VALUE-ADDED DECISION MAKING FOR MANAGERS

VALUE-ADDED DECISION MAKING FOR MANAGERS

Kenneth Chelst

Wayne State University
Detroit, Michigan, USA

Yavuz Burak Canbolat

Abbott Laboratories
Abbot Park, Illinois, USA

CRC Press
Taylor & Francis Group
Boca Raton London New York

CRC Press is an imprint of the
Taylor & Francis Group, an **informa** business

A CHAPMAN & HALL BOOK

CRC Press
Taylor & Francis Group
6000 Broken Sound Parkway NW, Suite 300
Boca Raton, FL 33487-2742

Version Date: 2011901

International Standard Book Number: 978-1-4200-7572-4 (Hardback)

Visit the Taylor & Francis Web site at
http://www.taylorandfrancis.com

and the CRC Press Web site at
http://www.crcpress.com

Contents

Preface

This book was developed from a course on decision and risk analysis that we have taught to hundreds of experienced technical managers over the last 18 years. Our primary thesis is that there is more to decision making than just picking the best alternative. A structured approach clarifies the strengths and weaknesses of the best alternatives and enables the decision maker to develop a plan for improving on the best by adding value and reducing risk. Throughout the book, we explore the important interaction between decisions and management action and clarify the barriers to rational decision making. Specifically, this book

- Provides a wide range of realistic decision contexts—routine, semi-routine, and strategic—for both industry and personal life
- Develops and illustrates the concept of value-added decision making
- Gives equal weight to modeling both multi-objective decisions and decisions in the presence of uncertainty, and comes packaged with Logical Decisions software for multi-objective decisions and Decision Suites software for probabilistic decisions
- Provides a comprehensive review of diverse challenges to "rational" decision making, including chapters on forecasting bias and decision bias
- Includes chapters on negotiated decisions, strategic decisions, and ethical decisions, covering alternate approaches and perspectives

The core of the book addresses decisions that involve selecting the best alternative from a distinct list of a handful or more of alternatives. The decisions include buying a car, picking a supplier or house contractor, selecting a technology, picking a location for a manufacturing plant or sports stadium, hiring an employee or selecting among job offers, deciding on the size of a sales force, making a late design change, and sourcing to emerging markets. More complex decisions that involve multiple dimensions simultaneously are covered in the later chapters on negotiations, strategy, and ethics.

There are numerous activities placed throughout the book. These activities are intended to encourage the reader to stop, think, and assimilate the ideas by finding similar examples in his or her work environment or personal life. These same activities appear in the exercise section.

Book Website

The book comes with two software packages, but we chose to not include discussion of the software commands in the book. We were concerned that any software changes would make those sections obsolete. Software downloads are described in the appendices at the end of the book. We will maintain on the website for the book some basic up-to-date tutorials developed around examples in the book. The website address is http://ise.wayne.edu/research/decision.php. In addition, we chose not to include basic probability in the book or appendices. If necessary, the basic probability concepts used in the book can be covered in a couple of lectures drawn from any introductory book. The website will also contain a password-secure section for instructors. This section will have PowerPoint presentations for each chapter and solutions to all of the numeric examples. To gain instructor access send a request to the author at kchelst@wayne.edu.

The book is divided into 5 parts and 17 chapters.

Part I: Structuring Hard Decisions

Chapter 1. The Case for a Structured Analytic Decision Process

The chapter begins with a discussion of common complaints about organizational decisions. It then explores the factors that complicate decision making and why gut feeling and instinct often produce bad decisions. Common symptoms of a poor decision-making process on both an organizational and personal level are detailed, concluding with a discussion of what constitutes a quality and efficient decision-making process.

Chapter 2. Influence Diagrams: Framing Multi-Objective and Uncertain Decisions

A wide range of decisions involves multiple objectives and/or uncertainty. This chapter introduces the reader to influence diagrams used in the early stages of decision making to obtain agreement on the critical values and uncertainties surrounding the decision. The chapter develops and explores this tool through a series of examples: investing in automation, planning a theater party, buying a used car, launching a new product that is late to market, and adding transmission power lines.

Chapter 3. Common Decision Templates

A primary concern with any structured decision process is that it is too complex and takes too much time. Chapter 3 presents decision templates that can serve as a foundation for efficiently beginning the decision-structuring process.

Part II: Decisions with Multiple Objectives

Chapter 4. Structure Decisions with Multiple Objectives

Multi-objective analysis begins by framing the decision with a hierarchy of objectives and sub-objectives in the form of a tree. The tree culminates with a list of measures used to characterize the objectives. The book discusses different types of measures. These include natural measures such as cost or miles per gallon as well as categorical measures. Categorical measures require the analyst to creatively define, quantify, and group the range of possible values for items that are often easier to describe qualitatively than quantitatively. The concepts are developed and illustrated with a series of examples involving lightbulbs, salesclerks, used cars, kitchen remodelers, and global facilities. The chapter concludes with a set of real-world case studies.

Chapter 5. Structured Trade-Offs for Multiple Objective Decisions: Multi-Attribute Utility Theory

This chapter presents a process for assigning weights that reflects the relative importance of each measure and objective. Weights are the primary mechanism for making trade-offs between objectives. The next step involves converting raw data on each measure into a scale between 0 and 1. Several methodologies for the conversion are presented. These two steps are combined in multi-attribute utility theory (MAUT) to calculate and compare the overall score for each alternative. The chapter uses Logical Decisions software to facilitate the process.

Chapter 6. Value and Risk Management for Multi-Objective Decisions

This chapter demonstrates the process of adding value by a thorough analysis of the strengths and weaknesses of the best alternatives. Ideally, this leads to creating an enhanced or hybrid solution that is even better than any of the original alternatives, by creatively improving the top alternatives, addressing areas of weakness within highly valued objectives, and reducing any significant risks. The chapter leverages the capabilities of Logical Decisions to identify and clarify strengths and weaknesses.

Chapter 7. Multiple Objective Decisions with Limited Data: Analytical Hierarchy Process

In this chapter, we introduce the analytical hierarchy process (AHP), a second analytical methodology for multi-objective decisions. AHP is less data intensive and structured than MAUT with regard to measures and scaling. It is less rigid in its data requirement and is built on a natural decision process of pairwise comparisons. The examples given include choosing a snowblower, selecting among job offers, and purchasing project management software. Classic decision analysis questions the validity of this methodology because of issues of rank-order reversals however. The chapter closes with a discussion of the relative merits of MAUT and AHP.

Part III: Decisions and Management under Uncertainty

Chapter 8. Value-Added Risk Management Framework and Strategies

This chapter begins with an exploration of the role that uncertainty plays in both day-to-day management and decision making. It presents a multistep risk management process that begins with risk identification and concludes with strategies to avoid, mitigate, or manage risks. The chapter includes a limited discussion of standard tools for identifying and prioritizing risks: fishbone diagrams, failure mode effect analysis (FMEA), and likelihood-impact maps.

Chapter 9. Spreadsheet Simulation for Decisions with Uncertainty

This chapter is a basic introduction to stochastic simulation used to model a collection of uncertainties. It is a descriptive tool that produces a risk profile, but unlike decision trees it does not identify the optimal decision. The software used is @Risk, an Excel add-on. Investment in project acceleration, forecasts of drug development profits, and sourcing to emerging markets are the examples used here.

Chapter 10. Decisions with Uncertainty: Decision Trees

This chapter presents decision trees as an analytic tool for making decisions involving uncertainty. It is a normative decision-making tool that identifies the optimal decision based on expected value or expected utility. The software package Precision Tree, an Excel add-on, is used to carry out the decision tree analysis and perform sensitivity analysis. The examples in this chapter include capacity planning, design change, automation investment, make or buy, and choice of technology.

Chapter 11. Structured Risk Management and the Value of Information and Delay

This chapter presents a structured approach to risk management that is developed around decision trees. It begins with the demonstration of two graphical tools, a tornado diagram and a spider plot, used to highlight critical variables. The chapter develops the concepts of expected value of perfect control, expected value of perfect information, and imperfect information. It illustrates the role that contingent contracts and real options can play in risk management.

Chapter 12. Risk Attitude and Utility Theory

This chapter introduces the concept of translating an individual's attitude toward risk into a risk utility function. This function is then used in decision trees with the objective of maximizing the expected value of the utility function. The value of utility theory is demonstrated with examples involving insurance and risk-sharing partnerships. The chapter explores a number of paradoxes that challenge key assumptions of utility theory.

Part IV: Challenges to "Rational" Decisions

Chapter 13. Forecast Bias and Expert Interviews

This chapter addresses a core issue in decision trees, the accurate specification of subjective probabilities. All decisions require data, estimates, and forecasts, and the chapter explores a range of biases that undermine forecasting accuracy. The biases discussed include motivational bias, overconfidence, availability bias, representative bias, confirmation bias, and errors in probabilistic reasoning. The chapter also describes an expert interview process designed to reduce these biases.

Chapter 14. Decision Bias

The goal of this chapter is to develop an understanding of how to overcome cognitive decision biases that are antithetical to good decision making. These biases produce tendencies to select non-optimal alternatives and to reject good ideas. Discussed here are biases of sunk cost, escalation and de-escalation of commitment, framing, status quo and omissions, regret, and groupthink. The chapter also discusses how mood affects decision making.

Part V: Decisions with Multiple Perspectives

Chapter 15. Value-Added Negotiations

This chapter develops the knowledge and analytic skills for making negotiated decisions that culminate in value-added agreements, providing a systematic framework for improving negotiation outcomes. One of the key recommendations is that negotiators focus on the interests of opposing parties and not on opposing parties' positions. The chapter describes a negotiation process intended to create value that the parties can then divide according to their needs. Cross-border negotiations affected by different cultural behaviors are also discussed.

Chapter 16. Ethical Decisions

This chapter is designed to raise awareness regarding a wide range of ethical issues that routinely arise when making decisions. It focuses on day-to-day situations that often involve conflicting ethical issues, presents a list of common ethical values, and identifies those that are most often in conflict. The chapter discusses a number of biases, barriers, and pressures that often affect our ability to see all the ethical issues involved in decisions we make. It concludes with a series of small cases that include some of the more common ethical conflicts.

Chapter 17. Strategic Direction, Planning, and Decision Making

This chapter develops a broad perspective as to how companies and individuals should plan, develop, and refine their strategies. It starts by describing a strategic planning process and the critical decision elements of a strategy. The chapter presents an overview of the concept of SWOT analysis (strengths, weaknesses, opportunities, and threats) and also introduces two basic tools, a decision hierarchy and a strategy table. Strategic planning often requires the development of alternative scenarios in order to assess the robustness of various strategies. The chapter concludes with a description of the dialogue decision process used in large organizations to keep key stakeholders engaged and aligned with the developing strategy.

Kenneth R. Chelst, PhD
Wayne State University
Detroit, Michigan

Yavuz Burak Canbolat, PhD
Abbott Laboratories
Abbot Park, Illinois

Acknowledgments

In 1992, the Department of Industrial and Systems Engineering (ISE) of Wayne State University launched the Engineering Management Master's Program (EMMP) at Ford Motor Company. Faculty from the Leaders for Manufacturing Program at MIT mentored the department in designing a graduate program that integrated engineering and business principles for Ford engineers. Ford has continued to support this program with fresh nominations of high-potential engineers each year since, even through the darkest days of the U.S. automotive industry from 2007 to 2009.

Ford management originally requested that a course in decision and risk analysis be included in the curriculum, but it offered little guidance on specific content. I was asked to develop and teach the course, having taken standard courses in decision analysis as a graduate student at NYU and MIT. Those courses emphasized technical modeling aspects, and I was unsure of how to make the topics relevant to experienced engineers. The decision tree examples seemed either simplistic or relevant only to the oil industry, while the multi-objective concepts seemed most relevant to major public policy decisions.

At the time, there was no widely used up-to-date text book that integrated relevant software packages. Reading the collection of articles in *The Principles and Applications of Decision Analysis*, edited by Ronald A. Howard and James E. Matheson, was a turning point in broadening my understanding of the way decision analysis might be applicable to engineers on the path to technical leadership. That book was complemented by *Decision with Multiple Objectives: Preferences and Tradeoffs,* by Ralph L. Keeney and Howard Raiffa. Together, these volumes led me to a fateful decision: the course would give equal time to multiple objective decisions and uncertain decisions. In addition, a significant portion of the course would be allocated to softer issues such as bias. Logical Decisions was an integral part of the course from the very beginning and was complemented by a variety of decision tree packages that have changed over the years before settling on Precision Tree. The Ford engineers would be required to complete and present team projects in the course. The breadth of the real-world decisions they tackled in a short span clearly demonstrated the relevance of the material to their day jobs.

For more than a decade, the primary book for the course was *Making Hard Decisions* by Robert Clemen; a new edition co-authored with Terence Reilly includes the Decision Suites software. This was supplemented by a large collection of outside articles. As time went on, however, I found I was replacing many of the examples in the book with problems more relevant to the Ford engineers. More important, MAUT was the lead modeling tool in the course and was supported by Logical Decisions software, whereas the Clemen and Reilly book places far more emphasis on decisions with uncertainty than on multiple objective decisions. In addition, since it does not use multi-objective software, the analysis it presents is limited. As a result, I began the process of converting my lecture notes and examples into a book, and along the way decided to broaden the scope to include chapters on negotiations, ethics, and strategy. Because I have few skills in negotiations, I asked Professor Hal Stack, the former director of the Center for Labor Relations at Wayne State University, to write that chapter. Dean Pichette, a retired Ford engineering manager and lecturer at Wayne State, accepted the challenge to develop the ethics chapter, which focuses on day-to-day ethical dilemmas.

Dr. Yavuz Burak Canbolat is co-author of this book. He completed his PhD at Wayne State under my direction. During his years of study, he was the teaching assistant for the course as well as a technical consultant for a number of major projects at Ford. Upon receiving his doctorate, he entered the pharmaceutical industry as an internal decision analysis consultant. Currently, he is a manager at Abbott Laboratories in the decision analysis group. His practical experiences over the last several years have further enriched the material covered in this book.

I would like to acknowledge a number of individuals who have contributed to and encouraged this work. David Strimling mentored me as I began to develop this course. As a senior manager at General Dynamics, he had both used and taught the tools of decision analysis. A number of my current and former doctoral students contributed examples and ideas along the way. These include Mustafa Sefik, Ali Yassine, Gang Wang, and Saman Alaniazar. Other friends and colleagues who have helped and offered encouragement along the way are Sam Bodily, Robert Bordley, Jay Johnson, Azriel Chelst, Kenneth Riopelle, and Philip Lanzisera. A special debt of gratitude is owed to the main editors of the book, Eric Schramm and Pessie Novick. Bob Stern of Taylor & Francis was enthusiastic about the book from the moment I approached him and stuck with us even though the writing took much longer than anticipated. Tom Edwards has worked with me to develop a series of examples that can even be taught to high school students. Without Gary Smith, the developer of Logical Decisions, the MAUT and AHP sections would have been limited to small, unrealistic examples. I owe a special debt of gratitude to the legions of engineers at Ford, Visteon, and ACH who challenged me every step of the way to ensure that the concepts taught were relevant. Their outstanding projects were proof of the relevance of decision analysis to working engineers. Without the continued support for EMMP from the technical leadership of Ford, especially Derrick Kuzak, I would never have developed the practical knowledge needed to write this book. Last, I owe an unrepayable debt to Tamy, my wife of more than 40 years, for supporting me through the excruciatingly long challenge of bringing this work to conclusion.

Kenneth R. Chelst
Director of Engineering Management Masters Programs
Department of Industrial and Systems Engineering
Wayne State University
Detroit, Michigan

I would like to acknowledge my colleagues at Merck and Abbott who helped me work on a broad range of decision analysis problems. I owe a special debt of gratitude to Kerime, my wife, for her support while working to complete this book.

Yavuz Burak Canbolat
Senior Manager, Decision Support Group
Abbott Laboratories
Abbott Park, Illinois

Authors

Kenneth R. Chelst received his BA from Yeshiva College, his MS from NYU, and his PhD in operations research from MIT. He is currently a professor of operations research and director of engineering management programs in the Department of Industrial and Systems Engineering at Wayne State University. He also has rabbinic ordination from the theological seminary of Yeshiva University. Dr. Chelst is co-principal investigator on Project MINDSET, an NSF-funded project to develop and implement a new high school mathematics course designed around operations research examples. He is a senior consultant for the International City and County Management Association (ICMA) in the area of emergency service management. As part of his responsibility for educating the next generation of technical managers at Ford, he has overseen numerous major internal engineering management projects. Dr. Chelst is an active member of the INFORMS Roundtable.

Dr. Chelst's research interests include engineering management, emergency service management, global engineering, and the use of operations research to enhance K–12 mathematics education. He was an Edelman Prize Finalist for a team project he headed for Ford. He is also the author of *Does This Line Ever Move: Everyday Applications of Operations Research* as well as *Exodus and Emancipation: Biblical and African-American Slavery*.

Yavuz Burak Canbolat received his BS from Yildiz Technical University, his MS from Sakarya University, and his PhD in industrial engineering from Wayne State University under the supervision of Dr. Kenneth Chelst. He is currently a senior manager in the Decision Support Group at Abbott Laboratories. Prior to Abbott Laboratories, Dr. Canbolat worked as an associate manager in decision analysis for Merck & Co., Inc., and as an instructor in the Industrial Engineering Department of Qafqaz University.

Dr. Canbolat has applied decision analysis and operations research techniques to R&D portfolio evaluation and management, strategic planning, financial and economic analysis, global operations and logistics, risk analysis, and capacity planning. He has developed sales and financial models to predict commercial opportunities for R&D projects. He has received numerous performance awards from Abbott Laboratories and Merck & Co., Inc., and research and teaching awards from Wayne State University and Qafqaz University.

Dean W. Pichette received his BSEE and MBA from Michigan State University and his MS in electronics and computer control systems from Wayne State University. He is currently a senior lecturer in the Industrial and Systems Engineering Department at Wayne State University. Previous positions include various engineering and management assignments at Ford Motor Company, with much of them focused on the development and implementation of electrical and electronic architectures and systems.

His teaching and research interests include systems engineering, project and program management, and engineering economics. He also co-leads capstone and leadership projects, which provide students with the opportunity to apply classroom learning to solve real-world business issues or to improve methods and processes using a project management methodology.

Hal Stack received his BA from the University of Notre Dame and his MA from the University of Wisconsin. He recently retired as director of the Master of Arts in Industrial Relations (MAIR) program and the Labor Studies Center, a comprehensive labor education and research center at Wayne State University. The MAIR program prepares students for careers in labor and employment relations. The Labor Studies Center provides training and technical assistance to unions on labor relations and workplace issues, and assists unions and employers in the design and implementation of joint initiatives to enhance labor–management relations, employee involvement, and organizational effectiveness.

Hal Stack's teaching and research interests focus on negotiation, interventions to increase union effectiveness, the design and implementation of joint labor-management programs, and the impact of organizational restructuring on labor relations and work organization. In addition to his labor relations work, he has worked with cities, nonprofits, community-based organizations, and government agencies as a facilitator of strategic planning, community visioning, consensus building, and participative organizational redesign.

Part I

Structuring Hard Decisions

Chapter 1

The Case for a Structured Analytic Decision Process

1.1 Goal and Overview

The primary goal of this chapter is to motivate the need for a structured analytic decision process by discussing the limits and flaws of an intuitive unstructured process.

The chapter begins by describing common concerns with decision making as practiced in many organizations. It explores three major dimensions that limit an intuitive process. First, there are common cognitive biases that contribute to flawed intuition. Second, decision making is difficult because of problem complexity and organizational and personal pressures. Last, intuition is limited in its ability to consider multiple objectives and uncertainty. Some of the symptoms of a poor decision process include arbitrarily revisited decisions and personality driven decisions with few alternatives honestly evaluated. This chapter explores the elements of good decision making: the process should be high quality, efficient, easily updated, disciplined, transparent, and committed to implementation. The process proposed here incorporates both hard data and expert opinion and can work both for the individual decision maker and for a group in a complex organizational setting involving multiple stakeholders.

1.2 The Challenge

The best laid schemes o' mice an' men, Gang aft a-gley

Robert Burns

We never have enough time to make the right decision the first time but we always seem to have enough time to review the decision over and over again *after* beginning implementation!

Engineering, marketing, manufacturing, and finance never seem to reach a consensus. They all leave the meeting more frustrated than when they came, and then complain that they weren't heard!

We spend more time justifying the decision we made than analyzing what the best decision should be!

Every time a new manager is appointed, we have to review the decision and start over from scratch!

Finance keeps raising the bar for 'return on investment' and the corporate technologists keep exaggerating the potential value of new technology. And the game goes on!

We don't have enough data to nail the decision, yet, we have more data than we could possibly process in time to make a timely decision.

These laments are often heard as corporate decision makers tackle tough problems. These individuals face high levels of complexity as they attempt to integrate technical and marketing expertise to design, develop, launch, manufacture, deliver, and continuously improve new products and services. Difficult issues arise whether the context is an automobile, a washing machine, a critical subsystem, an integrated chip, a software product, or a power plant. The decision-making environment includes not only the uncertainty of the marketplace but also the uncertainty of new technology and the accuracy of projected product performance. These challenges abound in a service economy as well, whether you are a provider or consumer who is considering anything from the design of a new health insurance policy to a cell phone plan. Decision makers at all levels must weigh their priorities while also making trade-offs for the sake of what can be delivered at a specific price.

Uncertainty, trade-offs, and cross-organizational teams abound, whether your organization provides a physical product or delivers a service. Health care providers must constantly assess the need to upgrade and integrate new technologies. Hotel managers and airline industry executives struggle to find the right mix of service, facility and equipment upgrade, and price in the presence of brutal competition. The trade-offs are no less complex when similar decisions are made by state and local government officials under pressure to hold the line on taxes while meeting citizen expectations of service.

The same issues extend to personal decisions. Trade-offs among objectives arise when buying a home, a cell phone, a digital camera, an automobile, health insurance, or financial services. Deciding which contractor to hire for a kitchen remodeling project is as complicated as picking a supplier to maintain the corporate IT infrastructure, but the home decision is more emotionally laden. When evaluating a possible job or even career change, an individual will need to balance short- and long-term goals involving significant career path uncertainty. A company might face a dilemma when deciding whether to replace an existing technology with a state-of-the-art alternative that is said to have unlimited potential but is as yet unproven.

1.2.1   Decision-Making Environments

The decisions explored in this text are primarily one-time decisions. They range from classes of decisions that are semi-routine such as choosing a supplier to those that are once in a lifetime, such as when to stop extreme efforts to prolong life. These decisions arise in corporate organizations and governmental bodies as well as in our private lives. Multiple objectives, uncertainty, and concerns regarding decision dynamics are critical in each of these environments; yet, there are important distinctions between them.

Corporations often have access to large amounts of data and multiple technical experts. These experts may be drawn from marketing, finance, operations, product development, manufacturing, and a variety of support services. The long-term objectives will consist of some set of financial measures such as profit and return on investment (ROI). Nevertheless, it will be impossible to link every decision directly to the ultimate fiscal viability of the firm. Hard decisions will cut across multiple departments of an organization. Often, they will involve multiple levels of higher management reviews. Additionally, the corporate world is experiencing unprecedented global pressures on profitability. Consequently, there is an overall tendency to place too much emphasis on short-term instead of long-term goals. Last, there is an information explosion that is rapidly changing how we do business, and there is a premium on speed to market.

Governmental bodies routinely make decisions that impact lives. They have responsibility for and are responsible to multiple constituencies. As a result, they must constantly balance efficiency and equity. Members of elected bodies must deal with powerful public sector worker organizations that can influence elections and ultimately their jobs. Public sector managers may be more closely aligned with their workers than with the top-elected officials, who, in theory, set budgets and policies and oversee the managers' performance. These managers and workers are often shielded from accountability by rules, regulations, and practices that were originally designed to protect the integrity of the work environment, preventing cronyism and nepotism. The decisions of such governmental bodies as city councils and legislatures often involve public discussion and almost always public votes. These votes can require more than simple majorities to pass, as in California, which requires a two-thirds majority for budgetary votes. In addition, preliminary discussions and the information gathered can be subject to freedom of information rules. There are also checks and balances between the legislature, executive, and the judiciary built into the decision-making system. Today almost all such organizations are facing extreme budget constraints that necessitate compromises across different agencies and different levels of service cuts. These must be carefully balanced in order to maintain a productive working relationship among the various departments of a particular organization.

Personal decisions impact one's own life as well as those of family members, and, to a lesser extent, friends and co-workers. There will always be multiple objectives. Individuals generally have limited time to gather all of the information that could help in the decision; nor do they necessarily have access to it all. At other times, they face information overload; they do not have the ability or time to process all of the information that is obtainable. Thus, people tend to face significant uncertainty as they make decisions. There is a constant tension between short-term goals and needs and longer-term interests. Personal values and biases implicitly affect how one approaches decisions as do the people who most influence the decision maker. Finally, most people face substantive budget constraints when making decisions involving significant investments.

1.3 Decision Analysis Effectiveness

We offer here no quick fixes for professional and personal decision challenges, nor do we have access to a prophet or visionary who can resolve all uncertainty. Instead, we offer a structured, well-established, and well-researched process that begins with decision framing and proceeds through formal quantitative analysis while reducing biases and attempting to avoid common decision-making pitfalls. This methodology explicitly and formally incorporates two factors that complicate many decisions: multiple objectives and uncertainty. It is also designed to support group decisions involving multiple perspectives and concerns. We present two methods to address multi-criterion decisions: Multi-Attribute Utility Theory (MAUT) and the Analytic Hierarchy Process (AHP). For uncertain decisions, the primary modeling technique is decision and risk analysis with decision trees.

Throughout the text, we emphasize that decision making is a process and not just an event (Quinn and Rohrbaugh 1981); our goal is to facilitate the development of an inquiry-based collaborative decision process instead of an advocacy conflict-ridden contest of wills (Garvin and Roberto 2001). The process is facilitated by software packages that structure the decision, compare alternatives, and provide sensitivity analysis in graphic form. We claim that the approach is proven QED: quality driven, efficient, and disciplined. Equally important, this inquiry-based process adds value to the resulting decision, as we explain and demonstrate.

An obvious but difficult question is "Do these analytic techniques measurably improve decisions when compared to alternative, less analytic and structured processes?" Matheson and Matheson (2001) performed ground-breaking macroanalysis of corporate performance and the use of best practices. They identified nine aspects of best practices of corporate management and created an IQ test to measure a company's use of these practices. High IQ companies were almost five times more likely to be high-performing companies when compared to low IQ companies. Four of the nine aspects directly relate to decision processes presented here: embracing uncertainty, employing disciplined decision making, a culture that emphasizes value creation, and alignment and empowerment. Other research into decision aid effectiveness presented three levels of effectiveness: process, output, and outcome effectiveness (Schilling et al. 2007).

Process effectiveness is described as the quality of the decision and is the most often studied variable. Rohrbaugh (2005) describes key elements of a quality decision process. These include adequate information, clear and rational thinking, flexibility, and creative and sufficient participation.

Matheson and Matheson (1998) add clear value trade-offs and commitment to action. Timmermans and Vlek (1994) and Davison (1997) focus on the quality of communication between the team members making the decision.

Schilling et al. (2007) used a series of six complex cases in a controlled environment to assess the perceived effectiveness of multi-criterion decision analysis (MCDA) when compared to existing decision processes. MCDA was consistently perceived as being better on a multiplicity of measures. It increased strategic insights, creativity, and the quality and quantity of information exchange.

Output effectiveness is described as the ability of the decision process to achieve the organizational objectives (Dean and Sharfman 1996). Phillips and Costa (2007) demonstrated how MCDA improves decision output effectiveness. In both public and private organizations, this process increases communication that leads to a shared understanding and a sense of shared purpose that helps organizations achieve their objectives. In addition, the clear articulation of multiple objectives increases transparency that facilitates communication of the decision to stakeholders and provides a mechanism for auditing how decisions are made. This transparency has increasing relevance for public sector decisions that too often seem to be dominated by special interests.

Outcome effectiveness is hard to measure, especially in an uncertain environment when the best decision might still produce a bad outcome. A detailed study of the role of decision analysis at Kodak estimated that decision analysis added $1 billion in value across 178 project decisions evaluated over a period of 10 years (Clemen and Kwit 2001). The fact that many highly successful pharmaceutical and energy firms routinely use decision and risk analysis is further confirmation of their overall effectiveness.

1.3.1 Quality

The decision processes of this text incorporate the needs and values of both the primary decision maker(s) and various stakeholders and facilitate communication. They factor in both hard data and subjective judgment in a transparent process that is intended to reduce the impact of cognitive biases that undermine rationality. The methodologies are designed to produce robust decisions in the presence of uncertainty and to be less subject to the variability of personality inherent in intuitive decision making. They yield insights into the strengths and weaknesses of the various alternatives, resulting in greater decision defensibility and better buy-in for implementation and tracking.

1.3.2 Efficient

When time is not just money but also a competitive advantage, no decision-making process can afford to be burdensome or subject to long delays and over analysis. We cannot claim that the methods in this book are faster than intuition, gut feeling, or seat-of-the-pants decision making, only that they can be streamlined as time and context demand (Timmerman and Vlek 1994). It is for this reason that we provide numerous decision templates to enable decision makers to more rapidly frame decisions by analogy. Where data do not exist or are too time-consuming to gather, subjective expert opinion can fill the gap. However, true efficiency comes from the need in many contexts to explain and modify the decision, often more than once. There may be higher and higher level management reviews or reviews and modifications when management changes. Decisions should also be reviewed and updated when new information becomes available, the environment changes, or new options arise.

1.3.3 Disciplined

Corporate executives often establish, oversee, and strive for discipline in product development, manufacturing, or service delivery, but they fail to expect discipline when it comes to decision making. They embrace Six Sigma as a quality management tool that can bring discipline and reduce variability in processes but often do not see the parallel need when it comes to the process of decision making. The decision-making processes presented here offer a discipline that is replicable throughout a corporation across a broad array of decision contexts. They support decision making at all levels of the organization and across departments. The ability to replicate the processes also means that they can be continuously improved and organizational structures can be put in place to deliver and maintain critical data inputs as well as to streamline and update decisions. The discipline can also bring transparency to the process, thereby enhancing buy-in and speeding implementation.

1.3.4 Value-Added Decisions

The types of decisions we address involve a limited number of distinct alternatives. If our only goal is simply to find the best choice, it would be difficult to demonstrate that the proposed process is much better than a gut feeling. However, the approach of this text is geared toward clarifying the strengths and weaknesses of the strongest alternatives, not merely in finding the best one. This clarification enables a decision maker to develop more robust strategies and cost-effective contingency plans as uncertainty evolves. In addition, the analysis provides insights that enable creative individuals and decision teams to enhance an alternative or to develop a hybrid that is better than all the original options. Thus, the outcome of a particular decision may be enhanced in any or all of the following ways:

- Considering a more complete list of alternatives will result in outcomes that have higher values and entail fewer risks.
- Clarity of explanation will lead to a greater chance of implementation.
- Clarification of the strengths and weaknesses of a particular plan will enable more effective updating as the situation changes.

1.4 Do Not Trust Your Gut

"Two of every three (executive) decisions use failure-prone practices" (Nutt 2002). The goal of this chapter is to force you to take a hard look at how you and others around you in your organization, whether it is a company, government agency, or educational institution, make important decisions. The primary alternative to a structured decision-making process (leaving aside astrology, crystal balls, and prophecy) is intuition developed from experience. We develop the case *against* intuition from three vantage points: (a) representative debacles and patterns of bad decisions, (b) experimentally demonstrated biases, and (c) evidence of decision complexity.

It is common to find articles and even entire books in praise of intuition. *Harvard Business Review* published "When to trust your gut" (Hayashi 2001), but later followed up with "Don't trust your gut" (Bonabeau 2003). The primary case for experienced intuition has often been made by senior executives who, almost by definition, have had successful decision-making careers—sometimes relying heavily on gut feelings. One survey has reported that 45% of executives use intuition more than analysis to run their businesses (Bonabeau 2003). And yet, as Ralph S. Larsen, the former CEO of Johnson and Johnson, noted (Hayashi 2001) what works during the early phases of a career may not succeed when one climbs up the corporate ladder: "Very often, people will do a brilliant job up through the middle management levels, where it's very heavily quantitative in terms of decision making. But then they reach senior management, where the problems get more complex and ambiguous, and we discover that their judgment or intuition is not what it should be. And when that happens, it's a problem; it's a *big* problem." Larsen's point reaffirms the main thesis of this book; indeed, our primary audience is not the CEO, but the lower-level manager working his way through the ranks of middle management. It is he, or she, who must learn from the very start to integrate a proper balance between intuition—a specious quality at best—and quantitative analysis—a far more reliable, scientific approach to assessment of plans.

Larsen himself stresses the utility of intuition with regard to mergers and acquisitions: "When someone presents an acquisition proposal to me, the numbers always look terrific: the hurdle rates have been met; the return on investment is wonderful; the growth rate is terrific. And I get all of the reasons why this would be a good acquisition. But it's at that point—when I have a tremendous amount of quantitative information that's already been analyzed by very smart people—that I earn what I get paid. Because I will look at that information and I will know, intuitively, whether it's a good deal or bad deal." Part of the justification for the unprecedented growth in salaries for senior corporate executives in the United States is reflected in Larsen's self-assessment of his value. A board of directors must pay a premium for the experienced-based intuition retained in the memory bank of the unique and irreplaceable CEO—or so goes conventional thinking.

We cannot speak to Larsen's personal track record on mergers, but the literature is clear: mergers and acquisitions do *not* usually deliver the forecasted added value to the acquiring corporation, and, in fact, they often reduce its value (Cartwright and Schoenberg 2006). In one study of 53 high tech mergers, only 11 were considered successes (Chaudhuri and Tabrizi 1999). In a study of 131 mergers greater than $500 million, 60% reported negative returns at the end of 12 months (Eccles et al. 1999).* Strangely enough, Hayashi included in his list of star decision makers Bob Pittman of America Online. Pittman is quoted as saying that "probably more than half of [his] decisions are wrong" but that he does not worry because he routinely reviews his decisions, learns quickly from his mistakes and makes adjustments as needed. This self-proclaimed flexibility notwithstanding, the AOL Time Warner merger failed to live up to expectations, and Pittman was forced to resign in 2001. In 2009, AOL was spun off and became an independent company once again.

Let's take a closer look at mergers and acquisitions, since decisions in these areas are both highly visible and prone to intuitive decisions (for better or worse). The acquiring corporate management team always pays a premium on the value of the acquired corporation, on average 36% (Nutt 2002) under the belief that the combined organization can deliver synergies in the form of increased market potential and dramatic cost savings. This optimistic executive belief in their own abilities to create greater value and minimize the challenges of combining two distinct corporate cultures is representative of a variety of decision-making biases described by Lovallo and Kahneman (2003) in the aptly titled HBR article, "Delusions of Success, How Optimism Undermines Executive Decisions." To cite an extreme example consider the Daimler-Benz and Chrysler merger/acquisition of 1998. This merger involved two companies of similar size with very distinct national and corporate cultures. The stock value of Chrysler at the time was an estimated $38 billion; in 2007, when the merger was reversed and Chrysler was sold, its value was less than $10 billion.

When considering claims of the value of intuition from successful executives, imagine the findings from a study of successful lottery winners. What would you learn from asking a 100 lottery millionaires, "What strategy did you use to pick the winning number?" If you believe their answers would be of value, you probably should stop reading this book now and either ask for a refund or sell it. Use the cash to buy a lottery ticket. While it is an exaggeration to say that CEOs reach their exalted positions simply by luck, randomness will cause individuals with winning streaks of so-called good decisions to stand out.

---

* The high tech mergers study reported eleven successes, nine failures and 33 with zero or slightly positive returns. The study of large acquisitions noted that 59% of the time the stock market value of the acquiring company went down with the announcement. Clearly, the markets did not believe the majority of these mergers created value for the acquirer.

1. **Activity**: Coin flip: You are about to flip a coin eight times. Please choose which of the following outcomes you expect.

    1. An equal number of heads and tails

    2. The number of heads and tails differs by exactly two

    3. The number of heads and tails differs by three or more

Now flip a coin eight times.

Record the number of heads _____ and tails _____ and the net difference _____.

Did the outcome match your choice? _____ If the choice and outcome agree, does that mean that you made the best decision? _____

In a class of 30 students, the number of students whose choice matches the outcome could be as low as 4 or as high as 16. Did all of those students whose decision and results match make the "best" choice, even though their choices for the identical experiment were different?* Do you believe that they were better at controlling their destiny in achieving their predicted result?

Now consider a hypothetical scenario. Your organization has 500 workers, each making decisions with only a one-in-two chance of success. A manager who decided "correctly" seven times in a row would seem to have a much better understanding than the average manager. The likelihood of picking correctly seven times in a row on a 50–50 bet is only one in 128 ($1/2^7$). A decider this good is in the top 1% of the class. However, purely by randomness, an average of 4 managers out of every 500 should hit it right seven times in a row. Those who decide correctly at least six times out of seven will be in the top 7% of their peer group. Nevertheless, they need not have any special insights—just better luck. Bazerman (2006) offers a similar analogy with regard to placing your money with an investment agent. Every year just by randomness a percentage of brokers will outperform a stock portfolio indexed to the S&P 500, but the percentage keeps dropping dramatically as you take a longer and longer multiyear look at the data. Bottom line: it is wiser to go with an index with low fees than trying to pick a fund that seems to be an above-average performer.

2. **Activity**: Can you point to a situation in which you believe your organization made an extremely risky decision to save a buck, such as using an unproven technology or inexperienced supplier, that you did not think was justified but the results turned out satisfactorily? Were decision makers rewarded because of the outcome? Explain.

_____

3. **Activity**: Can you point to a personal situation in which you or someone you know made an extremely risky decision that in retrospect was not really justified but the results turned out satisfactorily? Explain.

_____

Conversely, it is not necessarily correct to focus on examples of bad or unsuccessful outcomes in order to uncover bad decision processes. Research and Development (R&D) projects are notable examples of low probabilities of success. A failed R&D project, one that did not result in a profitable product, does not necessarily mean that it was a bad decision to pursue the project initially. However, a company that has an 8% success rate of converting projects into marketable products or services

---

* This experiment follows the Binomial Distribution Probabilities discussed in Appendix A.
  (1) Equal number (70/256 = .273)
  (2) Differ by exactly 2 (112/256 = .438)
  (3) Differ by 3 or more (74/256 = .289)
  Individuals who made the best choice, "differ by exactly 2," will still experience a bad outcome more than half the time. However, individuals who selected 1 or 3 will experience a bad outcome almost 3 out of 4 times.

year after year as compared to only 3% for a company in a similar industry likely has a better process for evaluating projects and managing the transformation from laboratory idea to commercial product.

4. **Activity**: Can you point to situation in which you believe your organization made a reasonable choice in an uncertain world but the results turned out unsatisfactorily? Were decision makers punished because of the outcome? Explain.

---

5. **Activity**: Can you point to a personal situation in which you or someone you know took a realistically evaluated risk but the results turned out to be unsatisfactory? (e.g., you spent a good deal of time gathering available information about a new job offer, but the company went bankrupt a year later due to corporate executive misinformation.) Explain.

Individual examples of debacles do not make the case *against* intuition in the corporate executive suite any more than successful executive careers built on intuition make the case *for* intuition. It is valuable to recall Robert Burns's classic lines,

> But, Mousie, thou art no thy lane,
>
> In proving foresight may be vain:
>
> The best laid schemes o' mice an' men
>
> Gang aft-agley...

The mouse expended significant energy in designing its little home only to have it plowed under by the oblivious farmer.

The focus on infamous debacles can also misdirect research. It has been argued (Fuller and Aldag 1998) that too much of group decision research has been distorted by the groundbreaking study of the Kennedy Administration's Bay of Pigs fiasco that led to the popularization of the term Groupthink (Janis 1972). However, Paul C. Nutt, in *Why Decisions Fail*, uses a more rigorous approach to determine good and bad decision-making processes and to highlight "blunders and traps that lead to debacles." We present the results of this and related research in Chapters 13 and 14.

At the microlevel, Klein (1999) builds a case for intuition based upon examples involving chess masters, fire fighters, officers on the field of battle, or emergency room physicians, who successfully employ split second intuition to make decisions that are, for the most part, successful. Even retrospectively, they are unable to provide a structure for their decision-making process. We do not debate Klein's underlying thesis, but we reject the idea that these experiences can be usefully extrapolated to the vast majority of professional or personal decisions. Unlike the chess master, few of us play and replay in our minds thousands upon thousands of variations of chess board patterns, learning to recognize good strategies without exhaustively thinking and weighing the alternatives. Nor are we like the firefighters or soldiers who spend hundreds and sometimes thousands of hours training for a range of emergencies and studying other people's mistakes so that they can make the best split second decision possible. Nor are we like the emergency room physician with a decade of education supplemented by thousands of hours of direct mentoring along with constant review and debate regarding actions and decisions including things that went wrong.

Even within the medical field, the Society for Medical Decision Making has, since 1979, pushed for more structure for medical decisions so as to integrate physician experience and hard data. Of particular concern is the reality that physicians often function under extreme time pressure leaving inadequate time to assess fully the available data. The journal *Medical Decision Making* features articles that offer structure to both individual treatment decisions as well as public policy questions. We will draw on their examples to demonstrate the main methodologies of this text.

There are three primary difficulties in developing intuition for tackling complex problems. First, we seldom have the opportunity to learn directly from our experience, since we rarely face the same decision context over and over again. How many times in your life are you going to choose which college to go to, what subject to major in, or which first job to accept at the start of your career? This problem is compounded in the American corporate culture wherein managers expect to change jobs frequently as they rise up the corporate ladder. This situation precludes their developing the deep technical intuition that is common in, for example, the managers of German and Japanese, automotive companies.

Second, as Yogi Berra said, "When you come to a fork in the road, take it." After selecting a college, we can look back on the experience we had, but we cannot know how our lives and careers would have developed if we had picked the alternative. The same is true when we pick a supplier for our IT infrastructure or our home remodeling. We will know how well the choice worked out and experience the problems that arose, but we will never know what would have happened had we gone in another direction. As Robert Frost said, "I could not travel both and remain one traveler." (Frost, "The Road Not Taken").

Finally, the feedback loop leading to success or failure, especially when involving major corporate decisions, could take years to close. There will be numerous intervening factors, some controllable and some uncontrollable, that will affect the final success or failure of, for example, the launch of a major new product or service. Moreover, the individual who made the decision is likely to have long since moved on to another part of the organization. All of the aforementioned issues contribute to why decision-making experience does not readily translate into expertise (Bazerman and Neale 1992).

Meteorology is one context in which decisions are repeated daily; some feedback loops are closed in a day and others in 10 days. Weather forecasters do not use intuition, but rather sophisticated models that are continually being refined. Over time, they have been able to deliver better and better probabilistic forecasts over longer and longer periods of time. Theirs, however, is not a typical situation.

Nevertheless, despite education and extensive research, executives continue to cling—with no small amount of pride—to the unique value of their intuition. Each can personally recall one or more instances in which, he claims, following his intuition produced exciting results. The intermittent reinforcement of high profile success makes it extremely difficult to overcome what is still a dysfunctional behavior. Perhaps, it is time for executives and their organizations to stop relying on intuition and begin adopting and refining approaches outlined here so as to develop more consistent and higher quality decisions.

While criticizing the overuse of intuition, we do not want to minimize the value of in-depth *knowledge* in specialty areas such as engineering design or new product marketing. In 2007, Ford Motor Company significantly reduced its vehicle launch problems, thereby reducing dramatically their warranty costs, when they brought together a team of experienced assembly line workers and engineers to spend months reviewing the final design of a new model. They were able to identify numerous potential problems and suggest appropriate means through which to prevent them from occurring (*Detroit Free Press*, August 28, 2007). This was possible because sitting at the table was literally more than a thousand person-years of work experience, involving some of the company's best workers from both engineering and manufacturing.

1.5 Maximize versus Satisfice

The decision-making processes developed in this text include the following basic tasks:

- Identify as wide a range of alternatives as practically feasible
- Collect comparable data for each of the alternatives
- Select the best alternative based on some measurement scale

These processes are aligned with the approach first articulated by von Neumann and Morgenstern (1953) that includes axioms that characterize rational economic decision making (see also Simon 1955, 1956, 1957). However, extensive research in psychology and behavioral economics has documented a wide range of common behaviors that violate the assumptions of consistent rational economic man. These issues are explored in two later chapters, on forecasting bias and decision-making bias. In these chapters, we define the biases and suggest ways of overcoming them with the goal of being more consistently rational economically. However, there is one decision-making approach, called satisficing, that is at almost total variance with a strategy of utility maximization.

A satisficer seeks an alternative that is good enough. He has defined for himself an acceptable threshold and picks the first alternative that surpasses this threshold. His search is far more limited than the maximizer's because he is unconcerned with finding and picking the best. Simon (1956) argued that this type of behavior is still economically rational if the cost of additional search and analysis is less than the expected gain from considering more alternatives. However, it is well documented that the primary reasons for choosing to satisfice rather than maximize are psychological and not the result of an analysis of the benefit of additional search.

A satisficer scans what is playing in the local movie theaters and picks a movie he thinks he will enjoy. A maximizer identifies multiple movies that he is interested in seeing and then decides which movie he would prefer that day. A satisficer looking for a home to buy will stop his search when he finds the first house he likes that has generally what he wants and is in his price range. A maximizer will continue searching until he has seen a number of houses he is willing to buy and only then compare and contrast them to determine which is best.

Schwartz et al. (2002) developed a scale to determine whether an individual is primarily a satisfier or maximizer. Some of the self-descriptions used to characterize an individual as a maximizer are as follows:

1. Whenever I make a choice, I'm curious about what would have happened if I had chosen differently.
2. When I am in the car listening to the radio, I often check other stations to see if something better is playing, even if I'm relatively satisfied with what I'm listening to.
3. I'm a big fan of lists that attempt to rank things (the best movies, the best singers, the best athletes, the best novels, etc.).
4. I never settle for second best.

The second dimension of this research assessed the overall psychological status of the two groups of undergraduates in the study. In comparison, maximizers reported significantly less life satisfaction, less happiness, less optimism, and lower self-esteem. They tended to more frequently make social comparisons between themselves and others in a variety of contexts. Schwartz et al. (2002) hypothesized that the very nature of maximization involves not only considering a wider range of alternatives but also always wondering whether there is another alternative that is even better. When a final choice is made, there is a greater opportunity for regret since the maximizer has taken more responsibility and ownership for the choice he has made and for the choices he did not make. Parker et al. (2007) documented the greater tendency for regret among maximizers and also found that maximizers more frequently reported avoiding or postponing decisions.

Last, Iyengar et al. (2006) carried out an interesting study that involved tracking more than 500 soon-to-graduate students at 11 diverse universities as they pursued their post-university jobs. These students were surveyed as they began their search, once again in the middle of the search, and also at the end of the process when they accepted a job. Maximizers reported an average 20% higher salary and yet reported that, on average, they were less satisfied with the final job offer when compared to satisficers. In addition, they experienced more negative emotions during the job search process when compared to satisficers.

All of the earlier research was done with college students and involved personal decisions. What is not clear is how much of this phenomenon exists in organizational decision-making settings. Do business managers often utilize a satisfice approach to making business decisions? Do personal behavioral decision traits carry over into the work world?

1.6 Established Biases

There is extensive literature based on classroom experiments and observations of decision makers that demonstrate common, unconscious biases that arise when making forecasts and decisions. Bazerman (2006) provides an excellent summary of the psychological literature, while Keeney et al. (1998) provide a focused managerial perspective. We cover these issues in Chapters 13 and 14 but for now we will illustrate the following three biases.

1.6.1 Sunk Cost

Have you ever gone to an expensive show or movie and after 15 or 20 minutes found you disliked the show or movie, yet you stayed to the very end? Did you find yourself thinking, "Well, maybe it will get better" as you sat there longer and longer trying to remain focused? This scenario illustrates the concept of sunk cost. Having invested money and time in an enterprise, you are reluctant to admit that your investment was wasted. You may even be embarrassed if you brought a friend along. This is likely to cause you to proceed to invest even more time, thereby escalating your commitment further.

A clear-headed rational decision maker would focus solely on future value and not money and time already wasted—cost that has already sunk. The decision whether or not to stay should simply depend on your estimate of the future. However, the concept of sunk cost can cloud your judgment. Experience tells you that movies or plays rarely get better after a poor beginning; they are not like football games. Yet you hope, irrationally, for the best in order to redeem your initial decision to attend the movie. The same bias comes into play with life and death consequences in wars. Even when a clear consensus agrees that a war cannot be won, it is extremely difficult to start planning a strategic withdrawal. How often have we heard that we cannot leave a particular confrontation because to do so would suggest the lives sacrificed thus far were wasted? And yet, it is unspeakably wasteful to risk additional lives because of a stubborn unwillingness to focus on the future and recognize the sunk cost of the past.

1.6.2 Framing

Scenario A: Imagine you drive up to the gas station and see the price per gallon is $3.10. The sign says this is the price if you pay cash. However, if you want to pay with a credit card, you will have to pay a 10 cent premium per gallon.

Scenario B: The price on the pump is $3.20 if you use a credit card. However, if you choose to pay cash you will receive a 10 cent discount per gallon.

Drivers in scenario B are more likely to use the credit card than in scenario A. The framing of the credit card as a base value instead of a premium affects people's willingness to pay the $3.20 credit card rate.*

---

* This example also includes an anchoring bias. Anchoring occurs when an individual fixes on the first number he sees and assesses everything as a deviation from this base. Negotiators may start negotiations by using an initial value that is extremely in their favor. They might price a product at double its value recognizing that counteroffers are likely to be 10% or 20% less, not 50%.

Such a framing bias had serious consequences for Coca Cola in 1985 and contributed to the decision to change its recipe and launch New Coke. Management was concerned about recent declines in market share (Whyte 1991). As a result, they made a risky decision to change their formula even though this meant jeopardizing their very large base of still loyal customers. It is well documented that when decisions are framed as a potential loss, decision makers become more risk-prone. For example, people will take the risk of holding on to stocks that have lost value longer than they should as compared to holding on to stocks whose value has increased. In both instances, the only factor, aside from tax issues, that should govern the decision is future projections and not the original price paid (Shefrin and Statman 1985; Odean 1998).

1.6.3 Motivational

One of the toughest biases to overcome is when self-interest leads to overly optimistic predictions. Many companies require proposals for a new product or service to include an estimate of the ROI. To be approved, the new concept must have an ROI that surpasses some established hurdle value. Not surprisingly, the team that has generated and worked diligently on the new concept may consciously or subconsciously overestimate market potential while underestimating the time and money it takes to bring the concept to market. Light rail and other mass transportation projects, for example, often generate motivationally biased forecasts. In one study of 10 light rail projects, the projected ridership was 15%–75% above the actual. Construction cost overruns averaged 150% and operating costs averaged 200% above forecasts (Nutt 2002).

1.7 What Makes a Decision Difficult?

Decision making is difficult for reasons we have grouped into three categories: impact, problem complexity, and context. Decisions with a major impact may keep you awake at night worrying. If the problem is complex, it will be difficult to sort out the factors, account for all of the issues, and assess the likely impact of your decision. Last, many decisions are not made in a vacuum: they are complicated by the need for others to be involved in the decision and by a process compressed due to time pressures.

1.7.1 Major Impact

Decisions become tougher as more hangs in the balance. On a personal level, we all periodically face life-changing decisions. Where to go to college? Whether or not to get married, and if so, to whom? Whether or not to have children, and if so, how many? Which community to live in and which house to buy? No less difficult are the decisions we make at the end of our lives. When faced with a serious illness, for example, we often must decide—for ourselves or for those we love—on a treatment option that involves a trade-off between quality of life and projected length of life.

Many of our high-impact personal decisions are job related. A career choice sets the educational foundation for a wide range of future decisions, as may one's first job choice. However, as societies change and people switch jobs and even careers more frequently than in previous generations, the consequences of each decision decline in significance. Although money often seems to dominate the decision-making process, other factors related to job environment are, in fact, more important when it comes to job satisfaction. And toward the end of an individual's career, one of the more difficult life-changing decisions—one that seems to be occurring more frequently, especially in the biggest corporations—is whether or not to accept the offer of a buyout or early retirement.

At the corporate level, mergers, acquisitions, and bankruptcy impact the very essence of a corporation as well as its tens of thousands of workers. Major internal organizational restructuring options should be, but are not always, treated as tough decisions. The process is further complicated by the inability of management to quantify the short-term as well as the long-term effects of their decisions. In some instances, these decisions force a paradigm shift upon the organization as it attempts to redefine itself. But aside from obviously significant restructuring decisions, there are other seemingly smaller decisions that may ultimately have equally great impacts on an organization. Outsourcing one segment, such as IT, is an especially tough call, one that is usually extremely difficult to reverse. This only adds to the pressure to get it right the first time.

1.7.2   Problem Complexity

Among the factors that increase decision complexity are multiple objectives and uncertainty. The challenge of balancing and trading off multiple, often-conflicting objectives adds complexity to every service and product development project. There will always be financial objectives to consider alongside of performance objectives. In addition, minimizing time to launch is often an implicit objective, especially in today's highly competitive market.

Decisions related to the selection of plant equipment involve balancing objectives such as operational issues, space requirements, training requirements, and cost. Decisions related to selecting a supplier involve both multiple objectives and elements of uncertainty if the company has not worked with the supplier before. There are issues of cost, quality of work, timeliness, and responsiveness to concern. Uncertainty complicates the comparison of alternatives, especially when the level of uncertainty is not equal across the alternatives, such as when choosing between a proven and unproven technology. The unproven technology is surrounded by uncertainty regarding the time it will take to complete myriad engineering design tasks as well as the cost and resources required to implement and maintain it. Other concerns relate to whether or not the new technology can be implemented in the time allotted and whether or not the launch of a product or manufacturing start-up might need to be delayed.

Every new product or service faces uncertainty regarding market demand. This uncertainty creates ambiguity regarding predictions of revenue and profitability. Unknowns surrounding competitive actions compound the randomness. Every policy regarding the handling of customer complaints and whether or not to provide financial compensation faces uncertainty regarding a disgruntled or satisfied customer's future behavior and how that behavior might influence others in his social group.

The analytic techniques of this book are designed to assist in selecting the single best option among distinct alternatives when faced with multiple objectives and/or uncertainty. However, there are many other decision contexts and various other factors that affect decision complexity, and the recommended modeling approaches for these are beyond the scope of this book. We list some other modeling techniques in the appendix at the end of the chapter.

1.7.3   Personal and Organizational Context

The context in which a decision is made can also increase its difficulty. If you are a wealthy orphan deciding on which college to attend, the decision is yours alone. In contrast, if your parents are alive and well and are paying for your college education, the decision dynamic becomes more interesting. The level of involvement will also vary as a result of whether or not your parents are also college graduates. At the minimum, you will have to provide a valid justification and explanation of your decision.

The same concept applies in hierarchical organizations in which critical decisions must be reviewed and approved by multiple layers of management and possibly across organizational divides

TABLE 1.1:   Context complicates decisions.

Management turnover	Multiple organizational perspectives
Time pressure	Global cultures
Competitive pressures	Negotiated decision
Strong personalities	Poor quality and availability of data
Dynamic environment	Competing interests: equity versus efficiency
Long lead times to implement	

such as finance, marketing, and engineering. If the decision is a technical one, there may be an added challenge that can vary across national cultures. In an American company, the technical expert may find he has to explain the decision to higher levels of management who are not technically sophisticated and know significantly less about the technical issues and challenges. However, much of the time spent defending the decision will focus on estimates of time and resources required while management may downplay the project's complexity, since it neither comprehends nor appreciates this aspect of the endeavor. If there is frequent management turnover, the requirement to justify a decision can become a nightmare. In contrast, in a German or Japanese company, the individual will often have to spend more time justifying the technical elements of a decision, as the higher-level officials are likely to have more expertise than he.

Time pressure is perhaps the single most common factor that increases the difficulty in making high-quality decisions (Svenson and Mauke 1993). It forces managers to take shortcuts such that there may be only enough time to evaluate one or two alternatives. Time pressure precludes taking a step back to look at the big picture, including multiple objectives. Complex decisions are simplified to absurdity. When time runs short in a group decision, the influence of personality becomes even greater. Strong personalities may push through decisions that are inadequately evaluated and negate other alternatives without allowing a full hearing. Table 1.1 lists contextual factors that complicate decision making.

Decisions that cut across organizational boundaries will be more complicated, particularly when it comes to reaching a consensus. In today's global economy, products and supply chains are likewise global, and meetings and decision teams are likely to involve experts and managers from diverse national cultures. These experts bring different values to discussions, the dynamics of which are often culturally sensitive.

Public sector decisions involving power plants, resource allocation, public transportation, or governmental regulations face a special trade-off decision. They must juggle both what is best for the public and what is equitable for the various special interest groups affected by the decision.

Negotiated decisions represent a unique class. Two sides are approaching the same issue, each from its own perspective. Usually, each side's objective is to get the best deal for itself that the other side will accept. Analytic tools can help any one side evaluate a contract offer; however, other skills are required in order to obtain the best results, especially if the two sides expect to maintain a mutually beneficial long-term relationship.

1.7.4   Fuel Tank Example

Let's illustrate the decision context challenge with what may seem a relatively narrow automotive concern: the design and location of a fuel tank. It would be nice if one human being had all the knowledge and information required to make this technical decision, but that is extremely unlikely. The decision team must have knowledge of material science as it relates to the fuel tank itself, the sensing devices within the tank, and the impact of diverse gasoline additives that vary across gasoline retailers. The team must understand manufacturability, packaging, and survivability under various crash conditions, not all of which can be tested in the laboratory. They must also be up to date on all the latest environmental regulations, current and proposed. In all likelihood, they will also need the blessing of finance officials to certify that their investment and variable cost estimates are accurate and in line with those that have been authorized for the targeted vehicle program. Finally, they might need a marketing analyst to assess the importance of the capacity of the fuel tank. Smaller size might ease issues of packaging but may place the intended vehicle at a competitive disadvantage. The challenge to the team is to draw upon and integrate the narrowly focused expertise of individuals from broadly different backgrounds—engineering, marketing, and finance.

Now imagine the fuel tank is to be used on a vehicle to be marketed around the globe. Fuel mixtures vary from country to country. Vehicle operating conditions, which might stress the fuel system, also vary in terms of ambient temperature and road conditions.

If all these factors have not created enough pressure for the decision team, someone brings up the infamous Ford Pinto gas tank that could not withstand certain rear-end crashes and resulted in the loss of millions of dollars in lawsuit judgments. Then, someone else jumps in with the story of CBS reporters and producers staging a side impact crash on a GM truck and then artificially creating a fireball explosion. In the end, the decision maker must decide while facing multiple levels of management scrutiny and recognizing the fact that there is no one viable gas tank that can withstand every possible crash scenario.

1.8 Symptoms of a Poor Decision-Making Process

Reliance on intuition as the primary decision arbiter is but one symptom of a poor decision-making process. Other common symptoms include a tendency to consider just one alternative, failing to look at the big picture, frequently revisiting one's decisions, allowing strong personalities to drive decisions, ignoring uncertainty, overusing inexpert opinion, and establishing a weak link between decisions and implementation.

1.8.1 Narrow Focus: One Alternative and One Objective

One of the most common symptoms of poor decision making is a tendency to frame decisions around a single alternative.

Should we set up low-cost manufacturing in China?
Should we buy a new software technology to streamline product development?
There is a new high-tech gizmo that can be added to a car; should we design it into the next product?

These sorts of question should not be considered in a vacuum; too many factors hang in abeyance. And yet, strangely enough, four out of five decisions consider only one idea (Nutt 2002).

Alternatively, do you find yourself bombarded with these types of questions from upper management in your organization while you are trying to get your work done? Does every new idea, repackaged idea, technology, material, or business opportunity that a manager learns about generate a study as to what the company should do? Does every new product or service your competitor launches lead to the question of whether your company should offer this product or service as well? Worse yet, are these requests framed with a sense of urgency?

While rushing to set up low-cost manufacturing in an emerging market, a certain company went through a series of wasted initiatives. This company chose a candidate product and identified a suitable manufacturer only to find out months later that one of this product's components could not be manufactured at the quality level necessary for inclusion in a mainstream product in the United States. The mistake was repeated as the company considered product after product for manufacture in the emerging market. Sadly, but not atypically, this dynamic was triggered in part by a senior executive announcement that within 12 months the company would be importing hundreds of millions of components from this low-cost country. The powers that be did not take the time to frame the issue as a set of related decisions or to broaden the discussion.

The discussion should have included the following questions:

- Which of our company's products should be manufactured in China, Mexico, Eastern Europe, or somewhere else?
- What are the risks, and what strategies can be employed to mitigate those risks?
- How can the company streamline product development and how does global manufacturing play a role in this process?
- Can manufacturing in a low-cost country also add value to our products or only cut costs?
- Can low-cost manufacturing open up new market segments for our products?

On a personal level, do you tend to frame your decisions as "Should I buy this car that is on sale?" instead of "Which car should I buy?" Much of price-reduction marketing is designed to get the consumer to focus on the one alternative, here and now, rather than consider the big picture that includes a range of alternatives. Similarly, do you live in a city in which local officials suddenly decide that there is a need for a new city hall, a new high school, and a new library or fire station, without delineating the process through which this important decision came about? If you are lucky, you get to vote yes or no on a bond issue, but you are unlikely to be made aware of any serious debate with regard to the range of options, leasing versus buying, or facility size.

One factor that contributes to limiting the range of alternatives and narrowing the focus is time pressure (Svenson and Mauke 1993). If you often hear around the office statements such as "We do not have time to consider other alternatives" or "We need to make the decision now!" then the decision-making process is flawed.

6. **Activity:** Describe the last time a senior executive or boss came up with one specific new alternative and asked you to evaluate it. Was there time pressure? Alternatively, describe a yes or no facility decision of your local government. Would it have made sense to look at a broader range of alternatives?

Has your organization blindly pursued one overriding objective to the detriment of other factors that would make an organization healthy and successful in the long run? Senior executives of American corporations have been on a short-term stock price binge for almost a decade. How many groups have faced serious cost-cutting drives while paying lip service to quality and customer service? What about market share as the focus and "to hell with profitability?" "Thou shalt cut inventory to the bone" is another mantra that ignores the need for safety stock to handle fluctuations in demand or in the supply chain

7. **Activity**: Describe a specific decision that was heavily influenced by your company's pursuit of one objective to the detriment of an entire range of other measures.

1.8.2 Decision Arbitrarily Revisited

Have you ever watched as turnover in management resulted in a comprehensive review of issues that you thought had long ago been resolved? American organizations move their managers from job to job more frequently than their counterparts in Europe and Japan. The reason is to increase the breadth of these managers' experience as they move up the corporate ladder. The individual may stay less than 2 years before moving to a whole new area. He is unlikely to develop in-depth knowledge within his sphere of responsibility. However, rather than simply accepting the opinions of those with technical expertise and abiding by earlier decisions, a new manager may be driven to make an impact quickly. Thus, many prior decisions are up for review even though the circumstances surrounding the original decision have not changed significantly.

The staff regathers the data, updates the presentation, and begins the debate all over again. Unfortunately, the lack of a structured decision-making process means that the earlier so-called final decision had an irreproducible dose of gut feel. It was also influenced by the dynamics of the decision meeting that may have involved a different mix of experts.

1.8.3 Personality-Driven Decisions

Another challenge arises when decisions are heavily influenced by strong personalities in leadership roles. If an executive states his support for one position and the result is that few if any negatives are allowed to be presented, then your organization has a dysfunctional decision-making process. If much of your time is spent trying to find out what one executive is likely to think about an option, your decision-making process is misdirected. If meetings are dominated by individuals with the highest rank, your organizational structure is dysfunctional. If much of the time of your support staff is spent proving that a preconceived solution is the best rather than assessing the potential of the alternatives, you have a problem of misplaced analysis.

Worse yet, an organization might have a "multiple-personality disorder." Strong personalities and their supporters in various parts of the organization may strongly advocate their specific agendas. Instead of open discussion and debate on merits and weaknesses of different alternatives, each side only presents the positives for its preferred choice, hiding the negatives so as not to undermine its cause. In the end, one side "wins" and the other "loses," which inspires the latter group to work, either passively or actively, to undermine decision implementation (Garvin and Roberto 2001).

> **8. Activity**: Describe an instance in which a strong leader did not allow for adequate analysis and discussion of strengths and weaknesses.

1.8.4 Ignoring Uncertainty

Market demand is unknown, technologies are unproven, product development timelines contain many uncertainties, and competitors are unpredictable. Yet, too many organizations still use only single-point estimates to guide their decision making. As an admission of their planning or forecasting fallibility, they add modest buffers to budgets, timelines, or manufacturing capacity. Yet, they shy away from explicitly acknowledging and analyzing the depth and breadth of the uncertainty surrounding a decision or initiative.

One of the major themes of this book is that uncertainty should be articulated, communicated, analyzed, and anticipated. Forecasting uncertainty requires greater knowledge and experience than coming up with a single-point estimate. Uncertainty is neither an admission of ignorance nor evidence of weak, unfocused leadership (Shephard and Kirkwood 1994). It is recognition of an uncertain reality. Einstein may or may not have been right when he said, "God does not play dice with the universe." However, in the absence of prophets who might have an inside track into perceiving God's will, we live in an uncertain dynamic world, especially with regard to technology.

9. **Activity**: Describe an instance in which a decision was made while ignoring a broad range of obvious uncertainties that would affect the outcome of the decision.

1.8.5 Inexpert Opinion

Many decisions involve data collection and extensive analysis. In a technical environment, this could also include complex testing. And yet, in the final analysis, a significant amount of subjective judgment is used to complement the data collection. Few decisions are exactly the same as the one before; the data are not complete and cannot cover all situations. There are always new facets and challenges to consider when new ideas are integrated. Moreover, the egalitarian nature of many U.S.-style decision meetings and conferences allows everyone present to offer an opinion on every issue brought up. This can be counterproductive; not all people at a meeting are equally qualified to offer their opinion. With apologies to George Orwell, some are "more equal than others," namely those who have expertise in particular areas. This problem is compounded by the American philosophy of management, which posits that even the managers of technical groups do not need to be experts in their respective fields in order to be good managers.

This prevailing sense of egalitarianism dictates that experts with decades of experience in their specialty must cope with representatives from other specialties chiming in at meetings. Worse yet is when, for example, the finance staff asks, "Why can't you get by with less time, money, and personnel?" or "Why can't this object be made of a lighter, less costly material?" Conversely, engineers do not hesitate to offer their opinion on market trends and whether or not the product fits the customer niche, even if there is little overlap between the engineer's experience or expertise and the concerns of the targeted customers. One senior automotive executive declared his objection to a vehicle's sound system because he had just bought a $10,000 sound system for his home and claimed that this qualified him to know what a good car sound system should sound like. He was oblivious to the fact that the vehicle was targeted at the low end of the price spectrum, and he was 30 years older than the average purchaser and had different values.

10. **Activity**: Describe an egregious example of an individual offering an opinion on a technical issue outside his area of competence.

1.8.6 Decisions Poorly Linked to Later Management Actions and Little Accountability

Has your organization decided to make a particular process paperless, change suppliers, change a material of a critical component, add a major new facet to a new product, or redesign a manufacturing process? All these decisions require extensive follow-up, throughout and beyond the process of implementation. Rarely are complex decisions simple to implement, especially with limited resources and an environment in which staff are already stretched to their limits.

A decision that goes unimplemented is not much of a decision. Yet, ease of implementation is not necessarily factored in when evaluating alternatives. All too often, key stakeholders with primary responsibility for making things happen may feel that their experience was not adequately considered in the decision-making process and have therefore not bought into the decision. They hope that the decision will just go away so that they can get on with their regular jobs without the added hassle of one more poorly thought out top–down decision.

Complex decisions cut across multiple organizational functions. The interdisciplinary team that was involved in the decision may not have sorted out areas of responsibility for the implementation and issues of coordination. The high-level decision makers may not even have direct responsibility for the groups that will bear the major burden for implementation.

Oftentimes, implementation carries added costs that were not budgeted for at the decision-making stage. State and federal government agencies and decision makers are notorious for forcing significant change without considering how these changes will be paid for. The No Child Left Behind Act, for example, mandated significant testing, monitoring, and process improvement requirements in local school districts without providing resources to sustain this initiative.

11. **Activity**: Describe a context in which a decision did not adequately account for difficulty of implementation. Explain what was lacking. Were there any significant unbudgeted costs associated with implementation?

1.8.7 No Feedback Loop on Decision Quality

Most amazingly, once a decision has been made, organizations often do not have processes in place to provide feedback on the quality of the decision. Decision makers are thus not held accountable for the impact of their decisions, especially in the long term. Even though ROI is a major factor in a wide variety of organizations and decision contexts, few companies actually track the ROIs for each decision. In such cases, there is no way of knowing whether forecasts of ROI were reasonable to begin with, based on realistic forecasts, or whether they were artificially inflated to justify decisions. Nor could they possibly know what, if any, systematic biases were built into the ROI estimates. In one set of interviews of the top 10 leaders of a corporate division, each admitted that he used personal adjustment factors when hearing presentations involving forecasts. These adjustments depended on their personal experiences with the presenter or organization in question and were rarely discussed openly.

In an uncertain world, the need for decision and implementation tracking goes beyond assessing the final outcome. As implementation proceeds, new information is gathered, and the underlying uncertainty is resolved. If in the original decision, uncertainty was clearly explicated, a flexible risk management plan could have been concurrently developed and rolled out as needed. Unfortunately, if the original decision ignored uncertainty, there was no justification for investing time and energy in developing a risk management strategy. As events unfold, not necessarily according to the single scenario planned for, management can only tweak the decision implementation in order to reduce the negative impact of unplanned for contingencies in an uncertain world.

Many of these symptoms of a poor decision process, such as a failure to consider uncertainty, apply equally to personal decisions. The following activities below ask you to score your organization and your personal decision processes.

12. **Activity**: *Organization*—Score each of the symptom categories listed in Table 1.2 (0 = not a problem; 1 = occasional problem; 2 = recurring problem; 3 = major problem) as they arise in your organization. A total score of 9 or less is excellent, 10–12 is good, 13–15 is problematic, and 16 or more is poor.

13. **Activity**: *Personal*—Score each of the symptom categories listed in Table 1.3 (0 = not a problem; 1 = occasional problem; 2 = recurring problem; 3 = major problem) as they arise in your life. A total score of 6 or less is excellent, 7 or 8 is good, 9–12 is problematic, and 13 or more is poor.

TABLE 1.2: Symptoms of an organization's poor decision process.

Symptoms	Score
(1) Too few alternatives (often only one)	
(2) Multiple objectives often not considered	
(3) Uncertainty ignored	
(4) Decisions arbitrarily revisited	
(5) Strong personalities dominate	
(6) Inexpert opinion affects decisions	
(7) Decisions poorly linked to implementation	
(8) Lack of long-term accountability for decisions	
(9) Other—specify	
Total score	

TABLE 1.3: Symptoms of a person's poor decision process.

Symptoms	Score
(1) Too few alternatives (often only one)	
(2) Multiple objectives often not considered	
(3) Uncertainty ignored	
(4) Decisions arbitrarily revisited	
(5) Family and friends overly influence your decisions	
(6) Delay making decisions as long as possible	
Total score	

1.9 Transparent and Efficient Decision Making

A primary factor of quality decision-making is transparency, clarifying the rational basis for a decision so as to facilitate effective implementation. Participants and stakeholders should understand the basis for a decision even if they do not necessarily agree with the final viewpoint. The starting point for this process is developing the appropriate frame for the decision using a structured method. The tool we propose for representing the decision frame is an influence diagram. The decision frame defines the scope of the decision or decisions, the timeframe to be covered, the underlying assumptions, the key objectives, and the main uncertainties. As discussed in Chapter 2, this descriptive tool can be directly linked to analytic modeling techniques involving multiple objectives and decision trees.

The modeling framework also explicitly incorporates uncertainty. Experts are interviewed individually or in small teams to obtain reliable estimates of the range of values. The structured interviews are designed to reduce, as much as possible, the standard types of biases that arise in making forecasts. Each area of expertise is explicated separately to avoid having individuals offer opinions in areas outside their expertise. Conversely, this method enables all parts of the organization to contribute to the decision-making process by offering their views within their areas of expertise.

The openness and clarity of the process facilitates communication and consensus from multiple organizational perspectives. This should lead to a broader commitment to action. Even more important, the clarity of decision making leads to significantly greater efficiency, notwithstanding that the first time this structured process is employed, it will likely take longer than past decision making. Moreover, critical decisions often undergo repeated reviews. As new information becomes available or uncertainties resolve, the decisions will need to be updated. This structured process, built around a mathematical model, is easily updated as model parameters change. Finally, as executives come and go, changes can be captured in the weights assigned to various objectives. This is also easily accommodated without revisiting the entire decision-making process.

The analytic modeling tools that are integral to the decision-making process are computer based. The software is designed not only to identify the best alternative but also to facilitate an assessment of its strengths and weaknesses. The software enables the decision maker to assess the robustness of the best alternative to changes in key parameters. The explicit modeling of uncertainty and objectives also facilitates the development of a risk management strategy and the creation of hybrid alternatives that are better than any of the original set of alternatives. The computer-implemented structured model is simple to update as new information becomes available. It also generates a consistent review process.

Are you motivated to consider an alternative to your current decision-making dynamic? Or are you just confused? There is an alternative to the modeling tools and approach we are proposing here, and it has certainly withstood the test of time. This method involves using a consultant with expertise in stargazing or astrology. More widely used alternatives today are based on two parts of the human anatomy: gut-feel and seat-of-the-pants. Then, there is always the classic American strategy of aggressive debating until truth wins out. The process presented here is not designed to discourage healthy debate but rather to structure and focus this debate around specific strengths and weaknesses that underlie assumptions as well as the quality of data and expert opinion (Garvin and Roberto 2001). Finally, you can simply list the pros and cons of the various alternative methods and see which list is longer.

You may feel comfortable continuing with gut feeling/seat-of-the-pants decision-making process or heated personality-driven debates, if that is what your organization relies on. But if, instead, you are motivated to consider an alternative to your current method, or are simply confused by what passes for process in your organization and would like to change it, then consider the approach and modeling tools described in the succeeding chapters.

Appendix 1. A: Other Modeling Tools

In this text, we present modeling techniques that are applicable to a wide range of difficult decisions but by no means all situations. Our primary decision focus involves *discrete choice decisions* involving a *limited number of comprehensive alternatives*, generally no more than 10, that face significant uncertainty and/or require trade-offs among multiple objectives. The goal of the analysis is to identify the single best alternative and to understand the strengths and weaknesses of the best and near best alternatives. In studying this methodology, there is a danger that the reader will try to fit every decision challenge into the paradigms presented here. One may mistakenly believe that we have provided the manager with a hammer and that, from now on, every decision can be reduced to trying to hit the nail on the head. In fact, however, a particular problem context might not require a hammer at all, but rather a screwdriver or saw. In Chapter 3, we provide decision templates that should help the reader understand the decision contexts that can be appropriately addressed with the modeling tools of this text. However, the reader interested in learning about a broader array

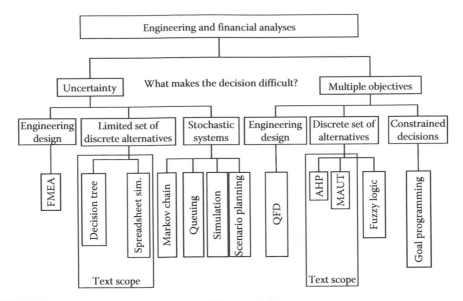

FIGURE 1.1:   Systems and decision modeling techniques.

of decision modeling tools may turn to any one of a large number of survey texts in the field of operations research and management science. The modeling techniques presented in Figure 1.1 broadly fall into two categories: probabilistic and deterministic.

1.A.1   Probabilistic Models

Randomness and uncertainty arise in a wide variety of decision and management contexts. One key element of many decision support systems involves forecasting models used to continually predict critical variables such as demand. Inventory management is a fertile area for the use of probabilistic models to cope with the projected and actual random demand. Probabilistic models are also essential to the field of financial engineering.

Designing and developing a new product always involves initial uncertainty as to how well the design will perform. Modeling techniques that are used to support the design, testing, and refinement of a new product include reliability, design of experiments (DOE), analysis of variance (ANOVA), and Taguchi methods. If the focus is to identify and quantify possible sources of component or system failures, failure modes effects analysis (FMEA) is an effective tool.

Decision trees incorporate a collection of random events, which are generally independent of one another or at most linked through conditional probability. However, randomness and uncertainty can pervade an entire complex system such as a production line or airport. Modeling tools for analyzing these interconnected *stochastic systems* include simulation, queueing theory, Markov chains, and hierarchical inventory models. For example, manufacturing plants and airport runways are often modeled through simulation. Queueing theory is used to model the performance of telecommunications systems, toll booths on a highway, or tellers in a bank. Markov chains are the basis for a number of inventory models and customer loyalty analysis.

1.A.2   Deterministic Models

Product mix planning and the associated production often involve thousands if not tens of thousands of decision variables in the presence of large numbers of constraints. These decision challenges are often addressed using mathematical programming models that maximize profit or minimize cost in the presence of thousands of constraints. Similar deterministic models are used to develop schedules for airline crews or process and blend petroleum products. This class of models can be used to address a number of logistics and supply chain operational decisions.

There is another whole group of deterministic models that have been developed around a graph-network structure involving arcs and nodes. These include routing vehicles or shipping product through a network or selecting an optimal subset of nodes for a network of facilities such as new car dealerships.

Multiple objectives are also a factor in every engineering design. Quality function deployment (QFD) is a technique that helps designers identify customers' most critical desires and then converts their concerns into design performance metrics while striving to maximize the quality assurance of the final product. Goal programming is a more structured but less widely used tool. This tool is an extension of mathematical programming in which the objective function is a weighted sum of deviations from each of a series of goals or objectives. The weighting reflects the relative importance of the goals to the decision maker.

Exercises

Complete Chapter Activities

1.1 Coin flip: You are about to flip a coin eight times. Please choose which of the following outcomes you expect.

    a. An equal number of heads and tails

    b. The number of heads and tails differs by exactly two

    c. The number of heads and tails differs by three or more

    Now flip a coin eight times.

    Record the number of heads _____ and tails _____ and the net difference _____.

    Did the outcome match your choice? _____ If the choice and outcome agree, does that mean that you made the best decision? _____

1.2 Can you point to a situation in which you believe your organization made an extremely risky decision to save a buck, such as using an unproven technology or inexperienced supplier, that you did not think was justified but the results turned out satisfactorily? Were decision makers rewarded because of the outcome? Explain.

1.3 Can you point to a personal situation in which you or someone you know made an extremely risky decision that in retrospect was not really justified but the results turned out satisfactorily? Explain.

1.4 Can you point to situation in which you believe your organization made a reasonable choice in an uncertain world but the results turned out unsatisfactorily? Were decision makers punished because of the outcome? Explain.

1.5 Can you point to a personal situation in which you or someone you know took a realistically evaluated risk but the results turned out to be unsatisfactory? (e.g., you spent a good deal of time gathering available information about a new job offer but the company went bankrupt a year later due to corporate executive misinformation.) Explain.

1.6 Describe the last time a senior executive or boss came up with one specific new alternative and asked you to evaluate it. Was there time pressure? Alternatively, describe a yes or no facility decision of your local government. Would it have made sense to look at a broader range of alternatives?

1.7 Describe a specific decision that was heavily influenced by your company's pursuit of one objective to the detriment of an entire range of other measures.

1.8 Describe an instance in which a strong leader did not allow for adequate analysis and discussion of strengths and weaknesses.

1.9 Describe an instance in which a decision was made while ignoring a broad range of obvious uncertainties that would affect the outcome of the decision.

1.10 Describe an egregious example of an individual offering an opinion on a technical issue outside his area of competence.

1.11 Describe a context in which a decision did not adequately account for difficulty of implementation. Explain what was lacking. Were there any significant unbudgeted costs associated with implementation?

1.12 Use Table 1.2 to assess your *Organization's* decision-making process—score each of the symptom categories listed in the following (0=not a problem; 1=occasional problem; 2=recurring problem; 3=major problem) as they arise in your organization. A total score of 9 or less is excellent, 10–12 is good, 13–15 are problematic and 16 or more is poor.

1.13 Use Table 1.3 to assess your *Personal* decision-making process—score each of the symptom categories listed in the following (0=not a problem; 1=occasional problem; 2=recurring problem; 3=major problem) as they arise in your life. A total score of 6 or less is excellent, 7 or 8 is good, 9–12 are problematic and 14 or more is poor.

Discuss the Factors That Made the Following Decisions Difficult

1.14 President Obama's decisions to send additional troops to Afghanistan. Clarify the components of the decision.

1.15 The decision whether or not to include a public option in the health care legislation of 2010.

1.16 The decision to fire Fritz Henderson as CEO of General Motors less than a year after he successfully led GM into and out of bankruptcy.

1.17 Identify a difficult decision at your organization that was made within the past 2 years by you, your manager, or a higher-level manager you interact with.

1.18 Identify a difficult decision at the local or state government level that was made within the past 2 years that could impact you.

1.19 Identify a difficult personal decision that you or a family member made within the past few years.

For questions 1.17, 1.18, or 1.19 discuss in 600 to 800 words all of the following:

a. The decision context and the specific decision that was made.

b. Major subsequent decisions, if any, influenced by this decision.

c. The primary objective and any secondary objectives that drove the decision.

d. Describe the factors that made it a hard decision. The factors should be grouped under the categories "major impact," "problem complexity," "personal or organizational context" as discussed in the text. The discussion of these factors and their categorization is the major focus of the write-up. Do not spend too much time describing the technical details that made the decision hard.

e. The dollar magnitude of the decision.

f. The risks associated with the decision.

g. Time pressures if any.

h. If the decision was revisited, explain the circumstances.

i. Constraints surrounding the selection of viable alternatives.

j. Globalization's potential impact or role in the decision, if any.

k. What concerns would you have with the quality of the process used to make the decision? (This is an important issue.)

References

Bazerman, M. H. (2006). *Judgment in Managerial Decision Making*, 6th edn., Hoboken, NJ: John Wiley.

Bazerman, M. H. and Neale, M. A. (1992). *Negotiating Rationally*. New York: Free Press.

Bonabeau, E. (2003, May). Don't trust your gut. *Harvard Business Review*, 81, 116–123.

Cartwright, S. and Schoenberg, R. (2006). Thirty years of mergers and acquisitions research: Recent advances and future opportunities. *British Journal of Management*, 17, S1–S5.

Chaudhuri, S. and Tabrizi, B. (1999, September). Capturing the real value of high-tech acquisitions. *Harvard Business Review*, 77, 123–130.

Clemen, R. T. and Kwit, R. C. (2001). The value of decision analysis at Eastman Kodak Company. *Interfaces*, 31(5), 74–92.

Davison, R. (1997). An instrument for measuring meeting success. *Information Management*, 32, 163–176.

Dean, J. W. and Sharfmann, M. P. (1996). Does decision process matter? A study of strategic decision-making effectiveness. *Academy of Management Journal*, 39, 368–396.

Eccles, R. G., Lanes, K. L., and Wilson, T. C. (1999, July). Are you paying too much for that acquisition? *Harvard Business Review*, 77, 136–146.

Fuller, S. R. and Aldag, R. J. (1998). Organizational tonypandy: Lessons from a quarter century of the groupthink phenomenon. *Organizational Behavior and Human Decision Processes*, 73(2/3), 163–184.

Garvin, D. A. and Roberto, M. A. (2001, September). What you don't know about making decisions. *Harvard Business Review*, 79, 109–116.

Hayashi, A. M. (2001, February). When to trust your gut. *Harvard Business Review*, 79, 58–65.

Iyengar, S. S., Wells, R. E., and Schwartz, B. (2006). Doing better but feeling worse. Looking for the "best" job undermines satisfaction. *Psychological Science*, 17, 143–150.

Janis, I. L. (1972). *Victims of Groupthink*. Boston, MA: Houghton Mifflin Company.

Keeney, R. L., Raiffa, H., and Hammond III, J. S. (1998, September–October). The hidden traps in decision making. *Harvard Business Review*, 84, 47–58.

Klein, G. (1999). *The Source of Power: How People Make Decisions*. Cambridge, MA: MIT Press.

Lovallo, D. and Kahneman, D. (2003, July). Delusions of success: How optimism undermines executives decisions. *Harvard Business Review*, 81, 1–9.

Matheson, D. and Matheson, J. (1998). *The Smart Organization: Creating Value through Strategic R&D*. Boston, MA: HBS Press.

Matheson, D. and Matheson, J. (2001, July–August). Smart organizations perform better. *Research Technology Management*, 44, 49–54.

Nutt, P. C. (2002). *Why Decisions Fail: Avoiding the Blunders and Traps That Lead to Debacles*. San Francisco, CA: Berrett-Koehler.

Odean, T. (1998). Are investors reluctant to realize their losses? *Journal of Finance*, 53, 1775–1798.

Parker, A. M., Bruin, W. B., and Fischhoff, B. (2007). Maximizers versus satisficers: Decision-making styles, competence, and outcomes. *Judgment and Decision Making*, 2, 342–350.

Phillips, L. D. and Costa, C. B. (2007). Transparent prioritisation, budgeting and resource allocation with multi-criteria decision analysis and decision conferencing. *Annals of Operations Research*, 154, 51–68.

Quinn, R. E. and Rohrbaugh, J. A. (1981). A competing values approach to organizational effectiveness. *Public Productivity Review*, 5(2), 122–144.

Rohrbaugh, J. (2005). Assessing the effectiveness of group decision processes. In: *The IAF Handbook of Group Facilitation*, Schuman, S., ed. San Francisco, CA: Jossey-Bass.

Schilling, M. S., Nadine Oeser, N., and Schaub, C. (2007). How effective are decision analyses? Assessing decision process and group alignment effects. *Decision Analysis*, 4, 227–242.

Schwartz, B., Ward, A., Monterosso, J., Lyubomirsky, S., White, K., and Lehman, D. R. (2002). Maximizing versus satisficing: Happiness is a matter of choice. *Journal of Personality and Social Psychology*, 83, 1178–1197.

Shefrin, H. and Statman, M. (1985). The disposition to sell winners too early and ride losers too long: Theory and evidence. *Journal of Finance*, 40, 777–790.

Shephard, G. G. and Kirkwood, C. W. (1994). Managing the judgmental probability elicitation process: A case study of analyst/manager interaction. *IEEE Transactions on Engineering Management*, 41, 414–425.

Simon, H. A. (1955). A behavioral model of rational choice. *Quarterly Journal of Economics*, 59, 99–118.

Simon, H. A. (1956). Rational choice and the structure of the environment. *Psychological Review*, 63, 129–138.

Simon, H. A. (1957). *Models of Man, Social and Rational: Mathematical Essays on Rational Human Behavior*. New York: Wiley.

Svenson, O. and Mauke, A. J., eds. (1993). *Time Pressure and Stress in Human Judgment and Decision Making*. New York: Plenum Press.

Timmermans, D. and Vlek, C. (1994). An evaluation study of the effectiveness of multi-attribute decision support as a function of problem complexity. *Organizational Behavior and Human Decision Processes*, 59, 75–92.

Von Neumann, J. and Morgenstern, O. (1953). *The Theory of Games and Economic Behavior*. Princeton, NJ: Princeton University Press.

Whyte, G. (1991). Decision failures: Why they occur and how to prevent them. *Academy of Management Executive*, 5(3) 23–31.

Chapter 2

Influence Diagrams: Framing Multi-Objective and Uncertain Decisions

2.1 Goal and Overview

The primary goal of this chapter is to develop skills in using an influence diagram to frame decisions involving uncertainty and multiple objectives in preparation for quantitative analysis.

The first challenge in tackling any decision is to create an appropriate frame around it. A decision frame defines the scope of the decision in terms of the factors to be considered, the time horizon, the organizational breadth, and a range of alternatives. In a group decision, it is especially critical to have the team reach consensus as to the appropriate frame before data collection and analysis begins. There is nothing more disconcerting than to proceed through a detailed analysis of several alternatives for sequencing next year's new product or new service launch and then have someone chime in halfway through the process that "we really need to plan the next five years all at once."

There are a number of brainstorming techniques and associated graphic representations that can be used to facilitate achieving a framing consensus. In this chapter, we explore the role of influence diagrams in framing decisions involving uncertainty and multiple objectives. The power of an influence diagram is its direct link to two analytic tools, decision trees and multiattributed utility theory, that move the decision maker from a descriptive statement of the decision problem to a prescriptive analysis. In Chapter 17, which focuses on strategic decision making, we introduce two additional framing tools: the strategy table and hierarchical decision pyramid.

Influence diagrams were first introduced in 1973 at the Stanford Research Institute as a tool to solve decision problems being studied by the Defense Intelligence Agency (Howard and Matheson 2005a, Howard et al. 2006). The decision analyst community soon recognized the value of influence diagrams as an excellent communication tool for solving complex problems. The initial applications were in the petroleum industry, where oil company executives needed to make a decision on whether or not to drill for oil in a particular location. Influence diagrams helped to form the basis for intelligently discussing the major factors that impact a decision and for representing these factors in the form of a diagram for easy understanding and evaluation. Researchers found that the executives would enumerate important variables and their relationships, after which the analysts would ask the executives to define important outcomes and values regarding the decision. Over the years, applications have spread from the oil industry to the pharmaceutical industry. Influence diagrams have also been utilized for medical issues, evaluation of military systems, and virtually the whole spectrum of decision-making problems. The journal *Decision Analysis* dedicated a special issue in 2005 to influence diagrams. Two of the articles focused on influence diagrams' impact on the fields of medical decision making (Pauker and Wong 2005) and artificial intelligence (Boutilier 2005).

A typical complex business decision involves representatives from different parts of an organization who bring differing perspectives to the problem. The major goal of a high-quality decision-making process is to communicate the issues, clarify the problems, and reach an action-oriented decision. An influence diagram is a simple but powerful descriptive tool that facilitates a common vision among decision makers surrounding the decision alternatives and context, whether this

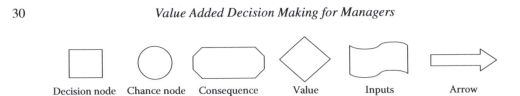

FIGURE 2.1: Influence diagram shapes.

involves a new automotive product at GM (Kusnic and Owen 1992), a medical decision (Nease and Owens 1997), a global product and manufacturing strategy at DuPont (Krumm and Rolle 1992), or a decision related to power company transmission lines (Borison 1995). Howard (1988) considers the influence diagram the best tool for transforming an opaque idea into a clear and crisp decision as well as the greatest advance he has seen in communication, elicitation, and detailed representations of human knowledge.*

In this chapter, we describe the building blocks of an influence diagram and the iterative process used in its construction. The visual modeling tool supports our major underlying premise for decision making: the need to include uncertainty and multiple objectives as integral elements. It also encourages the decision-making group to think about the essence of the decision in question. Is this really one decision or should it be viewed as a sequence of decisions?

2.2 Components of an Influence Diagram

An influence diagram provides a graphic map of the decision problem through six components, as illustrated in Figure 2.1: decision nodes, chance nodes, consequences and objectives, ultimate value, inputs, and arrows.

These elements are represented in an influence diagram by different shapes. Rectangles are used to represent decisions. Circles and ellipses represent the chance nodes that capture the uncertainties that the decision maker believes will influence the desired outcome of the decision. The rounded rectangles represent the consequences or subobjectives that may be of interest to the decision maker. There will be one-rounded rectangle for each of the subobjectives associated with the decision. Each of the objectives should be associated with either the term "minimize" or "maximize." The wavy box describes the deterministic inputs that are needed to support the decision. Finally, a single diamond restates in succinct terms the single ultimate value or goal of the decision. The consequences or multiple objectives lead toward the "overall objective."

The wavy box (inputs) is not a standard part of the literature of influence diagrams. However, we have found that it is important for the group to discuss the data needed to support the decision analysis and to assign responsibility for bringing data to the table. Unlike random events, these data may encompass little or no uncertainty.

These different shapes, referred to as nodes, are linked together by the last major component, arrows. An arrow connecting one node to another (see following text) describes the relationship between the two connecting nodes. By analogy, the nodes of an influence diagram are its vocabulary, with the arrows serving as its syntax—by connecting the nodes.

The influence diagram in Figure 2.2 describes a decision about developing a new late-to-market product scheduled to be launched several months after a competitive product is released. There is, first, the basic decision as to whether or not to develop this product. Two other related decision nodes involve what features to include in the product and the launch price. The major uncertainties are

---

* Glenn Koller (2005, p. 109) argues for a simpler style diagram when focusing on risk assessment. He calls his approach a contributing factors diagram. It includes only elements that contribute to uncertainty.

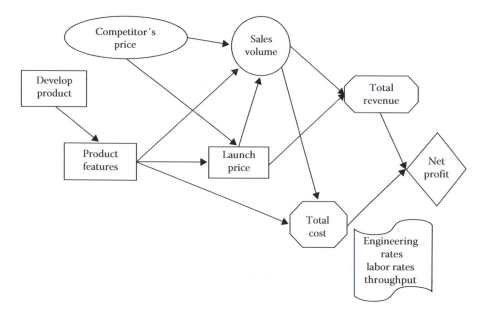

FIGURE 2.2: New late-to-market product development.

the competitor's price and the sales volume. The primary objectives are to minimize total cost and maximize total revenue, thus leading to the ultimate objective, to maximize net profit. The total cost is determined by gathering critical estimates of engineering rates, labor rates, and factory through-put. Later, in this chapter, we discuss this example in greater detail.

The construction of the influence diagram is an iterative process. Initially, it is more important to get the main issues down on the board in the form of nodes and only later worry about the relation-ships (arrows), the decomposition of the decisions, and the overall sequence of decisions and events. The process can be grouped into six steps:

1. Articulate decision(s)
2. Define objectives
3. Identify uncertainties
4. Step back and look at the bigger picture (considering time horizon, organizational breadth, and additional decision details)
5. Link nodes with appropriate arrows
6. Identify critical data needs

Step 1: Articulate decision(s)

The diagram construction process begins by articulating the basic decision, such as choosing a supplier, bidding on a house, selecting a material, selecting a medical treatment, specifying the number of service people, selecting a manufacturing technology, choosing a product design, or determining plant capacity. It is not critical at this stage to work out the details of the decision or whether this decision should be subdivided into multiple decisions. For example, the decision on plant capacity may be a multiyear expansion plan, with each year's expansion a separate decision. If the decision involves choosing a supplier, it is important to clarify whether this is to be a local sup-plier or a global supplier or possibly a pair of suppliers. If the decision is to define a buyout package for employees, will it be just one package or an array of packages?

TABLE 2.1:   Broad categories of common objectives.

Costs (variable and investment)	Time to complete
Profit—NPV, TARR, ROI	Risk of not meeting targets
Human resources required	Management issues
Long-term value	Operational issues
Performance	Sales and/or market share
	Training requirements

Step 2: Define objectives and ultimate value

The decision maker begins to articulate the objectives that are of importance and the final value such as picking the best supplier. When it comes to selecting a supplier, these objectives could include minimizing cost, maximizing quality, and maximizing engineering design capability. In the case of material selection, the objectives could be minimizing cost, maximizing manufacturability, and maximizing durability. When choosing a medical procedure, the objectives are likely to include life expectancy, quality of life, and cost. When buying a home to live in, the primary objectives might include cost, location, size, and amenities. When selecting a manufacturing technology, the major objectives could include minimizing total cost, minimizing space requirements, minimizing training requirements, and maximizing throughput. Table 2.1 summarizes broad categories of common objectives.

At this stage when constructing an influence diagram, it is not critical to think about how these objectives will be quantified or where data will come from. In Chapter 4, we present the concept of an objectives hierarchy, which provides a detailed expansion of these objectives and involves meticulous definitions of measures. To facilitate and stimulate discussion of objectives for a specific decision, we offer a list of categories that cover a broad array of decisions. In addition, in Chapter 3, we present influence diagram templates that can be used as starting points for various classes of decisions.

Step 3: Identify uncertainties

At this step, the team is challenged to define what factors cannot be known with certainty before the first decision is made. Are the projected costs accurate estimates, or can the actual costs vary by 5% or 10%, enough to possibly change the decision? The projections for demand are always uncertain variables, despite the salesmanship of the executive pushing for the development of a new product or service. When it comes to medical decisions, survival and side effects are common unknowns. When merging two companies, the synergies that will actually develop cannot be predicted with certainty. When investing in an emerging economy, political developments are a major uncertainty. Even within the personal domain, such as buying a home or a used car, there can be significant uncertainty regarding future repair costs.

Likewise, the element of time is a variable that must be considered in every decision and every action plan. However, except in the case of routine processes with long track records, time is a key uncertainty. In fact, time uncertainty could appear as a separate node for each activity if the decision maker wonders how long each task takes to complete. Alternatively, this uncertainty may be succinctly captured by a single node corresponding to "Will the project deadline be met?" This single-node form might be used when making or missing a deadline is one of the critical elements of the decision. If the medical decision involves a surgical procedure, the recuperation time the patient needs before returning to work or other daily activities is an uncertainty. On the other hand, a college student developing a semester schedule may wish to specify a separate node to represent the amount of time required by each subject.

Uncertainty with regard to cost is similar to uncertainty of time. The more experience a company has had with similar projects, the less the uncertainty is. Thus, there would be little uncertainty

TABLE 2.2: Common uncertainties.

Time needed to complete task or reach goal	Performance to specifications
	Warranty claims and quality control
Resources required	Competitive actions
Cost	Is task doable?
Market demand	Will some specific event occur, such
Revenue	as who will be elected president or
Throughput–productivity	will a pandemic occur?

surrounding the costs of a new warehouse or even the construction of a well-defined chip factory. In contrast, estimates of the variable cost of production for a totally new product can involve significant uncertainty, especially in the early design phase. In chemical processes and chip manufacturing, cost uncertainty starts with process yield variability. This uncertainty would be compounded if the technology of the production processes were unproven. In addition, cost estimates often fail to accurately factor in the impact of a learning curve (Wells 1993).

When selecting a college to attend, the starting tuition is known, but annual increases are a significant unknown factor. When selecting a health plan, the basic cost structure is known, but total costs for a year after accounting for deductibles will be a function of unforeseen medical problems.

As a design team is given a complex new design challenge, they are unsure as to whether they can deliver a design that will meet specifications within the given time and budget constraints. Additionally, the actual performance of the design upon release at a specified date in the future can be modeled as another uncertainty. This uncertainty arises for equipment, software, or pharmaceutical products. In software products, there will be uncertainty with regard to the number of undetected bugs at the time of release. For a car, the ultimate NVH (noise, vibration, and harshness) or ride and handling will be uncertain until physical prototypes are on the test track. And even after the first tests, there still will be uncertainty as to how much improvement can be achieved within the time allowed. For a drug, there may be years of uncertainty with regard to its effectiveness and possible toxicity.

Table 2.2 is a list of uncertain events, discussion of which can facilitate diagram construction. The terms are generally self-explanatory with one exception. The item, "Will some specific event occur?" refers to a wide range of uncertainties that involve whether or not something happens. For example, a company involved with significant environmental issues is directly impacted by the random event, the outcome of a presidential election. Bars and restaurants near Yankee Stadium see their revenues impacted by whether or not the New York Yankees play in the World Series. Other discrete events that may or may not occur: laid off from a job, Congress passes specific legislation, or a company declares bankruptcy.

While developing an influence diagram, confusion can arise when laying out the uncertainties and the objectives, since there can be overlap between the two entities. A cost can be both an uncertain variable and an objective in terms of the need to minimize. One option is to define total cost as the objective to be minimized and to specify as uncertain one or more highly volatile cost components such as energy cost. We will try to clarify this issue later in the chapter by citing examples. However, it is important to note that there is no single correct design for an influence diagram since the same decision frame can often be represented in multiple ways.

Step 4: Step back and review the big picture (and nodes)

The team should now pause and think in terms of the big picture. How broad a frame should be used for the decision? Can they isolate and decide on manufacturing capacity for a single plant for the coming year? Or must they look at multiple years and multiple plants at the same time? For example, if a company is considering opening an office in the near future in China, can it focus on just this decision or must the decision be integrated into a global strategy that covers all branch decisions in a variety of countries over the next 5 years? Can the company decide on one specific product or service or should the decision-making team look at the entire product strategy? Are they selecting an IT supplier for North America or a supplier who is to be a global partner? Do they need to select one piece of equipment for one plant or consider a common strategy for multiple plants, even if only one piece of equipment is to be bought this year?

There are three key questions with regard to framing or scoping the decision.

1. How many years are to be covered by the decision analysis?

2. How broad a geography must be included so as to capture the interaction between a local decision and a more global decision?

3. How broad a product line is to be considered?

The goal is not simply to define the "right" time frame or scope but for key stakeholders to openly discuss the decision and reach a consensus. It is always easy to suggest that a broad frame be used. However, the broader the frame of reference, the more data is required and the more complicated the analysis will be. Thus, more time will be needed to reach a decision and implement it. There will always be a trade-off between decision-making efficiency and breadth of analysis.

A direct corollary to the three questions is whether a single decision or a series of decisions is to be made. When deciding whether or not to launch a new product or service, is there much to discuss and decide with regard to the price of the service, or is price a given based on the competitive marketplace? In deciding whether or not to open an office in a particular country, will the decision be affected by the choice of city or is the basic decision independent of the exact final location?

After reviewing the decision nodes, the team should focus on the objectives to see if the list is complete. Is profit the primary objective, or is market share more important? Should the team include hard-to-quantify objectives such as the decision's impact on the company's image or reputation? If this is a public sector decision, such as where to locate a library or park, do concerns over the fairness to different population groups arise?

Finally, the list of key uncertainties should be debated. In global decisions, does the team need to worry about uncertain political events in the countries under consideration? Are currency fluctuations a significant factor? Is variability with regard to inflation a concern in the time horizon of the study? Are the costs of implementation well understood, or is there significant uncertainty? What will be the timeline for completion?

Step 5: Link nodes with arrows

Arrows are used to bring the picture together, as well as to define how uncertain nodes and decisions are related or sequenced. Interspersing decisions and random events with judicious use of arrows identifies which events are unknown today as well as which will be known in time for a later decision. For example, in planning a multistage capacity expansion, the decision maker will not know the demand for the coming year, but he may obtain more market information before having to make the final decision in year two for expansion or contraction. Similarly, a plan to launch a family of new products over a staggered timeframe will rely on new data that is unknown when the first product is launched, but becomes clear once the process has begun and before the latter part of the product strategy is rolled out.

The placement of arrows is the single biggest source of confusion in building an influence diagram. The difficulty lies in the fact that an influence diagram is not a flow chart. There are specific rules of arrow placement that, like rules of grammar, may seem counterintuitive. These rules are best discussed and explored through the use of examples in the next section.

Step 6: Data input

In every decision context, much data are required that can be ascertained accurately. Often, these data must be gathered from multiple sources with different parts of the organization responsible for certifying data accuracy. This last set of nodes is intended to clarify data needs and responsibility for collection. In planning for a new product, the demand is an unknown, but the size of the market segment may be known. It will be marketing's responsibility to bring that data to the decision analysis. Although yields and throughput may be random variables, the manufacturing staff will know with some certainty the cost of equipment and related facility costs.

2.3   Learn by Simple Example: Automation Investment

Consider the decision case at Boss Controls (BC). BC is gearing up to manufacture an option that would be made available over the next several years to a total of 1 million purchasers of new cars worldwide. Initial projections posit that the take-rate, the percentage of people who purchase the option, could be low or high. The plan calls for BC to deliver the option to the Original Equipment Makers (OEMs) at a price of $60 each. Timothy O'Leary, VP for imaginative products, is considering two alternatives that differ significantly in the level of investment in automation and the related variable cost of production. The decision is over which automation investment to choose.

The next step in framing the problem is to understand the values of the decision maker. In this simple case, Tim wants to maximize corporate profits by meeting demand for the product. Thus, the overall objective is to maximize the profits, and there is no need of intermediate objectives or consequence nodes. (The word maximize was omitted so as not to squeeze the wording in the figure.) The influence diagram for this case is given in Figure 2.3.

The arrow from the decision node to the value node shows that the automation investment decision influences profits. Similarly, the outcome of the uncertain take-rate also influences profits. The input data box makes reference to the need for data on the volume of the market segment, the variable cost, and the investment cost.

Note that there is no arrow connecting the chance node and the decision node. Absence of an arrow between the chance node and the decision node does *not* mean that the uncertainty in take-rate does not influence the decision. It only means that the probability that the take-rate is low or high is independent of the decision Tim makes with regard to automation. Everything in the diagram affects the optimal decision. We will discuss more about the properties of arrows as we proceed.

In the initial representation, it was assumed that the OEM's forecast of 1 million cars to be sold was on target. This would be true if the demand for the particular vehicle line(s) exceeded capacity

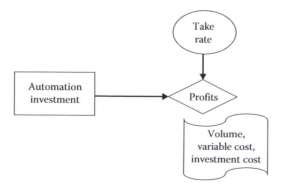

FIGURE 2.3:   Automation investment.

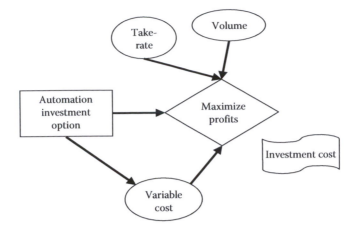

FIGURE 2.4: Automation investment expanded.

or if the company had a consistent policy of offering incentives to keep the lines running at full capacity. In other contexts, this total would also be a random variable and represented by a distinct random node, as seen from Figure 2.4. In theory, the two random nodes, take-rate and volume, could be combined into one, called sales. However, by keeping two separate nodes, the diagram better represents the sources of uncertainty. In addition, there may be uncertainty regarding the variable cost linked to the different investment alternatives. A random node reflects this uncertainty.

Notice that in Figure 2.4, there is an arrow from the decision box to variable cost but not to other random events. Why? The choice of investment options affects the value of the uncertainty associated with the variable cost of manufacturing. However, it does not affect the uncertainty surrounding the demand for vehicles or the take-rate.

2.4 Divide and Delay Decision: Plan an RSVP Theater Party

The Department of Industrial Engineering of Welcome State University is planning its first ever theater party for its faculty, staff, alumni, and special guests. They have purchased 100 tickets for the event and plan an afterglow. The total number of people on their first draft of an invitation list includes more than 500 names. Initially, the primary objective was to maximize the number of people actually attending the theater party. After some thought, however, it was determined that the primary objective was to maximize good will as depicted in Figure 2.5. The major difference in this reformulation is that good will is earned even by inviting people who choose not to attend. The primary uncertainty is the percent of people invited who would respond yes. A secondary uncertainty involves no-shows.

The chair of the department is concerned, since this is the inaugural event and he has no prior data on response rates. He is afraid that if he invites too many, the number of acceptances could exceed the number of tickets, with a resultant loss of goodwill. If too few are invited and the acceptance rate is low, there will be too few attendees, and the department will have missed out on an opportunity to build goodwill, primarily amongst the alumni.

A staff member with experience in planning events points out that the decision has been framed too narrowly. There is enough time before the date of the show to send out two waves of invitations, and, thus, there are two decisions, not one (see Figure 2.6). In the first wave, the number sent would be based on the most optimistic estimate of the percentage of people who will say yes, with a required RSVP window of 3 weeks. At the end of 3 weeks, they would know how many have said

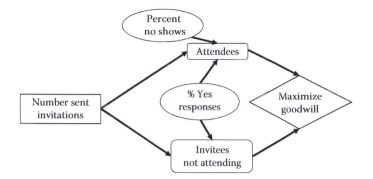

FIGURE 2.5: Theater party invitations.

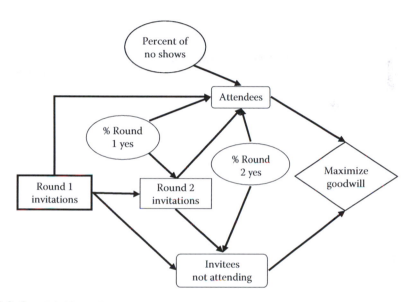

FIGURE 2.6: Divide and delay decision—Theater party invitations.

yes and also be able to estimate the positive response rate. They can use this information to decide how many invitations to send in the second round.

1. *Activity*—Theater party initiations: Expand on this example but without adding more rounds of invitees. What options exist if there are tickets left over? Is it possible to buy more tickets at the last minute?

Decision(s) _____

Uncertainties _____

2.5   Arrows in Complex Influence Diagram: New Product Late-to-Market

The creation of a good and meaningful influence diagram requires judgment and experience. Some of the rules mentioned in the following can make the process of construction easier. We will rebuild Figure 2.2 step by step.

Decision context: A company is considering developing a new multifunction cell phone that will be ready 3 months after its competitor introduces a similar product.

Articulate the decision: The basic decision is whether or not to develop the product knowing that the competitor has a head start. After a little thought, it is obvious that for the cell phone to succeed, it will need some competitive advantage(s) over the competitor's product, which will be released earlier. Therefore, key decisions involve product features and price.

Start at the value node and work back to the decision nodes: It is a good idea to start by recognizing what this decision maker values most and working toward achieving those values (Nease and Owens 1997). For this case, a single objective, Net Profit, describes what the decision maker wants. He has to decide whether launching a product that will be 3 months late to the market can help him maximize net profits. Working backward, he sees that Net Profit is a natural result of two factors: Total Cost and Total Revenue. Each of these is also an objective, because he would like to maximize total revenue and minimize total cost. The number one uncertainty for the company is sales volume. In this example, the company has a clear understanding of the features of its competitor's product but is not sure how the product will be priced. Competitor's price, therefore, is another uncertainty node. The nodes are shown in Figure 2.7.

At this stage, one should not unnecessarily crowd the model by including chance nodes and decision nodes. A complex and crowded influence diagram is both difficult to understand and hard to explain. Include a node only if it will have a significant impact on the decisions and values or if it helps clarify the context. In this example, we have chosen to add the consequence nodes Total Revenue and Total Cost to clarify the issues.

Arrows are used to link different node types, and there are subtle differences of interpretation depending upon the types of nodes that are linked.

Arrow between two decision nodes: This arrow use reflects either a time delay in the sequence of decisions or the presumption that one decision influences the second decision. For example, the pricing decision may depend on the decision on the features designed into the product. If the product will have more features than its competitor, the company may decide to price it higher. Thus, an arrow pointing from the Product Features decision to the Price decision in Figure 2.8a indicates the influence of the product feature decision on the price decision.

However, it may be that product pricing will be driven by market conditions or that the features under consideration will in no way directly influence a pricing decision. In that case, an arrow will *not* be placed between product features and price (see Figure 2.8b), since the pricing decision will not change on the basis of the product features.

Arrow from random node to a decision node: An arrow from a chance node to a decision node shows that the outcome of the chance node will be known before the decision even though that information is not known at present. For instance, in this example, the company's product will be launched after its competitor's. The decision maker would place an arrow pointing from the

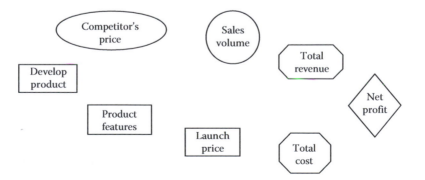

FIGURE 2.7:   Nodes of the new product late to market case.

FIGURE 2.8: (a) Arrow from a decision node to a decision node. (b) Alternative: Arrows from a decision node to two decision nodes.

FIGURE 2.9: Arrow from a chance node to a decision node.

Competitor's Price chance node to the Launch Price decision node in Figure 2.9 to show that the competitor's price will be known at the time of the Launch Price decision.

Never use an arrow from a circle to a decision to represent the fact that the decision is affected by the random event. This is the most common error made when constructing an influence diagram. Remember that everything in the diagram has an impact on all decisions. The mere fact that the random event, Competitor's Price, is in the diagram means that a forecast of what the competitor is likely to do will influence the decision whether or not to develop and launch the product. Nevertheless, there should NOT be an arrow from the Competitor's Price node to the initial decision node Develop Product, because such an arrow would indicate that the decision maker knows the competitor's price before he decides whether or not to develop the product.

Arrow from a chance node to a chance node: An arrow can be used to connect a chance node to another chance node to show the probabilistic conditional dependence of the two. The uncertainty in sales volume may depend on the actions the competitor takes. The competitor may price the product at a level that will make it difficult for others to compete and thus harm the new product's sales. In another scenario, the competitor's product may not be as cost efficient, which would be reflected in a higher price than necessary. This would increase potential sales for the newer product, even though there is a 3 month lag in the launch. Thus, the outcome of the chance node Sales Volume will depend on the actions the competitor takes. This relationship is shown in Figure 2.10.

Arrow from decision node to random node: An arrow can also show conditional dependence between a decision and a random event. Here, the arrow signifies that the unknown sales volume of the product will depend on its price (see Figure 2.11). The relation is that the higher the price, the lower the sales volume and vice versa. The specifics of this relationship are not important in constructing the diagram, just that there is a relationship. In later analysis, the nature of this relationship will be critical.

Arrow from several consequences to the overall objective: An influence diagram may include one or more consequences that influence the overall objective. In Figure 2.12, Total Revenue and Total Cost contribute to Net Profit. Total Revenue and/or Total Cost are influenced by random events, and, as a result, Net Profit will be an uncertain value. However, once the other two values are known, Net Profit will no longer be uncertain.

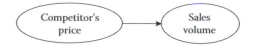

FIGURE 2.10: Arrow from a chance node to a chance node.

FIGURE 2.11:   Arrow from decision node to a chance node.

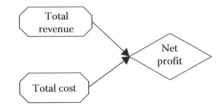

FIGURE 2.12:   Consequences combine to create overall value.

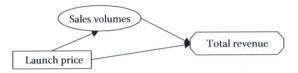

FIGURE 2.13:   Arrows from a chance node and a decision node to a consequence.

Arrow from decision or chance node to consequences: Use arrows to show which decision and chance nodes affect a particular consequence. One or more nodes can influence a consequence. Similarly, a single node can influence many consequences. For example, in Figure 2.13, the arrows show that the total revenue from the product will depend directly upon the sales volume and the launch price. This results in two arrows feeding into Total Revenue.

In this example, the Launch Price decision will also influence sales and as a result indirectly affect Total Revenue. It will also directly influence Total Revenue, since Sales Volume times price represents Total Revenue.

Notice that Total Revenue and Total Cost are not portrayed as chance nodes even though they are obviously uncertain variables, as is Total Profit. Anytime a chance node leads into a consequence or value node, it is automatically assumed that these variables are uncertain. However, once the uncertainty surrounding all the other chance nodes is resolved, the assumption is that these consequences can be calculated accurately.

Review the set of nodes for completeness: Are there any critical issues not reflected by nodes? Is there any uncertainty regarding the features of the competitor's product? Will the comparative quality of the two products be a factor and is this an uncertain variable? Will the sales volume in any way be affected by uncertainty in economic conditions? Are there any other decisions to be made, such as possibly delaying the product development to see how well the competitor's product does in the marketplace?

Review the independence of nodes not connected by arrows: After drawing all the necessary arrows, review the influence diagram to assure that any absence of arrows between two nodes was intentional. Thus, if there were no arrow between two chance nodes, Competitor Price and Sales Volume, that might mean that the decision maker does not envision competing on price. However, if that is the case, there is no need to include a chance node for Competitor Price.

Ensure that there are no cycles in the influence diagram: There should be no cycles in an influence diagram, because a node cannot influence its own outcome. This is an important point to remember, because there are other diagrammatic tools in which cycles are common. For example, a workflow or process diagram will often have cycles to represent that the process is iterative. An influence diagram, however, is not a flow diagram, although they are often mistakenly viewed

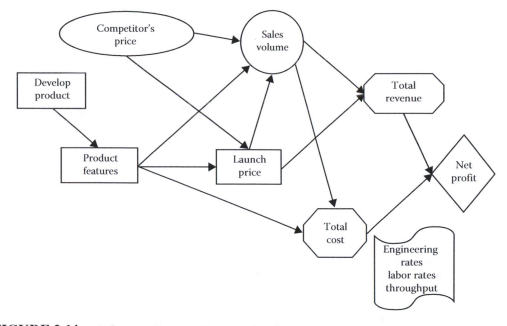

FIGURE 2.14:   Influence diagram: New product late-to-market.

that way. A flow chart may be used to indicate the sequence of events and activities in a decision-making process. Influence diagrams, on the other hand, are structured displays of decisions, uncertain events, and outcomes, providing a snapshot of the decision environment at a single point in time.

List data inputs: Last, the decision makers should identify other key data that will be needed for the analysis. These data are primarily linked to estimating the cost of product development and production. These include engineering rates, labor rates, and throughput.

2.5.1   Summarize Diagram

The complete influence diagram for the late-to-market case was originally given in Figure 2.2. Note that the diagram (the same influence diagram that is shown in Figure 2.14) consists of three decisions in sequence. The first decision is whether or not to develop the product. The second decision is regarding the product features to be included in the product. This decision will be made after the first, and an arrow between the two decisions shows this relationship. The third decision in the sequence is the launch price. Before this decision is made, the outcome of the chance node—Competitor's Price—and the decision—Product Features—will be known. Uncertainty in Sales Volume will be dependent on Launch Price, Competitor's Price, and Product Features. A change in the value of any of the nodes will affect the uncertainty in Sales Volume. Finally, Sales Volume and the Launch Price will affect the consequences Total Cost and Total Revenue. Total Cost and Total Revenue combine together into the value node Net Profit.

2. *Activity*—Late-to-market cell phone: Does the influence diagram in Figure 2.14 describe the elements that would be of concern to you if you were the decision maker? How would you modify this diagram? Here are some questions to consider.

   a. Why might you create separate nodes for product development and manufacturing costs?

b. Once a decision on price has been made, what uncertainty might still exist with regard to the actual price that will be paid by purchasers?

---

c. What other uncertainties or consequences might you add to the diagram?

2.6 Multiple Objective Influence Diagram: Buying a Used Car

Pete is a freshman in college looking to buy a used car both for social functions and to drive back and forth to his part-time job 20 miles away. Pete would like to minimize total cost. At the same time, he wants to maximize accessories and aesthetics. Pete found that deciding which car to purchase was not a trivial task and asked his close friend Isabel, who had taken a Value Added Decision Making course the previous semester, to help him structure his thoughts in a consistent manner. Isabel recommended that Pete create an influence diagram to frame the problem.

The overall objective is identified as maximizing the value to Pete. Because he is on a tight budget, he wants to minimize the purchase price. However, he realizes that with a used car, repair costs could be significant. He considers reliability of a used car a main concern, since he cannot afford to miss work. When evaluating cars, he also wants to take into account the car's accessories and the image it will project to potential dates as well as to his peers. Uncertainty about maintenance affects the assessment of both his total cost and the car's reliability. In addition, longevity primarily affects total cost.

Pete begins to lay out his basic framework in the influence diagram. He places his ultimate objective to maximize value and then adds the primary subobjectives, reliability, total cost, accessories, and aesthetics. Isabel asks him to discuss his primary uncertainties. The two things he is concerned about are the regular repair costs, and how long the car will last before a major system failure would force him to simply junk it. Working from right to left, he adds two uncertainty nodes. Isabel reminds him that gas prices are fluctuating wildly and that uncertainty leads into the cost as well.

Isabel then guides Pete to think more specifically about each of the major objectives. Purchase price and miles per gallon obviously influence total cost. These are considered two separate subobjectives. Pete is concerned about his immediate cost of buying the car, because his cash reserves are limited. He is also concerned about weekly expenditures. However, reliability is a little harder to define. There is the significant uncertainty regarding repairs that has already been noted. One measure to be minimized, he decides, is odometer mileage, but he is uncomfortable that this does not fully capture the issue of the reliability of the car to transport him to work on a regular basis. Isabel reminds him that *Consumer Reports* rates the long-term reliability of used cars and so that measure is added to the influence diagram as well.

Pete is very much interested in the impression the car will make on passersby and includes three measures of aesthetics: color, exterior body, and interior condition. Isabel asks him if he is also interested in whether or not the car is fun to drive. He responds that he really is not a car guy and driving is not an exciting experience for him.

With regard to accessories, Pete would like to have a functional air conditioner and heater, of course, but the quality of the sound system is also especially important. He expects to use his vehicle on dates but would like to be able to offer friends a ride as well. Thus, he would like to maximize seating capacity. The complete influence diagram is shown in Figure 2.15. In this context, the primary focus of the framing process has been capturing the multiple objectives (Edwards and Chelst 2007).

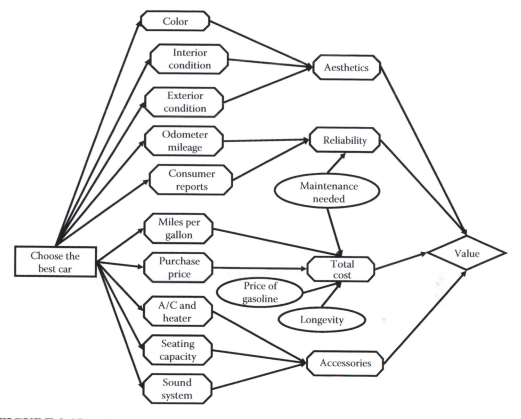

FIGURE 2.15:   Buying a used car.

2.6.1   Information Seeking: Buying a Used Car

In many contexts, there are opportunities to seek additional information before making the primary decision (Howard and Matheson 2005). Usually, there is a cost associated with the decision to seek additional information. This cost might be directly linked to the cost of gathering the information, or it might be the cost of lost opportunity as a result of delaying the decision. For example, when buying a used car, you may hire a mechanic to check it out or pay for a report about its accident history. In choosing a medical treatment, you may wait for further tests. In deciding on the design and equipment for a full-scale manufacturing plant, a company might decide to build a pilot plant. In launching a new product, a key decision involves the extent of test marketing. Often, the information gathered is not a perfect predictor, but it does update the assessment of the probability distribution of a key variable.

In the used car example, Pete is concerned about whether a particular car has been involved in a serious accident or damaged in a flood, or if the car may even be a salvaged title, a car that the insurance company had once declared a total loss. He decides to purchase a 30 day unlimited access to car history reports as he searches for a car. His other major concern is the need for repairs. He would like to have the car checked out by a mechanic, but each assessment will cost him $75. He is therefore planning on inspecting only the two best cars he finds.

These additions lead to a revision of his influence diagram. The decision is now split into two decisions as illustrated in Figure 2.16. The first decision identifies the two best cars. An uncertain event, the results of a vehicle history report, feeds into this decision. This uncertainty will be resolved before he finalizes his two best choices. The decision as to which car is preferred will be made later after he has received a mechanic's report for each of his finalists. The mechanic's report

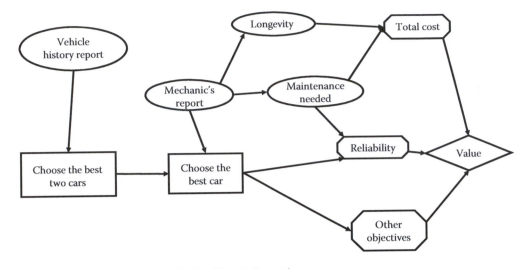

FIGURE 2.16:   Used car revised—New information.

will also change the probabilistic estimates of longevity and cost of maintenance. Figure 2.16 contains the revised influence diagram. In this discussion, the primary concern has been reliability, so the influence diagram has compressed most of the multiple objectives into a single node.

2.7   Oglethorpe Power Corporation: Actual Case

Oglethorpe Power Corporation (OPC), a generation and transmission cooperative, provides wholesale power to consumer-owned distribution cooperatives in Georgia (Borison 1995). They delivered 20% of the power in the state, with Georgia Power Company (GPC) dominating the remainder. In late 1990, OPC learned that there was an opportunity to expand their business by investing in an additional 1000 MW transmission line. This opportunity was available through Florida Power Company (FPC), which was planning to tap the surplus power generated in Georgia. OPC had to decide whether to invest in this project.

The process started with developing a clear statement of the problem and identifying the values, objectives, uncertainties, and associated decisions. An influence diagram was generated based on this information. The diagram was cleaned and modified in consultation with key experts and analysts in a series of meetings. Figure 2.17 illustrates the final version of the influence diagram.

Net present value of the savings was chosen as the ultimate objective for evaluating the decision. The influence diagram showed that there were three decisions involved in the process: whether to build a transmission line, whether to upgrade associated transmission facilities, and the nature of control over the new facilities. All told, the diagram represented a total of 18 decision policies. Five major uncertainties were identified: the cost of building new facilities, the demand for power, the competitive situation, OPC share, and spot price.

The arrows in the influence diagram pointing to the consequence nodes show how the savings were calculated. The influence diagram helped decision makers visualize and understand the complexity of the decision process. Framing the problem was the initial stage in the process, followed by a comprehensive debate and analytical evaluation of the decision at hand.

A comprehensive decision analysis was completed in less than 2 weeks. This analysis helped persuade OPC to change the preferred joint venture strategy to an independent strategy of direct negotiations.

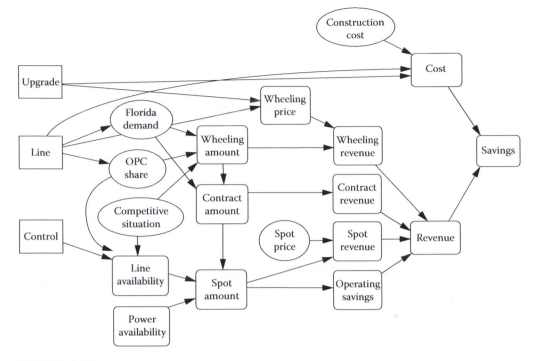

FIGURE 2.17:   Oglethorpe influence diagram.

2.8   Influence Diagram Construction: Review

How do you begin creating an influence diagram? Start with nodes but without arrows. The first sets of nodes are decision nodes. The key question to ask is the following: Does the diagram represent only one decision or will a current decision lead to more than one or more subsequent decisions? The criteria and justifications for subdividing a decision are as follows:

Distinct decisions: You should consider picturing distinct decisions if there is a time lag between sequential parts of the decision and additional information will be obtained in the meantime. The decision is split in two if a current decision affects future decision options. For example, installation of flexible automotive assembly capacity enables decision makers to change production schedules as the demand mix changes. Decisions should be subdivided if separate decisions are to be made with regard to geography or product classes. You should leave out a subsequent decision if it cannot be realistically analyzed now or if the choices will not change the relative merits of the earlier decisions.

Objectives: Brainstorm with key decision makers or confidantes as to the goals and key components of the overall objective. For clarity, separate an objective into its components (e.g., revenue and cost lead into profit or return on investment, or variable and fixed cost combine into total cost). Do not shy away from including objectives that are difficult to quantify, such as supplier reliability or craftsmanship.

Uncertainty: Specify what you do not know. Do not be shy or frightened by how much you do not know. However, do not nitpick about random variables that have relatively small amounts of uncertainty.

Arrows: Add the arrow links carefully. Remember that there are specific "grammatical" rules governing placement and interpretation of arrows. Be especially careful about placing an arrow from a random event to a decision. Check to make sure there are no cycles.

Restructure diagram: In almost every instance of a complex diagram, many arrows and lines will cross in the first draft. Alter nodes and arrows to add clarity as well as to reduce overlap and confusion.

Input: Discuss who is responsible for what additional data needs to be obtained in support of the decision.

Revisit: Do not be afraid to modify the diagram by adding or eliminating objectives or random events. Your goal is to create an efficient and effective influence diagram that contributes to communication and is not overwhelmingly complex.

2.9   Solving Influence Diagrams

In this text, we have explored the role of influence diagrams in decision structuring and communication. The actual analysis of the decision will be discussed later in the context of two classes of analytic tools: one focuses on decisions in which the primary source of complexity is multiple objectives, and the other is used when uncertainty is the major complicating factor. There is, however, extensive literature and research on how to carry out an analysis by "solving" the influence diagram to which we refer the interested reader.

In general, solving an influence diagram for a complex problem is complicated (Clemen and Reilly 2001) and may require an enormous computational effort (Cano et al. 2006). Howard and Matheson (2005a) explore in depth the relationship between influence diagrams and decision trees and provide several examples of the transformation between the two. Several algorithms have been developed to solve the influence diagrams involving symmetric decision structures (Shachter 1986; Zhang 1998). (In symmetric decision structures, each and every decision encounters the same set of random events and random variables.)

However, the solution of influence diagrams has a serious drawback when it comes to dealing with asymmetric decision problems. One approach involves converting the asymmetric problem into an equivalent symmetric representation. This process can significantly expand the size of the influence diagram and create a considerable amount of unnecessary computation. Several methods have been proposed to cope with this difficulty. For example, Shenoy (2000) proposed to use valuation networks to represent and solve asymmetric decision problems. Cano et al. (2006) proposed an approximate inference algorithm to handle very large models. Interested readers can find a review of asymmetric decision problems in Bielza and Shenoy (1999).

2.10   Recent Articles on Influence Diagrams

Detwarasiti and Shachter (2005) use influence diagrams to evaluate a team decision situation under uncertainty and incomplete sharing of information. The authors assume that all team members agree on common beliefs and preferences and hence represent the team as a single rational individual with imperfect recall. Since the optimal solution with perfect recall might not be achievable for most such problems, the authors introduce Strategy Improvement and its variation, Uniform Strategy Improvement, as solution methods. They show that the notions of strategic irrelevance and the requisite influence diagram allow the use of all available information to improve decision quality and find the joint strategy that is maximally stable over the largest sets of decisions possible with the Uniform Strategy Improvement algorithm.

Charnes and Shenoy (2004) develop a Multistage Monte Carlo (MMC) simulation method to solve influence diagrams using local computation. The MMC method samples only a small set of chance variables for each decision node in the influence diagram, and this reduces the complexity of solving an influence diagram that has many variables. The MMC sampling technique proposed in this paper draws independent and identically distributed observations to solve multiple-stage decision problems. This method is designed to compute an approximately optimal strategy, rather than to calculate the optimal one. The approach uses information about the domains of probability conditionals and utility functions that are coded in influence diagrams and allows them to obtain the same degree of precision while calling for less computation as compared to global simulation.

Demirer and Shenoy (2006) propose sequential valuation networks that are a hybrid of sequential decision diagrams and valuation networks. Sequential valuation networks use the graphical ease of sequential decision diagrams to represent the asymmetric structure of a problem and attach value and probability valuations to variables as in value networks. The hybrid method adopts the best features of sequential decision diagrams and valuation networks while overcoming many of their respective shortcomings. This method breaks down a large asymmetric problem into smaller subproblems and then uses a fusion algorithm of valuation networks to solve the subproblems.

Exercises

2.1 Critique an influence diagram—global location

ABC Systems is one of the largest global manufacturers of personal printers. It is evaluating three potential facility location sites in Malaysia, Taiwan, and China to maximize sales in this region as well as to maximize global profits. An influence diagram for the decision is given Figure 2.18.

a. Why do you think it is important to include the objectives Labor Force Skills and Political Stability?

b. What additional objectives or consequences might you want to consider?

c. Would you classify government policies as uncertain?

d. What additional uncertainties might you want to consider?

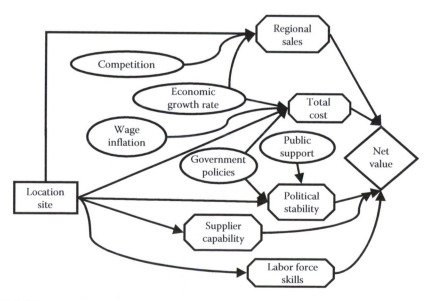

FIGURE 2.18: Global facility location.

2.2 Modify influence diagram examples in text

a. Theater party initiations: Expand on this example in Figure 2.6 but without adding more rounds of invitees. What options exist if there are tickets left over? Is it possible to buy more tickets at the last minute? What new decision(s) and new uncertainties would you add?

b. Late-to-market cell phone: Does the influence diagram in Figure 2.14 describe the elements that would be of concern to you if you were the decision maker? How would you modify this diagram?

Here are some questions to consider for the cell phone example.

Why might you create separate nodes for product development and manufacturing costs? Once a decision on price has been made, what uncertainty might still exist with regard to the actual price that will be paid by purchasers? What other uncertainties or consequences might you add to the diagram?

2.3 Construct an influence diagram—software: late design change
Make a design change in hopes of solving a software bug. It is only 6 weeks before a software product's launch. On rare occasions, commercial users who are beta testers have run into a problem that caused the software to lock up. However, software engineers are not able to reproduce the crash in a controlled environment. They are fairly sure that the problem stems from a series of three subroutines. They have a new design that can be implemented quickly and that should solve the problem.

In constructing the diagram, identify all the key issues and inputs. Use your judgment to specify whether or not a key variable is simply a deterministic input or will represent significant uncertainty that should be taken into account. Are there any downstream decisions?

2.4 Construct an influence diagram—supply new engine: yes or no
Your organization has been offered an opportunity to supply a complex new engine for a recreational vehicle. Your current capacity is nearing its limits. To take on this new product, you might have to add workers and/or increase investment in new equipment as well as possibly outsource some of the work to a new supplier. Currently, you are in the midst of a major productivity improvement effort whose benefits are not yet discernible. You are concerned about maintaining your overall reputation, given the time pressures associated with the project and the manufacturing complexity of the new engine. What should you do in response to this opportunity? Think broadly but remember that you will need to compress your ideas into a one-page diagram. Do not include downstream decisions or random events that will have little or no impact on the decisions at hand. However, make sure to include those decisions or random events that could have significant impact on the starting point for the current key decision regarding whether or not to supply the new engine.

2.5 Construct an influence diagram—children's movie to produce
The Erstwhile Production Company is considering a number of scripts that can be turned into a movie targeted at children under the age of 12. The company can only produce one movie this year. What are the key decisions that follow closely upon the decision as to which movie to pick? What are the major uncertainties? In setting up the overall profit objective, be sure to explicitly note various sources of profits. Are there any elements of the decisions that can be delayed?

2.6 Construct an influence diagram—President Obama's sequential decisions to send more troops to Afghanistan in 2009
Shortly after assuming office in January 2009, President Obama quickly decided to send 17,000 more troops to Afghanistan. Later that same year, he took a much longer time deciding to send an additional 34,000 troops. In constructing an influence diagram be sure to include at least three components to the second decision. Describe what objectives might

have been present in the second decision that were not significant in the first. Identify any new information he was able to ascertain before the second decision was made that he was uncertain of at the time of the first decision.

2.7 Construct an influence diagram—job offer

It is a good year in the economy and you have just graduated college. You have been given several job offers in your field of study, but these offers come from different cities around the country. Construct an influence diagram of this decision.

What additions or subtractions would you make to the diagram if you were an individual in mid-career with a spouse, 2 children aged 6 and 9, and a home? If the children were teenagers, would you add anything to the diagram?

References

Bielza, C. and Shenoy, P. P. (1999). A comparison of graphical techniques for asymmetric decision problems. *Management Science*, 45, 1552–1569.

Borison, A. (1995). Oglethorpe Power Corporation decides about investing in a major transmission system. *Interfaces*, 25(2), 25–36.

Boutilier, C. (2005). The influence of influence diagrams on artificial intelligence. *Decision Analysis*, 2, 229–231.

Cano, A., Gómez, M., and Moral, S. (2006). A forward–backward Monte Carlo method for solving influence diagrams. *International Journal of Approximate Reasoning*, 42, 119–135.

Charnes, J. M. and Shenoy, P. P. (2004). Multi-stage Monte Carlo method for solving influence diagrams using local computation. *Management Science*, 50, 405–418.

Clemen, R. T. and Reilly, T. (2001). *Making Hard Decisions with Decision Tools Suite*, 2nd rev. edn. Belmont, CA: Duxbury Press.

Demirer, R. and Shenoy, P. P. (2006). Sequential valuation networks for asymmetric decision problems. *European Journal of Operational Research*, 169, 286–309.

Detwarasiti, A. and Shachter, R. D. (2005). Influence diagrams for team decision analysis. *Decision Analysis*, 2, 207–228.

Edwards, T. and Chelst, K. (2007). Purchasing a used car using multiple criteria decision making. *Mathematics Teacher*, 101(2), 126–135.

Howard, R. A. (1988). Decision analysis: Practice and promise. *Management Science*, 34, 679–695.

Howard, R. A. and Matheson, J. E. (2005a). Influence diagrams. *Decision Analysis*, 2, 127–143.

Howard, R. A. and Matheson, J. E. (2005b). Influence diagram retrospective. *Decision Analysis*, 2, 144–147.

Howard, R. A., Matheson, J. E., Merkhofer, M. W., Miller, A. C., and North, D. W. (2006). Comment on influence diagram retrospective. *Decision Analysis*, 3, 117–119.

Koller, G. (2005). *Risk Assessment and Decision Making in Business and Industry*. Boca Raton, FL: Chapman & Hall/CRC.

Krumm, F. V. and Rolle, C. F. (1992). Management and application of decision and risk analysis in Du Pont. *Interfaces*, 22(6), 84–93.

Kusnic, M. W. and Owen, D. (1992). The unifying vision process: Value beyond traditional decision analysis in multiple-decision-maker environments. *Interfaces*, 22(6), 150–166.

Nease, R. F. and Owens, D. K. (1997). Use of influence diagrams to structure medical decisions. *Medical Decision Making*, 17(3), 263–275.

Pauker, S. G. and Wong, J. B. (2005). The influence of influence diagrams in medicine. *Decision Analysis*, 2, 238–244.

Shachter, R. D. (1986). Evaluating influence diagrams. *Operations Research*, 34, 871–882.

Shenoy, P. P. (2000). Valuation network representation and solution of asymmetric decision problems. *European Journal of Operational Research*, 121, 579–608.

Wells, W. (1993). *Unified Life Cycle Modeling: A Framework for Manufacturing Cost Extimating and Analysis*, Ph.D. Dissertation, Industrial and Manufacturing Engineering, Wayne State University, Detroit, MI

Zhang, N. L. (1998). Probabilistic inference in influence diagrams. In: *Proceedings of the 14th Conference Uncertainty in Artificial Intelligence*, Cooper, G.F. and Moral, S., eds. San Francisco, CA: Morgan Kaufmann.

Chapter 3

Common Decision Templates

3.1 Goal and Overview

The goal of this chapter is to facilitate the application of a structured decision process by providing influence diagram templates for broad classes of decisions.

A common complaint that impedes the wider adoption of structured decision modeling tools is the claim that they take too long to apply. This chapter is intended to facilitate the start-up process by providing a skeletal structure to build upon for a wide range of corporate, governmental, and personal decisions. The previous chapter presented influence diagrams as a tool for framing discrete choice decisions from among a limited number of alternatives. The diagram brings into focus and discussion the primary uncertainties and major objectives as well as the link between the decision in question and subsequent decisions. In this chapter, we illustrate how this model can be applied to broad classes of decisions listed in the following.

1. In-house or outsource: make or buy (a) simple and (b) strategic
2. Change or keep status quo: (a) upgrade and (b) late design change
3. Products: (a) launch and (b) portfolio
4. Project management: product development tasks
5. Capacity planning: (a) basic and (b) flexible
6. Technology choice: (a) large scale and (b) personal
7. Personnel or organization selection: hire research faculty
8. Facility location: sports arena
9. Bidding: make offer
10. Personal decision: University
11. Information gathering decisions: market research, medical tests, prototypes, and pilot plants

For each class of decisions, we discuss the most common uncertainties and objectives and we provide at least one related influence diagram. We also point the reader to several research papers on each topic as well as to additional examples that appear elsewhere in the text. Corner and Kirkwood (1991) provide a survey of decision analysis applications.

3.2 In-House or Outsource (Make or Buy)

Organizations are generally constrained in terms of their capacity to deliver products and services. These constraints may be in the form of physical limitations on a manufacturing or service facility. Analogously, human resources may be constrained with regard to trained and experienced personnel needed to perform critical tasks. As a result, private and governmental organizations must decide which products and services they are going to make and deliver themselves and which they will buy from a supplier. Often, it is not desperation but economic competitiveness that drives the decision to buy from an emerging market or outsource IT and backroom services to a country such as India. The classic make or buy decision is only part of a broader class of decisions we label as in-house or outsource.

In the simplest context, the primary objective is to minimize the total cost of manufacturing the product or providing the service. If this involves a significant capital investment, the decision will also affect corporate ROI (return on investment). The primary uncertainty relates to demand for the product or service. Additional uncertainty could relate to wage inflation both internal and at the supplier. Another uncertainty involves transportation costs. This latter uncertainty has taken on additional prominence as the price of oil has fluctuated wildly.

Productivity gains are key to long-term uncertain reductions in cost. If anticipated productivity gains are dependent on the choice of in-house or outsource, these gains should be included in the diagram. Some companies seek to write into contracts with suppliers a specific annual cost reduction target. However, the major potential savings result from a systems perspective that requires consistent feedback between, for example, manufacturing and product development. This is harder to achieve when the product is outsourced, especially to a distant land.

Perhaps the most significant uncertainty often not considered is the level of communication and oversight required with an outsourced product or service. When dealing with a new supplier, the need for almost continuous communication and the cost of oversight are factors that are likely to weigh heavily in the overall budget. However, if the selected supplier is extremely experienced in delivering the service or product, the level of management oversight and involvement might not be as costly.

If the supplier of the product or service is foreign based, numerous other uncertainties become of concern. Most importantly, currency uncertainty could be significant irrespective of recent trends. Too often, this uncertainty is ignored by executives seeking quick fixes to internal wage pressures. For example, sourcing jobs to Canada looked like a great idea when the Canadian dollar was worth less than $0.70 U.S. but became far less attractive in 2010, when the dollars were almost equal in value. Analogously, although the Chinese RMB is strictly controlled, wage pressures in cities such as Shanghai rapidly drive up the cost of experienced labor (Figure 3.1). A broad array of issues associated with the production in emerging markets are discussed in a case study in risk management in Chapter 8 and another case study on facility location in Chapter 12. These include uncertainty regarding political and economic stability, risks of pandemics that affect travel, other natural catastrophes such as earthquakes and tsunami, or a general increase in criminal lawlessness.

Until now, we have focused on one objective: minimizing cost. Forward thinking companies reflect more broadly before outsourcing, as characterized in Figure 3.2. At the minimum, they take a close look at the potential impact on quality, a fundamental objective of most companies. In a manufacturing context, when considering the purchase or production of particular components, it is relatively straightforward to incorporate a measure of quality into the decision-making process. Unfortunately, too many companies have neglected to factor in quality when considering outsourcing customer service or IT services to overseas companies. They underestimate the importance of obtaining information regarding the training and skills of the workers of the outsourcing company as well as the rate of turnover within the company. This neglect can have far reaching negative consequences.

Manufacturing companies have begun reviewing all of their activities to determine what is core and what is not (Fine and Whitney 1996). A supplier is often viewed as more than a low-cost manufacturing alternative. The company may choose to require the supplier to assume extensive design and testing responsibilities. In making this decision, however, the company must consider the strategic value of maintaining this capability internally. It might be in the company's best interest to remain "hands on" when it comes to quality control. Conversely, there may be strategic value in building a relationship with a company that has design capabilities that the purchasing company

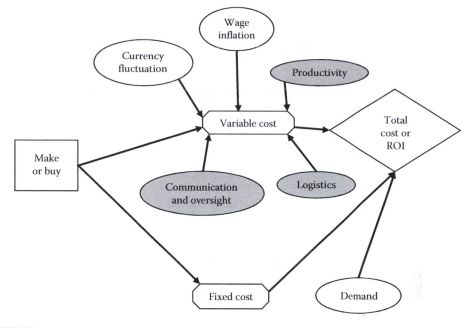

FIGURE 3.1:   Make or buy decision—minimize cost.

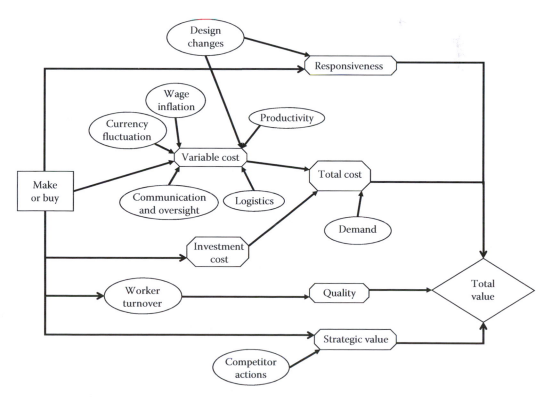

FIGURE 3.2:   Make or buy decision—multiple objectives.

lacks. This is often true in sectors that have come to rely on many highly sophisticated subsystem suppliers. Of special concern in the automotive industry, for example, are emerging technologies for components such as batteries that are linked to an emerging generation of hybrid and electric vehicles. Similarly, pharmaceutical companies must determine which suppliers are at the cutting edge of technology when it comes to the design of new drugs as well as which delivery systems are most efficient in the human body. This objective is influenced by the uncertainty regarding how competitors will evolve their respective networks of supplier relationships.

Depending upon the product, the purchase decision may need to be made early in the design process because of long lead-time required in setting up manufacturing facilities. Consequently, the technical success of the current planned design may still be uncertain while setting up facilities for its manufacture. Thus, one fundamental objective concern is maximizing the responsiveness of the manufacturing system to design changes. Is the physical plant set up in such a way that it can accommodate design changes without having to be rebuilt? This uncertainty creates variability in the cost of manufacture both inside the company and at the supplier. With regard to suppliers, it is a well-known phenomenon that a late change in design is likely to result in a disproportionate raise in price (Walton 1997).

3.3   Change (Upgrade) or Keep Status Quo

Decision makers often face decisions with just two basic alternatives: maintain status quo or change in attempt to improve or in response to new information. Staying with the status quo may call for an additional decision: to revisit the situation a day, a week, or some other specified time in the future. Some examples of decisions include whether or not to upgrade software or equipment, incorporate a late design change, shut down a plant, upgrade a piece of equipment, adopt a new technology (Weber 1993) (Bhattacharya et al. 1998), or accelerate a new product development effort (Hess 1993). This class of decisions is confounded by a well-documented decision bias that leans toward maintaining the status quo. This has been demonstrated in decisions involving whether to sell or keep a stock or change the portfolio mix (Samuelson and Zeckhauser 1988).

Many of the aforementioned examples have personal decision parallels such as replacing a car, computer, or cell phone. More difficult is the decision to move into a larger house or, in the case of an elderly parent, into an assisted living accommodation. Another example involves whether to cancel a scheduled activity. This activity might be a seminar with low enrollment or an extended family activity that faces an uncertain weather forecast. In the public sector, a classic decision involves when to order the evacuation of a region in response to the forecast of a hurricane or whether to evacuate a building because of a bomb threat.

The decision regarding whether or not to upgrade is another never-ending organizational and personal challenge in a rapidly advancing technological society. This issue arises with regard to both software and hardware. We examine a university decision to install a major upgrade of its course management software such as Blackboard.

The major immediate uncertainty relates to all of the changeover issues and compatibility issues. Compatibility is a major concern since faculty have prior semester course material to which they wish to refer and, in many cases, copy into subsequent offerings of the same course. There is also a question of the effort that would be required to implement this changeover. The primary motivation to make the change is usually the additional performance features. Purchase or licensing cost is a major consideration; moreover, further costs are incurred in training the IT support staff. Any major corporate upgrade will temporarily strain the capacity of the IT staff by increasing their immediate workload.

In every upgrade context, there is significant long-term uncertainty regarding three related issues; when will the next upgrade come, what will be its performance, and how much will it

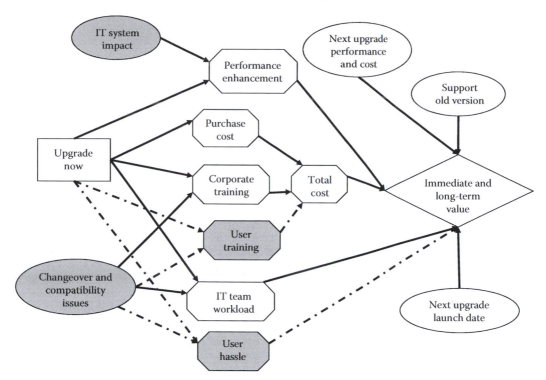

FIGURE 3.3: Corporate software upgrade.

cost. In addition, one must ascertain how long the software developer will continue to support the earlier version. The ultimate goal is to determine both the immediate and long-term value of this upgrade.

A midwestern university recently adopted an upgrade to its course management software that led to disastrous consequences. In fact, many faculty members cut back their usage of the software, or stopped using it altogether, as a result of this decision to upgrade. There were several primary values and uncertainties that were not adequately considered by those who made the decision to upgrade. We have shaded these uncertain events and objectives along with making those arrow connections dashed lines (see Figure 3.3).

The latest upgrade introduced a totally different look and style associated with all aspects of course management. The decision makers overlooked or underestimated the significant difficulties and training costs involved. In addition, they had not adequately developed an understanding of this software's use of IT system infrastructure. The entire university's computer system often ground to a snail's pace during peak usage of this software, namely the weeks surrounding the start-up of class. Faculty routinely received messages stating that there were too many users, suggesting that the user in question try later. Other routine university business processes were also severely hindered.

3.3.1 Late Design Change

Another common decision in product development involves incorporating a late design change. This type of decision arises both in hardware and machine design as well as with the launch of a new software product. There could be several reasons for considering last-minute design changes that occur long after the design freeze date and critically close to the launch date. The motivation for the change might be performance related, to include another feature, enhance a performance

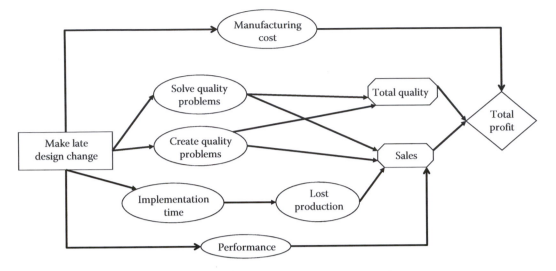

FIGURE 3.4:   Late design change.

measure (e.g., speed, torque, and hp), or adopt a new technology. The changes might have the benefit of allowing the organization to produce a product closer to customer preferences at the time of launch. Alternatively, the motivation might be the recognition of a design weakness that could generate warranty costs or liability. For example, all software has unresolved bugs at release. There is always a debate regarding which bugs to fix before the release.

Figure 3.4 describes the change decision and related uncertainties. It includes two fundamental objectives that drive many late design changes: sales and total quality.

The decision to introduce the design change so late in the process is sensitive. While addressing a particular concern, the change might also introduce numerous others. The first uncertainty is whether or not the late change will actually achieve its intended goal of improving performance, reducing warranty, or eliminating a software bug. Next, there is uncertainty as to how long it will take to implement this change. A delay in launch might result in lost production. Last, nearly all rushed changes can yield unintended consequences. Tackling a warranty problem in one subsystem might produce problems in another subsystem. Removing a software bug can introduce a new one.

Nevertheless, design teams often implement late changes, in spite of change freezes that have been established. The automotive industry, for example, averages more than one late change for each of thousands of automotive components. This applies to both U.S.- and foreign-based car companies. If not carefully managed, these changes can add millions of dollars to the cost of manufacturing. They also increase the risk of an extremely expensive massive vehicle recall at a cost that is likely to exceed a 100 million dollars.

3.4   Products: Launch, Portfolios, and Project Management

The launch of a new product was discussed in the preceding chapter. The problem context focused on the launch of a late-to-market product in which there were still decisions to be made regarding product features. In the following diagram, the stage is set for launching a product or service in an already well-defined format (Figure 3.5). This diagram can be used to represent the final decision before moving ahead with a new vehicle program. Although development costs have all been paid for, the really large investment costs involve set up of the manufacturing process and production lines. Alternatively, the situation could involve decisions as routine as approving a major

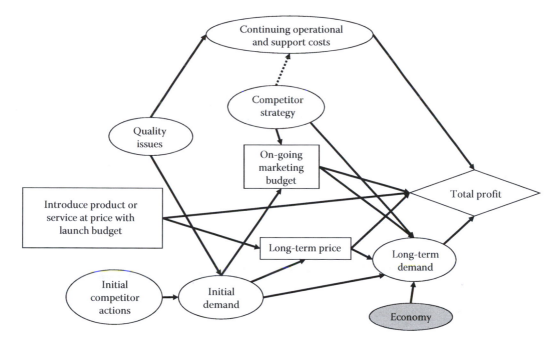

FIGURE 3.5: Launch product.

new menu group for a fast food chain, or as radically new as launching a completely new product or service, such as Kindle—an invention that integrates both product and services. Until now, we have discussed decisions that involve designs, but even after all issues regarding the design of a product have been resolved, it may prove necessary to cancel its launch for issues that have nothing to do with design.

In considering the primary "go or no-go" decision, it is important to think not only in terms of the immediate dynamics but also to consider a second round of longer term uncertainties and decisions. First and foremost, there should be an explicit recognition of uncertain competitor actions that might impact initial demand for the product or service. Additionally, in this initial phase, there are often quality issues that are difficult to resolve until the product or service is actually launched.

The initial demand will affect estimates of long-term demand. After seeing how the product or service is initially received, the company has an opportunity to adjust its price to support long-term market demand. In addition, as time progresses, the competition's long-term strategy should become clearer. This too will influence total demand.

The long-term ongoing costs are uncertain at the time of launch and are likely to be influenced by quality concerns since this would drive need for support services. The other major cost will be the ongoing marketing effort to sustain the product or service over the long haul. This is distinct from any marketing launch costs. It is a decision that will be influenced by the as-yet uncertain initial demand and the competitor's unfolding strategy. There is a dotted line between the competitor strategy and ongoing cost. This symbol was used to reflect the fact that in some situations the competitors' policies could affect the company's need to provide comparable support services in order to remain competitive. In other cases, competitors' actions may have no affect on operating costs.

One uncertainty that may or may not be a factor is the economy (shaded in gray). The key to whether or not to include this element lies with whether or not this uncertainty plays a role in any part of the decision to launch the product. This is of special concern as this text is being written at a time when the United States is experiencing its worst economic period in more than half a century.

When including economy as an element, it may be appropriate to limit the uncertainty only to the U.S. economy or, in some cases, even more narrowly to the local economy.

3.4.1   Product Portfolio

Companies have come to realize the need to evaluate and optimize a portfolio of products rather than designing each product as if it existed in isolation (Sanderson and Uzumeri 1997). By planning an entire portfolio as an entity, companies can achieve efficiencies of design and manufacturing by (a) minimizing overlap and cannibalization, (b) maximizing commonality of components that are invisible to the end-user, and (c) providing coverage of all segments and global markets. Mathematical programming models have been used to optimize an entire portfolio as a single entity (Young 1998). Often, decisions regarding the content of a product portfolio are made sequentially with one or more products added while others are dropped. It is this type of decision we address here.

In Figure 3.6, we illustrate a basic decision with two product classes: luxury and standard. The main decisions involve establishing a set of price points for the family of products and defining the characteristics of each product so as to provide increasing value consistent with higher price points. This class of decisions arises whether the product family is microprocessors, cameras, or automobiles. It even arises in the service industry as exemplified by the portfolio of warranties available with an expensive purchase. It also can be applied to medical insurance plans.

As in every product-planning problem, the major uncertainty revolves around product demand. That uncertainty influences the pricing and feature decisions. In the portfolio problem, there is an intervening random event between demand for the standard and luxury products. That uncertain event is the degree of cannibalization of the luxury product by the standard product. The closer the two products are in features and the farther apart they are in price, the more cannibalization there will be between the two product categories. There are also a number of other random events that affect all product decisions that are relevant here as well, but for diagram simplicity, we have left out. These include uncertainty regarding competitors' actions and the economy.

Du Pont (Krumm and Rolle 1992) used a complex influence diagram as an integral part of the decision analysis process to address product family planning in the broadest possible global frame. In their context, demand uncertainties can be dramatically influenced by uncertainties in the global

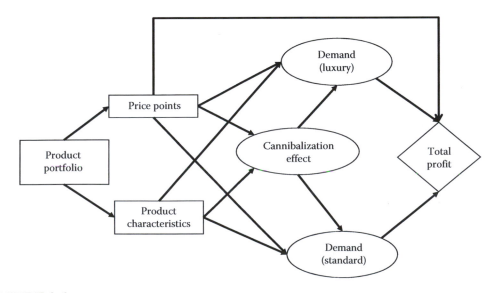

FIGURE 3.6:   Product portfolio: two product classes.

economy. The Du Pont decision problem consisted of evaluating three product planning strategies: the current strategy, cost leadership strategy, and product differentiation strategy. The major uncertainties considered in the analysis included competitors' strategies, market size, market share, and prices. The uncertainty assessments were performed for several different product types in each of three global regions. The decision model compared the worldwide NPV (Net Present Value) for each of the strategies.

3.5   Project Management: Product Development

Complex projects contain inherent risks and uncertainties. The traditional method of combating these uncertainties is to factor contingency into the time and cost estimates of a project (Rosenau and Moran 1993). Unfortunately, even with contingency factors, most projects miss their target schedule, cost, or scope. The contingency approach to project risk management treats the symptoms of project management problems but does not allow managers to identify and understand the sources of risk involved in the project that are responsible for schedule delays and cost overruns (Browning 1998). This makes DRA (Decision and Risk Analysis) an attractive tool to help managers make critical upfront decisions that consider uncertainty in the completion times, cost, and performance of project activities (Booker and Bryson 1985; Bhuta 1992). We use a product development project to illustrate several project management decisions (Figure 3.7).

Designing products for manufacture requires several design iterations before achieving a set of design specifications (Reinertsen 1997). The key set of initial decisions involves the execution strategy. The sequencing of project activities is a major issue in engineering project management (Kusiak and Wang 1993; Yassine et al. 1999b). Concurrent engineering advocates the parallel execution (or overlapping) of development activities as a tool for faster product introduction (Kusiak and Park 1990; Krishnan et al. 1997). Sequencing of tasks is concerned with the possibility of executing sequential tasks concurrently or with some degree of overlapping and the associated risk involved. The degree of overlap is a critical upfront planning decision.

Manufacturing feasibility will be one of the first uncertainties the product development team must confront (Yassine et al. 1999a). This will directly influence the overall uncertainty regarding the feasibility of initial designs to meet stated objectives. Next comes an iterative design process.

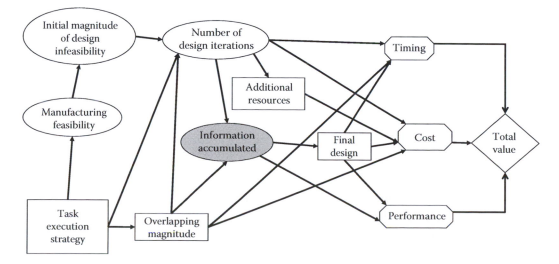

FIGURE 3.7:   Project management: product development.

The number and magnitude of these iterations are seldom known with certainty at the outset of the design process. The single most critical aspect of project planning is to project the impact of iterations on the schedule and budget of the development process. Without this, there will be a large discrepancy between the baseline plan and the actual duration and cost of the process.

The single most critical uncertainty that is often not easy to measure is the rate and amount of information accumulated as the team cycles through the design phase. The amount of information accumulated is a function of both the iterations and the overlapping strategy. These iterations directly impact the later decisions regarding additional resources. At some point, the team leader must decide that they have reached a final design and move on to actual product implementation and manufacture. In the end, there are three major objectives that need to be balanced in every major project: time to complete, product performance, and cost.

3.6 Capacity Planning

A fundamental element of long-term planning is capacity. This issue arises whether the focus is a power plant, a manufacturing facility, or a service facility such as a hospital or sports stadium. Capacity planning decisions generally involve large capital investment. They can take years to plan and implement and their effects can linger on for decades, as in the case of power plants and stadiums.

In planning capacity, companies almost always balance estimates of short-term demand and long-term potential. This is certainly true in a manufacturing environment and applies equally to public facilities, such as hospitals. The length of the planning horizon automatically increases the surrounding uncertainty as the decision maker must peer deeper into the crystal ball to predict the future. The uncertainty surrounding overall demand for a product or service can be influenced by changes in overall economic conditions. Over the long term, there may be changes in market taste, market structure, or technological breakthroughs that dramatically influence the overall demand. The risks are compounded by uncertainty about the company's share of that demand that will be influenced by actions of direct competitors offering the same or similar product as well as by competitors offering a substitute (natural gas for oil).

In addition to these externalities, there can be significant uncertainty regarding the actual operating capacity or throughput of a particular facility (Spetzler and Zamora 1989). The initial yield for manufacturing a new computer chip, for example, can be highly uncertain. A primary reason for declining prices over the lifetime of a new chip is that companies such as Intel dramatically improve manufacturing yield from year to year. Over the long term, there may be technological breakthroughs that will also affect the yield. Likewise, any time a new manufacturing process or new technology is used, there will be uncertainty. A complex assembly could face unanticipated bottlenecks. All new operations face a learning curve that affects both short-term productivity and long-term capacity (Wells 1993).

Globalization of markets and supply chains complicates capacity planning even more. In a global plan, plant capacity must both service local demand as well as contribute to the global supply chain. This leads to a critical second question, "Where to build the capacity?" (Canbolat et al. 2007; MacCormack et al. 1994).

In Figure 3.8, we focused on the capacity decision and two primary objectives. If demand is high, capacity will limit annual sales. In addition, plant capacity directly affects the cost of operations. In Figure 3.9, we expand the basic diagram to reflect the context in which product mix is a major uncertainty. The company's ability to rapidly respond to changes in a mix can be a substantial competitive advantage and will affect market share. Japanese automaker plants, for example, are significantly more flexible than those of Ford, GM, and Chrysler. This is the result of a better control of the vehicle design process that has enabled them to allow for flexibility within their assembly plants.

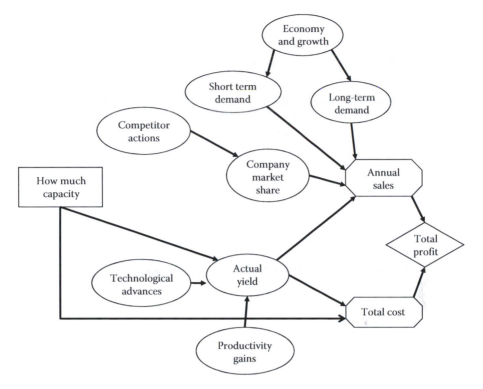

FIGURE 3.8: Capacity planning.

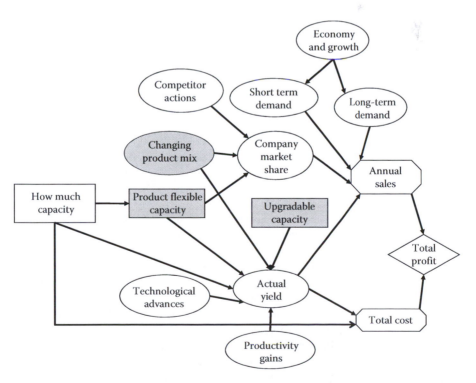

FIGURE 3.9: Capacity planning with flexibility.

In this instance, there is a subsidiary decision regarding how to divide up this capacity and, in particular, how much flexibility to include (Graves and Jordan 1995). Flexibility allows production mix to adjust as product demand shifts, with minimal incremental investment. Demand uncertainty drives the need for and the financial value of flexibility investments. Flexibility mitigates the risk of demand uncertainty in return for some extra investment.

Figure 3.9 includes the decision to design in product flexibility (The extra elements are shaded). This decision will likely influence the plant's overall yield as well. In this diagram, we also included another element of upfront capacity planning. Will the design of the facility include the capability for upgrading the capacity and possibly expanding its cost effectively?

3.7 Technology Choice

Individuals and organizations routinely face decisions as to which technology to choose. Do we buy an Apple or PC type computer? Which software should we buy to manage the enterprise or to schedule school buses. In designing a product, the decision may relate to the choice of materials or imbedded microprocessors. In most cases, this decision includes an implicit decision as to the corporate provider of the technology. Decision complexity will be primarily linked to the need to consider multiple objectives with uncertainty an important, though less significant, concern. In Chapter 4, we present a technology choice example with a detailed breakdown of objectives and related measures.

In every technology choice decision, there is trade-off between cost and performance (Hammond and Keeney 1999; Keeney et al. 1986). The costs usually entail purchase costs as well as a range of other costs: operating, training, and integration. If, for example, the technology is energy intensive, there would be significant uncertainty regarding its cost, as energy prices are notoriously volatile.

The difficulty of systems integration may be uncertain, a factor that would influence the integration costs. Imagine the choices automotive companies face as to which battery technology to use as they develop electric vehicles. It is often difficult to anticipate all of the potential problems especially when dealing with a relatively new technology. This will influence both operating costs and performance. In particular, reliability could affect the downtime of the technology or the need for rapid replacement.

Many major technology purchases involve significant training of on-site personnel. The ease of operating the system will determine the level of education required of operating and maintenance personnel as well as the need for start-up and continuing training. In addition, as random problems arise, the quality and quantity of support services will be a critical, albeit initially uncertain, concern. Last, as we enter a more environmentally conscious time, an additional objective might be minimizing environmental impact (Figure 3.10).

In a large-scale decision, there also may be a concern as to how quickly the technology can be implemented and integrated into existing systems. Many hospitals and health care systems, for example, are driven by federal mandates to move quickly to fully integrate computerized patient record systems.

Embedded in the earlier diagram is a performance objective that covers a multitude of sub-objectives relevant to the particular technology. This leaves room for significant uncertainty. Let us expand this performance objective by exploring a selection of alternative airport screening systems (Clemen and Reilly 2001). The performance objective is primarily linked to detection capabilities: (a) detection of metal weapons and (b) detection of explosives. Since this process interferes with the free flow of the public, another objective is to maximize "passenger acceptance." This measure could involve a subjective assessment of customer acceptance or an objective criterion, such as passenger-processing time as a surrogate measure (Figure 3.11).

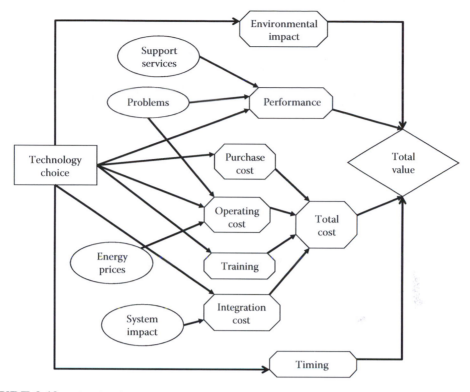

FIGURE 3.10: Technology selection—large scale.

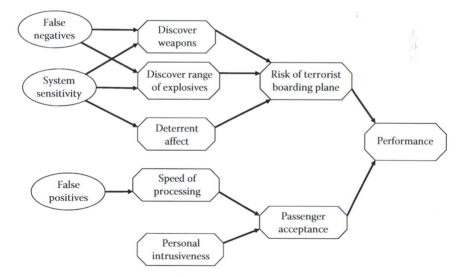

FIGURE 3.11: Detection technology—performance objectives.

The primary objective of any detection technology is to accurately identify a range of possibilities. In this case, one objective relates to detecting explosives, which can exist in a range of compositions. The other focuses on weapons, which are primarily metal. Moreover, a good system also has a deterrent effect. Ideally, the system should be so consistently effective that terrorists will not even attempt to board a plane while carrying a weapon or explosive.

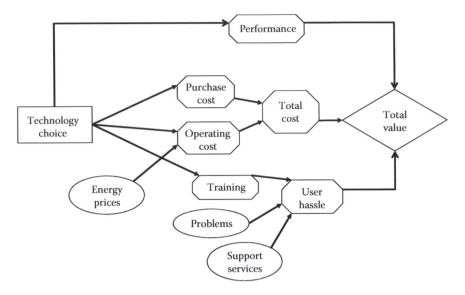

FIGURE 3.12:   Technology selection—personal.

The primary uncertainties associated with any detection device are the rate of false positives and false negatives that occur under real operating conditions. Another uncertainty is the overall sensitivity of the detection system under operating conditions. This concept applies analogously to the personal choice of a hearing aid. The device must distinguish sounds that are noise and therefore not to be amplified, from those that are voices of someone talking.

There is second set of objectives that is relevant to the overwhelming majority of passengers, and these objectives affect the customers' willingness to accept the burden of security screening. The speed of processing is of special interest, but speed is influenced by the uncertain rate of false positives. Each false positive involves a significant delay. In addition, there is a concern over the intrusiveness of the body scanning devices, so much so that Congress originally objected to their use.

Individuals choosing a technology will be interested in many of the same objectives as an organization and will face the same uncertainties regarding problems and support services. In an organization, it is most appropriate to assess the impact of these objectives on cost. However, for the individual, the need to minimize the inconvenience associated with using a new technology takes on greater importance (Figure 3.12).

3.8   Personnel and Organizational Selection: Hire Faculty

We are all exposed to a variety of multifaceted objective rankings of people and organizations. A complicated formula is used to rank quarterbacks, based on a variety of passing measures. College rankings are used to make decisions as to whom to invite to the NCAA basketball tournament. In Chapter 4, we present a detailed discussion of objectives and measures for evaluating job performance. These would be primarily used to determine bonuses and pay raises.

The public is similarly bombarded with rankings of organizations. Universities and departments within universities are ranked in a popular issue of *U.S. News & World Report*. There are lists for the best companies to work for and the best cities to live in. All of these are primarily interesting pieces of information that may help some individuals decide where to apply for college or a job.

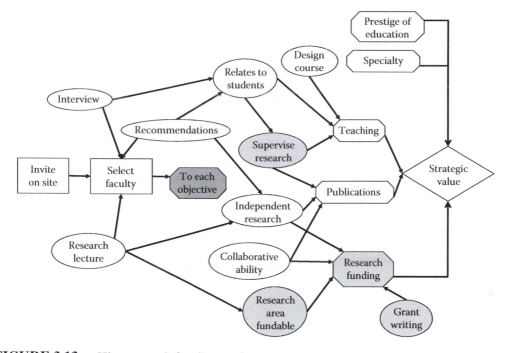

FIGURE 3.13: Hire research faculty member.

At the early stage of information gathering, multiple objectives and measures of past experience matter the most. However, the decision whom to hire is filled with significant uncertainty with regard to future performance especially when hiring someone directly upon graduating university. Experience has demonstrated that work performance is only moderately correlated to academic performance. Even when hiring an individual who has prior work experience, the uncertainty is significant. Laws and common practice regarding privacy generally limit the hiring company's ability to obtain accurate information about the candidate's performance in a prior job.

In the current environment, it is not uncommon to have more than 100 applications for a single position. The first step is to sort through the applications and decide on a handful of priority candidates to invite for an on-site visit. This visit primarily consists of interviews.

In Figure 3.13, we illustrate the factors involved in the decision to hire a new faculty member at a research university. (To limit the clutter, we have left off the arrows from the decision to each of the five objectives.) For this class of job, the on-site visit also includes delivering a seminar. The results of the interviews and seminar are uncertain when the candidate is invited to campus. However, these uncertainties are resolved prior to the final decision. With regard to faculty hires, two or more letters of recommendation are also part of the final hiring process.

In selecting a candidate, the obvious primary goals are to maximize teaching performance and research performance. If the field is engineering or science, the faculty member will also be required to obtain research contracts. The shaded elements in Figure 3.13 relate to the need for funded research. This would generally not apply when hiring faculty for a business school or many liberal arts fields such English or History.

Typically, a department's faculty recruitment focuses on one or more areas of specialization that are priorities for hiring. Thus, the candidate's specialty is a major element of the evaluation. Additionally, universities often pride themselves on the prestige of the universities where faculty members earned their PhD degrees. This information is often listed alongside the faculty name on a department website. It is believed that the more prestigious the university degree, the more likely the new faculty member will excel in research.

The hiring process is fraught with uncertainty about future performance. The single biggest concern involves the ability of the individual to conduct independent research once he has parted with his advisors. In the first year or two after hire, the new faculty member is typically continuing or fine tuning the research for his PhD and publishing the results. However, within a couple years, the new hire must develop an independent identity, while working with little or no supervision. Success in this area is difficult to predict, although the letters of recommendation can help.

More and more research is being done collaboratively, cutting across specialties. In addition, research university students often play a critical role in increasing the productivity of the faculty member. Thus, the candidate's ability to collaborate with colleagues and supervise students, as yet an uncertainty, is critical in projecting the candidate's future research productivity. Again, interviews and recommendations are useful tools when making this assessment.

The competition for funded research is brutal; success in obtaining funds is even more uncertain than success in publishing one's research in quality journals. Few graduate students receive much experience in grant writing while pursuing their doctorates. In addition, the potential for obtaining funding varies in different areas of research. The most generous grants tend to go to multifaculty efforts that cut across several universities. The individual's ability to collaborate with other educational institutions and become part of one of these initiatives often affords him the best opportunity for obtaining initial funding.

The same uncertainty applies to a new faculty's teaching ability. Even if the candidate has been a graduate teaching assistant, there is more to actual teaching than preparing the lectures of an already designed course. Of special concern is how successfully the candidate will relate to a wide variety of students. This is of particular importance when hiring for middle tier public universities that have diverse student populations. The interview can shed some light on this uncertainty but interviews are limited in their predictive value when it comes to ongoing performance; this is especially the case with regard to predicting the individual's ability to grow and mature in a new environment. Letters of recommendation can also help but these must be read with care and taken with a grain of salt, since the writers are motivated to help the candidate obtain a job. In addition, the aforementioned privacy laws tend to inhibit the writer's efforts at providing a balanced picture.

Each faculty hire is expected to have a realistic chance of earning lifetime tenure. Once achieving tenure, few faculty members move to other universities except for periodic sabbatical leaves. Thus, each hire can affect the long-term strategic direction of the department.

3.8.1 Supplier and Contractor Choice

The days are gone when important supplier decisions simply involve a comparison of costs adjusted for quality (Nydick and Hill 1992). With the increasing awareness of supply chain management, organizations seek strategic alliances with global, full-service suppliers. These suppliers need to support the lead organization in all business aspects including quality, cost, delivery, and global presence. In addition, manufacturers of complex products are passing on more and more of the subsystem design responsibility to their tier one suppliers. Suppliers must be evaluated through measures that capture their ability to contribute to the design of the subsystem. Boeing's development and manufacture of the Dreamliner, for example, involved supplier selection and integration of unprecedented scope. It also involved significant risk that, as of 2011, remains unresolved.

3.9 Facility Location: Sports Arena

The decision where to locate a major public facility is almost always multifaceted. This is true whether the facility is a library (Clemen and Reilly 2001), a power plant (Kirkwood 1982; Wenstop

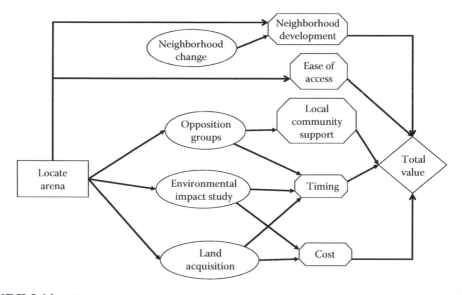

FIGURE 3.14: Locate sports arena.

and Carlsen 1988), a hazardous waste site (Merkhofer and Keeney 1987), a sports arena (Carlsson and Walden 1995), or a service terminal (Hegde and Tadikamalla 1990). One of the earliest applications of MAUT was in deciding where to locate a new Mexico City Airport (de Neufville and Keeney 1972).

The primary effort in structuring facility location decisions involves developing the multiple objectives and the associated measures. Issues of uncertainty tend to be of secondary significance. In Chapter 4, we present detailed discussion of multiple objectives associated with a decision to locate a global manufacturing facility. Here we will limit the description to some high level objectives and uncertainties (Figure 3.14).

The decision context we use here is the location of a new sports arena in a city. There is a wide range of values that sports owners and public officials consider in deciding where to build a facility of this type. As a result, the final decisions are quite diverse. For example, New York City built the new Yankee Stadium alongside the old one in the Bronx. The owner of the New Jersey Nets decided to build its new arena in a highly developed, densely populated area in Brooklyn. In contrast, the New York Giants built their stadium on generally vacant land in New Jersey alongside the New Jersey Turnpike. Similarly, the New England Patriots situated their stadium far from Boston.

In every location context, there is an objective to minimize cost. However, in an urban context, there is generally significant uncertainty as to how much it will cost to acquire the needed property. This uncertainty can contribute to delays in starting and completing the project. Another common uncertainty involves the results of an environmental study. It is not uncommon to find unanticipated hazardous materials on the site that warrant costly special handling.

An important objective in picking a location is the general level of support from the local community for the project. Often, there will also be a subgroup that is vehemently opposed to having the project take place in their backyard. It is hard to predict the extent of this opposition and its uncertain impact on delaying the project and undermining its overall support. An important factor in selecting a stadium location involves ease of access, both by car and mass transit. Poor access may lead to added costs, a factor we did not show in this diagram. Last, the reason why cities and neighborhoods seek such projects is the hope that the project will spur development in and around the facility location. This synergy effect is the most uncertain aspect of them all.

If the facility were not a stadium but a high tech plant, then maximizing the availability of highly skilled human resources would be another important objective to add to the diagram. For example, Toyota built its U.S. Engineering Research Center close to the University of Michigan, so that it could take advantage of the ready availability of new hires.

3.10   Bidding: Make Offer

With the growth of online auctions, bidding is becoming an increasingly dynamic and critical part of both business and personal life. Transportation companies routinely respond to requests for quotes to pick up and deliver goods for companies and individuals. Additionally, many people stay glued to their computers as the deadline approaches on an e-Bay auction.

Figure 3.15 illustrates a simple, one-time bid on a business opportunity. There is only one major decision: how much to bid? The key uncertainty relates to the competitor's bid. The short-term objective is to win the bid. However, some companies forget that the ultimate goal is to make a profit. They live and die by the "winner's curse" that results from outbidding everyone else, but ultimately paying more than the contract is worth. In complex business situations, there will be uncertainty regarding both the total cost and the ultimate value of a particular contract. A company such as Schneider International uses sophisticated operations research models to determine the cost of integrating a specific transport job into their overall delivery network schedule. They often choose not to bid on specific jobs.

The same issues of cost and value uncertainty apply when considering a bid to purchase a home. There may be hidden costs to repair the home. Moreover, many recent home purchasers discovered, to their chagrin, that house values can decline precipitously. In addition, the long-term value of residential property will be affected by uncertain community development as well as by personal lifecycle events.

In a bidding context such as e-Bay, there is the added decision regarding the timing of the bid. Timing is also critical in many large-scale acquisitions. Over time, the purchase price and the value of a company targeted for acquisition are likely to vary significantly as a function of the economy and the state of the company's business niche.

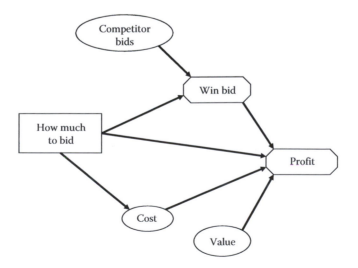

FIGURE 3.15:   Simple bid.

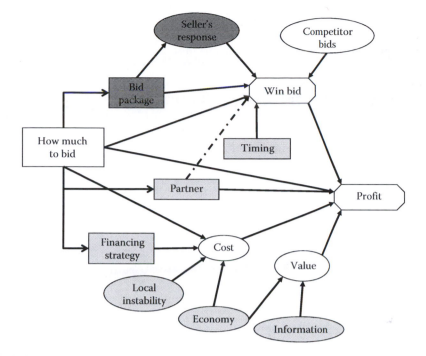

FIGURE 3.16: Large scale bid with multiple decisions.

The decision regarding how much to bid on a large-scale development project such as the Iraqi oil fields is extremely complex. In oil lease decisions, there are large amounts of data, but with the geological information, there is still significant uncertainty regarding the amount of oil that will be ultimately retrievable as technology improves. Furthermore, the total value of the resource in question is uncertain, since this value will be influenced by economic developments in the world at large. The cost of development is further influenced by the economy and the instability of the local region (Figure 3.16).

One key decision involves whether to include partners and how these partners should be selected. Deciding to include partners can reduce the risk of investment but can also diminish the potential profit. The dotted line reflects the situation in which the choice of partner also influences the chance of winning the bidding the contest. In global investment decisions, it may be essential to have a local partner on the scene, because this will make your proposal more attractive to local decision makers. This also applies in very large scale projects, such as when bidding on the development of major weapons systems for the U.S. government; in such cases, partners who bring special expertise are likely to increase the lead organization's chance of winning the bid.

Another decision involves how to finance the bid. Every home purchaser decides how much money to put down and how large a mortgage to take. In recent decades, a number of companies have financed corporate acquisition binges by assuming large amounts of debt. These debt levels eventually proved unsustainable and have, in some instances, forced the purchasing company into eventual bankruptcy. Similarly, many home buyers have been unable to sustain debt loads as they lost their jobs. Warren Buffett, a major investor in Kraft, expressed serious concerns in 2010 as to Kraft's plan to fund the purchase of Cadbury by issuing stock. Kraft decided to go ahead as planned, and it is too soon to say whether or not Buffet's concerns will be realized.

Another related decision involves the bid package and not just the amount. The bid composition often influences the way that the potential seller responds to an offer. A company being purchased will be interested in the amount of cash offered as well as in acquiring stock in the purchasing company. The top management may be interested in their future role in the acquiring company. This, for example, was a major issue in Boeing's purchase of McDonnell Douglas.

The bidding process has interesting applications in sports. Multiple teams may bid for the services of a highly prized coach or a highly prized athlete. In bidding for a coach, the scope of authority will be an important factor in addition to the salary and benefits. The movie *Blind Side* offers an interesting take on the package offered to a promising high school football prospect. In this case, the package even included benefits for the player's younger brother as an inducement. Teams seeking to acquire a star through a trade, in order to increase their immediate chances for a championship, are challenged to come up with a set of players who can help the team develop over the long-term.

For each bid placed, there is a mirror decision on the part of the seller as to whether or not to accept a particular bid or which bid to accept. This applies equally to the sale of a house and the acquisition of a company. Likewise, it applies to both a high profile coach and an unrestricted free agent.

3.11   Personal: University Selection

We all make numerous decisions on a daily basis. Although many of these decisions involve some form of multiple objective trade-off, we rarely need a formal structured approach in our personal lives (Keeney 2004). For example, every time we go out to eat, we consider cost, quality and type of meal, the restaurant's location, and its ambience. If we intend to take along our children, we also assess the restaurant's level of family friendliness. When making other decisions, such as shopping for food and additional staples, we establish our preferences through similar processes, and, afterward, we tend routinely to patronize the same stores. There are, however, other decisions that we might make as infrequently as once a year, once every few years, or once in a lifetime that warrant the time and energy associated with more careful thought and structure. These decisions include which job to accept, which university to attend, which car or house to buy, and which organizations to join. (See Chapter 4 for a used car decision.) We also make regular technology upgrade decisions for our computers, cell phones, and digital cameras.

All of the readers of this book have already made a university selection decision for themselves and may have even helped someone else wrestle with this decision. It would be interesting to ask whether or not the reader considered all of the objectives and recognized the associated uncertainties. The primary objectives in choosing a university relate to social life, academics, and cost. In this time of uncertainty, few people embark on college or graduate school without also thinking about career. Additionally, the location of the university may be of interest in and of itself. Weather factors or the proximity to family may factor into the decision (Chelst and Edwards 2005). The choice between an urban school versus one that is located in a college town would also speak to the preferred quality of social life (Figure 3.17).

There are a number of other critical uncertainties. Most notable of these, perhaps, is the state of the economy throughout the student's school years, a factor that is likely to influence family finances and the ability to pay the cost of education. The economy will also play a significant role in defining the most attractive career choices at the time of graduation. Universities differ in the level of support they provide as students shape the paths that their careers will take. While all universities offer support services, the actual help these services provide is uncertain. For this reason, the picture includes a dashed line between support services and academics to represent only those students who feel they may need significant guidance when choosing their courses. Although a student can see the list of course and faculty credentials, there will still be significant uncertainty regarding the quality of instruction at the time the student selects a college.

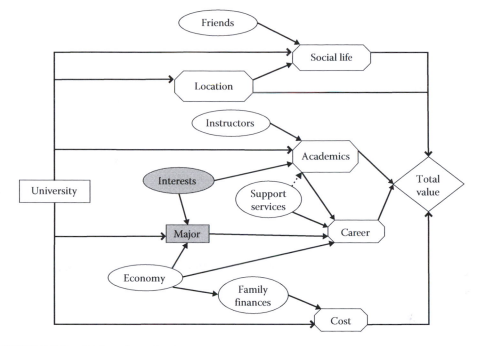

FIGURE 3.17: University selection.

The figure includes a shaded box with regard to deciding on a subject major. Some entering students know with near certainty the area they want to major in, but the literature indicates that the majority of students do not make this decision until later in their academic careers. Many students will change their decision multiple times. The key uncertainties that will influence this choice are the students' developing interests over time and the state of the economy at the time of their decision.

3.12 Information Gathering: Market Research, Prototypes, and Pilot Plants

In a number of the previous examples, the primary decision could be broken down into a sequence of smaller decisions. Often, in the process of making these decisions, information is obtained that clarifies some of the uncertainty. For example, in Figure 3.5, the decision to launch a product is followed by a long-term price decision. Before making the second decision, the company will gain competitive and market demand information that will help them decide on the long-term price. Similarly, the selection of a university is followed by the selection of a major. By the time the second selection must be made, the student will have learned more about his interests and be able to make a better choice. In these instances, knowledge can be expected to accumulate naturally during the interim between the two decisions.

One of the early critical contributions of the field of decision analysis was to formalize and quantify the expected value of gathering information in an uncertain environment. This expected value can be determined even in situations in which the information improves a forecast but does not completely resolve the uncertainty. Information of this sort has been termed *sample information* or *imperfect information*. In many instances, the first decision is whether or not to invest money in gathering information.

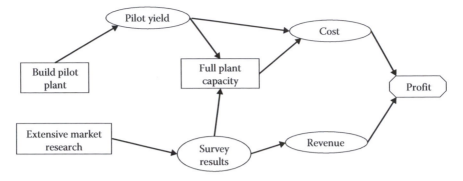

FIGURE 3.18:　Gather imperfect information.

The value of imperfect information has engendered major debates regarding several medical testing policies. The latest recommendations have suggested that annual mammograms for women should begin at age 50 instead of 40 and that the start of routine pap smears should be similarly delayed because of high rates of false positive results. Of greater concern has been the widespread use of full-body scans as a diagnostic tool. One estimate claims that this will significantly increase the cancer rate, leading to the deaths of tens of thousands of people.

The classic example brought in business textbooks involves the use of a market test to better estimate demand. In an engineering management context, this issue arises when evaluating the potential benefit of constructing a pilot plant. In the chemical industry or chip-making industry, a prototype plant may be built as a means through which to estimate initial manufacturing yields. In the auto industry, they may use soft tooling (or experimental manufacturing processes) to evaluate alternative manufacturing strategies or pieces of equipment in order to determine the most reliable way to manufacture a specific component and build a quality product. However, the information gathered from these tests is imperfect at best—in part, because it is impossible to predict the learning that will take place as a process matures and workers become more experienced. We have inserted both information gathering decisions in Figure 3.18.

Analogous decisions arise when assessing the value of building a product prototype to reduce the risk and magnitude of costly design iterations (Klein et al. 1994). For example, building and testing a prototype for a molded part may reveal a problem with the mold, thus eliminating the expense of developing and building an injection mold (Ulrich and Eppinger 1995). The three major uncertainties associated with this decision are (a) the time needed for design activities prior to the prototype build decision (i.e., upstream), (b) the time needed for prototype development, and (c) the impact of the prototype on subsequent downstream tasks. The build prototype decision itself can be further subdivided into multiple decisions. There may be a choice between spending the whole budget on a comprehensive, reliable (but expensive) prototype toward the end of the development process, or developing a series of less reliable (and supposedly cheaper) prototypes at various points within the development process (Thomke and Bell 1998).

3.13　Summary

In this chapter, we presented and framed a wide array of decisions and highlighted the prevalence of uncertainty and multiple objectives. These examples were drawn from both the manufacturing and service sectors. They also included public sector decisions as well as personal ones. In each instance, we offered an influence diagram as a template to build upon when decision makers are faced with a particular application of any of these classes of decisions. To emphasize the point that these templates are just starting points, the homework encourages the reader to critique specific diagrams by adding or deleting nodes and arrows.

This chapter serves another purpose. It attempts to clarify the types of one time decisions that are the major focus of the modeling and analysis tools developed in this text. Many of the decision contexts will be revisited in later chapters as we demonstrate the analytic tools that can be used to rank order decision alternatives and determine their respective strengths and weaknesses.

Exercises

I For each of the following examples, modify the decision context and suggest changes to the influence diagram. The changes could be subtractions or additions.

    3.1 Make or Buy—Strategic Decision: Figure 3.2

    3.2 Technology upgrade: Figure 3.3

    3.3 Product Launch: Figure 3.5

    3.4 Technology choice—Personal: Figure 3.12

    3.5 Personnel Selection: Figure 3.13

    3.6 Facility Location: Figure 3.14

II Construct an Influence Diagram

    3.7 Bid Acceptance—different scenarios

        a. You have put your house up for sale and received one offer to buy your home.

        b. You have put your house up for sale and received multiple offers to buy your home.

        c. Your company has received one hostile offer to purchase your company.

        d. Your company has received multiple purchase offers.

        e. Your government has received multiple corporate offers to develop a natural resource.

    3.8 Personnel decision—different jobs

        a. You are in charge of making the decision whom to promote from assembly line worker to first line supervisor.

        b. You are responsible for selecting the manager of a supermarket from among current employees.

        c. You are responsible for selecting the department chair of from a pool of faculty.

References

Bhattacharya, S., Krishnan, V., and Mahajan, V. (1998). Managing new product definition in highly dynamic environments. *Management Science*, 44 (11: Part 2), S50–S64.

Bhuta, C. (1992). Management of risk in projects. *Engineering Management Journal*, 4(1), 15–20.

Booker, J. and Bryson, M. (1985). Decision analysis in project management. *IEEE Transactions on Engineering Management*, 32, 3–9.

Browning, T. (1998). Sources of schedule risk in complex system development. *Proceedings of the 8th Annual International Symposium of INCOSE*, Vancouver, British Columbia, Canada, July 26–30, pp. 187–194.

Canbolat, Y. B., Kenneth Chelst, K., and Garg, N. (2007). Combining decision tree and MAUT for selecting a country for a global manufacturing facility. *Omega*, 35, 312–325.

Carlsson, C. and Walden, P. (1995). AHP in political group decisions: A study in the art of possibilities. *Interfaces*, 25(4), 14–29.

Chelst, K. and Edwards, T. (2005). *Does This Line Ever Move?: Everyday Application of Operations Research*. Emeryville, CA: Key Curriculum Press.

Clemen, R. T. and Reilly, T. (2001). *Making Hard Decision with Decision Tools Suite, 2nd rev. ed.* Belmont, CA: Duxbury Press.

Corner, J. and Kirkwood, C. (1991). Decision analysis applications in the operations research literature, 1970–1989. *Operations Research*, 39, 206–218.

De Neufville, R. and Keeney, R. L. (1972). Use of decision analysis in airport development for Mexico city. In: *Analysis of Public Systems*, Drake, A. W., Keeney, R. L., and Morse, P. M., eds. Cambridge, MA: MIT Press.

Fine, C. and Whitney, D. (1996). Is the make-buy decision process a core competence? MIT Center for Technology Policy and Industrial Development working paper.

Graves, S. and Jordan, W. (1995). Principles on the benefits of process flexibility. *Management Science*, 41, 577–594.

Hammond, J. and Keeney, R. (1999). Making smart choices in engineering. *IEEE Spectrum*, 36(11), 71–76.

Hegde, G. and Tadikamalla, P. (1990). Site selection for a 'sure service terminal'. *European Journal of Operations Research*, 48, 77–80.

Hess, S. (1993). Swinging on the branch of a tree: Project selection applications. *Interfaces*, 23(6), 5–12.

Keeney, R. L. (2004). Making better decisions. *Decision Analysis*, 1, 193–204.

Keeney, R., Lathrop, J., and Sicherman, A. (1986). An analysis of Baltimore Gas and Electric Company's technology choice. *Operations Research*, 34, 18–38.

Kirkwood, C. (1982). A case history of nuclear power plant site selection. *Journal of the Operational Research Society*, 33, 353–363.

Klein, J., Powell, P., and Chapman, C. (1994). Project risk analysis based on prototype activities. *Journal of the Operational Research Society*, 45, 749–757.

Krishnan, V., Eppinger, S., and Whitney, D. (1997). A model-based framework to overlap product development activities. *Management Science*, 43, 437–451.

Krumm, F. and Rolle, C. (1992). Management and application of decision and risk analysis in Du Pont. *Interfaces*, 22(6), 84–93.

Kusiak, A. and Park, K. (1990). Concurrent engineering: Decomposition and scheduling of design activities. *International Journal of Production Research*, 28, 1883–1900.

Kusiak, A. and Wang, J. (1993). Efficient organizing of design activities. *International Journal of Production Research*, 31, 753–769.

MacCormack, A., Newmann, L., and Rosenfield, D. (1994). The new dynamics of global manufacturing site location. *Sloan Management Review*, 35(4), 69–80.

Merkhofer, M. and Keeney, R. (1987). A multi-attribute utility analysis of alternative sites for the disposal of nuclear waste. *Risk Analysis*, 7, 173–194.

Nydick, R. and Paul Hill, R. (1992). Using analytic hierarchy process to structure the supplier selection procedure. *International Journal of Purchasing and Materials Management*, 28(2), 31–36.

Reinertsen, D. (1997). *Managing the Design Factory*. New York: Free Press.

Rosenau, M. and Moran, J. (1993). *Managing the Development of New Products*. New York: Van Nostrand Reinhold.

Samuelson, W. F. and Zeckhauser, R. (1988). Status quo bias in decision making. *Journal of Risk and Uncertainty*, 1, 7–59.

Sanderson, S. and Uzumeri, M. (1997). *Managing Product Families*. Burr Ridge, IL: Irwin Professional Publishing.

Spetzler, C. and Zamora, R. (1989). Decision analysis of facilities investment and expansion problem. In: *The Principles and Applications of Decision Analysis*, Vol. I, Howard, R.A. and Matheson, J.E., eds. Menlo Park, CA: Strategic Decision Group.

Thomke, S. and Bell, D. (1998). Optimal testing under uncertainty. Harvard Business School Working Paper #99-053 (October).

Ulrich, K. and Eppinger, S. (1995). *Product Design and Development*. New York: McGraw-Hill.

Walton, M. (1997). *Car: A Drama of the American Workplace*. New York: W. W. Norton.

Weber, S. (1993). A modified analytic hierarchy process for automated manufacturing decisions. *Interfaces*, 23(4), 75–84.

Wells, W. (1993). Unified life cycle modeling: A framework for manufacturing cost estimating and analysis, PhD dissertation, Industrial and Manufacturing Engineering, Wayne State University, Detroit, MI.

Wenstöp, F. E. and Carlsen, A. J. (1998). Ranking hydroelectric power projects with multicriteria decision analysis. *Interfaces*, 18(4), 36–48.

Yassine, A., Chelst, K., and Falkenburg, D. (1999a). A decision analytic framework for evaluating concurrent engineering. *IEEE Transactions on Engineering Management*, 46, 144–157.

Yassine, A., Falkenburg, D., and Chelst, K. (1999b). Engineering design management: An information structure approach. *International Journal of Production Research*, 37, 2957–2975.

Young, M. R. (1998). A minimax portfolio selection rule with linear programming solution. *Management Science*, 44, 673–683.

Part II

Decisions with Multiple Objectives

Chapter 4

Structure Decisions with Multiple Objectives

> Itel Corp. is planning the installation of a new V-chip plant. Meanwhile, the U.S. Environmental Protection Agency is in the midst of developing new regulations that will significantly tighten the discharge limits from industrial wastewater treatment plants across a variety of industries. The new regulations are scheduled to be finalized at the same time as the launch of the new chip program. Therefore, Itel Corp. needs to purchase and install a wastewater treatment plant that will meet these new regulations. How will Itel Corp. select the wastewater technology? What are the objectives of the wastewater technology selection problem?

> Danielle and Daniel Lyons have decided to remodel their kitchen. They have spoken to their friends and neighbors who shared the names of two contractors they have used. In addition, there are two local companies with good reputations that advertise heavily. The Lyons invited each of the four potential contractors to their homes and each has submitted designs, cost estimates, timelines, and references. How can they utilize the information provided to make a rational decision? In addition, they are concerned about the quality of the work and timeliness of project completion. What measures should they use and how do they obtain data?

4.1 Goal and Overview

The goal of this chapter is to develop skills for constructing a multiple objectives hierarchy that includes both quantitative and qualitative measures.

Business, political, and personal decisions often involve a number of objectives that may conflict (Dyer et al. 1992). Consider the kitchen remodeling decision. The Lyons want to balance the quality of material and possibly the physical expansion of the kitchen area against the cost and time needed to complete the job. Analogously, multiple-objective trade-offs arise every time a manager has to select a supplier for a complex product or component. Purchasing a digital camera is a personal technology choice decision that is likely to involve trade-offs between cost and performance that parallel the Itel decision mentioned earlier. Where to live and where to locate a stadium or a factory are, likewise, decisions that always involve multiple objectives. In fact, it is multiple objectives and complex trade-offs that make our personal decisions so hard. For example, imagine the agonizing choice of whether to place an elderly, incapacitated parent in a nursing home or in a room in one's own home. There are obvious cost considerations as well as numerous quality of life issues that face both the family and the incapacitated parent. As long as no single alternative is best with regard to each and every factor, a decision maker must carefully determine the objectives and develop a strategy for assigning relative weights.

There are a number of methods to deal with multi-objective decision problems. These include the Multi Attribute Utility Theory (MAUT), the Analytical Hierarchy Process (AHP) (Saaty 1980), goal programming, and fuzzy logic. Both MAUT and AHP can effectively incorporate quantitative and qualitative factors into their multiple objective frameworks and deal with uncertainty. These methods are primarily used in a decision context that involves selecting the *single best option* among a limited set of distinct alternatives. Both are supported by software that is easy to use and flexible, such as Logical Decisions® for Windows (LDW). In this book, we use MAUT as a primary tool and AHP as a secondary tool to deal with multiple-objective problems.

In this chapter, we examine several concepts related to framing multi-objective decisions: objectives, measures, objectives hierarchy, and alternatives. We add detail to the framing tool introduced earlier, influence diagrams. An influence diagram is used to reach agreement among key stakeholders as to the major objectives. Analysis, however, requires a more detailed approach to clarifying multiple objectives. Ultimately, there need to be clear measures that facilitate quantification and analysis. We begin with two simple examples: light bulbs and service level. The choice of a light bulb illustrates a technology decision that is likely to trade off performance and cost. The manager of a facility must make a decision regarding the illumination of the light bulb he installs. At what point does the cost of electricity outweigh the usefulness of extra wattage? Likewise, in establishing a service level, a retail store must often trade off the number and quality of the cash register attendants and the hourly cost of service. We will elaborate on these concepts as we progress to several more complex choices: a building contractor, facility location, used car, performance evaluation, power plant, watershed, and wastewater technology choices.

4.1.1   Other Multi-Objective Tools

MAUT was developed to incorporate both multiple objectives and the principles of utility theory. Utility theory captures an individual's attitude with regard to risk and may be a nonlinear function of the dollar magnitude of the risk. It is based on principles of rational consistent decision making. If the analysis does not include the formal assessment of risk attitude, the more appropriate term is Multi-Attribute Value Theory (MAVT), which is commonly used in Europe instead of MAUT (Belton and Stewart 2002). The broad field of multi-objective analysis is often labeled Multi-Criterion Decision Analysis (MCDA; Belton and Stewart 2002). Other methods include goal programming, fuzzy logic, and ELECTRE. Goal programming, proposed by Charnes et al. (1955), is an extension of linear programming designed to handle problems with multiple, sometimes conflicting, objectives. It attempts to combine the logic of constrained optimization with the decision maker's desire to satisfy several goals. Each measure is given a goal or target value to achieve. The ultimate goal is to minimize deviations from this set of target values. The weighted sum of these deviations is combined into in an achievement function to be minimized. Jones and Tamiz (2002) provide a bibliography of diverse goal programming applications between 1990 and 2000.

Fuzzy Logic was developed by Zadeh in 1965 to deal with problems involving vagueness and ambiguity that are often associated with the processes of human thinking, reasoning, and cognition (Zadeh 1965). Fuzzy Logic is a superset of conventional logic (Aristotelian two-valued logic) that has been extended to handle the concept of "partial truth." Classical logic is equipped to deal only with "completely true" and "completely false" situations. In contrast, Fuzzy Logic is multi-valued logic that allows intermediate values to be defined between conventional evaluations like yes/no, true/false, black/white, and 0/1. Notions like rather warm or pretty cold can be formulated mathematically and processed by computers. Fuzzy Logic modules have been developed for process control, flexible manufacturing systems, flexible automation, and multi-criteria decision making. There are over 2000 commercially available products that employ Fuzzy Logic, ranging from washing machines to high-speed trains. Metaxiotis et al. (2004) presents recent applications of Fuzzy Logic to decision support systems in various sectors.

Another family of multi-objective decision models uses "outranking relations" to rank a set of alternatives. These were developed in France and have attracted users and researchers primarily in Europe. The outranking methods seek to eliminate alternatives that are, in a particular sense, "dominated." ELECTRE and its derivatives are the most prominent techniques in this class of models. It was first introduced by Benayoun et al. (1966). Today, the most widely used versions are known as ELECTRE II and III (Wang and Triantaphyllou 2006). ELECTRE has been widely used in civil and environmental engineering (Hobbs and Meier 2000). TOPSIS is a variant of ELECTRE (Hwang and Yoon 1981).

4.2 Description of the Overall MAUT Process

The fundamental principles of MAUT were developed by Keeney and Raiffa (1993) to help decision makers structure complex problems as well as evaluate and select alternatives under conflicting multiple objectives and uncertainty. One of their earliest case studies involved the selection of the location for a new airport to service Mexico City (de Neufville and Keeney 1972). The objectives included multiple measures of cost, airport traffic capacity, access time to the airport, safety, social disruption, and noise pollution. Historically, many of the early applications involved major public policy decisions as these naturally involve multiple objectives and often reflect the interests of different constituencies. These applications included studies of air pollution control, instructional program budgets, fire department operations, nuclear power plant locations, aircraft landings, blood bank policies, and sewage sludge disposal (Keeney and Raiffa 1993; Hokanen and Salminen 1997). Environmental issues and land use planning continue to be the major areas of application (Merrick et al. 2005), and executives in the power industry are leading supporters of using MAUT to make strategic decisions (Kidd and Prabhu 1990; Keeney and McDaniels 1992; Keeney et al. 1995).

The MAUT process is illustrated in Figure 4.1 (Strimling 1996). The application of MAUT involves four main tasks: structuring decisions, creating and describing alternatives, clarifying preferences, and analyzing alternatives.

These tasks can be further delineated as follows:

4.2.1 Structuring Decisions

The process of structuring decisions is highly creative and requires expert knowledge of the issues relevant to the decision (Keeney 1996). Three elements of the process are listed as follows:

- Identify and clarify requirements, goals, and objectives and define the problem scope. The requirements specify what needs to be accomplished and the basic capabilities that must be designed into a new product or service.
- Define relevant measures affecting the decision outcome and structure them into a hierarchical form called an objectives hierarchy.
- Create measure scales for hard-to-quantify variables such as customer satisfaction, implementation difficulty, and overall risk.

4.2.2 Describe Alternatives

This is often the most time consuming task. First, the decision maker working with stakeholders must develop a range of alternatives and avoid the common mistake of narrowing this range too quickly. Then—with the help of diverse experts—data and expert opinion must be gathered for each

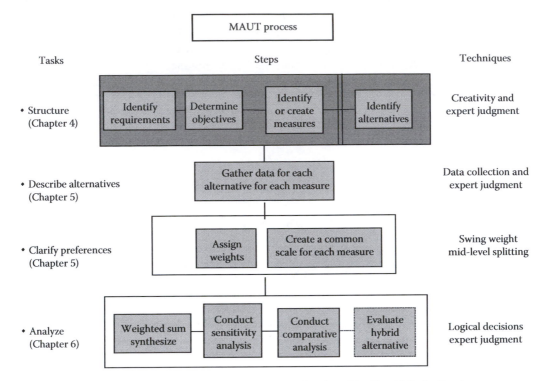

FIGURE 4.1:   MAUT process 1. (This figure was developed by David Strimling.)

measure of each alternative. Depending upon the range of options and data availability, this task can take weeks or several months.

4.2.3   Clarify Preferences

In order to determine the overall score or ranking of an alternative, the disparate measures and objectives must be combined into a single aggregate score. The planners must

- Elicit preference information from the decision maker(s) concerning the measures that will be used. This will help determine the relative importance of the various measures and is expressed in weights.
- Develop the decision maker's set of utility functions by establishing a scaling function within each measure. Each function converts the score on a performance measure into a value on a 0–1 scale. For numeric measures, the default assumption is linearity.

4.2.4   Analyze Alternatives

The analysis begins after the data are collected and appropriately scaled, and measures and objectives are weighted.

- Calculate the overall score for each alternative. All scores will be between 0 and 1. An overall score of 1 means the alternative scores the highest possible value on each and every measure and, thus, represents a theoretical ideal.

- Compare alternatives as to their relative strengths and weaknesses. Determine which measures contribute the most and the least to each alternative's total score.

- Determine the impact of uncertainty on the relative rankings.

- Perform sensitivity analyses on the weights assigned to assess the robustness of the solution to changes in the weights to determine if small changes in weights result in a different preferred alternative.

- Create improved hybrid alternatives by brainstorming ways to address the weaknesses of the highest ranked alternatives.

MAUT uses functions to transform the diverse criteria to a single, common, dimensionless scale, or utility. The decision maker acts to maximize a utility function that is itself a function of the multiple objectives and the relative importance placed on each objective. In addition, a separate single utility function (SUF) scales the relative preference for different scores on a specific measure. This helps the decision makers evaluate the impact of various differences on multiple measures. If, for example, there is 5 minute time differential between how quickly two people with different skill levels can service a customer's needs, what significance does this difference represent to the overall customer experience? Is it worth incurring additional costs in order to save these 5 minutes? The overall utility score of an alternative is a weighted sum of n different SUFs. It takes on the following form for simple cases that assume independence between measures and objectives (Figure 4.2).

Let

x_{ij} is the raw score of alternative j on measure i.

$u_i(x_{ij})$ is the decision maker's utility function for measure i that transforms the raw score into a utility value between 0 and 1.

w_i is the decision maker's weight assigned to the ith measure.

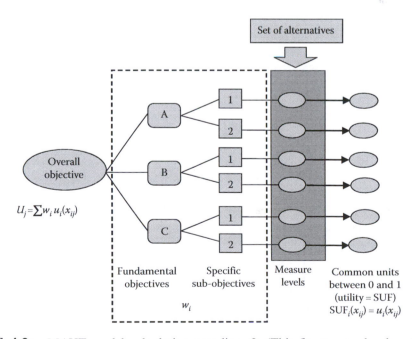

FIGURE 4.2: MAUT model calculation paradigm 2. (This figure was developed by David Strimling.)

It is conventional to normalize the weights to unity so that the following condition holds:

$$\sum_{i=1}^{n} w_i = 1$$

U_j is the overall utility score of the jth alternative and is determined by taking a weighted sum.

$$U_j = \sum_{i=1}^{n} w_i u_i(x_{ij})$$

4.3 Basic Terminology

Objective: An objective is defined by Keeney and Raiffa (1993) as a statement of the desired state of the system under consideration. Thus, when stating objectives, we use the terms minimize or maximize. Typical objectives are to minimize cost or maximize profit. In our context, an objective reflects a *direction* of improvement (e.g., minimize cost) and not a specific target value.

Goal: The term *goal* is often used interchangeably with the term objective. However, many view these terms as distinct. In colloquial speech, the term *goal* often identifies a *specific level or target* of achievement toward which to strive. In this sense, a goal is different from an objective in that it is either achieved or not. A goal regarding cost may be "to reduce cost by 10% in one year." President Kennedy's stated goal in 1961 was to reach the moon by 1970. For our purposes, we treat the terms *goal* and *objective* as distinct.

Decision context: A decision context is identified as the setting in which a decision occurs. Each decision context affects both the objectives to be considered and their relative importance. The MAUT model used to select a supplier will vary significantly if the decision context involves suppliers of flashlight batteries or six cylinder engines.

Measure: A measure is used to characterize performance in relation to an objective. For example, if an objective is to minimize cost, then the measure scale might be defined in terms of dollars, euros, or yen.

Attribute: This term is used interchangeably with the term measure.

Cutoff value: The cutoff value is a minimum (or maximum) acceptable score on a measure. If an alternative falls below a minimum cutoff value on even one measure (or above a maximum), that alternative is likely to be dismissed regardless of how well it scores on other measures.

Objectives hierarchy: An objectives hierarchy is a structured representation of a set of objectives and measures that organize them from most to least general. It is similar to an organization chart. Objectives will often have sub-objectives that are more specific. For example, in selecting a library location, maximize accessibility for the public is critical. Accessibility may be subdivided into three distinct population subgroups: elderly, general adult, and children. (The software package Logical Decisions uses Goals Hierarchy as its default term.)

Utility: Utility is a standardized measure of the relative desirability of a score on a particular measure or total score for an alternative. Since attributes may have different scales such as dollars, seconds, or miles per gallon (mpg), it is very difficult to evaluate alternatives without creating a commensurate scale. Utility is used to convert the levels of attributes into a common scale that ranges from 0.00 to 1.00 in order to compare and/or combine qualitative and quantitative attributes. Zero corresponds to the worst level and one corresponds to the best. Just as the utility scale for a single measure is called an SUF, the score for an alternative that involves weighted sums of objectives and measures is called a multi-attribute utility function (MUF). Utility scales were originally used for uncertain variables to reflect the decision maker's attitude toward risk.

Alternatives: Alternatives are choices to be evaluated and ranked by analysis.

4.4 Fundamental Objectives

Intelligibly articulated objectives help clarify the driving forces behind a decision (Keeney 1996). The fundamental objectives qualitatively define all the concerns in the decision context and provide guidance and a foundation for evaluating the relative desirability of the alternatives. If objectives are vague and inadequately defined or, worse yet, even missing, time and resources may be wasted collecting unnecessary data while useful information is ignored. All organizations and individuals have objectives that guide their decisions even if they are not explicitly articulated. Decision makers often express the desire to minimize cost, maximize quality, maximize market share, or maximize customer satisfaction. Some common objectives and sub-objectives are listed in Table 4.1.

Too often, however, diverse parts of the same company are measured differently and driven by alternative objectives that are not aligned with those of their colleagues next door.

Almost every decision has a financial objective either to minimize cost or maximize profit. In theory, there is no need for sub-objectives or multiple measures, since all of the financial measures, such as transportation costs and production costs, can be combined into a single measure such as net present value (NPV) or return on investment (ROI). However, in practice, decision makers do not treat all costs the same. For example, variable cost and investment costs are treated separately because decision makers routinely trade off annual operating costs against investment dollars. Similarly, although headcount also contributes to cost, decision makers who are concerned about the number of workers might use the objective of minimizing headcount. This would be applicable only if management is not constrained from reducing its workforce. However, in some contexts—as with unions and guaranteed work—a specific project might have the reverse objective, namely to maximize labor content. If the decision involves a choice of equipment, the decision maker may have to consider operating costs, maintenance costs, and training costs separately since each is likely to come from a different budget category.

In many decision contexts, one objective is to minimize time, a critical variable in today's globally competitive business environment. Companies strive to reduce product development time and launch their products as soon as possible. If the decision alternatives involve the choice of a new process or technology, implementation difficulty will have a direct impact on the speed with which the company can achieve its goal and this could serve as a key differentiator.

Every company strives to maximize its competitive advantage in order to stay in business and thrive. In certain situations, a company may strive to increase market share and be willing to give up some profit in exchange. Competitiveness may also include factors that are not as easy to quantify, such as competitive styling. Then again, a company may choose one alternative over another because it is less likely to lead to copying and loss of intellectual property. This measure is important in technical decisions, especially when countries such as China do not actively protect the intellectual capital of others.

Maximizing performance is the broadest category of all; it is most dependent on context. Performance measures for different classes of products, such as computer chips, materials, or software, will vary dramatically. For equipment purchase decisions, performance may be defined by such factors as ease of use or ease of maintenance. Upgradeability is a critical factor in new

TABLE 4.1:   Common objectives.

Major Objectives		Sub-Objectives	
Direction of Preference	Objectives	Direction of Preference	Objectives
Minimize	Cost	Minimize	Investment cost
			Variable cost or operating costs
			Number of workers
			Warranty cost
			Training cost
			Maintenance cost
		Maximize	Production rate
	Time	Minimize	Development time
			Implementation difficulty
Maximize	Profit	Maximize	ROI
			NPV
	Competitiveness	Maximize	Market share
			Overall style and appearance
			Difficulty in copying
	Performance	Minimize	Weight
			Power consumption
		Maximize	On-time delivery
			Ease of maintenance
			Durability
			Ease of use
			Upgradeability
	Safety	Minimize	Minor injuries
			Serious injuries
			Fatalities
	Quality	Minimize	Rework or scrap rate
			Failure rate
		Maximize	Life cycle
			Labor skill
			Craftsmanship
	Stability	Minimize	Corruption
		Maximize	Economic stability
			Political stability
	Customer satisfaction	Minimize	Waiting time
		Maximize	Problem resolution
			Personalized service
	Location	Minimize	Environmental impact
		Maximize	Access
			Expandability

technology decisions. Two of the sub-objectives of maximizing performance involve minimizing power consumption and weight. There is no contradiction between stating that the overall objective is to maximize performance while defining its sub-objective as minimization.

Quality might involve obvious measures such as number of defects per 1000 parts or total warranty costs or might be defined by something more esoteric such as craftsmanship. In public sector decisions, such as governmental regulations or power plants, safety concerns will play a significant role. In global location decisions or supplier decisions, the overall stability of the relationship will be of significance for any long-term investment decisions.

TABLE 4.2: Questions drive identification of objectives.

Requirements: What are the requirements of the decision situation? What are the environmental social, economic, or health and safety requirements?

Program objectives: What are the ultimate objectives? What values are absolutely fundamental? What are the objectives for customers, employees, shareholders, and decision makers? What environmental, social, economic, or health and safety objectives are important?

Customer guidance: What does the customer want? What does he or she value? What should a customer want?

Technical performance measures: How do you measure achievement with regard to this objective? Which objectives are the most important and why?

Different perspectives: What would competitors or one's own constituency be concerned about? At some time in the future, what would concern decision makers? Where would the company like to be in 10 years?

Alternatives: What would constitute a perfect alternative, a terrible alternative, or some reasonable alternative? What is good or bad about each?

Problems and shortcomings: What is wrong or right with the organization? What needs to be fixed? How can the organization be improved?

4.4.1 Identify Objectives

Objectives must be clear, specific, measurable, agreed upon, and realistic in order to facilitate decision making. In complex decisions, the objectives should represent the interests of the various organizations affected by the decision. For instance, when choosing a supplier, representatives from purchasing, logistics, manufacturing, and engineering should participate in the creation of the fundamental objectives hierarchy.

Eliciting objectives requires creativity and hard thinking about a decision situation. The most obvious way to identify the objectives is to ask a group of decision makers or stakeholders first to recapitulate the decision context and then individually provide a written list of objectives. The decision maker or stakeholders then move to a group discussion of the lists. One goal of this method is to provide a variety of examples so as to facilitate a decision-making team's thought process regarding the primary objective, sub-objectives, and specific measures. Table 4.2 introduces techniques used by Keeney (1994) to elicit objectives. He suggests asking a variety of questions from multiple vantage points so as to stimulate thinking and make sure that the list of objectives is comprehensive.

4.5 Objectives Hierarchy: Examples

An objectives hierarchy is a diagram of relationships between objectives, sub-objectives, and measures. This hierarchy aids in communicating the results of framing the decision. The higher levels represent more general objectives, which are often vaguely stated and, hence, not operationally defined. Typically, two or more sub-objectives are associated with objectives at the next level of the hierarchy; these provide more specific statements regarding desirable characteristics of alternatives and also help define the objectives in greater detail. As we go down the hierarchy, objectives at the lower level become more specific and more operational, describing the important elements of the more general levels. The lowest levels of the hierarchy represent measures. A measure by definition is quantifiable.

We start with a simple example in order to explore the basic concept of fundamental objectives. One of the simplest technology choices we make on a regular basis is which light bulb to buy and install in a fixture. In every instance, there is a trade-off of *performance* (the amount of light) and *cost*, just as Mr. Frail sees with his bulb selection problem in the case that follows.

Case: Bill Frail has recently been promoted to a product development manager position and will soon move to his new office, which is being repaired. He will select light bulbs for the office. In the office, there are 10 fixtures. His goal in replacing the bulbs is to maximize lighting performance and minimize lighting costs. He decides to evaluate incandescent bulbs using two criteria: bulb performance and annual cost, including operating cost and purchase price. Though bulb performance or output is measured in lumens, he preferred watts as a performance measure, since he evaluated only incandescent bulbs, and wattage is more frequently used as a measure of comparison by buyers. Imagine now that the responsibility covers a network of warehouses and factories involving the purchase of tens of thousands of bulbs. The same objectives would be relevant, but purchase price would now become a more significant factor.

The objectives hierarchy for the bulb selection example is given in Figure 4.3 as it would appear in the software package LDW. In the objectives hierarchy, squares represent objectives while ovals surround the measures. In the bulb choice problem, Mr. Frail's decision is to select the best bulb in order to maximize performance and minimize cost.

It is a relatively straightforward task to determine an annual cost based on kilowatt usage per year and the number of bulbs to be purchased each year. The latter is a function of the average lifespan of the bulb and the average number of hours per year of usage. As a result, there is just one measure, annual cost. However, it often makes sense to keep the two measures distinct, *operating cost* and *purchase price*. In major decisions, managers usually treat investments and operating cost separately for the following reasons:

- Separate budgets
- Different tax ramifications
- Affect ROIs
- Different time dimension—capital investment costs may be amortized over 5 or 10 years based on corporate accounting rules

In the bulb example, the capital cost is so small that normally you would combine it with operating cost to calculate an annual cost. However, Figure 4.4 illustrates the objectives hierarchy for the bulb selection example when Mr. Frail divides annual cost into two separate measures: annual operating cost and annual purchase cost.

In the service economy of the twenty-first century, many companies face critical questions regarding the level of service to provide and at what cost. Mod Stack is a medium priced retailer of fashionable clothing. The store is laid out in such a way that it can have three, four, or five cashier stations spread throughout the store that can be staffed with sales people. The company is in the process of deciding how many stations to have and the level of qualifications for the staff.

FIGURE 4.3:   Objectives hierarchy for choosing the best bulb.

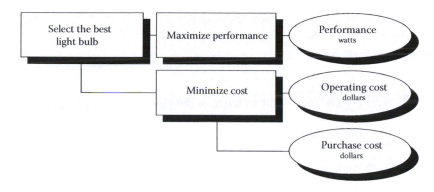

FIGURE 4.4: Objectives hierarchy for "best bulb" example with three measures.

Activity 1: Construct an objectives hierarchy for Mod Stack. It should include a high level objective, at least two sub-objectives, and at least one measure for each of the sub-objectives

1. Highest level objective: *Maximize store performance*
   a. Sub-objective _____
      i. Measure _____
      ii. Measure _____
   b. Sub-objective _____
      i. Measure _____
      ii. Measure _____

4.6 Top-Down Approach: Global Facility Location

There are two complementary starting points from which to develop the elements that make up an objectives hierarchy for large decisions: a top-down approach and a bottom-up approach. A top-down approach starts from the overall (most general) objective, which is then successively decomposed into sub-objectives that describe their higher level parent objectives. A decision maker is asked, "What do you mean by that upper-level objective?" in order to reveal the lower-level fundamental objectives. The bottom-up approach, on the other hand, often starts with specific measures that can be used to differentiate between existing alternatives. These measures are then grouped into a sub-objective that shares a common goal as described in the next section.

The top-down approach is summarized as follows:

1. Identify the overall fundamental objective
2. List three to five major objectives that specify and clarify the intended meaning of the objectives in terms of more specific objectives.
   a. Ask the decision maker to state what aspects of the higher-level objectives he considers important.

For example, when the goal is to maximize safety, the major objective can be decomposed into different subpopulations. For example, one would want to consider the respective needs of the public and personnel when looking at a power plant or, similarly, the needs of respective adults and children when considering features of a car.

- Subdivide a specific objective to develop lower tier objectives in successively greater detail. For example, cost is divided into (a) investment and (b) variable.

- Continue until the lowest level is sufficiently well defined that a measure can be associated with it.

We illustrate the top–down approach with our example of a facility location decision. Planet Inc., a growing global toy company, is planning to establish a new manufacturing facility in Asia or Eastern Europe in an attempt to minimize cost and maximize regional sales. This ultimate, or overall, objective can be achieved by maximizing regional sales, minimizing total cost, maximizing labor utility, maximizing country stability, and maximizing performance. These five objectives help to define the overall objective and specify the meaning of "maximize net value" in more detail. Figure 4.5 depicts the structuring of the facility location selection problem into an objectives hierarchy.

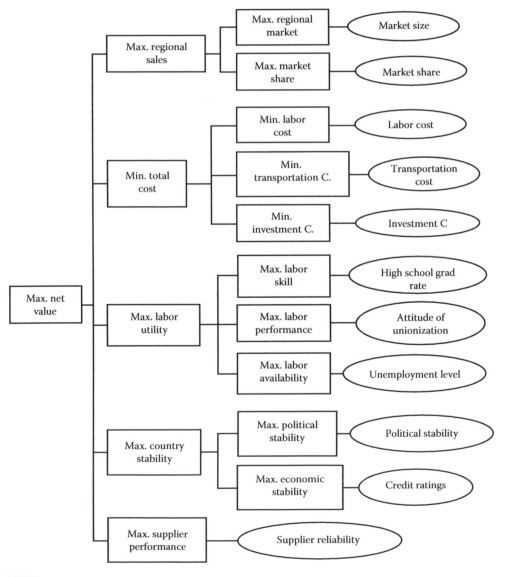

FIGURE 4.5:   Objectives hierarchy for facility location selection.

Except for supplier performance, none of the second level objectives (major objectives) is sufficiently well defined that a single measure can be associated with it. To identify sub-objectives associated with "maximize regional sales," the decision maker is asked what is meant by maximizing regional sales. In our example, maximizing regional sales has two aspects, each corresponding to one sub-objective: maximizing market share and maximizing regional market. The corresponding measures are market share and regional market size or projected sales. The measures are illustrated in the fourth level of the objectives hierarchy. We introduce measures in more detail at the end of this section.

Minimizing total cost consists of three readily understandable components: labor, transportation, and investment. Although it is possible to combine these three values into NPV, decision makers tend to treat these as three separate objectives that can be traded off. In light of dramatic fluctuations in fuel costs, most companies will also want to treat minimize transportation costs as a distinct objective.

The nature of the available workforce is of major concern to any executive setting up a factory in a foreign country. Maximizing labor utility is linked primarily to a number of surrogate factors that describe the labor skill, labor availability, and labor performance. One measure of the labor skill of the country's workforce is the percent of the population that has completed high school. The performance of the workforce is further affected by the strength and attitude of the unions in the country. (Later, we describe a three-point scale used to characterize the union environment.) Labor availability can be measured by using the unemployment rate.

The fourth objective relates to country stability. A company making a long-term investment in a plant that is located abroad is concerned about the long-term political and economic stability of the surrounding region. Political stability is identified in terms of the impact of changes in government on business conditions as well as any threats to the stability of the current government. (Later, we describe a three-point scale used to characterize political stability.) Economic stability can be captured by the S&P currency issuer credit rating.

Finally, every factory must be supported by a local supply chain. The decision makers will need to take a close look at the range of local suppliers and make a subjective assessment of the overall supply chain infrastructure, a detail that factors into the measure issues of supply chain product quality and delivery performance. We will discuss these kinds of constructed measures in Section 4.8.

Activity 2: Top-down—identify fundamental objectives for a lighting system in a manufacturing plant. Identify three high-level fundamental goals and for each goal identify at least two subgoals

---

4.7 Bottom-Up Approach: Kitchen Remodeling

The bottom-up approach often begins with projected alternatives and some data or knowledge about the strengths and weakness of the various alternatives. With this information in hand, the decision maker questions why he should care about each measure and searches for ways to group the measures into meaningful categories.

For example, the Lyons have decided to remodel their kitchen and invited four potential contractors to their homes. Each contractor submitted designs, cost estimates, a timeline, and a list of references. What measures should the Lyons use and how do they obtain relevant data? They talked with the contractors, received background information about each contractor, discussed their ideas about the kitchen, and received an estimate with the details of price, timeline, and cabinet source. They also spoke with three references for each candidate and visited at least one finished job by each builder. They found differences with regard to these contractors in a number of areas as follows:

- Number of years in business
- Total labor cost
- Total material cost
- Duration of job
- Creativity
- Cabinet brand reputation
- Use of subcontractors

There were obvious differences among the candidates when it came to such questions as years of experience, costs, and estimates of the time needed to complete the job. Interestingly, the Lyons found one of the contractors to be particularly creative in coming up with ideas to address their needs. The candidates did not all use the same brand of cabinets. In addition, one of the contractors subcontracted out the majority of the work.

After following up with references, the Lyons found additional differences with regard to: quality of references, history of cost overruns, delays in completing the job, cleanliness of the working environment, follow-up and resolution of problems as they arose, and the fit and finish of the kitchen.

The Lyons were interested in trying to organize these differences in a more structured approach. They started by asking themselves the question: why is each of these differences important? It was clear to them that cost was a fundamental objective since they were working on a limited budget (see Table 4.3). Another major category was the ultimate quality of the rebuilt kitchen. They struggled with grouping such factors as cleanliness, on time delivery, and the efficiency with which each contractor resolved problems. Soon they realized that all of these reflected a concern over the hassle of the entire project. This factor was especially important because they were both employed full-time outside of the home.

Table 4.3 summarizes the Lyons' goals hierarchy and measures. It was obvious which sub-objectives to assign to cost. However, some of the others could have been categorized either as quality or hassle. Responsiveness to concerns was listed under hassle but could have alternatively been included with quality. The use of subcontractors was listed under quality, but it could also have been listed under hassle. There is no "right" objectives hierarchy structure, so long as all critical issues are included and not double counted.

TABLE 4.3:   Lyon's goals hierarchy.

Goal	Objectives	Sub-Objectives	Measures
Best contractor	Minimize cost		Total labor cost
			Total material cost
			Cost overrun history
	Minimize hassle	Time	Duration kitchen unavailable
			Weeks of delay
		Cleanliness	Cleanliness created scale
		Responsiveness	Follow-up and resolution scale
	Maximize quality	Creativity	Creativity scale
		Cabinet quality	Brand and store reputation scale
		Subcontracting	Use of subcontractors
		Craftsmanship	Fit and finish scale
		Corporate experience	Years in business
			Quality of references scale

4.7.1 Summary Guidelines for an Objective Hierarchy

The objectives hierarchy is the key to multi-objective decision making. It specifies what is important to the decision makers and therefore defines the measures and the data that must be collected. The objectives hierarchy includes all relevant aspects of a decision and concerns of the decision maker. If an important objective is inadvertently left out, decision makers will be reluctant to accept the results of the analysis. In a classic personal story, a leading MAUT theorist applied his art to a decision regarding which academic position to accept. He was unhappy with the results until he realized that he had forgotten to include a measure that reflected the social scene of the respective cities. This was a particularly grievous error since was unmarried at the time.

A common mistake in designing a hierarchy involves developing an excessively long list of measures. This complicates the data collection process and makes it difficult to assign meaningful weights to the various measures. In addition, with long lists, there is a tendency for measures to overlap. Thus, weights assigned to overlapping measures will result in double counting the same factor. For example, the weight and fuel economy of an engine are closely linked. So are production cycle time, production cost, and throughput. An alternative problem arises when an objective is divided more and more finely into smaller and smaller measures that can easily be combined. This often happens with the objective to minimize cost. Different measures of cost should be included only if they can be traded off and not if they are simply going to be added together.

Last, the decision maker may decide to prune the objectives tree only after specifying the alternatives and collecting the data. If there is little difference among all the alternatives with regard to one measure, it is wise to delete that measure. Also, an objective may be eliminated if it is impossible to collect data or create a meaningful scale that can be used to characterize differences between alternatives.

WARNING—Missing or incomplete data are a major problem for MAUT

MAUT does not have a mechanism for handling situations in which one or more of the alternatives do not have values for specific measures. This can arise when comparing tried and true alternatives, such as suppliers or technologies, against new untested options. Leaving a measure blank for an alternative is equivalent to assigning a zero utility. Ideally, the decision-making team should consider using expert judgment to input a probabilistic range that will fill in the gap.

After an objectives hierarchy is created, it is helpful to check that the hierarchy and the associated measures are both complete and concise. The objectives should:

1. Span the full range of considerations
2. Contain as few measures as possible:
   a. Ten or fewer measures for most problems
   b. Fifteen to twenty measures at the bottom for major studies
3. Consist of nonredundant or overlapping measures
4. Specify meaningful differentiation between alternatives
5. Be directly relevant to decision at hand
6. Outline significant concerns
7. Be measured or estimated at reasonable cost
8. Not be decomposable into more meaningful measures
9. Be easily understood by stakeholders and decision makers
10. Not distinguish between measures that may be added together (e.g., two types of variable cost)

4.8 Measures

In the building contractor example, the objective of minimizing total cost can be measured in dollars. Similarly, the objective of minimizing time can be measured in weeks. However, some objectives are not easily measured on a single, natural numerical scale. For example, what kind of measures can be used to assess craftsmanship, cleanliness, or responsiveness?

A measure should be comprehensible; an expert has to be able to look at each alternative and assign the alternative an appropriate score. In addition, the decision maker should clearly understand how a score on a particular measure relates to the overall objective. If the decision context is related to health, for example, a scale that registers only in terms of *healthy* or *sick* is insufficient, because this measure would have different meaning to different people. Similarly, a quality measure should be more specific than poor, acceptable, and good. Not all measures lend themselves to natural numeric representations. In addition, some measures involve significant subjectivity. In the next section, we discuss both natural and constructed measures.

4.8.1 Natural Measures

In many cases, measures lie on a natural scale, such as time measured in weeks, mpg for fuel economy, NPV in dollars, or defects per 1000 for quality. If a measure has an obvious interpretation, we call it a natural measure. Table 4.4 shows common objectives with their corresponding natural measures.

4.8.2 Zero or One Measures

In some instances, the alternatives are described by whether or not they have a specific attribute. In this case, the measure will take on the value zero if the attribute is not present and one if it is. The examples of such measures and their context include "a vehicle has ABS brakes or not," "a technology is computer controlled or not," "a house has central air conditioning or not," or a "chemical is classified as toxic by the EPA or not." The zero-one scale can also be used if the measure reflects two distinct alternative characteristics in which one is clearly preferred over the other for the target population. For example, a light bulb may be fluorescent or incandescent, a transmission automatic or manual, and a flowering plant perennial or annual.

4.8.3 Constructed Measures

Many important objectives, however, do not have natural measures. Examples include maximizing customer satisfaction, minimizing union conflict, minimizing opposition to a project, maximizing the craftsmanship of a product, or maximizing the ease of using a particular machine. We are

TABLE 4.4: Common objectives and their natural measures.

Sub-Objectives	Natural Measures
Cost, profit, or NPV	Dollars or any currency
Time: such as cycle time, life cycle, development time, or durability	Seconds, minutes, hours, days, and weeks
ROI	Percentage
Weight	Grams, tons, and pounds
Proximity	Miles and minutes
Market share	
On-time delivery	
Failure rate	

all exposed to surveys that include arbitrary numeric scales ranging from 1 to 10 or word scales with measures such as poor, good, and excellent. This subjective approach may work with large surveys used to track measures over time. However, when dealing with a limited number of expert opinions in an organization, it is critical that all respondents understand the meaning of the scale before rating various alternatives. We therefore recommend that a scale be anchored with meaningful descriptions associated with the different values on the numeric scale in question. Note, however, that we do not rigorously define in excruciating detail the specific meaning of each value. The wording of the descriptions is designed to guide an expert in a process that is, nonetheless, highly subjective.

When selecting a country for a new plant, a major concern is the level of conflict and agitation that the local trade union might create, a factor that would affect plant operations. Table 4.5 illustrates three possible levels of union conflict or agitation, ordered from most preferred to least preferred. Note that the last column provides the detailed descriptions that define what is meant by union attitude, which enables the analyst to assign a score to each country under consideration.

Similarly, customer satisfaction can be defined as a customer's overall experience to date with a product or service (Johnson and Fornell 1991). To measure customer satisfaction, a company can use an index such as the American customer satisfaction index (Fornell et al. 1996) or the Swedish customer satisfaction barometer (Fornell 1992), or develop its own index, as depicted in Table 4.6. A survey can be used to elicit customers' perception of the performance and quality of the product as well as the level of their loyalty. Measuring loyalty in the financial services industry was so critical that IBM Consulting created the Customer Focused Insight Quotient (CFIQ), a concept they have trademarked.

Ease of use is often a factor in the choice of technology. Table 4.7 illustrates a five-point scale to measure ease of use in the case of a personal computer. In this example, ease of use reflects the degree to which a person thinks that using the computer will be hassle-free.

TABLE 4.5: Trade union measure of conflict.

Level	Words	Preference	Description
1	Low	Most preferred	Cooperative, rational in demands: work stoppages fewer than once a year
2	Mid		Demanding: work stoppages more frequent than once a year; infrequent short strike during contract negotiations
3	High	Least preferred	Highly organized and aggressive union; frequent work stoppages and strikes during negotiations

TABLE 4.6: Customer satisfaction measure.

Level	Words	Preference	Description
1	High	Most preferred	More than 80% of the customers have no complaint about the quality and performance of the product and would consider buying the product again
2	Medium		Customers are not totally satisfied with the product (20%–40% of the customers have complaints), but more than 70% of the customers would consider buying the product again
3	Low	Least preferred	More than 40% of the customers are dissatisfied with quality and performance of the product. At least 30% of the customers state that they will buy the product from the competitors

TABLE 4.7: Ease of use of a computer.

Level	Words	Preference	Description
1	Very high	Most preferred	Users rarely need to reboot. Easy to install/uninstall software and hardware. Automatically performs system maintenance and optimization for user tasks. Failures are automatically fixed
2	High		Users rarely need to reboot. Plug and Play usually works. Does not require routine maintenance. Does not optimize for user tasks. Only 90% of the failures are automatically fixed
3	Medium		Users sometimes need to reboot. Easy to install/uninstall software and hardware. Users need to do some of the routine maintenance. Half of the failures are automatically fixed
4	Low		Several specific applications or PC system problems require a reboot. Software is difficult to install. The user must open the cabinet to install hardware. Users need to fix half of the problems
5	Very low	Least preferred	Users need to do maintenance. Hardware and software fail often and errors are difficult to diagnose

The Lyons' decision regarding their choice of a remodeling contractor required the creation of a number of significant subjective scales.

Cleanliness scale

1. Clean—workers clean up every day and owners can walk through kitchen most days
2. Messy—kitchen not routinely cleaned up but rest of house OK
3. Dirty—dirt and dust often spread to adjacent rooms

Responsiveness scale—follow-up

1. Highly responsive—responds to concerns within 24 hours and usually resolves issue in 48 hours
2. Adequately responsive—responds to concerns within 48 hours and often, although not always, resolves issue within a week
3. Needs improvement—can take longer than a week and sometimes 2 weeks to get back to client to discuss concern

Creativity scale

1. Highly creative: Came up with three or more ideas Lyons had not considered
2. Creative: Came up with one or two new ideas
3. Mundane: Did not come up with any new ideas. Just planned to do what was asked.

Cabinet quality

1. Top of the line and used by designers
2. Second best brand
3. Common brand found in many stores such as Home Depot

Craftsmanship—fit and finish

1. Excellent: Up close inspection showed no alignment problems or gaps
2. Good: No visible problems from 5 ft away but closer inspection indicated some minor alignment problems in two cabinets
3. OK: Minor alignment problems were visible from 5 ft away

The Lyons would not consider any contractor whose craftsmanship scored below three

Corporate experience—references

1. Excellent: All three references were very satisfied
2. Good: One of the three thought everything went very well and two saw opportunities for improvement, but all three would use the company again
3. OK: Each reference reported some problems, but ultimately, after much discussion, all issues were resolved. However, one of three would not use company again

Activity 3—Create measures for house purchase

- Describe a *constructed* measure _____
- Describe a *yes or no* measure _____

4.8.4 Group Numeric Ranges to Make It Meaningful

Some measures have a natural scale, but the actual value is too specific to reflect general attitudes about that measure. The analyst may choose to group the measure into ranges that reflect differences that are significant to the decision makers. For example, with regard to a building contractor's experience level, a three-level scale, as depicted in Table 4.8, would be sufficient. Notice that the Lyons would not consider a contractor with fewer than 5 years experience.

If a decision maker were selecting a supplier of components for his plant, the proximity of the supplier's factory would be a significant concern. The actual mileage between the two could be used as a natural measure. However, his concerns might be better reflected by a five-level scale that represents how quickly the supplier can respond to urgent requests as illustrated in Table 4.9.

For most measures, the direction of preference is obvious. Fewer defects and higher customer satisfaction are better. However, in some instances, the order of preference is not clear. A student selecting a college may be interested in the college's proximity to his home. Some may prefer to be nearer to their immediate family (minimize) while others may prefer to be farther away (maximize). Table 4.10 presents the constructed measure scales for the "nearness to family" measure from both perspectives.

How would you handle the scale if the student prefers to be not too close to home but not too far? In that case, preference with regard to nearness to family is not monotonic, relative to actual

TABLE 4.8: Contractor's level of experience.

Level	Words	Preference	Description
1	Excellent	Most preferred	More than 20 years
2	Good		10–20 years
3	OK	Least preferred	5–10 years

TABLE 4.9:   Time to deliver: grouped measure.

Level	Words	Preference	Description—Time to Deliver
1	Immediate vicinity	Most preferred	Less than 20 min
2	Nearby		20 min to 1 hr
3	Same shift		1–4 hr
4	Same day		4–8 hr
5	Next day	Least preferred	More than 8 hr

TABLE 4.10:   Constructed scale: nearness to family.

Nearer to Home Is Better	Farther Away Is Better	Nonmonotonic
1. Within 0.5 hr drive	1. More than 5 hr	1. 1–2 hr
2. 0.5–1 hr	2. 2–5 hr	2. 0.5–1 hr
3. 1–2 hr	3. 1–2 hr	3. 2–5 hr
4. 2–5 hr	4. 0.5–1 hr	4. Within 0.5 hr drive
5. More than 5 hr	5. Within 0.5 hr drive	5. More than 5 hr drive

distance. The student might enjoy the convenience of being able to go home without much effort to get a home-cooked meal or for a long weekend during which someone else might do his laundry. However, he may not want to be so close that he feels his parents are interfering with his independence. The last column in Table 4.10 illustrates how to create a monotonic scale from a nonmonotonic measure of distance with "one hour to two hours" the student's preferred level while "more than five hours" is the least preferred level.

It is important to remember that the purpose of creating measures and scales is to facilitate decision making by clarifying differences. There is little value in a scale that describes the quality of a supplier on a range of 0–100 when all suppliers are rated between 95 and 97. This predicament arises when a company has high standards on certain measures for prescreening companies to be considered for the contract. In that case, the decision maker should focus and expand the description of the narrow range of 95–97 or eliminate the measure entirely if there are no substantive differences that would affect the decision.

In summary, we recommend the following when developing constructed measures:

1. Three to five levels
2. Descriptive phrase(s) should be used to describe each level
   a. The words used must have specific meanings
   b. Levels must be unambiguous and carefully defined

4.9   Example: Buy a Used Car

This section demonstrates how to structure a multi-objective problem and develop measures using the "buy a used car" (Edwards and Chelst 2007) example that we introduced in Chapter 2. In this case study, Pete, a college freshman, was looking to buy a used car both for social functions and to drive back and forth to his part-time job 20 miles away. Pete's overall objective was to maximize the car's value in terms of his personal needs. He identified four major objectives that impact

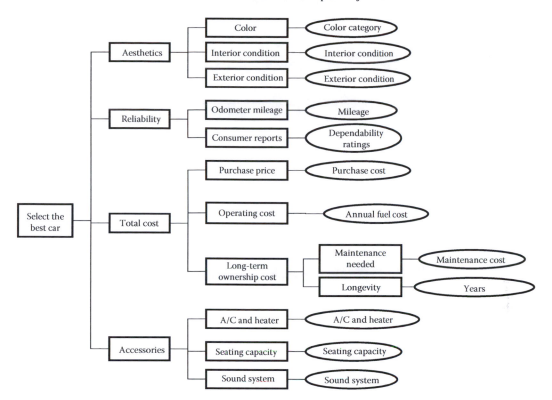

FIGURE 4.6:   Objectives hierarchy for buying a used car.

the overall objective of maximizing this value: reliability, total cost, accessories, and aesthetics as illustrated in objectives hierarchy (Figure 4.6).

Odometer mileage and consumer reports were combined into the reliability objective. While Pete uses a natural measure for odometer mileage, he uses a constructed measure for dependability ratings, as developed by J.D. Power (see the J.D. Power's website at www.jdpower.com for more information). Dependability ratings measure problems experienced by original owners of 3 year old vehicles using a five-circle rating system as noted in Table 4.11.

Total cost objective consists of purchase price, operating cost, and long-term ownership cost. Operating cost is primarily the annual fuel cost, which is the product of three uncertainties: fuel economy (mpg), miles driven, and price of gasoline. Long-term ownership cost is derived by considering maintenance cost and longevity. Maintenance cost is easily measured in terms of dollars. Pete measured the longevity in terms of years of the car's expected remaining life.

Pete developed constructed measures for sub-objectives of the aesthetics objective. He specified three categories for color: light (most preferred), neutral, and dark (least preferred). He used four-level scales developed by Kelly Blue Book (www.kbb.com) for interior and exterior body condition, best, good, fair, and poor.

Pete developed a four-level constructed measure for A/C and heater measure, ranging from most preferred level (both systems work) to least preferred level (neither system works). The preference for the second best level may vary in different geographic locations. For example, on the southwest coast where the weather is always warm or hot, people may prioritize A/C overheating, whereas people in cold geographic areas are more likely to be concerned about the car's heating system. Regarding seating capacity, Pete expects to use his vehicle on dates but would also like to be able to offer friends a ride; hence, he would like to maximize seating capacity. Pete also considers the sound system a high priority and created a four-level constructed measure for this objective.

TABLE 4.11:   Measures for buy a used car example.

Measure	Scale/Levels
Odometer mileage	Miles
Dependability ratings	Five circles: Among the best
	Four circles: Better than most
	Three circles: About average
	Two circles: Worse than average
Purchase cost	Dollar
Annual fuel cost	Dollar
Maintenance cost	Dollar/year
Longevity	Years
Color	Light, neutral, or dark
Interior condition	Excellent, good, fair, or poor
Exterior condition	Excellent, good, fair, or poor
A/C and heater	Both work, A/C only, and heater only or neither works
Seating capacity	Six or more
	Four or five
	Two
Sound system	Radio and CD player
	Radio and cassette player
	Radio only
	None

4.10   Identify Alternatives

First executive: "I think A is a good idea—make it happen or prove me wrong!"

Second executive: "I know B is a bad idea—find an alternative."

Managers often identify too few alternatives for a given decision context. They tend to create and evaluate alternatives that are obvious, readily available, or have been used before in similar decision contexts. Sometimes, an alternative has been inspired by a hot new concept that an executive recently heard about but has not fully digested. When, as is generally the case, the number of alternatives is limited (often to a single option), the generation of new alternatives tends to consist of nothing more than tweaking the alternatives that already exist. The failure to consider a wide range of alternatives is often cited as the single most common mistake in decision making. Some simple guidelines for defining alternatives are:

- They are comparable in completeness
- Adequate details enable decision maker to judge relative worth
- They are wide ranging
- There are at least two but no more than seven
- They are nondominated by others on the list

Completeness: The alternatives should be comparable in terms of the completeness with which they address a particular concern or need. For example, when choosing between a totally integrated software package and one that includes multiple packages, the latter alternative should spell out the combination(s) of packages under consideration. The same can be said for deciding between contracting with one full-service supplier who can deliver a total system to your plant or a combination of suppliers, each of whom will provide individual components. In a global organization, completeness is reflected in the ability of a technology or supplier to support a company's needs worldwide. Some technologies or suppliers can be used everywhere, while other choices might require a different solution for each locale.

Adequate detail: A corollary to completeness is that the alternatives should be described with enough detail, so that hard data can be gathered for each measure or experts can be consulted to evaluate each alternative on each measure.

Wide ranging: The alternatives should represent the broad range of viable alternatives realistically available. An analysis of a wide range of alternatives enables the decision maker to understand better their individual strengths and weaknesses. This will facilitate the construction of hybrid alternatives that may outperform all alternatives on the original list.

Two to seven: The decision-making process is flawed if decision analysis becomes focused on a single alternative—one that is, perhaps, presented as the latest and greatest idea—so that the only question considered is whether or not this idea will work for your company. When faced with an overwhelmingly large number of alternatives, however, the list should be screened down to no more than a wide ranging set of seven that covers the entire spectrum. If, after analysis, one or two alternatives stand out, the decision maker may then want to take a closer look at some of the screened out alternatives that are comparable to the best.

Nondominated: After putting together the list of alternatives and gathering some of the data, it is useful to quickly review the list. One or more of the alternatives may outperform or dominate another alternative on every measure of significance. The weaker alternative can then be removed from the list.

It stands to reason that a final decision can only be as good as the best alternative that was considered. The development of a wide range of good alternatives requires creative and systematic thinking on the part of diverse individuals. The process should incorporate all of the following:

- Start with the elements of the objectives hierarchy
- Define an ideal alternative by removing constraints
- Incorporate multiple perspectives
- Think independently
- Benchmark best practices

Objectives hierarchy: The objectives hierarchy offers a good starting point for brainstorming. It consists of a collection of objectives to be minimized or maximized. Participants in the decision making can be asked to identify individual alternatives that are best with regard to one or more specific objectives. Next, they should be encouraged to identify or create alternatives that might better balance the objectives.

Remove constraints: What would an ideal alternative look like if there were no practical constraints or limitations? What would be the best alternative if cost were not an issue? Then consider the various practical limitations one at a time to create ideal alternatives that are within the realm of reality. Continue the process until consensus is reached regarding which options are feasible.

Multiple perspectives: The decision-making team should include individuals from different parts of the organization and even outsiders who have a stake in this decision. Ask each to articulate what would be the best alternative from his or her perspective. Use these responses to create additional alternatives that balance any competing interests.

Think independently: Often, a group is brought into a room to brainstorm alternatives. Research, however, shows that creativity of individuals often exceeds that of a group; that is, the sum of the parts is greater than the whole. In a group setting, some new ideas may be suppressed if they are subject to the criticism of others before these ideas have been fully formed or articulated. It is therefore preferable that individuals submit their personal thoughts regarding possible alternatives before coming to a group discussion. The initial long list can then be "reality checked" by experts with diverse backgrounds. This should curtail the staggering amount of wishful thinking that tends to direct the creation of new alternatives, especially when the time horizon for completion is several years down the road.

Benchmark best practices: An important source of alternatives can come from looking outside your organization by benchmarking other companies as to their best practices. The literature available can also be a source of innovative ideas. However, do not allow what already exists to restrict conceptualization of visionary alternatives. Professor Norbert Wiener, one of the most creative geniuses of the twentieth century, always thought about a new scientific problem on his own before reading the literature.

Words of caution: Do not go too far down the path of evaluating alternatives before subjecting the alternatives to a substantive reality check. Executives with limited technical knowledge are notorious for coming up with ideas that are impossible to implement—that is, without the kind of scientific breakthrough that occurs once in a lifetime. They also routinely minimize the projected time and resources required to implement radical alternatives. Any choice that is buttressed by comparison to President Kennedy's success in starting the United States on a journey of less than 10 years to reach the moon should be dismissed. Likewise, consider with *extreme* caution alternatives that do not fit within the culture of your organization. Many of the failed attempts to transfer the best practices of Japanese automotive companies to U.S. companies resulted from the lack of understanding that in order for these practices to succeed, there would need to be a fundamental change in organizational behavior from top to bottom.

4.11   Real-World Applications

4.11.1   British Columbia Hydro

The directors of British Columbia Hydro (BC Hydro), a major electric power corporation with billions of dollars in annual sales, realized that they would face many complex strategic decisions in the 1990s. They needed to make decisions regarding additional resources to generate electricity, construct transmission lines, negotiate power agreements, and evaluate the environmental impact of their facilities and activities. The directors wanted to make these decisions in a consistent and effective way, using quality information and sound logic in a coordinated manner (Keeney and McDaniels 1992).

Various departments play key decision-making roles at BC Hydro. To facilitate the coordination of their decisions, Ken Peterson was appointed director of planning, with the goal to make BC Hydro the best-planned utility in North America. He realized that all decisions should contribute to achieving a set of long-range objectives, and, with the help of consultants, he worked toward identification of the organization's strategic alternatives and the construction of a MUF. The process began with interviews of key decision makers to identify the objectives and continued through the steps necessary to define and assess the utility function. The list included six major objectives: economics, environmental impact, health and safety, equity, quality of service, and public service, with a total of 18 measures as depicted in Figure 4.7. Most of the major objectives have sub-objectives.

Maximizing economics was subdivided into minimizing cost of electricity, maximizing funds transferred to the British Colombia government, and minimizing economic implications of natural resources. They identified a measure for each of the lowest level objectives in the objectives hierarchy (see Table 4.12). Minimizing cost of electricity use is measured in terms of mills per kilowatt-hour in 1989 Canadian dollars. Minimizing cost of electricity use conflicts with the objective to maximize the annual dividend paid to the government.

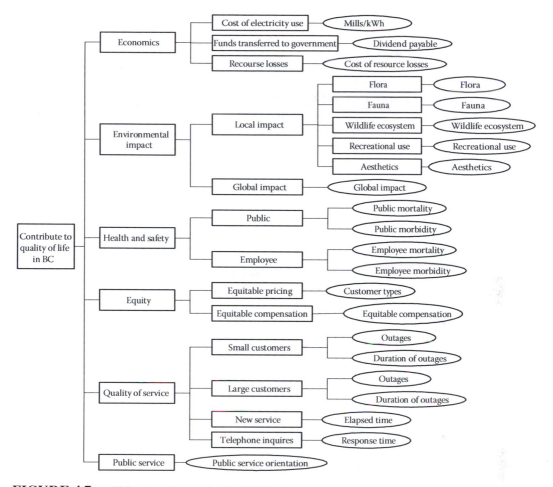

FIGURE 4.7: Objectives hierarchy for BC Hydro case.

The environmental impact objective is broken into local environmental impacts and global environmental impacts. For local environmental objective, five distinct measures were developed. For example, flora is measured in terms of number of acres of mature forest that would be lost.

The health and safety objective uses mortality and morbidity measures both for the public as well as for BC Hydro's employees. The mortality measure weights the death of a younger person at a greater value than the death of an older person, since the younger person loses a longer expected lifetime. The morbidity measure specifies the employees' lost work time caused by either injuries due to accidents or illness induced by emissions from the power plant.

The assessment of the weights provided important insights to the executive leadership team. It was observed that four of the objectives—economics, environment, health and safety, and service quality—mattered the most. The team assessed relative weights for the various measures through questions about trade-offs. For example, in Peterson's view, two outages per year of 2 hours' duration each, to 20,000 large customers, are equivalent to an increase in energy costs of 1.7 mills/kWh. A 1.7 mills/kWh increase in cost is equivalent to $83,300,000 in net earnings. Thus, if BC Hydro had an opportunity to reduce expected outages of that nature at a cost of less than $83 million, it would be a good investment. The measures and their scales are listed in Table 4.12. The last measure in the table, Public Service Orientation, uses a constructed scale ranging from 1 to 4.

TABLE 4.12: Measures for BC hydro case.

Measure	Scale/Levels
Cost of electricity use	Mills per kilowatt-hour in Canadian dollars
Funds transferred to government	Annualized dividend payable in Canadian dollars
Recourse losses	Cost of resource losses in Canadian dollars
Flora	Hectares of mature forest lost
Fauna	Hectares of wildlife habitat of Spatsizi Plateau quality lost
Wildlife	Hectares of wilderness of Stikine Valley quality lost
Recreational use	Hectares of high-quality recreational land lost
Aesthetics	Annual person-years viewing high-voltage transmission lines in quality terrain
Public mortality	Public person-years of life lost
Public morbidity	Public person-years of disability equal in severity to that causing employee lost work time
Employee mortality	Employee person-years of life lost
Employee morbidity	Employee person-years of lost work time
Customer types	Residential, commercial, and industrial
Equitable compensation	Number of individuals who feel they are inadequately recompensed
Outages to small customers	Expected number of annual outages to small customers
Duration of outages to small customers	Average number of hours per outage to small customers
Outages to large customers	Expected annual number of outages to large customers
Duration of outages to large customers	Average number of hours per outage to large customers
Elapsed time	Elapsed time until new service installed
Response time	Time until human answers the telephone
Public service orientation	Level 4—Very public-service oriented
	Level 3—Moderately public-service oriented
	Level 2—Somewhat public-service oriented
	Level 1—Minimally public-service oriented

4.11.2 Upham Brook Watershed

Merrick et al. (2005) used MAUT to assess the quality of the endangered Upham Brook Watershed, Richmond, VA, and identify future programs to improve the quality of watershed. The watershed has multiple stakeholders, including the residents of the watershed, community organizations, industry, and government. The MAUT approach allowed the consideration of multiple stakeholders' views.

They interviewed the project committee to elicit the committee's objectives for the watershed. The project committee described "maximize the quality of the watershed" as the overall objective. The authors asked the group to specify 10–15 action verbs and modifiers that define the quality of the watershed. Similar verbs were grouped together to form affinity groups. They identified an objective that best described each affinity group. The objectives were then structured in an objectives hierarchy, the first three levels of which are illustrated in Figure 4.8.

The overall objective of maximizing the quality of the watershed is divided into "maximize quality of wildlife habitat" and "maximize quality of human habitat" objectives. The wildlife habitat objective is then broken down into five classes of wildlife species that reside within the watershed boundaries, each with the same five objectives. The human habitat objective is disaggregated into improving watershed quality for residential stakeholders and improving watershed quality for commercial/industrial stakeholders, each with the same four objectives.

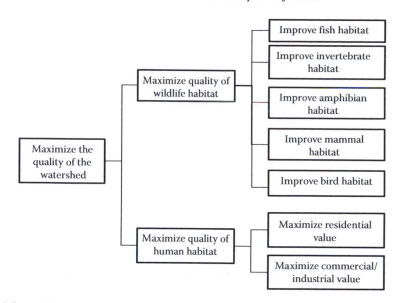

FIGURE 4.8: Objectives hierarchy for the watershed quality.

TABLE 4.13: Measures for watershed quality.

Measure	Scale/Levels
Water quality for amphibians	% of acidity readings below 4.5 pH
Water quality for all others	% of dissolved oxygen readings below 5 ppm
Riparian zone and floodplain	% of riparian zone with underdeveloped or natural vegetation
Land use	% of Upham Brook with underdeveloped or natural vegetation
Natural steam channel for all others	% of nonredirected channel
Natural steam channel for amphibians	EPA rapid bioassessment metric
Natural stream flow	% of impervious surface in watershed
Water safe for designated usage	% of fecal coliform measurements in violation of regulations
Access and recreation	% of stream length accessible to public
Flood safety	% of riparian zone and floodplain with construction
Aesthetics	% of stream length with trash present

The project committee chose the measures that would be used to monitor and report on the attainment of the watershed objectives as shown in Table 4.13. The results of the project were used to guide the Upham Brook restoration programs.

4.11.3 Wastewater Treatment Plant Technology Selection

Itel Corp. is planning the installation of a new V-chip Plant. Meanwhile, the U.S. Environmental Protection Agency is in the midst of developing new regulations that will significantly tighten the discharge limits from industrial wastewater treatment plants across a variety of industries. This case is an adaptation of a real world decision that was modified to preserve confidentiality. The new regulations are scheduled to be finalized at the same time as the launch of the new chip program. Therefore, from the very onset of program planning, Itel Corp. needs to plan for the purchase and

TABLE 4.14:   Objectives for selecting wastewater technology.

Objective	Measure
Minimize investment cost	Dollars
Minimize maintenance cost	Dollars
Minimize operating cost	Dollars per gallon
Minimize floor space	Square feet
Minimize time	Months
Maximize supplier potential	Constructed (high, medium, and low)
Maximize system capacity	Gallons per day
Maximize metallic water quality	Metallic parts (mg/L)
Maximize oily water quality	Oily parts (mg/L)

installation of a wastewater treatment plant that will meet new regulations. How will Itel Corp. select the wastewater technology? What are the objectives of the wastewater technology selection problem?

Currently, it is unclear what the pretreatment standards will stipulate. Nor is it known which EPA regulations will apply, because the machinery V-chip process involves multiple operations that overlap two different classifications of the regulations. The two wastewater discharge regulations are the Metal Finishing Wastewater Regulations (MFWR) and the Oily Wastewater Regulations (OWR). It is anticipated that Itel will be required to negotiate with the EPA to determine the wastewater pretreatment discharge regulations that best apply to their manufacturing process. For legislative and enforcement reasons, the EPA will permit the Itel plant to apply only one of two wastewater regulations. David Edison, director of the new plant, formed a team to research the available technologies that meet the requirements of newly proposed EPA regulations and evaluate alternative technologies.

The challenge is to determine which wastewater treatment technology Itel Corp. should select. Because of uncertainty regarding regulations, Itel is seeking to select the best wastewater treatment technology that meets the new proposed regulations for *both* EPA classifications at a minimum cost. Once a technology is selected, the company would like to investigate the potential of outsourcing the ownership and operation of the new wastewater treatment plant to a third-party supplier.

In order to identify alternatives and clarify objectives, the team conducted a study to examine existing industry wastewater treatment technologies and also conducted an extensive wastewater effluent sampling study at plants with identical processes as the Itel Plant site. The results of these studies revealed that there are four different wastewater treatment technologies that can meet the new proposed regulations for both EPA classifications with various quality ramifications. Each of these four technologies has advantages and disadvantages that complicate the decision. After an extensive study and discussion, the team came up with a list of objectives and measures (in no special order), as shown in Table 4.14.

The objectives and measures can be represented in an objective hierarchy (Figure 4.9). The overall objective is to select the best wastewater technology that keeps Itel Corp. in compliance with all applicable regulations, while allowing the Itel plant to operate in a cost effective manner. Minimizing investment cost, minimizing maintenance cost, and minimizing operating cost objectives are combined into the overall objective of minimizing cost. Maximizing metallic water quality and maximizing oily water quality objectives are grouped under the maximizing water quality objective. Several other objectives in Figure 4.9 are not classified as part of a major objective(s), because no combination of these factors will clarify the meaning of any upper level objective except for the overall objective.

The company would like to explore the possibility of outsourcing the ownership and operation of the new wastewater treatment plant to a supplier in order to reduce company-owned capital assets. The choice of technology affects the potential number of suppliers who are capable of

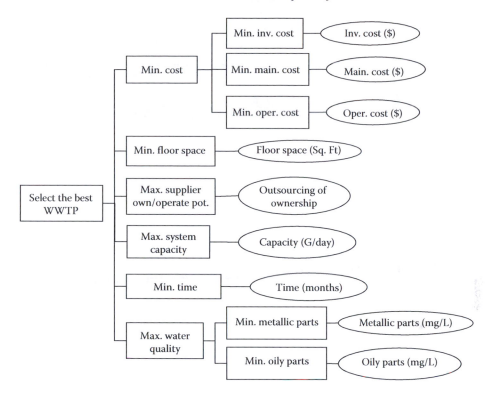

FIGURE 4.9: An objectives hierarchy for the wastewater technology selection.

TABLE 4.15: Constructed measure scale for outsourcing of ownership.

Measure Level	Order of Preference	Description
High	Most preferred	Three or more suppliers are available
Medium		One to three supplier are available
Low	Least preferred	Supplier is not available

running the facility. The team created a three-level constructed measure scale for outsourcing of ownership as illustrated in Table 4.15. The most preferred level corresponds to three or more third-party suppliers.

4.11.4 Evaluate Employee Performance

All of the examples until now use multiple objectives to frame alternatives so as to determine the best among a limited set of alternatives in a specific context. However, there are other situations in which it is appropriate to use multiple objectives and measures as a performance evaluation tool. Many sports enthusiasts, for example, routinely peruse rankings of sports teams and individual athletes. The vote for most valuable player in various professional sports implicitly involves combining multiple measures of performance.

One organization used MAUT to create a structured and consistent process for evaluating worker performance. There was no intent to rank order all of its employees to find out who was best or most valuable. Rather, the ultimate objective of the employee performance evaluation system was to assist employee professional development and increase organization effectiveness through increased personal performance. This process was also used to place workers in performance bands

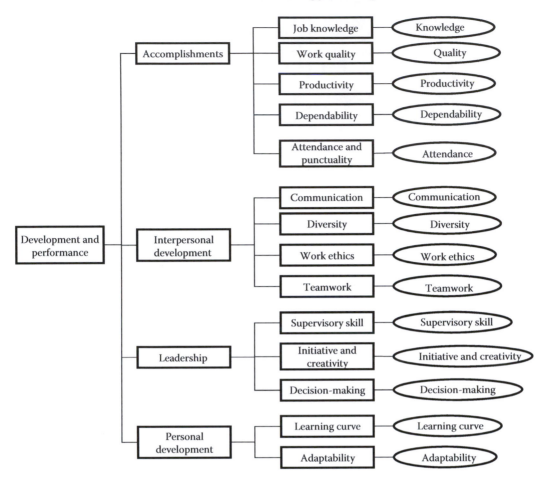

FIGURE 4.10: Objectives hierarchy for employee performance evaluation.

to facilitate fair and consistent allocation of bonuses and pay raises. Figure 4.10 depicts an objectives hierarchy for employee performance. Accomplishment of annual objectives; interpersonal development; and leadership and personal development are combined into the overall objective of maximizing employee development and performance.

Job knowledge, work quality, productivity, dependability, attendance, and punctuality specify the accomplishments of annual objectives, and, therefore, they are grouped under the accomplishments objective. A three-level or five-level measure was constructed for each sub-objective under the accomplishments objective, as illustrated in Table 4.16. The table orders each measure level from the most preferred to the least preferred level and provides a meaningful description of each level.

Communication, diversity, work ethics, and teamwork are classified under interpersonal development as they clarify the meaning of the interpersonal development objective. Table 4.17 depicts a three-level constructed measure for each sub-objective under the interpersonal development objective.

The leadership objective consists of three sub-objectives, supervisory skill; initiative and creativity; and decision-making and problem-solving. Measure levels and their descriptions are given in Table 4.18.

Personal development is linked to learning curve and adaptability. Table 4.19 shows a three-level constructed measure for each of these two sub-objectives.

TABLE 4.16: Measure levels under the accomplishments objective.

Measure	Most Preferred				Least Preferred
Job knowledge	Exceptional: Consistently exhibits exceptional knowledge and outstanding skills in even the most complex aspects of the job	Good: Frequently demonstrates better than average knowledge and skills in all aspects of the job	Adequate: Has adequate knowledge and skills to handle job duties	Need improvement: Requires considerable assistance to handle job duties	Limited: Application of knowledge is limited. Required skills are poorly demonstrated
Work quality	Highest: Consistently produces work of highest quality	High: Produces high quality work; makes minimal errors	Acceptable: Produces acceptable work with some errors	Marginal: Often produces unacceptable work, makes frequent errors	Unacceptable: Produces marginal to unacceptable work; makes excessive errors
Productivity	Highest: Consistently completes work ahead of schedule; seeks additional tasks; highest output level	High: Completes most work ahead of schedule; above average output level	Average: Completes the majority of work within specified deadlines; acceptable output level	Marginal: Does not usually complete work within time limits; output level is below average	Unsatisfactory: Does not complete work within time limits; generally unsatisfactory output level
Dependability	Exceptional: Carries out work assignments with exceptional degree of independence and efficiency		Average: Carries out work assignments with expected degree of independence and efficiency		Limited: Carries out instructions and responsibilities only with close supervision
Attendance and punctuality	Consistent: Is consistently present and on time		Good: Demonstrates adequate attendance and punctuality		Poor: Has erratic rate of attendance or punctuality

TABLE 4.17:   Measure levels under the interpersonal development objective.

Measure	Most Preferred		Least Preferred
Communication	Exceptional: Exceptionally articulate in verbal and written communication	Good: Communicates clearly	Poor: Communicates poorly
Diversity	Diligent: Works assiduously to foster an open and inclusive environment; actively involved in diversity initiatives; always displays behavior that respects and values individual differences	Acceptable: Contributes in promoting an open and inclusive environment; participates in diversity initiatives; generally displays behavior that respects and values individual differences	Limited: Needs encouragement to support a diverse and inclusive environment; fails to display behavior that respects and values individual differences
Teamwork	Exceptional: Is very effective interpersonally; works extremely well with others	Good: Works well with others; facilitates cooperation	Unacceptable: Has difficulty in relating to others; is not readily cooperative
Work ethics	Exceptional: Practices exceptional work ethics; demonstrates scrupulous integrity in all work	Good: Practices good work ethics; demonstrates integrity in all work	Poor: Fails to practice good work ethics; does not demonstrate integrity at work

TABLE 4.18:   Measure levels under the leadership objective.

Measure	Most Preferred		Least Preferred
Supervisory abilities	Excellent: Excels in supervision and leadership of subordinates; exceptional ability to build and lead a team	Acceptable: Provides effective supervision and leadership of subordinates	Unsatisfactory: Provides little or no supervision and leadership of subordinates
Initiative and creativity	Excellent: Consistently exceeds requirements for independent action and resourcefulness; highly motivated and creative	Good: Meets requirements for independent action and resourcefulness; diligent worker	Unsatisfactory: Rarely initiates independent action as required by the job; requires constant supervision
Decision making/ problem solving	Excellent: Consistently demonstrates outstanding problem-solving skills; consistently able to handle complex problems creatively	Good: Demonstrates good problem-solving skills; occasionally able to handle complex problems	Unsatisfactory: Has difficulty recognizing and solving routine problems; does not show evidence of requisite analytical skills

TABLE 4.19 Measure levels under the personal development objective.

Measure	Most Preferred		Least Preferred
Learning curve	Exceptional: Exceptional in learning new techniques and tools; is very interested in learning	Good: Learns new techniques in specified time; demonstrates interest in learning	Poor: Fails to learn new techniques in specified time; does not demonstrate interest in learning
Adaptability	Excellent: Exceptional contribution to efficient operation of unit; consistently seeks ways to improve work methods	Good: Develops better methods of completing work; occasionally provides constructive suggestions	Inadequate: Does not provide constructive suggestions

Exercises

Complete Chapter Activities

4.1 Create two different types of measures for a house purchase

    a. Describe a constructed measure

    b. Describe a yes or no measure

4.2 Construct an objectives hierarchy for Mod Stack. It should include a high level objective, at least two sub-objectives, and at least one measure for each of the sub-objectives

    a. Highest level objective: Maximize store performance

        i. Sub-objective and one or more related measures

        ii. Sub-objective and one or more related measures

4.3 Top–down—Identify fundamental objectives for a lighting system in a manufacturing plant. Identify three high-level fundamental objectives and for each objective identify at least two sub-objectives

4.4 Measures from two perspectives—Car performance

Specify and describe how the following measures may reflect two distinct perspectives: (a) fuel economy and (b) warranty. What measures would each perspective use? It may or may not be the same measure.

Cases

4.5 Supplier selection—team assignment

Part A. "Luxury" seat supplier (2–3 page discussion)

    You and a colleague are responsible for selecting a supplier for the design, development, and manufacture of seating modules for the next generation luxury SUV. Your company has established specific targets, and each supplier will respond with a general proposal.

    Create an objectives hierarchy to structure your decision. You should have at least two levels, not counting the overall goal. At the lowest level, you should have measures. For some objectives that are not easy to quantify, you will need to "construct" your own measure scale (numeric or nonnumeric) along with a word description to clarify the meaning of different values. See Tables 4.5 through 4.8 for examples of constructed measures.

a. Discuss the meaning of the objectives and measures. Discuss in detail any numeric or nonnumeric scales you created.

b. Discuss problems that arose as you tried to achieve a single agreed upon objectives hierarchy and the associated measures.

Part B. Car horn (one page discussion of key differences in objectives hierarchy).

c. Now assume that you have the responsibility for selecting the supplier for a car horn for a specific model vehicle. Create a new objectives hierarchy. This should be somewhat different from the hierarchy created for the seat supplier decision. The differences could be in the objectives and/or measures. Discuss what changes you made and why. Include your new objectives hierarchy.

(Assigning weights to the different objectives are not part of this assignment. Needless to say, the weights for the two decisions would be different. However, we are looking for differences in the structure of the objectives hierarchy.)

For both Part A and Part B do not have too many objectives and sub-objectives. Remember that in an actual decision, you would ultimately need to (a) obtain data for each measure and each supplier and (b) assign weights to each objective, sub-objective and measure

4.6 Library location

A city is looking for a solution to overcrowded and inadequate conditions at its public library. The library was built in 1942, when the city had a population of 18,000. The city's population has since tripled. The city could conceivably expand the current library building, but this would not be cost effective; the operating costs of an augmented building would be higher than those of a new one. Other alternatives include building a brand new building or buying a big building.

Create an objectives hierarchy to structure your decision. You should have at least two levels, not counting the overall objective. At the lowest level, you should have measures. For some objectives that are not easy to quantify, you will need to "construct" your own measure scale (numeric or nonnumeric), along with a word description to clarify the meaning of different values.

Submit a write-up that discusses the following issues:

a. Discuss the meaning of the objectives and measures. Discuss in detail any numeric or nonnumeric scales you created.

b. Discuss problems that arose as you tried to achieve a single agreed upon objectives hierarchy and the associated measures.

4.7 Automotive engine

An automobile manufacturer is developing a new vehicle. The automobile is to be a sporty, small vehicle. Three engine types have been identified as alternatives for use in this new vehicle: 2.8, 3.1, or 3.8 L. Which engine should be used in this new vehicle? Specify three to five major objectives you would use in choosing between engines. One obvious objective is to maximize performance. For your performance objective, specify two or (at most) three measures you would use to capture all aspects of this objective. Specify the type of measures to use: natural or constructed?

In framing the aforementioned decision regarding the choice of engine, be sure to consider multiple perspectives. When selecting a major system for a complex product, the objectives and measures should take into account the interests of the purchaser as well as the concerns of the internal organizational units such as product development, manufacturing, and purchasing. The purchaser is primarily concerned with performance, price, and styling effects. Manufacturing is concerned about the fit within its manufacturing capacity and the cost of manufacture. Product development has an interest in how this will affect other product design issues, such as how difficult it will be to fit the engine and Powertrain package within the front section of the vehicle.

References

Belton, V. and Stewart, T.J. (2002). *Multiple Criteria Decision Analysis: An Integrated Approach.* Boston, MA: Kluwer Academic.

Benayoun, R., Roy, B., and Sussman, N. (1966). Manual de Reference du Programme Electre, *Note De Synthese et Formaton*, No.25, Direction Scientifque SEMA, Paris, France.

Charnes, A., Cooper, W.W., and Ferguson, R. (1955). Optimal estimation of executive compensation by linear programming. *Management Science*, 1, 138–151.

De Neufville, R. and Keeney, R.L. (1972). Use of decision analysis in airport development for Mexico City. In: *Analysis of Public Systems*, Drake, A.W., Keenet, R.L., and Morse, P.M., eds. Cambridge, MA: MIT Press.

Dyer, J.S., Fishburn, P.C., Steuer, R.E., Wallenius, J., and Zionts, S. (1992). Multi-criteria decision making, multiattribute utility theory: The next ten years. *Management Science*, 38, 645–654.

Edwards, T.G. and Chelst, K.R. (2007). Purchasing a used car using multiple criteria decision making. *Mathematics Teacher*, 101(2), 126–135.

Fornell, C. (1992). A national customer satisfaction barometer: The Swedish experience. *Journal of Marketing*, 56, 6–21.

Fornell, C., Johnson, M.D., Anderson, E.W., Cha, J., and Bryant, B.E. (1996). The American customer satisfaction index: Nature, purpose, and findings. *Journal of Marketing*, 60, 7–18.

Hobbs, B.F. and Meier, P. (2000). *Energy Decisions and the Environment: A Guide to the Use of Multicriteria Methods.* Boston, MA: Kluwer Academic Publishers.

Hokkanen, J. and Salminen, P. (1997). Choosing a solid waste management system using multicriteria decision analysis. *European Journal of Operational Research*, 98, 19–36.

Hwang, C.L. and Yoon, K. (1981). *Multiple Attribute Decision Making: Methods and Applications.* New York: Springer-Verlag.

Johnson, M.D. and Fornell, C. (1991). A framework for comparing customer satisfaction across individuals and product categories. *Journal of Economic Psychology*, 12, 267–286.

Jones, D.F. and Tamiz, M. (2002). Goal programming in the period 1990–2000. *Multiple Criteria Optimization: State of the Art Annotated Bibliographic Surveys*, Ehrgott, M. and Gandibleux, X., eds. Boston, MA: Kluwer, 129–170.

Keeney, R.L. (1996). *Value-Focused Thinking: A Path to Creative Decision Making.* Cambridge, MA: Harvard University Press.

Keeney, R.L. (1994). Creativity in decision making with value focused thinking. *Sloan Management Review*, Summer, 35, 33–41.

Keeney, R.L. and McDaniels, T.L. (1992). Value-focused thinking about strategic decisions at BC Hydro. *Interfaces*, 22(6), 94–109.

Keeney, R.L., McDaniels, T.L., and Swoveland, C. (1995). Evaluating improvements in electric utility reliability at British Columbia Hydro. *Operations Research*, 43, 933–947.

Keeney, R.L. and Raiffa, H. (1993). *Decisions with Multiple Objectives: Preferences and Value Tradeoffs.* New York: Wiley.

Kidd, J.B. and Prabhu, S.P. (1990). A practical example of a multi-attribute decision aiding technique. *Omega*, 18, 139–149.

Merrick, J.R.W., Parnell, G.S., Barnett, J., and Garcia, M. (2005). A multiple-objective decision analysis of stakeholder values to identify watershed improvement needs. *Decision Analysis*, 2, 44–57.

Metaxiotis, K., Psarras, J.E., and Samouilidis, J. (2004). New applications of fuzzy logic in decision support systems. *International Journal of Management and Decision Making*, 5, 47–58.

Saaty, T.L. (1980). *The Analytical Hierarchy Process*. New York: McGraw-Hill.

Strimling, D. (1996). Private communication.

Wang, X. and Triantaphyllou, E. (2006). Ranking irregularities when evaluating alternatives by using some ELECTRE methods. *Omega*, 36, 45–63.

Zadeh, L.A. (1965). Fuzzy sets. *Information Control*, 8, 338–353.

Chapter 5

Structured Trade-Offs for Multiple Objective Decisions: Multi-Attribute Utility Theory

> Dawn chemicals will purchase and install a wastewater treatment plant to meet new U.S. Environmental Protection Agency regulations. Dawn has identified the following objectives: minimize investment cost, minimize maintenance cost, minimize operating cost, minimize floor space requirements, maximize operating pattern, maximize system capacity, minimize timing, and maximize water quality. What are the feasible alternative technologies? How can Dawn evaluate and compare the different alternatives? What wastewater treatment technology should it select?

5.1 Goal and Overview

The goal of this chapter is to present a process for assigning weights that reflect the relative importance of given measures and objectives. It includes methodologies for converting raw data into a scale between 0 and 1. These two steps combine to determine and compare the overall score for each alternative. The set of skills developed here are as follows:

- Assign consistent weights for measures and objectives.
- Create a utility score for distinct types of measures.
- Combine these skills to determine the overall score for each alternative.
- Utilize Logical Decisions software to facilitate this process.

Many managerial problems involve multiple conflicting objectives. In Chapter 4, we presented numerous examples of the need for trade-offs. The simplest trade-off involves only the two objectives of cost and performance and is similar to the light bulb decision of Mr. Frail. The example involving wastewater treatment technology and Dawn Chemicals, however, is significantly more complex. It involves investment cost, maintenance cost, operating cost, system capacity, time for implementation, water quality, and so on. In this chapter, we present a methodology for making efficient, logical, and defensible trade-offs. The light bulb example is our baseline case, and we expand upon it as the chapter proceeds.

There is a need for a structured approach to multiple objective decisions because rarely does one alternative dominate all the others with regard to every objective under consideration. We saw this in the light bulb selection decision, where the 100 W bulb is best in terms of performance but highest in cost.

Making trade-offs is one of the most important and difficult challenges that face decision makers. The methodology for doing so involves four steps:

1. Assign a weight to each measure that reflects its relative value to the decision maker. The weights must sum to 1.

2. Scale every measure to between 0 (worst) and 1 (best).

3. Convert each alternative's score on each and every measure to the common scale for that measure.

4. Determine the overall score for each alternative. Each alternative's score will be between 0 and 1.

First, the decision maker assesses the relative importance of the measures. Are performance and cost of equal importance? If so, then the two measures will be given equal weight: 0.5 and 0.5. This is a subjective question, and the answer depends on the decision context and personal preference of the decision maker(s). In this chapter, we introduce the swing weight method to assess the relative importance of measures. In Chapter 6, we use the hierarchical weights method to assign weights for the problem that has a large number of objectives. In Chapter 7, we introduce Analytic Hierarchy Process (AHP), which offers a different approach to determining weights.

Next, this chapter presents two techniques to scale measures: proportional scoring and mid-level splitting (MLS). Both methods generate a utility score between 0 and 1. For consistency, 0 is always considered the worst score and 1 the best. Proportional scoring is easy to use but assumes a linear relationship between a measure level and its utility score. The MLS method is a structured interview process that guides the subject matter expert (SME) or decision maker when developing a nonlinear relationship. We also introduce a third approach, Direct Assessment, which is used for constructed measures. In this method, the decision maker assigns a utility score to every possible value.

Finally, the decision maker compares the alternatives by taking a weighted sum of the utility scores for the relevant performance measures. In this chapter, we present Multi-Attribute Utility Theory (MAUT) to deal with conflicting objectives. MAUT is a comprehensive theory that addresses all aspects of the four-step process described earlier, including how to conduct interviews to determine the respective preferences of the decision maker and SME.

One of the strengths of MAUT is that it is not restricted by linearity assumptions. The utility scale for a measure need not be linear, that is, improving performance by 50% may increase the utility score by more or less than 50%. In addition, the weighted sum of scores across measures need not be a linear additive function. Later, in the chapter, we illustrate contexts in which the linear additivity assumption breaks down and present a multiplicative formula used to calculate utility scores. MAUT also easily incorporates probabilities when describing the uncertainty that surrounds the performance of an alternative on a specific measure.

To facilitate the application of MAUT, we introduce the reader to Logical Decisions® for Windows (LDW), a software package developed for dealing with multiobjective problems. In this chapter, we limit our illustrations to the basic features of LDW. You will learn to structure an LDW analysis by defining alternatives, goals, and measures; entering the data; scaling the data; and assessing the relative importance of the measures. Finally, you will learn to interpret the results of the analysis. On our book website, we offer a brief tutorial on how to get started using LDW.

5.2   Concepts and Terminology

Utility: A utility value is the abstract equivalent of a measure converted from natural units such as days, dollars, or watts to a scale from 0 to 1. A constructed measure such as high, medium, and low can be similarly converted. Thus, utility represents the desirability or satisfaction of a measure, or of an entire alternative, on a scale from 0 (worst) to 1 (best).

SUF: A Single-measure Utility Function (SUF) is a function that converts the specific value for a measure into a utility value between 0 and 1.

MUF: A Multi-measure Utility Function (MUF) is a formula that combines the utilities for the individual measures computed by the SUFs into the total utility for an alternative.

Additive utility function: A weighted sum of *n* different utility functions takes on the following form for simple cases that assume linear additive independence between measures and objectives:

$$U = \sum_{i=1}^{n} w_i u_i(x_i)$$

where

w_i is the weight of the *i*th measure

$u_i(x_i)$ is the individual utility function of the *i*th measure

Normalizing the weights to unity results in the following condition:

$$\sum_{i=1}^{n} w_i = 1$$

Proportional score: A linear utility function that calculates a score for an intermediate value that corresponds to its relative distance between the best and worst possible values of the measure.

Mid-level splitting method: A procedure for determining a decision maker or expert's SUF. The methodology begins by identifying the measure level that is exactly half-way in preference between the lowest and highest levels for a measure.

Weight: Weight represents the relative importance of the objectives or measures.

Range: The range of a measure is determined by the most preferred and least preferred levels of the measures. It should represent realistic values for a specific decision context, not merely the extreme values for the alternatives under consideration. For example, although we would prefer a zero cost, zero is an unrealistic value in almost every situation, and the preferred cost is almost always expressed as a dollar amount that is greater than zero.

SMART swing weight method: A method that allows a decision maker to assign weights that reflect the relative importance of the different measures by explicitly considering the "swing" or range of each measure from its least preferred to its most preferred level.

Cutoff value: The maximum or minimum value of a measure that a decision maker considers acceptable for the specific decision. An alternative that does not meet the cutoff value for even one of the measures may be cast aside entirely as unacceptable.

5.2.1   Multi-Attribute Utility Theory

MAUT is a methodology originally developed in the 1970s by Keeney and Raiffa (1993) for aiding decision makers when comparing and selecting among complex alternatives. The MAUT process is given in Figure 5.1. The application of MAUT involves the following broad categories of steps: structure decision, describe alternatives, clarify preferences, and analyze alternatives. These tasks can be further broken down as follows:

5.2.2   Structure Decision (Discussed Earlier in Chapter 4)

- Identify and clarify requirements, goals, and objectives and define the problem scope.
- Define relevant measures affecting the decision outcome and structure them into a hierarchical form called an "objectives hierarchy." Create constructed measures where necessary.

5.2.3   Describe Alternatives

- Describe feasible alternatives and gather data for each alternative, including any uncertainty. In many decisions, the data for a new concept are incomplete. Data often come from interviews of experts who use judgment to characterize the anticipated performance of a particular alternative on a specific measure.

5.2.4   Clarify Preferences

- Elicit preference information concerning the measures from the decision maker(s) in order to determine the relative importance of the various measures. This information is captured by weights.

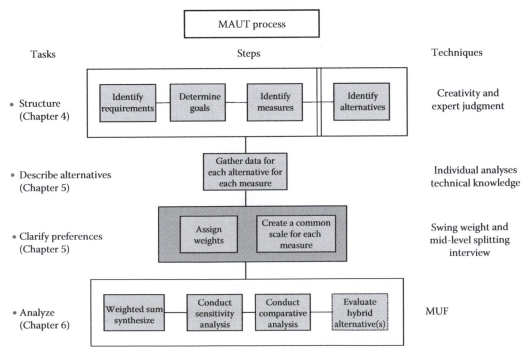

FIGURE 5.1:   MAUT process.

- Develop the decision maker's or expert's set of utility functions by establishing a scaling function for each measure. Each function converts the score on a performance measure to a 0–1 scale.

5.2.5   Analyze Alternatives

- Calculate the overall score of each alternative. All scores will fall between 0 and 1.

- Compare alternatives as to their relative strengths and weaknesses. What measures contribute the most and the least to each alternative's total score?

- Perform sensitivity analyses on the weights assigned to assess the robustness of the solution. Can it withstand change without optimal solution changing?

- Create improved hybrid alternatives by brainstorming ways to address the weaknesses of the highest ranked alternatives.

5.3   Compare Alternatives

Ideally, managers would be able to maximize performance at no added cost but, more often than not, one must give up something on one objective in order to achieve more in another. Making trade-offs is challenging; as the number of measures and alternatives grows, the decision complexity increases exponentially. The light bulb example seems almost trivial; most of us can handle this type of decision informally. However, Dawn Chemicals' decision regarding the waste treatment plant involves several measures of cost, as well as numerous measures of performance, and is beyond easy intuitive analysis. We are not just trading off apples for oranges in this case; it seems that we are juggling apples and watermelons—and hardwood trees as well.

Every manager, from his or her first day on the job to the last, lives and breathes trade-offs. However, these trade-offs are primarily technical decisions that affect how products and processes are designed and the way that services are delivered. When contemplating how to reduce the weight of a particular component, for example, a technical manager considers different materials that may be stronger and lighter, but he must also bear in mind the fact that these materials cost more. When adding durability to a product, an engineer considers how much stronger and heavier the various parts of its components or system must be while also considering the costs that these changes would incur. Managers and executives grow in their understanding of these trade-offs as they become more experienced. Sometimes, they can even discover, for example, ways of improving quality or performance without sacrificing any other measure.

Over time, managers of nontechnical functions also develop intuitive performance-value trade-offs. Managers in a wide array of organizations understand the relationship between experience and performance along with the corresponding cost of hiring more experienced staff. For example, there is an enormous difference between a 1st year teacher and one who has 5 years of experience. The same applies to sales people or technical support staff as well as to people in managerial positions. There may also be a trade-off between salary and turnover rate. Some retailers have chosen to operate at the low end of the value cost trade-off by setting salaries slightly above minimum wage. Their customers can expect little service, and the store managers can expect high turnover, since the staff does not feel greatly beholden to the store. High-end retail chains pride themselves on the knowledge and expertise of their sales clerks, who do not only process transactions but also help consumers make wise choices.

An understanding of these trade-offs is critical for anticipating the performance of various alternatives and suggesting new alternatives. This knowledge of technical or operational trade-offs is a prerequisite for deciding where the most preferred location on the trade-off curve might be. The aforementioned are examples of value trade-offs. A technical trade-off question, on the other hand, might be "How much *will it cost* to improve the durability or reduce the physical weight of the product?" An example of a value trade-off question is "How much are we *willing to spend* to improve the durability of a product and how much are we willing to spend to reduce its weight?" The answer to this question represents the relative value that the decision maker places on durability as compared to the physical weight. Another example is "How much are we *willing to spend* to reduce waiting time for the customer?" This represents the relative value the decision maker places on the customer's convenience as compared to the company's cost. The difference between these two types of trade-offs is subtle and, at first, often confusing for decision makers. The technical trade-offs concern the objective question, "How much will it cost to improve performance," for example, whereas the value trade-offs concern the subjective question, "How much is it *worth (to us)* to improve performance?"

The essence of this chapter is a structured process for aiding managers to make a rational and easily explainable value trade-offs, thereby reaching a defensible decision.

Technical Relationship Trade-offs
Are NOT
Value Trade-Offs

Example: Weight, Cost, Service

Technical trade-off
How much will it cost to achieve a one pound reduction in weight?
How much decrease in waiting time is achieved by adding one more cashier?
How much will customer satisfaction improve if waiting time is reduced by 1 minute?

Value trade-off
How much would we be willing to spend to reduce the weight by 1 lb?
How much would we be willing to spend to reduce waiting time by 1 minute?

5.4 Trade-Off Conflicting Objectives

In this section, we begin with the light bulb selection example, a case that involves only two objectives. Later, we discuss a retail staffing decision, a used car purchase, selection of a kitchen remodeler, and a real-world case involving the selection of a coating process for electronic circuitry. Last, we provide several case studies from relevant journal articles.

Bill Frail decided to buy the bulbs from a local hardware store and made his calculations according to the residential electricity rate charged by a local utility company. He estimated that he would use each bulb for an average of 12 hours a day for 250 days a year, a total of 3000 hours a year. The average life of the 60 and 75 W bulbs he considered was 1500 hours. For the 100 W bulb, he considered a longer lasting bulb that was projected to last an average of 3000 hours. On average, he would need two 60 or 75 W bulbs per year for each of his 10 lighting fixtures or only one 100 W longer-life bulb. (He had decided if he were to use the 100 W bulb, he would buy those that last twice as long as the cheaper bulbs since they were on sale.) The annual operating cost calculation assumed an electric rate of $0.10 per kWh. The data are summarized in Table 5.1.

TABLE 5.1: Incandescent bulb alternatives.

Bulb Watts	Initial Bulb Cost ($)	Kilowatts/ Year	Total Annual Operating Cost ($)	Total Bulb Purchasing Cost ($)	Total Annual Cost ($)
A	B	C=A * 3000 h/1000	D=C * 10 * 0.10	E=B * number of bulbs	F=D+E
60	0.45	180	180	9	189
75	0.50	225	225	10	235
100	1.50	300	300	15	315

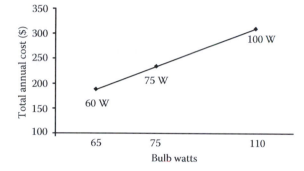

FIGURE 5.2: Graph of cost vs. bulb performance.

Figure 5.2 contains a plot of the alternatives with performance on the horizontal axis and price on the vertical axis. The 60, 75, and 100 W bulbs show up on the graph as three points. Notice that the data points almost fit a straight line because the annual operating cost is a linear function of the wattage which accounts for over 95% of the total cost.

As seen in Table 5.1 and Figure 5.2, the bulb with the best performance costs the most. In fact, none of the bulbs under consideration is best on each and every measure. Mr. Frail cannot easily select the best bulb by looking at either Table 5.1 or Figure 5.2. The question of which is the "best" bulb will depend on how much Mr. Frail is willing to pay to obtain better bulb performance. To answer the question, we will start with the 60 W bulb and assume that he will purchase it if others are not better. The basic question becomes, is it worthwhile to switch from 60 to 75 W while incurring added costs? The switch would be worthwhile if he is willing to pay $46 ($235−$189) more per year to increase bulb performance from 60 to 75 W. In other terms, is he willing to pay $3.07 for each additional watt? This is a subjective assessment. While the extra cost may be worthwhile for one person, it may not be worthwhile for another.

Let us continue with the example. Assume that Mr. Frail is willing to pay $46 to increase the bulb performance from 60 to 75 W. Would he be willing to pay $80 ($315−$235) more to switch from 75 to 100 W bulbs? Is he ready to pay $3.20 for each extra watt? Does the light supplied by an extra 25 W per bulb warrant the added expense? If so, he makes the switch to the 100 W bulbs. If not, he purchases the 75 W bulbs.

This procedure of considering direct trade-offs of performance and cost can be used with decisions that involve two or three measures and a few alternatives. However, as the number of measures increases, the trade-offs become more complicated, and graphical interpretation becomes more difficult. Therefore, we introduce the concept of weights that capture analytically the trade-off described in words.

5.4.1 Determine Weights: SMART Swing Weight Method: Two Objectives

In the following activity, you are asked to assign weights to two measures: performance and cost. First, think in terms of which measure is more important to you, and assign a higher weight to that measure. The two weights should add to 1.

Activity 1

Measure	Assign Weight
Performance	_____
Cost	_____

The simplest (and usual default) assumption is that the two measures are equally important. When working with two measures, that means each is assigned a weight of 0.5. In reality, however, this is rarely the case. In the aforementioned activity, we asked the reader to directly assign relative weights to performance and cost. As simple as this task is, we do not recommend it as the standard procedure. It is critical that before answering the question "Which is more important, performance or cost?" the decision maker pay careful attention to the range of values for each measure. We argue that the appropriate question is "Which *range of values* is more significant?" Later, we demonstrate the pitfalls of ignoring ranges when assigning weights.

Range Specification

Minimum range: Difference between the best and worst values for the stated alternatives.
Realistic range: Pick a range that is realistic for the problem and easy to work with. Allow for the possibility of other realistic alternatives that may fall outside the initial range.

First, we need to specify a realistic range for each measure. In the light bulb example, the 60 W minimum and 100 W maximum are the two extreme values of the alternatives and are obvious choices for specifying the range. However, it is often preferable to use a range that is wider than just the minimum and maximum values of the specific alternatives. This allows us to add alternatives at a later date without having to redo the assessment of weights and rescale all the single utility functions. Thus, the stated range for cost is not set at the exact cost of the respective bulbs, that is, $189–$315, but rather at $150–$350 to allow for consideration of a somewhat less expensive bulb or one that is more costly.

With the ranges specified as in the Table 5.2, the SMART swing weight process proceeds as follows.

- Bill Frail reviews the ranges and considers which improvement would be of greatest significance to him. Would Bill rather increase wattage performance to 100 W per bulb while keeping the $350 cost or would he prefer to reduce total cost to $150 per year and use all 60 W bulbs?

TABLE 5.2: Swing weights for simple bulb selection example.

Measure	Least Preferred	Most Preferred	Rank	Points	Final Weight
Performance	60 W	100 W	____	____	____
Cost	350	150	____	____	____
			Total	____	

TABLE 5.3:   Swing weights for simple bulb selection example.

Measure	Least Preferred	Most Preferred	Rank	Points	Final Weight
Performance	60	100	1	100	100/165=0.61
Cost	350	150	2	65	65/165=0.39
			Total	165	

- He ranks improvement in performance (rank order 1) ahead of cost reduction (rank order 2).
- Bill assigns 100 points to the highest ranked measure. Now, relative to the range for the highest ranked measure-range, how significant is the range for the second ranked measure? The answer to this question is expressed as a relative percentage.
  - He values the improvement on the cost range (from 350 to 150) as only 65% as important as spanning the range on performance. Bill assigns 65 points to cost. He puts these values into Table 5.3.
- Calculate weights. Add up the points which total 165 and rescale the values to 1.
  - The weight for performance is now 0.61 and for cost is 0.39.

If you were in Bill Frail's situation, which range would you value most? Fill out Activity 2 according to your preferences. Rank order the measures and assign points. Then, calculate the final weights. Compare your answers here to the answers you gave in Activity 2. Do not be surprised if there are differences.

Activity 2: Own Preference 10 Light Bulbs: Performance vs. Cost

Measure	Least Preferred	Most Preferred	Rank	Points	Final Weight
Performance	60	100	____	____	____
Cost	3500	150	____	____	____
			Total	____	

Now, we will modify the decision context, while still dealing with light bulbs. Assume you are the manager of a hotel and have responsibility for purchasing bulbs for 1000 bedside lamps. In the new decision, the annual cost ranges from a worst $35,000 per year to a best $15,000 per year. Activity 3 asks you to determine weighting the preferences for this decision context. In thinking about the trade-off, the decision maker must keep in mind two perspectives. Cost is an internal priority and of concern to management. Light bulb performance is primarily the concern of the customer who will be renting the hotel room. The need to consider a dual perspective regarding cost and performance commonly arises. Such a need would apply likewise in an example cited earlier, wherein store managers who were deciding the number of checkout counters to staff had to trade off cost and customer waiting time.

Two Perspectives
Cost (Internal) and Performance (Customer)

- Rank order your measure ranges in Activity 3.
- Assign 100 points to the most significant range.

- Assign a relative number of points, less than 100, to the second ranked measure range.
- Calculate the total points and rescale so the weights sum to 1.
- Compare your weights for the two different decision contexts.

Activity 3: Hotel Preference Light Bulbs: Performance vs. Cost

Measure	Least Preferred	Most Preferred	Rank	Points	Final Weight
Performance	60	100	___	___	___
Cost	$35,000	$15,000	___	___	___
			Total	___	

Are these weights the same as those you assigned in Activities 1 and 2? Too often weights are assigned on the basis of vague claims that cost is worth twice as much as performance. Suppose you are looking at houses, all of which cost about the same, ranging between $290,000 and $310,000, but whose features differ widely. Why should price be the highest rated measure in your decision, especially if mortgage rates are low? An extra $20,000 may represent only a slight increment in your monthly payments. In this case, price should have a relatively low weight in the overall score unless you are especially concerned over a difference of $20,000 in the purchase price and subsequent mortgage.

In the aforementioned activity, we provided no details regarding the nature of the hotel. How would your responses change if you were manager of a low end, extended stay hotel as compared to the manager of a high end downtown hotel?

This broader application of the light bulb problem highlights the importance of both decision context and range of values. It reminds the reader to think hard before assigning weights and to bear in mind that the process should not be reduced to the simple question, "How much more or less important is performance relative to cost?" In addition, if you carry out this activity in the presence of others, it is obvious that preferences will differ. The varying perspectives can serve as an important tool for arriving at an effective decision.

5.4.2 Interpretation of Weights

Mr. Frail's responses resulted in a weight of 0.61 assigned to the range for performance. Wattage ranged from 60 to 100, a difference of 40. The weight assigned to a range has the following meaning: A set of 10 bulbs each with 60 W, the least preferred value, earns 0 units of utility or utile. A set of 100 W bulbs, the most preferred value, earns 0.61 units of utility. This means that every additional watt beyond 60 W earns approximately 0.015 units of utility (0.61/40). Similarly, there was a $200 range for cost: from $350 down to $150. A set of bulbs costing $350 per year earns 0 utility units and a set costing $150 per year earns 0.39 units. Thus, every dollar in cost reduction is worth approximately 0.002 utility units (0.39/200). Alternatively, one might say that the value of each additional watt for this set of 10 bulbs is seven and half times as large as the value of each dollar saved annually: 0.015 as compared to 0.002.

Let us highlight one last time the importance of range. Assume that Mr. Frail's assignment of points and their resultant weights of 0.61 and 0.39 would have been the same had we used the cost range of the actual bulbs. The annual costs ranged from a high of $315 to a low of $189, a difference of $126. As a result, every dollar in cost reduction would be worth 0.003 utile, 50% more than in the earlier calculation.

Activity 4: Significant and Insignificant Price Ranges for Automobiles

Imagine that a friend is considering purchasing an automobile. Describe under what circumstances you might recommend assigning a relatively

(a) Large weight to variations in the purchase price _____
(b) Small weight to variations in the purchase price _____

5.5 Single-Measure Utility Function: Proportional Scores

In order to compare light bulbs, cost and performance measures need to be converted to equivalent scales before taking a weighted sum. MAUT uses a 0–1 scale, in which 0 reflects the worst level of a measure and 1 reflects the best level in the range under consideration. The simplest method for doing this (and the default assumption) is a linear proportional scale. Each point between the extreme values is assigned a value between 0 and 1 based on how far it is along the line connecting the two extreme values.

The general formula for a proportional calculation of the utility score of a value x_i is:

$$U_i(x_i) = \frac{x - Worst\ Value}{Best\ Value - Worst\ Value}$$

We will use the light bulb example to illustrate the calculation. Mr. Frail set 60 W as the worst possible value and 100 W as the best.

$$U_p(100\ W) = 1 \quad U_p(60\ W) = 0$$

For the 75 W bulb's performance,

$$U_p(75\ W) = \frac{75 - 60}{100 - 60} = 0.375$$

For the cost measure, the scale is reversed compared to the wattage measure. A higher value of cost is worse than a lower value.

$$U_c(\$150) = 1 \quad\quad U_c(\$350) = 0$$

The 75 W bulb's utility for cost is

$$U_c(75\ W) = \frac{235 - 350}{150 - 350} = \frac{-115}{-200} = 0.575$$

Because we used a cost range that goes beyond the specific bulbs under consideration, there is also a need to calculate the utility score for the costs associated with the 60 and 100 W bulbs.

$$U_c(\$189) = (189 - 350)/(150 - 350) = 161/200 = 0.805$$

$$U_c(\$315) = (315 - 350)/(150 - 350) = 35/200 = 0.175$$

TABLE 5.4: Utility scores for three bulbs on two attributes.

	60 W Bulb	75 W Bulb	100 W Bulb
Performance	0	0.375	1
Cost	0.805	0.575	0.175

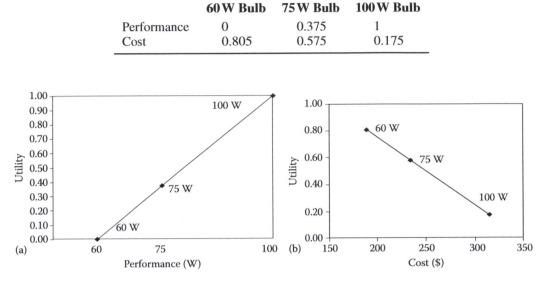

FIGURE 5.3: Utility functions for the (a) performance and (b) cost.

The utility scores for each attribute for all alternatives are given in Table 5.4. The utility curves for measures are usually graphed, and, in this case, they are represented by two straight lines, as depicted in Figure 5.3. These lines must be monotonic. The wattage scale is monotonically increasing (a positive slope), and the cost scale is monotonically decreasing (a negative slope).

5.6   Aggregate Utility: Total Score for Each Alternative

Once each of the measures has been rescaled to between 0 and 1 and weights have been assigned to the respective measures, a total utility score, defined as a weighted sum, can be calculated for each of the alternatives.

$$U = w_p U_p(Performance) + w_c U_c(Cost)$$

Using Mr. Frail's preference for weights and the assumed linear proportionality of the SUF for cost and performance, we calculate the score for the 60 and 100 W bulbs as follows:

$$U(60 \text{ W}) = 0.61*(0.00) + 0.39*(0.805) = 0.31$$

$$U(75 \text{ W}) = 0.61*(0.375) + 0.39*(0.575) = 0.45$$

$$U(100 \text{ W}) = 0.61*(1.00) + 0.39*(0.175) = 0.68$$

The 100 W bulb is ranked first, with a utility of 0.68. This value means that the 100 W bulb is 68% of an ideal alternative: a 100 W bulb that would cost only $150 per year. The 60 W bulb's utility score is less than half that of the 100 W bulb's utility score. The 75 W bulb falls in the middle with

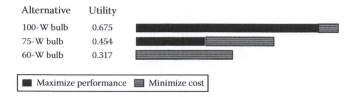

Alternative	Utility
100-W bulb	0.675
75-W bulb	0.454
60-W bulb	0.317

■ Maximize performance ▨ Minimize cost

FIGURE 5.4: Stacked bar results for bulb selection.

a score somewhat higher than the 60 W bulb. The results are consistent with our intuition, since Mr. Frail assigned most of the weight to performance.

Now imagine that Mr. Frail's preferences with regard to the two measures were not as previously noted, but that his weights were 0.50 and 0.50. The utility scores for each of the alternatives would become 0.40 for the 60 W bulb, 0.48 for the 75 W bulb, and 0.59 for the 100 W bulb. Although the absolute scores have changed, the relative ranking remains unchanged.

IMPORTANT: It is not uncommon to find that the rank ordering is not sensitive to changes in the weights. It is for this reason that we recommend, when working with groups, that the project leader not invest too much time upfront in ironing out differences in weight assignments. He should, instead, accept the initial differences and run the MAUT model for each of the different weights that have been suggested. Only when the rank orderings are significantly different among the various decision makers is there a need to struggle to achieve a consensus on the weights.

The software package Logical Decisions presents the results of the analysis in an easy to read, insightful manner. Figure 5.4 depicts the stacked bar ranking results as provided by LDW. It rank orders the alternatives and illustrates graphically how each alternative obtains its score. The most preferred alternative, the 100 W bulb, draws strength almost exclusively from performance. The 75 W bulb is ranked second on both measures.

5.6.1 Customer Service Staffing

In this section, we will demonstrate trade-offs by using a different example. Nancy Chicila is a customer service manager at MONEYMARK, a local financial planning and advising company. Nancy has four servers (staff) who handle the customer service hot line. She has recently received many complaints from customers about being placed on hold for a long time. She decided to analyze the problem to determine whether MONEYMARK needed to hire more employees in order to reduce waiting time. She planned to use a queuing model to estimate the impact of additional staff on the length of waiting time.

After studying historical data, she found that the time between customer calls and service time was characterized by an exponential distribution. Data showed that the arrival rate (λ) averages 28 calls per hour, and service rate (μ) averages 8 customers per hour. Average cost, including salary and training per employee per hour, was $18.

Nancy has sufficient staff available to handle their customers on average. Nevertheless, waiting lines form because customers do not call at a constant, evenly paced rate; nor are all situations resolved within an equal amount of time. There are random periods of time, wherein the number of customers exceeds capacity and the line grows. Eventually, however, the line stops growing and begins decreasing and is sometimes empty.

If Nancy hires new staff for customer service, customers will, on average, spend less time on hold. However, employing these additional servers will increase her costs. The problem she is facing is how to minimize customer waiting time while still minimizing costs. These are conflicting objectives, and Nancy cannot optimize both of them simultaneously. She needs to make a trade-off between minimizing cost and minimizing waiting time.

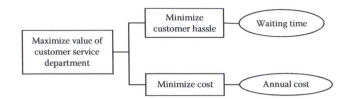

FIGURE 5.5:   Objective hierarchy for customer service staffing.

The simplest objectives hierarchy consists of minimizing waiting time and cost as shown in Figure 5.5. Waiting time is measured by the number of minutes a customer spends on the phone waiting for customer service. Cost includes annual salary and benefits and is expressed in terms of US dollars.

MONEYMARK's cost for a customer service staff member is $18/hour. A customer service representative works 8 hours a day for 200 days a year. Nancy calculated the annual cost and average waiting time on the phone using M/M/s queuing process for 4–6 staff members, as shown in Table 5.5. We recommend that interested readers review any management science or operations research book for additional information on queuing modeling.

Under the current staffing strategy (four staff), a customer spends on average more than 11 minutes on the phone waiting for a customer service representative. If Nancy hires an additional representative, the waiting time dramatically declines to less than 2 minutes. On the other hand, hiring the additional customer service representative would increase the company's annual cost from $115,200 to $144,000. Adding a sixth representative would reduce waiting time further still to a 0.5 minutes.

Table 5.6 shows rank orderings and weights for this example. The ranges specified were set only slightly beyond the actual values. The cost range was set at $115,000 for most preferred and $175,000 for least preferred. The range for waiting was from 0 to 12 minutes. Nancy considered the two ranges. She was more interested in reducing waiting time from 12 to 0 minutes than in reducing cost from $170,000 to $115,000. She assigned 100 points to the first ranked measure (waiting time). She assigned two-thirds as many points, or 67 points, to reducing cost. The last column of Table 5.6 illustrates a final weight for each measure. Waiting time accounts for 60% of the total and annual cost 40% of the total weight.

She created linear utility functions for waiting time and annual cost as shown in Figure 5.6.

TABLE 5.5:   Annual cost and waiting time for different staffing levels.

Strategy	Annual Cost ($)	Waiting Time (min)
4 staff	115,200	11.1
5 staff	144,000	1.9
6 staff	172,800	0.5

TABLE 5.6:   Rank ordering and weights for customer service staffing.

Objective	Measure	Least Preferred	Most Preferred	Rank Order	Points	Weight
Minimize waiting time	Waiting time (min)	12	0	1	100	0.6
Minimize cost	Annual cost ($)	175,000	115,000	2	67	0.4

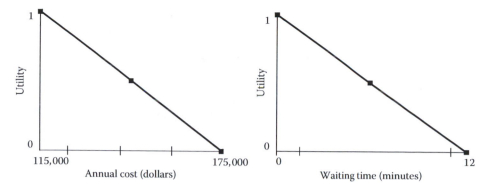

FIGURE 5.6: Utility functions for annual cost and waiting time measures.

TABLE 5.7: Utility scores for customer service staffing.

	Wait		**Cost**		**Total Score**
Servers	**Time**	**Utility**	**Dollars**	**Utility**	**Utility**
4	11.1	0.075	$115,200	0.997	0.444
5	1.9	0.842	$144,000	0.517	0.712
6	0.5	0.958	$172,800	0.037	0.590

For the waiting time measure, the range is 12 minutes. A higher numeric value of time is less desirable than a lower value and has a lower utility (Table 5.7).

$$U_T(11.1) = (11.1 - 12)/(0 - 12) = 0.075$$

$$U_T(1.9) = (1.9 - 12)/(0 - 12) = 0.842$$

$$U_T(0.5) = (0.5 - 12)/(0 - 12) = 0.958$$

For the cost measure, the range is $60,000. Again, a higher numeric value is less desirable and has a lower utility.

$$U_C(115,200) = (115,200 - 175,000)/(115,000 - 175,000) = 0.997$$

$$U_T(144,000) = (144,000 - 175,000)/(60,000) = 0.517$$

$$U_T(172,800) = (172,800 - 175,000)/(60,000) = 0.037$$

The total utility for each alternative is calculated as a weighted sum of the time utility and cost utility.

$$U_S(4 \text{ servers}) = 0.6(.075) + 0.4(.997) = 0.444$$

$$U_S(5 \text{ servers}) = 0.6(.842) + 0.4(.517) = 0.712$$

$$U_S(6 \text{ servers}) = 0.6(.958) + 0.4(.037) = 0.590$$

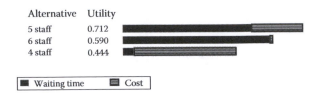

Alternative	Utility	
5 staff	0.712	
6 staff	0.590	
4 staff	0.444	

■ Waiting time   ▨ Cost

FIGURE 5.7:   Ranking of the alternatives for customer service staffing example.

Figure 5.7 illustrates the LDW output on a stacked bar chart. The five-staff strategy is ranked first with a utility score of 0.712 and is significantly better than either of the other alternatives. The six-staff strategy is second best. The five-staff strategy gains utility by reducing both waiting time and cost and is especially strong in terms of waiting time. The four-staff strategy is worst in terms of waiting time, and the six-staff strategy is worst in terms of cost.

These results can be interpreted as follows: Nancy Chicila is willing to spend an extra $38,300 to reduce customer waiting time by more than 9 min, from 11.1 to 1.9. However, she is unwilling to spend an additional $38,800 to reduce the wait by an additional 1.4 min. Even without a structured weight assignment approach, Nancy would likely have come to the same conclusion for this simple example.

5.7   Assessing Weights Revisited: Large Set of Measures

In this section, we present a more complex example that involves more objectives and measures in order to demonstrate the swing weight method and in order to assess the relative importance of measures. In these contexts, it is preferable that someone interview the decision maker rather than have the decision maker sit down and perform a self-assessment. The interviewer should encourage the decision maker to provide a rationale for every step of the process. This improves the transparency of the decision-making process and makes it easier to go back and revise weights should the decision context change.

Recall that the swing weight process asks the decision maker to consider swinging a measure from its least preferred value to its most preferred value when considering how to assign weights. The first task is to rank order the ranges in terms of significance. When ranking the measures, the interviewer asks the decision maker, "Which measure would you most want to change to its most preferred value in this decision context?" The second task is to assign points to a sort ordered list so as to quantify the relative importance of the ranges. This process proceeds as follows:

1. Establish realistic ranges for every measure and list all measures with their least and most preferred values.

2. Rank order the measure ranges based on decision maker preferences.

    a. Interview a decision maker or SME and identify the measure whose range is most significant. Provide a brief rationale.

    b. Place this measure at the top of a table alongside its extreme values.

    c. Identify the second most significant measure range. Provide a brief rationale.

3. Repeat step 2 until all the measure ranges are ranked and a sort ordered list has been created.

4. Assign points relative to the highest ranked measure range.

a. The highest ranked measure range is assigned a value of 100.

b. Assess the relative importance of swinging the second highest ranked measure from worst to best and score it as a percentage of the 100 for the highest ranked measure range.

c. Repeat step 4b for each measure as compared to the highest ranked measure. As you move down the ordered list of measures, the points assigned should decrease.

5. Normalize the weights. Determine the relative weights by adding up the total points and dividing each measure's assigned points by this total.

Let SW_i be the swing weight initially assigned to measure i.
The weight for each measure w_i is just:

$$w_i = \frac{SW_i}{Sum\_of\_all\_SW_i}$$

5.7.1   Used Car

In this case study, Pete, a college freshman, was looking to buy a used car both for social functions and to drive back and forth to his part-time job 20 miles away. Pete's overall objective was to maximize the car's value, according to his priorities. He identified four major objectives that impact the overall objective of maximizing the value: reliability, total cost, accessories, and aesthetics as illustrated in the objectives hierarchy in Chapter 4. Pete also developed twelve measures by which to evaluate the alternatives, as presented in Table 5.8.

Pete looked at many cars and, after his initial prescreening, decided to evaluate four cars: Honda Civic, Chevrolet Cavalier, Ford Ranger, and Mazda Miata. He collected data for each measure for each alternative, as depicted in Table 5.9. The data source includes the owner of the car, visual inspection of the car, Kelly Blue Book online, and JD Power online. In reviewing the data, it is important to note that Pete was not comparing generic used cars of each brand, but rather actual used cars that were for sale. He used the data to set reasonable ranges for each measure. In general,

TABLE 5.8:   Objectives and measures for used car example.

Objective	Measure	Preferred		Rank Order	Points	Weight Measure
		Least	Most			
Reliability	Mileage	130,000	80,000			
	Dependability ratings	2 circles	4 circles			
Total cost	Purchase cost	$6,500	$2,500			
	Mpg	20 mpg	30 mpg			
	Maintenance annual	$600	$400			
	Longevity	3 years	5 years			
Aesthetics	Color	Dark	Light			
	Interior	Poor	Excellent			
	Exterior	Poor	Excellent			
Accessories	A/C and heater	Neither works	Both work			
	Seating capacity	2	6 or more			
	Sound system	None	Radio and CD			

TABLE 5.9: Data for used car example.

Measure	Honda Civic	Chevrolet Cavalier	Ford Ranger	Mazda Miata	Least Preferred	Most Preferred
Odometer mileage	125,000	100,000	85,000	125,000	130,000	80,000
Dependability ratings	3 circles	3 circles	3 circles	4 circles	2 circles	4 circles
Purchase cost	$5,000	$3,000	$4,000	$6,000	$6,500	$2,500
Mile per gallon	30	29	25	24	20 mpg	30 mpg
Maintenance cost	$400	$500	$500	$600	$600	$400
Longevity	3 years	4 years	4 years	3 years	3 years	5 years
Color	Neutral	Neutral	Dark	Light	Dark	Light
Interior condition	Good	Fair	Good	Excellent	Poor	Excellent
Exterior condition	Good	Good	Fair	Excellent	Poor	Excellent
A/C and heater	Both work	Heater only	A/C only	Both work	Neither works	Both work
Seating capacity	4 or 5	4 or 5	2	2	2	6
Sound system	Radio and cassette player	Radio and CD player	Radio only	Radio and CD player	None	Radio and CD player

he increased the ranges slightly to account for the possibility that he might find another car. The actual range for purchase price was $3000–$6000, but he expanded this range by $500 above and below. The range was, therefore, $2500–$6500. He did the same for mileage, setting the range from 80,000 to 130,000 miles.

Isabel, a friend of Pete's, recommended that he use a structured interview process to ensure that the weights he had assigned to the measures were appropriate. Isabel then interviewed Pete. She asked him to rank order his set of measures, assign points across the range of each measure, and calculate a weight for each measure. Isabel had Pete rank the measures by having him pay explicit attention to the range for each measure. She asked Pete to identify the measure that would be most important for him to increase from its least preferred to its most preferred value. After looking at a list of measures with their respective most and least preferred levels, Pete said that he would prefer to lower the purchase cost from $6500 to $2500. Isabel asked Pete whether lowering the purchase cost to $2500 would represent the most important improvement in the list. After Pete's confirmation, they identified the second most important measure. Pete reviewed all measures except the purchase cost and thought that improving the dependability rating from two to four circles was important because he did not want to spend a lot of money on repairs or deal with the hassle of getting to work when his car was in the shop. Therefore, he ranked dependability second. Pete felt the dependability range was 80% as important as the cost range. Isabel and Pete continued the interview until all the measures had been ranked, as illustrated in Table 5.10 and relative points were assigned.

Table 5.10 shows points and weights for each measure. Purchase price ranked highest with a 0.15 weight. Interior condition was ranked lowest with a weight of 0.02. Isabel decided to carry out one additional check to see if Pete was comfortable with the final weights. She aggregated the measure

TABLE 5.10: Rank ordering and weights for buy a used car example.

Objective	Measure	Least Preferred	Most Preferred	Rank Order	Points	Weight Measure	Weights Objectives
Reliability	Odometer mileage	130,000	80,000	5	70	0.11	0.23
	Dependability ratings	2 circles	4 circles	2	80	0.12	
Total cost	Purchase cost	$6,500	$2,500	1	100	0.15	0.41
	Mile per gallon	20 mpg	30 mpg	3	75	0.11	
	Maintenance cost	$600	$400	11	25	0.04	
	Longevity	3 years	5 years	4	70	0.11	
Aesthetics	Color	Dark	Light	10	30	0.05	0.12
	Interior condition	Poor	Excellent	12	10	0.02	
	Exterior condition	Poor	Excellent	9	40	0.06	
Accessories	A/C and heater	Neither works	Both work	8	50	0.08	0.25
	Seating capacity	2	6 or more	7	55	0.08	
	Sound system	None	Radio and CD player	6	60	0.09	
Sum					665	1	1

weights within each objective. Total cost objective ranked first, collecting 41% of the total weight, and was followed by accessories objective, with 25% of the total weights. Reliability objective ranked third, accounting for 23% of the total weight. Finally, aesthetics objective ranked fourth, with 12%. The only change Pete considered making was to increase aesthetics to 15% and reduce reliability to 20%. In the end, he left the numbers as they were.

Pete proceeded to use a proportional scale to convert all of the vehicle values into a corresponding utility score. He began with odometer mileage, wherein lower is better. The range was between 80,000 and 130,000 miles, a total of 50,000 miles. The Honda Civic with 125,000 miles scored 0.1 as compared to the Ford Ranger with 85,000 miles, which received a 0.9. An argument could be made that the mileage scale should be adjusted for different vehicles, but Pete chose not to do so. Next, he considered the values for fuel economy, wherein higher is better. The range was from 20 to 30 mpg. The Honda, at 30 mpg, scored a utility score of 1. The Chevy Cavalier was close behind at 0.9. After completing Table 5.11, Pete was ready to calculate a total score. He multiplied each utility value by its corresponding weight. The Chevy Cavalier ranked highest with 0.68.

Isabel recommended that Pete use LDW to calculate total scores and rank the alternatives; this would provide more insight as to the strength and weaknesses of the various alternatives. Figure 5.8 shows a stacked bar chart. The specific Chevrolet Cavalier that was for sale ranks first with a utility score of 0.68, followed by the Honda Civic. The Chevrolet is extremely strong on total cost, while it is only average on the other objectives. The Honda Civic for sale draws its strengths from accessories, but is weak on cost and on reliability because of its high odometer mileage. Based on the results of the MAUT process, Pete decided to buy the Chevrolet Cavalier.

TABLE 5.11:   Utility scores for each vehicle and measure.

Measure	Honda Civic	Chevrolet Cavalier	Ford Ranger	Mazda Miata	Weight Measure
Odometer mileage	0.1	0.6	0.9	0.1	0.11
Dependability ratings	0.5	0.5	0.5	1	0.12
Purchase cost	0.375	0.875	0.625	0.125	0.15
Mile per gallon	1	0.9	0.5	0.4	0.11
Maintenance cost	1	0.5	0.5	0	0.04
Longevity	0	0.5	0.5	0	0.11
Color	0.5	0.5	0	1	0.05
Interior condition	0.67	0.33	0.67	1	0.02
Exterior condition	0.67	0.67	0.33	1	0.06
A/C and heater	1	0.33	0.67	1	0.08
Seating capacity	0.75	0.75	0	0	0.08
Sound system	0.67	1	0.33	1	0.09
Total score	0.55	0.67	0.49	0.48	

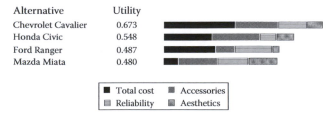

Alternative	Utility
Chevrolet Cavalier	0.673
Honda Civic	0.548
Ford Ranger	0.487
Mazda Miata	0.480

■ Total cost   ▨ Accessories
▨ Reliability   ▨ Aesthetics

FIGURE 5.8:   Rankings of the alternatives for the used car example.

5.7.2 Card Method: Weight More than 10 Measures

When the number of measures is more than 10, it becomes difficult to look at the entire list, try to rank order them, and specify swing weights. In general, people are better at making pairwise comparisons than sorting long lists. Thus, to facilitate the interview, we suggest recording on a separate card each measure with its least and most preferred level, as illustrated in Table 5.12. The decision maker should then select any two cards and compare their respective ranges. The card bearing the more significant measure is placed on the top of the deck of cards and the other is placed below it. A third card is selected and its range is compared to the previous two so as to determine where within the sorted deck of cards to place it. He repeats this process until each of the cards has been inserted into the deck in its appropriate location. He then records the rank of each measure and proceeds as before to determine the swing weight. Starting with the second highest ranked measure, he compares the significance of its range to that of the highest ranked and assigns a swing weight that is less than 100. He repeats the process for every card further down in the deck, assigning smaller and smaller weights each time. The used car example in the preceding text might have benefited from this approach, especially if Pete had not had a willing interviewer in his friend Isabel.

5.7.3 Hierarchical Top-Down Approach

The bottom-up approach starts at the measure level and involves repeated pairwise comparisons against the highest ranked measure. To ensure consistency when assigning points, the measure is also compared to the measure ranked immediately above it. A hierarchical structure of major objectives with groups of measures suggests the possibility of an alternative approach.

- Rank order the major objectives in order of total importance.
- Allocate 100% points among these objectives

When ranking these major objectives, it is important to consider all of the measures and their associated ranges. The number of points allocated to an objective is equal to the amount that would be earned if each and every measure within that objective moved from its least preferred value to its most preferred value.

- For each objective, apply the swing weight method to the measures within the objective. The measure weights within each objective should total to 1.
- Multiply the overall objective weight by the measure weight in order to obtain the global weight for that measure.

In Table 5.12, we recreate the used car example, employing a hierarchical approach. Total cost was ranked highest and assigned 40% of the weight. Accessories were ranked second, with 25% of the weight. Reliability and aesthetics were ranked third and fourth, respectively, with correspondingly smaller weights. The measures within each objective were then ranked. For example, dependability was ranked higher than odometer mileage but was given only slightly more weight (0.55) than mileage (0.45). The global weight for each measure was determined by multiplying their weights by 0.2, the weight of reliability.

Within cost, the purchase cost range was most important and given 40% of the total cost objective's weight. Longevity and miles per gallon were tied for second place, and each was given 25% of the weight. Each global weight was determined by multiplying by 0.4, the total cost weight. Similar calculations were carried out for measures within accessories and aesthetics.

TABLE 5.12: Hierarchical rank ordering and weights for buy a used car example.

Objective	Measure	Least Preferred	Most Preferred	Rank	Weights Objective	Rank	Weight Measure	Weight Global
Reliability	Odometer mileage	130,000	80,000	3	0.20	2	0.45	0.09
	Dependability ratings	2 circles	4 circles			1	0.55	0.11
Total cost	Purchase cost	$6,500	$2,500	1	0.40	1	0.4	0.16
	mpg	20mpg	30mpg			2	0.25	0.1
	Maintenance	$600	$400			4	0.1	0.04
	Longevity	3 years	5 years			2	0.25	0.1
Aesthetics	Color	Dark	Light	4	0.15	2	0.4	0.04
	Interior condition	Poor	Excellent			3	0.15	0.015
	Exterior condition	Poor	Excellent			1	0.45	0.045
Accessories	A/C and heater	Neither works	Both work	2	0.25	3	0.3	0.075
	Seating capacity	2	6 or more			2	0.3	0.075
	Sound system	None	Radio and CD player			1	0.4	0.1
Sum					1			

This hierarchical approach has an added benefit in decision contexts that involve different levels of executives and managers. The top-level executives might focus only on the relative importance of the fundamental objectives. SMEs or mid-level managers might assess weights of the subobjectives or measures. There may even be different groups of managers asked to weight measures within each objective. Marketing would focus on customer-related attributes. Finance or engineering might focus on product or cost issues. In this way, each aspect of the decision would benefit from the input of expert evaluators.

5.8 Assess Individual (Single-Measure) Utility Function: Nonlinear Utility Functions and Constructed Measures

A linear utility function assumes that the desirability of an additional unit of a measure is constant for any level of that measure. This assumption is not correct for some real-world measures. For example, suppose you are purchasing office space. One important measure is the floor area of the office. You think that about $1500\,ft^2$ is adequate for your working conditions, but, if necessary, you could manage with as little as $1000\,ft^2$. The square footage of the alternative offices is between 1000 and $2000\,ft^2$. It may be very important to you to increase the space from 1000 to $1500\,ft^2$ because this increase would significantly improve working conditions. However, an increase from 1500 to $2000\,ft^2$, although attractive, is of marginal value to you. In this example, the relation between floor area and utility score is not linear. A utility function as depicted in Figure 5.9 may represent the nonlinear relation between floor area and utility score. The concave curve is much steeper in the range from 1000 to 1500 when compared to the range of 1500–2000. The utility score for $1500\,ft^2$ would be 0.5 rather than 0.75 if the measure were strictly proportional. The same issue of nonlinearity might arise when using the number of bedrooms as a measure for houses that you are considering to buy. Similarly, you may apply a nonlinear utility function to waiting time on the phone.

Activity 5: Which Measures Would You Scale with a Nonlinear Utility Function?

- Home choice: three, four, or five bedrooms
- Car acceleration time 0–60 mph: range 6.5–9 seconds
- Waiting time on phone with customer service: range 0–12 minutes
- Gas Mileage: range 20–30 mpg on a highway

Suggest a measure in your work environment that would have a nonlinear utility function

FIGURE 5.9: Utility function for office floor area.

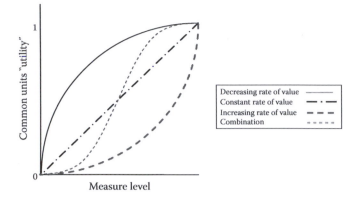

FIGURE 5.10: Possible utility functions.

There is a wide variety of nonlinear utility forms. The most common used utility functions take on the shapes depicted in Figure 5.10. These graphs apply to measures for which higher values mean higher utility.

Below we provide a short discussion and motivation for each curve.

Decreasing rate of value (concave): Small increments in the measure from the least preferred level add "significant" value. Less and less value is added as the most preferred level is approached. The floor space example was typical of this curve type. An increase in the measure from 1000 to 1100 ft² was significantly more important than an increase from 1900 to 2000.

Increasing rate of value (convex): Small increments from least preferred level add "little" value. As the level improves, however, each additional fixed increment has progressively greater value. The largest incremental value occurs as the measure approaches its most preferred value. For example, imagine the selection process in the first round of the NBA draft. The difference between the 24th and the 23rd pick is small when compared to the difference between the second and first pick.

S-Shaped combination: Small changes occur in the utility score as you move away from the least preferred value. Similarly, small changes occur in the utility score as you approach the most preferred. Assume that most customers are not bothered by waiting on line for a few minutes. However, once the wait is longer than 5 minutes, customer satisfaction declines significantly. In this case, reducing the wait from 12 to 11 minutes gains very little in customer satisfaction. In addition, once the wait is reduced to only 1 minute, there is little additional value in reducing it further.

Targets: The shape of the utility function depends on the decision context and personal preferences. There is no right or wrong utility function for a measure. However, there is a bias that can distort the shape of the curve inappropriately. Often, managers have a target value in mind for a specific measure. In some instances, these are official targets established by senior management. They perceive almost any substantive drop below the target value as driving the utility score to near zero. They feel that there is little added value in exceeding the target score. As a result, their utility function is S shaped but extremely steep around the target value. The thinking that leads to this type of function that overemphasizes a target value tends to make it difficult to make trade-offs between measures and should generally be avoided.

Consider a computer company that wants to increase the number of laptop sales from 1,000,000 to 1,200,000 units over the next year. There is pressure on the marketing department to reach the sales target. The department predicts that laptop sales will vary between 1 and 1.4 million units, depending on the strategy they follow. As laptop sales approach the target value (1.2 million units), more incremental value is added, as depicted in Figure 5.11. However, once the target is achieved, less incremental value is added as it approaches the most preferred level. The solid line depicts an extreme target bias. The dotted line would be more appropriate if there were a need to trade-off sales and marketing cost.

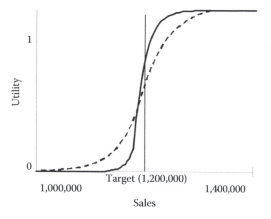

FIGURE 5.11:   Extremely nonlinear utility function for laptop sales.

A public announcement of a corporate goal of profitability is an example of a target that can distort decisions especially if the company has been unprofitable for several years. As the company approaches profitability, the utility score increases dramatically. Therefore, executives will make decisions that over emphasize immediate benefits in order to surpass this hurdle and not disappoint the stock market. However, once the company achieves profitability after a long absence, the amount of profit in that year is not that important.

In activity 6, you are asked to identify a measure and a relevant decision context in which a utility function is concave and another in which it is convex.

Activity 6: Concave or Convex Measures

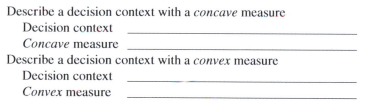

Describe a decision context with a *concave* measure

    Decision context   _____

    Concave measure   _____

Describe a decision context with a *convex* measure

    Decision context   _____

    Convex measure   _____

5.8.1   Mid-Level Splitting (MLS) Technique

How does one create a utility function? In this section, we introduce the MLS method to generate utility functions.

If the decision makers are uncomfortable with the abstract concept of a utility, they should not spend too much time trying to specify a nonlinear curve. In addition, if there are multiple decision makers and experts involved, spending too much time obtaining consensus on nonlinearity could undermine support for the decision-making process. Becoming bogged down in minutiae in order to improve modeling accuracy may undermine the broader goal of gaining support for a process. Ideally, the decision-making process should be perceived as adding value without being too complex. Thus, the exploration of nonlinearity should be limited to only the two or three most highly weighted measures and carried out in an efficient manner. However, for relatively rare decisions of major magnitude, in which decision-making speed is not important, the MLS method for a critical, highly weighted measure might proceed to greater levels of detail. The U.S. military, which develops major new weapons systems with total life-cycle costs in the hundreds of billions of dollars, must also balance life and death issues. For this reason, it often proceeds to carry out an in-depth exploration of nonlinear utility functions.

When developing a single utility function, the decision maker or SME should review the four forms presented and decide qualitatively which of the forms best represents his thinking about the measure. This will help him consistently apply a procedure such as MLS to determine specific values along the curve. MLS is a process that determines the measure levels that correspond to three specific points on a utility function: 0.25, 0.50, and 0.75. After assigning 0 and 1 to the range of least preferred and most preferred levels, the interviewee is asked to determine the level of preference that is midway between least preferred and most preferred. The mid-preference level is identified by establishing two changes in a measure level that have equal utility to the decision maker or SME. Once the mid-preference level is established, 0.5 is assigned to that point. Note that mid-preference level does not necessarily represent the average of the two ends of the range. The measure levels for 0.25 and 0.75 are identified in a similar way. In the following, we present the process of the MLS method.

- Divide utility range between 0 and 1 into equal intervals.
- Determine measure level with 0.5 utility.

In most instances, we recommend stopping at this point. This is the point that determines whether the curve is concave or convex. If necessary, however, proceed to the following steps:

- Determine measure level with 0.25 utility.
- Determine measure level with 0.75 utility.

When using the MLS method, the SME or decision maker answers a series of questions about changes in measure until mid-level is established. We introduce this technique with an example in which Nancy Chicila of MONEYMARK is deciding on the number of customer service representatives to hire. One of the measures is customer waiting time. The range is from 0 to 12 min.

- Let $M = $ (Best level + Worst level)$/2 = 6 = $ midpoint of total range
- $U(0) = 1$ and $U(12) = 0$
- Ask the decision maker or SME which change produces a greater value improvement.
  - Change 1: Improve from 12 to 6 min
  - Change 2: Improve from 6 to 0
- If, for example, the answer is that Change 2 has a greater impact, this implies
  - $U(6) - U(12) < 0.5$ and $U(0) - U(6) > 0.5$
  - Because $U(0) = 1$ and $U(12) = 0$ then
  - $U(6) < 0.5$ and the value whose utility is equal to 0.5 must be a number less than 6 min $\rightarrow X_{0.5} < 6$.

In Nancy's opinion, unless the waiting decreases to less than 5 minutes, the utility score does not reach 0.5. She sets 4 minutes as the 0.5 level. This means that an improvement from 12 to 4 minutes is equivalent to an improvement from 4 minutes to zero waiting time. Thus, $X_{0.5} = 4$. This point is then added to the utility curve as illustrated in Figure 5.12.

The decision maker should review his defined curve and decide if this characterization adequately captures this nonlinear utility function. If not, he would repeat the process to identify the values of X with utilities equal to 0.25 and 0.75. To specify the first and third quartiles of the utility function, we repeat the same process we used to determine the mid-preference level, whose utility is 0.5. To determine $X_{0.25}$, we first look at the mid-point between 12 and 4 min, namely 8 min. If he views the 4 min reduction from 12 to 8 of approximately equal value to an improvement from 8 to 4 min, then 8 would be the first quartile value: $X_{0.25} = 8$. However, Nancy feels that this is not the case. The waiting time would need to drop to 7 min before the two changes would be of equal value. The reduction

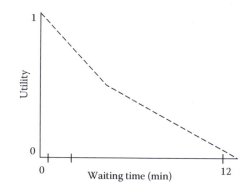

FIGURE 5.12: Nonlinear utility function for waiting time.

from 12 to 7 minutes would be of the same value as a reduction from 7 to 4 minutes. She assigned a utility of 0.25–7 minutes. Following is a summary of the process to specify the first quartiles:

- Determine $X_{0.25}$
- In this example, $X_{0.5}=4$
  - Calculate $(12+4)/2=8$
    - Change 1: Is there greater value in improving from 12 to 8?
    - Change 2: Or is there greater value in improving from 8 to 4?
  - Nancy preferred Change 2. Thus, $X_{0.25} < 8$, and she set $X_{0.25}=7$.

She also wanted to specify $X_{0.75}$, the waiting time with a utility of 0.75.

- Again, $X_{0.5}=4$
  - Calculate $(4-0)/2=2$
    - Change 1: Is there greater value in improving from 4 to 2?

Or

    - Change 2: Is there greater value in improving from 2 to 0?

Nancy viewed change 2 as more significant, since it eliminated almost all waiting time. She then sets the midpoint at 1.5 min, which means that $X_{0.75}=1.5$.

After identifying the cost levels corresponding to utility scores of 0.25 and 0.75, the utility function is created as illustrated in Figure 5.13.

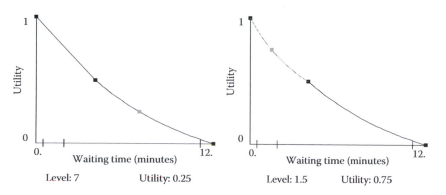

FIGURE 5.13: Nonlinear utility function for waiting time.

TABLE 5.13:   Nonlinear utility scores
for customer waiting time.

| Servers | Time | Utility | |
		Linear	**Nonlinear**
4	11.1	0.08	0.03
5	1.9	0.84	0.70
6	0.5	0.96	0.91

Ranking for overall goal: nonlinear utility waiting time

Alternative	Utility
5 staff	0.626
6 staff	0.558
4 staff	0.419

■ Waiting time □ Cost

FIGURE 5.14:   Nonlinear utility and customer service staffing example.

The nonlinear utility for waiting time is summarized in Table 5.13. For example, the utility of 1.9 minutes was 0.84 with a linear function and 0.70 with nonlinear utility function. The nonlinearity changed the overall score for each of the alternatives (Figure 5.14). The score for the five-staff option declined from 0.71 (Figure 5.7) to 0.63. The second best option went from 0.59 to 0.56. Nevertheless, the overall ranking was not affected by the introduction of a nonlinear utility function for waiting time.

5.8.2   Direct Assessment Method for Constructed Measures

The MLS method can be used only for continuous measures. But many measures have constructed scales or discrete numerical measures. In cases of this sort, the direct assessment method is used to create a utility function.

Consider the aforementioned simple bulb selection example. Bulb performance has three levels: 60, 75, and 100 W. Mr. Frails assigns 0 and 1 for the worst (60) and best level (100), respectively. He needs to identify the utility score for 75 W. Mr. Frail must ask himself, "Is an improvement from 60 to 75 more important than the improvement from 75 to 100?" If the answer is that the improvement from 60 to 75 is more important, the utility score of 75 must be greater than 0.5. He then assigns a numeric value keeping in mind this stated preference. Conversely, if Mr. Frail feels that the improvement from 75 to 100 is more important, the utility score of a 75 W bulb must be less than 0.5, and he then proceeds to assign a specific utility score (less than 0.5) to the 75 W value, again keeping in mind this stated preference.

Activity 7: Direct Assessment of Bedrooms

Directly assess the utility of the number of bedrooms in a house purchase decision. The possible values are three, four, or five bedrooms. More bedrooms are preferred. Since five bedrooms are best, it has a utility of 1. Three bedrooms are least desirable and have a utility of 0.

- SUF(5) = 1
- SUF(3) = 0
- SUF(4) = ?

- Which change produces a greater value improvement?
  - Change 1—Improve from three bedrooms to four
    - $SUF(4) > 0.5$
  - Change 2 – Improve from 4 bedrooms to 5
    - $SUF(4) < 0.5$
- Which increase is of greater value to you? _____
- Specify $SUF(4) =$ _____

5.9   Group Decision Making

Group decision making is common in corporate, government, and engineering management. It often involves multiple stakeholders from different parts of the organization or different segments of the community (Edwards 1977). Product design choices within one part of product development, such as the power supply module team, often affect other product development teams whose systems must use the power supply. These choices also directly impact manufacturing and final assembly. In many companies, the decision-making team would also include representatives from finance, who critically review all investment costs. Representatives from marketing are included as well to make sure the product is aligned with customer expectations and desires.

Similarly, city council members represent different constituencies from diverse socioeconomic groups as well as from different geographies. They must take on challenging decisions that will affect their constituencies in diverse ways. These decisions can include facility location, neighborhood development, resource allocation, and taxes. Different individuals will generally have different objectives and measures as well as preferences as to how much weight to assign these variables. As a result, conflicts arise naturally in the decision-making process. It is not surprising that the early applications of MAUT involved public sector decisions and that public power companies were early organizational adopters.

Based on their experience, the authors of this text claim that influence diagrams and MAUT facilitate the process of decision making by clarifying explicitly the differences between individual decision makers. In contrast, decisions made based on gut feelings and personal experience provide only a limited foundation for discussion and do not foster a drive toward consensus. Such decision making amounts to little more than a jumble of personal expert assessment of different factors, as well as personal preferences as to how to weight these factors. The most the group can hope for is to refute each other's best arguments and obtain a significant majority in favor of one decision or the other.

When it comes to making decisions, a core problem with groups is that the human mind tends to focus on one or at most two measures—generally, those measures that the individual considers in his best interests. As such, purchasing will focus on piece price, finance on capital investment, product engineers on high-tech performance, and so on. However, when a structured approach is taken and the group must consider a long list of measures to weight, one rarely hears someone to say, "I want to put all of the weight just on piece price," or on quality, or on marketing. For example, with a list of 10 measures, the range of weights assigned to piece price might vary from 2% to 25%; this is a significant difference, but no one will assign 50% to piece price and leave the other nine measures to divide up the remaining 50%.

The following experience is typical. A company was considering three alternative suppliers. The existing supplier tested well, with high quality, but the other two were lower priced. However, the quality capabilities of the other two suppliers were marginal because of their lack of experience with the complex component at hand. No one wanted to be pinned down to assign a specific dollar value to the losses that poor quality might incur. However, when the analysis was carried out, everyone agreed that at least 20% of the weight should be assigned to quality. As a result, it became obvious that the existing supplier was preferred.

This illustrates an important point. Even if the group cannot agree on the exact weights for a measure, they might agree on a range. With the existing software tools, it is possible to run a model with multiple alternative weights to see whether the rankings are highly sensitive to specific weights. In addition, sensitivity analysis can assess the robustness of the rankings to changes in the weights.

MAUT has two important elements that facilitate group decisions. First, it separates the data collection and expert judgment from the weighting process. The performance of each alternative with regard to every measure is addressed by experts for that measure. Manufacturing costs are specified by experts in manufacturing, market forecasts are determined by marketing, and product performance is determined by product engineers. This approach does not automatically eliminate all forms of disagreement, but it does focus the disagreements within very specific parameters that can be easily understood. Moreover, the data collection derived from diverse experts creates a natural tendency for the various members of the team to buy into the results of the final analysis.

Second, the decision makers, without focusing on the specific strengths and weaknesses of their preferred alternatives, explore in more abstract terms their respective motivating values for the specific decision. The crux of their disagreements will be captured in the weights they assign to the various measures. It will become obvious to all what is driving the lack of consensus. It is common in our experience to find that in a group of eight decision makers, there will be broad agreement among six of those present with no more than two members of the group placing significantly more weight on a specific measure. For example, finance might be the only division of the group to assign an unusually high weight to investment cost. However, even significant variability in weights does not necessarily presage disagreements on the ranking of the best alternative.

It is important to incorporate the perspectives of all stakeholders—even those who may not be part of the formal decision-making group—into the decision-making process. There are three distinct contexts that allow for potential individual input.

1. *Data input:* Different experts assess the performance of each alternative on each specific measure by gathering hard data or by offering subjective assessments. This affords the different organizational groups an opportunity to provide input into the data matrix. These individuals are unlikely to be part of the formal decision-making group.

2. *Single utility function for each measure—nonlinear:* For example, if a particular measure were related to overall product performance, the technical expert on product performance might assess the shape of the utility function. If, on the other hand, the measure were a factor in a customer's purchasing decision, a marketing expert might shape this utility function. These individuals may also be part of the final decision-making group that establishes overall weights.

3. *Weights:* This is the primary place for all key stakeholders and organizational groups to present their differing perspectives as to how to make the decision.

The process proceeds as follows:

1. *Data matrix:* Ask specific experts, and not decision makers, to assess the performance of each alternative for each measure. You might have one expert on fuel economy or you may require distinct experts who understand the performance of the individual technologies under consideration. Similarly, if you were considering a range of materials you may need multiple experts, each knowledgeable as to the characteristics of the specific materials under consideration.

2. *Utility function:* Assume linearity or ask an expert to shape each individual utility function. The expert may be a technical individual, or someone who can reflect customer, citizen, or organizational preference.

3. Ask decision makers to assign their respective weights along with an explanation of their reasoning.

    a. Clarify the differences in weights.

    b. Strive to achieve consensus, but be satisfied with a result that yields two or even three sets of disparate weights.

    c. Also, create a set of weights that reflects the average of the group.

4. Run the MAUT model using each set of weights as well as the average, and rank order the alternatives.

    a. Often, despite differences in weights, two alternatives out of a long list will appear at the top of everybody's list.

    b. Determine how the weights affect the rank ordering of the top two alternatives.

5. Discuss in depth the strengths and weaknesses of the top two alternatives

6. Explore the possibility of developing a new hybrid solution that would be best on both weighting scales, or use the average to break any deadlocks.

Alternatively: Consider a case in which the top two alternatives earn scores of 0.72 and 0.70, with the scores reversed depending upon the weighting preferences. Assume all other alternatives score 0.65 or less regardless of the different perspectives. There is a natural tendency, especially among quantitative thinkers, to want to determine the absolute best alternative. However, given the subjective nature of the MAUT process and the use of expert judgment, the proper response is to consider the top two alternatives as equally good essentially. There really is no need to argue endlessly over which is better. Instead, the decision maker should consider some other organizational factor that was unquantifiable to make the final selection or decide by a coin toss. The more important issue at this stage is that everyone agrees that the choices are essentially equally desired. They should then be willing to implement the final decision without feeling their preferred alternative was dismissed.

5.9.1 Mathematical Aggregation (No Consensus on Weights)

Although an average weight does not necessarily reflect the group's preferences, it is an easy starting point for discussion and analysis. The average can be determined easily and without a group meeting. Ferrell (1985) suggested that if the individual group judgments are unbiased, the simple average of the individual estimates is the best way of aggregating the preferences. LDW allows one to keep the judgments of individual group members as well as to incorporate a separate function for the group average. If the rank ordering of the top two or three alternatives among individuals is not extremely disparate, the group may be comfortable agreeing to use the average to break the deadlock. People understand the concept of average and are comfortable with its usage, although Nobel laureate Kenneth Arrow has observed that the use of averages can violate certain postulates of good decision making. (See Section 5.9.2.)

Another strategy is to strive for consensus early on while setting the weights. Companies that use MAUT may bring in a facilitator to drive the weighting toward consensus. The Delphi method is a well-documented process for striving for consensus among experts albeit it is very time consuming (Hasson et al. 2000). It was originally designed to develop consensus forecasts from a group of experts but more recently has been used in the medical field and for other public policy issues (Powell 2003; Angus et al. 2003).

The Delphi process follows a cycle of anonymous individual statements and aggregated feedback. Members do not initially meet face to face, and equal participation is structured by the use of written questionnaires. A problem is identified, and members are asked to anonymously provide their solutions through a questionnaire. Questionnaires are repeatedly administered to group members for revision and are intermixed with feedback from questionnaire summaries until consensus is reached. The absence of group discussion avoids voice dominance and reduces group pressure to conform, although the benefits of group synergy are sacrificed in this process. Experience indicates that this process can produce significant convergence.

In the context of MAUT, we suggest the process first begin with a group meeting to obtain agreement on the measures to be used as well as to explain the weighting procedure. Only afterward should a form be used by each decision maker to assign initial weights, along with an explanation of his preferences. The Delphi process can then be implemented to provide group feedback and to ask for revisions.

Alternatively, the group can work toward consensus in a meeting at which individual ideas are gathered and combined. Each member is first asked to make suggestions; critiques are only allowed after the first round of statements is completed. Group members provide written and individual judgments, followed by score aggregation, discussion, and possible new judgments until consensus is reached.

5.9.2 Arrow's Impossibility Theorem

We just argued in favor of a rather loose process for reaching consensus, but is there no simple formula that can be used to aggregate the various preferences in order to create a *group* preference? The answer Nobel laureate Kenneth Arrow (1951) found is that there is no mathematical way to create a consistent group preference. He studied various strategies for determining group preferences when the preferences of individual members are expressed as orderings. He identified the properties that a satisfactory aggregation procedure should have: universal domain, completeness and transitivity, positive association, independence of irrelevant alternatives, nonimposition, and nondictatorship.

In Arrow's impossibility theorem (or Arrow's paradox), Arrow proved that if the decision-making group consists of at least two members and there are at least three alternatives to decide among them, it is impossible to design an aggregation procedure that satisfies all these conditions at once. This means that there is no aggregation rule combining several individuals' rankings of alternatives to obtain a group ranking that will simultaneously satisfy all six properties.

We demonstrate Arrow's impossibility theorem with an example using a simple average method and majority rule vote. Suppose three SMEs in a group, SME1, SME2, and SME3, must agree on the selection of a particular technology. Three technologies, A, B, and C, are available, and the SME's preferences are depicted in Table 5.11. For example, SME1 prefers A to B and B to C.

A is the first choice of SME1, second best alternative for SME3, and third for SME2. On average, A is the second best alternative. When a simple average technique is used, all alternatives are ranked as second best, as shown in Table 5.14. None of the alternatives is ranked first or third in this example.

A group might then consider deciding by majority rule. When the majority rule vote is used, SME1 and SME3 prefer A to B and hence A gets two votes and B only one. This implies that $A > B$ (note that $>$ means is "preferred to"). When B and C are compared, B gets two votes and C only one, which implies that $B > C$. Finally, if we compare A with C, C gets two votes and A only one, which implies that $C > A$. So not only do we have $A > B > C$ but we also have $C > A$, which implies

TABLE 5.14:   Arrow and average of preferences.

	SME 1	SME 2	SME 3	Average
A	1	3	2	2
B	2	1	3	2
C	3	2	1	2

TABLE 5.15:   Arrow and majority rule vote.

	Yes	No	Result
A > B	2	1	A > B
B > C	2	1	B > C
A > C	1	2	A < C

that the preferences of the group are not transitive. This result is known as Condorcet's paradox (Table 5.15).

Arrow's theorem states that any aggregation scheme will have shortcomings. Given that no method is perfect, it is important to consider reasonable alternative approaches that recognize group differences. The good news is that standard software packages allow the analysis to be carried out from multiple perspectives without complicating the analysis.

5.10   Uncertainty

Uncertainty is a fact of life in many, if not most, substantial decision contexts. The early development of MAUT included formal procedures for assessing trade-offs in the presence of uncertainty. In a multi-objective decision, uncertainty may apply to a single hard-to-predict measure for one or more of the alternatives. For example, in a choice of technology alternatives, a measure that describes how long it will take to test out and fully implement the technology would be uncertain. When choosing among suppliers, the projected warranty costs would also be uncertain. Uncertainty can be represented in two distinct ways.

Probabilistic data: The data for an uncertain measure can be represented as a probability distribution. If the utility score is proportional, then the expected value is sufficient. For a non-linear utility function, the analysis would use the expected utility for that measure. LDW has an option that simulates the uncertainty and presents a score range for each alternative that involves uncertainty.

Risk measure: An alternative approach captures uncertainty with a separate risk objective. This single objective is defined as the total risk associated with each alternative. This objective will have a constructed measure. The measure can have levels such as low, moderate, and high risk. In general, the risk measure is likely to reflect a lack of information. For example, when considering alternative suppliers, the decision maker might rate his current supplier as a low risk, a supplier with much experience with other companies as a moderate risk, and a relatively new supplier as a high risk. When considering the appointment of a new manager, a current employee may pose less risk than an outside hire.

5.11   Contractor Selection for Kitchen Remodeling

Let us consider once again the Lyons' selection of a contractor for kitchen remodeling. The Lyons contacted four different potential builders. In the process of obtaining bids and talking to friends, they came up with three broad categories of goals: minimize cost, minimize hassle of construction and follow-up, and maximize quality of the kitchen. At the bottom, they ended up with 15 measures, as listed in the Table 5.16. The Lyons planned on asking for five references from each candidate and following up with three of each set of references by phone and, if possible, with a visit to one of the houses to see the kitchen. They eventually received visits and quotes from four kitchen remodeling contractors. However, one of the four's references did not check out well and he was dropped from the list. Because they had three separate sets of references, they had three data points on the issues of cost overrun and on-time delivery. They then input these data points into the model as three equally likely values. (Logical Decisions requires the probabilities add exactly to one. Therefore, we used 0.33, 0.34, and 0.33.)

The Lyons collected data on each measure for each contractor by reviewing the contactors' proposals, interviewing the references, and seeing the contractors' previous works, as shown in Table 5.16. After gathering the data, the Lyons did not feel the warranty range of differences, 7–9 years, was significant and deleted this measure from the analysis. They also were not sure whether the data on worker experience was particularly valuable or reliable, and this too was excluded. As such, they used only 13 measures to rank the alternatives.

To determine the relative importance of each measure, the Lyons specified the least and most preferred levels of each measure (see Table 5.17). They ranked the measures by reviewing the measures

TABLE 5.16:   Measures and relevant data for kitchen remodeling.

Measure	Build Rite	Quality Build	Cost Conscious
Total labor cost	$34,000	$26,000	$25,000
Total material cost	$20,000	$12,000	$10,000
Cost overrun history	0% ($p=0.33$)	2% ($p=0.33$)	6% ($p=0.33$)
	2% ($p=0.34$)	5% ($p=0.34$)	9% ($p=0.34$)
	7% ($p=0.33$)	9% ($p=0.33$)	15% ($p=0.33$)
Duration kitchen unavailable	13 weeks	10 weeks	9 weeks
Weeks of delay	On time ($p=0.33$),	1 week late($p=0.33$)	2 weeks late ($p=0.33$)
	1 week late ($p=0.34$)	2 weeks late($p=0.34$)	3 weeks late ($p=0.34$)
	2 weeks late ($p=0.33$)	3 weeks late ($p=0.33$)	4 weeks late ($p=0.33$)
Cleanliness scale	Clean	Messy	Dirty
Follow-up and resolution scale	Adequately responsive	Highly responsive	Adequately responsive
Creativity scale	Highly creative	Creative	Mundane
Brand and store reputation scale	Top of line	2nd best brand	2nd best brand
Percent use of subcontractors	25%	40%	65%
Fit and finish scale	Excellent	Good	Good
Years in business (grouped in ranges)	12 (good)	8 (ok)	22 (excellent)
Quality of references scale	Excellent	Good	OK

TABLE 5.17: Relative importance of measures for kitchen remodeling.

Objective	Measure	Least Preferred	Most Preferred	Rank Order	Swing Weight	Final Weight	Objective Weight
Cost	Total labor cost	35,000	25,000	4	80	0.10	0.27
	Total material cost	20,000	10,000	4	80	0.10	
	Cost overrun history	15%	0%	7	55	0.07	
Hassle	Duration kitchen unavailable	15	7	6	60	0.08	0.22
	Weeks of delay	5	0	9	40	0.05	
	Cleanliness scale	Dirty	Clean	11	20	0.03	
	Follow-up and resolution scale	Need improvement	Highly	8	50	0.06	
Quality	Creativity scale	Mundane	Highly creative	1	100	0.13	0.51
	Brand and store reputation scale	Common brand	Top of line	3	90	0.12	
	Percent use of subcontractors	70%	20%	12	10	0.01	
	Fit and finish scale	OK	Excellent	2	95	0.12	
	Years in business, but grouped	OK	Excellent	10	30	0.04	
	Quality of references scale	OK	Excellent	5	70	0.09	

FIGURE 5.15:   Stacked bar results for kitchen remodeling: multiple objectives.

FIGURE 5.16:   Stacked bar results for kitchen remodeling: uncertainty.

with the least and most preferred levels. The Lyons ranked the creativity measure first and assigned it 100 points. Their most preferred goal was to improve creativity from mundane to highly creative. The fit and finish measure was ranked second and assigned 95 points, followed by brand and reputation. Reducing material cost from $20,000 to $10,000 and reducing labors cost from $35,000 to $25,000 was ranked fourth. (Note that both measures have a range of $10,000.)

The seventh column in Table 5.17 illustrates final weight of each measure. The creativity measure collects 13% of the total weight; while the percent use of subcontractors measure accounts for only 1% of the total weight. The quality objective collects half of the total weight (51%) as depicted in Table 5.17. The remaining weight is shared by cost (27%) and hassle (22%) objectives.

The Lyons used linear utility functions for all numeric measures. They used the default assumption also for nonnumeric measures that were categories. The default assessments approach assigns 1 to the most preferred category, 0 to the least preferred category, and 0.5 to the middle category for a three-level measure.

Figure 5.15 illustrates the total scores of alternatives and their rankings. Build Rite is ranked first with a utility score of 0.651 and closely followed by Quality Build (0.630). Built Rite is extremely strong on quality, but the weakest on cost. Quality Build is the best in terms of hassle and ranked second on cost and quality objectives. Cost Conscious is ranked third with a utility score of 0.462. Cost Conscious is the best in terms of cost, but weak on quality and hassle objectives.

There was uncertainty associated with two measures: cost overrun and weeks of delay. The uncertainty can affect the rank ordering of the best two contractors, Build Rite and Quality Build, as shown in Figure 5.16. There is overlap in the two ranges and, as a result, Quality Build could be ranked first. However, a careful review of the uncertainties suggests that Build Rite should still be preferred. For example, with regard to cost overruns Build Rite stochastically dominates Quality Build. Its three possible values, 0%, 2%, and 7%, are each better than the corresponding Quality Build values, 2%, 5%, and 9%. Nevertheless, since the two random measures are independent, Quality Build could still come out ahead with regard to cost overrun. A similar pattern applies to weeks of delay.

5.12   Real-World Application: Multi-Attribute Risk Analysis in Nuclear Emergency Management

Hamalainen et al. (2000) implemented MAUT in nuclear emergency management and planning, to deal with conflicting objectives, multiple stakeholders and uncertainties. The MAUT was used

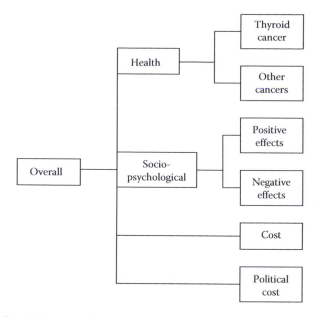

FIGURE 5.17:   Goals hierarchy for nuclear emergency management case.

to evaluate and rank nuclear emergency management strategies for protecting the population after a simulated nuclear accident. The study was part of Real-time On-line Decision Support (RODOS), a European Union project on developing a support system for nuclear emergency management. The RODOS project was designed to assess and predict the consequences of an accident, and to support decision makers in choosing proper countermeasures.

Four meetings, half a day long each, were organized in Finland in 1997 to develop early-phase countermeasure strategies for protecting the population. National nuclear authorities and technical experts who assess the situation and advise higher level political decision makers attended the meetings. The meetings focused on urgent protective actions such as iodine prophylaxis, providing shelter, and evacuation. The primary goals were to test the RODOS system as well as to study and extend the applicability of decision support system for different situations.

During the meetings, the participants used brainstorming to identify all factors to consider when making decisions regarding countermeasures after a nuclear accident. After eliminating irrelevant factors and combining similar factors, they constructed the final goals hierarchy as shown in Figure 5.17.

The team identified five (unranked) countermeasure strategies, as defined in the following:

0. No additional countermeasures taken.

1. Distribute iodine tablets and provide shelter in Rauma, a city of 30,000 inhabitants and 12 km south of the NPP.

2. Provide shelter in Rauma and the closest areas around that city, and distribute iodine tablets within a radius of 100 km away from the site.

3. Provide shelter in the same area as in Strategy 2, but distribute iodine tablets to all areas affected by the accident, including areas beyond 100 km away.

4. Evacuate Rauma after the cloud has passed the area, with provision of shelter and distribution of iodine tablets during the plume passage.

TABLE 5.18:   Relative importance of measures for nuclear emergency management case.

Measure	Least Preferred	Most Preferred	Final Weight
Thyroid cancer	240	0	0.33
Other cancers	320	0	0.26
Positive effects	0	100	0.03
Negative effects	100	0	0.10
Costs (millions)	180	0	0.03
Political cost	100	0	0.26

SMART swing weight method was used to specify relative importance of the measures. (See Table 5.18.) Thyroid cancer measure received one third of the total weight, as the decision makers are more interested in improving this measure (number of incidents) from its least preferred level (240) to the most preferred level (0). The decision makers were also very concerned about other cancers and political cost measures. Each of these measures accounted for 26% of the total weight. In this case study, reducing cost from 180 million Euros to 0 Euro received only 3% of the total weight.

Nonlinear utility functions were used for two cancer measures as well as for the cost measure, as depicted in Figure 5.18. The decision makers are very concerned about thyroid and other cancers. The attractiveness of a strategy rapidly diminishes as the number of cancer incidents increases, as shown in Figure 5.18a and b. In contrast, the decision makers are less worried about the increase in cost from its least preferred level. As cost exceeds 130 million Euros, its utility declines rapidly.

Figure 5.19 shows total scores and rankings of the nuclear emergency management strategies for base case scenario. Strategy 0 was worst in terms of thyroid cancer and strategy 4 was worst in costs. The other strategies scored about equally on the cost and cancer measures. Strategy 3 was ranked first with a utility score of 0.783 and was closely followed by strategies 2, 1, and 4. Strategy 0 was ranked fifth with a total score of 0.566.

The decision makers and experts assessed the impact of a nuclear accident on measures under each nuclear emergency management strategy. In the base case scenario analysis, they used their realistic (likely) estimates with respect to impact magnitude. As the decision makers were also concerned about the worst possible impact of an accident on each measure, they ranked the strategies for the worst case scenario as shown in Figure 5.20. In the worst case scenario,

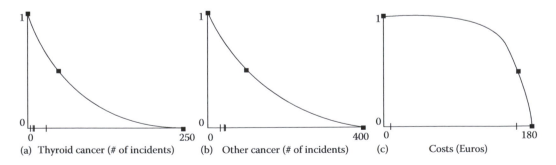

(a) Thyroid cancer (# of incidents)   (b)   Other cancer (# of incidents)   (c)   Costs (Euros)

FIGURE 5.18:   Nonlinear utility function for nuclear emergency management case.

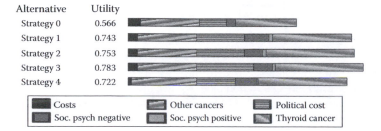

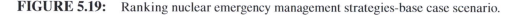

FIGURE 5.19: Ranking nuclear emergency management strategies-base case scenario.

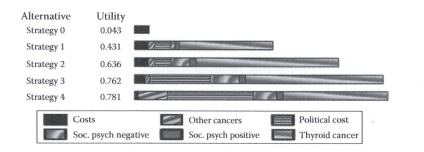

FIGURE 5.20: Ranking nuclear emergency management strategies-worst case scenario.

strategy 4 was ranked first with a utility score of 0.781, followed by strategy 3. Strategy 0 received a total score of only 0.043 in the worst case scenario while its score was 0.566 in the base case scenario. Irrespective of scenario analysis, strategies 3 and 4 received higher total scores in this analysis.

5.13 Selection of Best Conformal Coating Process

Global Electronic decided to install a new conformal coating process in 2002. (Corporate name and data changed.) Conformal coating is applied to electronic circuitry to protect it from moisture, dust, chemicals, abrasion, and temperature extremes that, if uncoated, could result in a failure of the electronic system. John Smith, vice president of Global Electronic, created a five-person team, the coating process selection team (CPST), to identify and evaluate possible processes. The CPST prescreened a number of available processes and reduced the number of viable candidates to three processes: selective spray coating, sil-gel potting, and conformal coat and extract. The processes are described in the following text box.

In the coating process selection case study, the team identified nine measures. The team realized that these processes vary widely in flexibility, weight, coating control, foreign material, facilities and tooling cost, labor cost, material cost, scrap cost, and process development time. They identified four major objectives: maximize performance, maximize reliability, minimize cost, and minimize development time. Since the problem involves conflicting objectives, they decided to use MAUT to evaluate alternatives and select the best coating process.

Conformal Coating Process Selection

Global Electronic must install a new conformal coating process due to upcoming Powertrain Control Module (PCM) design requirements that are incompatible with the existing process at the plant. These coatings are applied to the printed wiring boards to protect circuitry from environmental exposure after the installation of all surface mount devices, but before final assembly of the module. The process should ideally be capable of selectively applying the coating to various areas of the circuit board, coating some areas while avoiding others.

Prescreening of a wide variety of available processes has reduced the number of viable candidates to three processes: Selective Spray Coating, Sil-Gel Coating, and Conformal Coating and Extract. A description of these follows:

Selective spray coating: This process involves using spray equipment. After surface mount devices and in-circuit test processes, the PWB is sprayed with the coating. The equipment has the capability of applying a bead of conformal coating that can protect keep-out areas (such as the heat sink region) from being sprayed. This equipment also has the capability of spraying both sides of the circuit board, so that separate lines are not required. The coating is cured in an Ultra Violet oven after completion of the spraying.

Sil-gel potting: With this process, PWB is laminated to the housing after service mount devices and in-circuit test processes. Sil-Gel is injected into the bottom cover under the circuit board using slots located at the end of the connector. Sil-Gel is also placed on the top side of the circuit board using an amount sufficient to cover all exposed pads, leads, and connector pins.

Conformal coating and extract: Similar to Sil-Gel, the PWB is laminated to the housing after service mount devices and in-circuit test processes. The conformal coating process is used, but this is followed by a process that extracts as much conformal coat out of the bottom housing as possible, to minimize the amount of material used per module. The extracted material is mixed with virgin material and reused.

How can Global Electronic evaluate the available coating processes? What process should Global Electronic select?

They identified eight objectives and nine measures, as illustrated in Figure 5.21. The measures associated with weight, cost, and time have natural metrics such as grams and dollars. Coating control has only two levels, while flexibility and foreign material have three. Since coating control has two levels, a binary metric is used to scale coating control. Flexibility is a measure of how readily the chosen process can accommodate mechanical design changes that may occur in the future. The constructed scale refers to how adaptable the equipment is to design changes and the amount of original investment that could be preserved if a change were required. Foreign material has an adverse effect on the reliability of the module, especially if the material is metallic based. Nonmetallic particles are somewhat less of a concern, but still must be considered when selecting a coating process. Coating control provides an assessment of the process' ability to apply coating where it is needed, as well as its ability to prevent coating bleed into undesirable areas of the printed wiring boards. Table 5.19 presents the constructed measure levels and their definitions.

The relevant data are presented in the following Table 5.20. The table also indicates the least and most preferred levels of the measures. The production volume is predicted to be 250,000 units per year, as the actual costs for the three processes under consideration are calculated. Only development time involves uncertainty. For example, the development time of sil-gel potting may be 16, 20, or 24 weeks with probabilities of 0.40, 0.50, and 0.10, respectively.

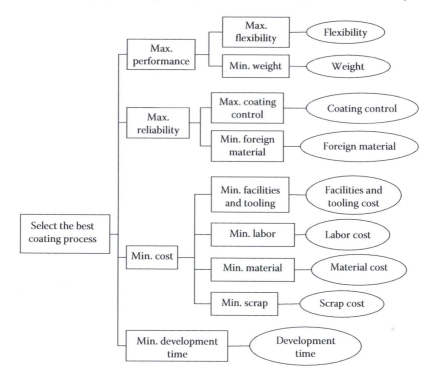

FIGURE 5.21: Goals hierarchy for coating process selection.

TABLE 5.19: Constructed measure levels for coating process selection.

Measure	Most Preferred		Least Preferred
Flexibility	High: 75%–100% reusable	Medium: 50%–75% reusable	Low: 25%–50% reusable
Foreign material	Superior: No particles	Excellent: 1–2 non-metallic	Good: 3–5 nonmetallic or 1 metallic
Coating control	1: Problem areas unlikely to affect function		0: Problem areas may affect function

5.13.1 Assigning Weights and Identifying Utility Functions

The CPST identified measures and collected the relevant data. They then interviewed three SMEs (the advanced manufacturing manager, the electronics line manager, and the process engineering supervisor) to identify the relative importance of these measures. They used the swing weights method to assign weights. Each SME was interviewed separately, followed by a group discussion among the team members to reach a consensus. Table 5.21 summarizes the weights assigned by the SME group. Foreign material, flexibility, and material cost (each has more than 15% of total weights) have the highest weights. In terms of objectives, minimize cost is ranked first and collects 32.4% of the total weight. This was closely followed by maximize reliability, which accounts for 31.5% of the total weight. The maximize performance objective is ranked third and accounts for 21.3% of the total weight. Finally, minimize development time is ranked fourth, with 14.8%.

TABLE 5.20: The relevant data for coating process selection.

Measure	Selective Spray		Sil-Gel Potting		Coat and Extract		Least Preferred	Most Preferred
Flexibility	High		Medium		Low		Low	High
Weight (g)	6		230		20		250	5
Coating control	0		1		1		Average	Good
Foreign material	Superior		Excellent		Good		Good	Superior
Facilities and tooling cost ($)	315,000		25,000		110,000		350,000	20,000
Labor cost ($)	40,000		10,000		20,000		45,000	9,000
Material cost ($)	17,000		615,000		63,000		650,000	15,000
Scrap cost ($)	95,000		0		11,000		100,000	0
Development time (Weeks)	DT[a]	Pr[b]	DT	Pr	DT	Pr	50	15
	28	0.15	16	0.40	28	0.10		
	32	0.45	20	0.50	30	0.20		
	36	0.35	24	0.10	34	0.60		
	48	0.05			40	0.10		

[a] DT is development time.

[b] Pr is probability.

TABLE 5.21:   Relative importance of measures and objectives.

Objective	Measure	Least Preferred	Most Preferred	Swing Weights	Weights	Objective Weights
Min. development time	Development time	50	15	80	0.148	0.148
Min. cost	Facilities and tooling cost	350,000	20,000	30	0.093	0.324
	Labor cost	45,000	9,000	20	0.037	
	Material cost	650,000	15,000	90	0.167	
	Scrap cost	100,000	0	15	0.028	
Maximize reliability	Coating control	Average	Good	70	0.130	0.315
	Foreign material	Good	Superior	100	0.185	
Maximize performance	Flexibility	Low	High	85	0.157	0.213
	Weight	250	5	30	0.056	

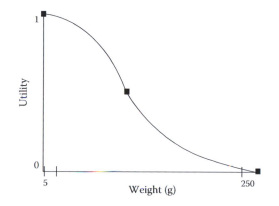

FIGURE 5.22:   Utility functions for weight.

After the team assigned the weights, they constructed the utility functions. Figure 5.22 illustrates a utility function for weight. The utility score of the weight measure is an s-shaped curve. As the weight of a coating material exceeds 74 g, the performance of a process grows rapidly worse, resulting in a correspondingly rapid decrease in the utility function. If the weight is more than 135 g, there is a small decrease in the utility function, due to the small difference between the performance of 135 and 250 g coatings.

5.13.2   Ranking the Alternatives

The results are presented in Figure 5.23. Selective spray is ranked first with a utility score of 0.702, followed by Sil-Gel Potting. Selective spray is extremely strong on performance, while average on the other objectives. Sil-Gel Potting draws its strength from development time and reliability. On the other hand, it is weak on performance and cost. Coat and Extract is ranked third with a utility score of 0.596. Coat and Extract is extremely strong on cost while weak on the other objectives.

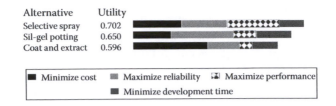

FIGURE 5.23:   Stacked bar ranking for the coating processes.

FIGURE 5.24:   Impact of uncertainty on ranking coating processes.

As depicted in Figure 5.24, uncertainty has an impact on the rank ordering of the best two processes. Selective Spray involves significant risk, and there is a possibility that it may be ranked second. Sil-Gel Potting involves the least uncertainty, with a utility range of 0.014. Since the uncertainty may change the rank ordering of the best two alternatives, the team cannot recommend either Selective Spray or Sil-Gel Potting without a risk management strategy. We introduce risk management in Chapter 6.

5.14   Nonlinear Additivity: Multiplicative Form

MAUT applications commonly assume a linear additive function for the total utility.

$$U = \sum_{i=1}^{n} w_i u_i(x_i)$$

where
  w_i is the weight of the ith measure with the weights summing to 1
  $u_i(x_i)$ is the individual utility function of the ith measure

In this formula, each attribute contributes an independent value to the total score. However, there are contexts in which interaction between measures affects the total score. Consider the case of maximizing the craftsmanship of a car instrument panel or craftsmanship of kitchen cabinetry. Two measures of craftsmanship are gaps and misalignment. If an alternative scores poorly on either measure, the overall craftsmanship score would be low. This is labeled destructive interaction (Smith 2007) and the attributes complement each other (Keeney and Raiffa 1993). In contrast, consider the goal of maximizing competitiveness of a highly stylized product. Two aspects of competitiveness are styling and price. A product can succeed in the marketplace if it has a substantial competitive advantage in either category. This is labeled as constructive interaction (Smith 2007) and the attributes substitute for each other (Keeney and Raiffa 1993).

If the attributes satisfy the condition of preference independence, the utility function has a multiplicative form. Preference independence means that the order of preference on one attribute is not influenced by the value of another attribute. This condition is generally met. However, this assumption might be violated if a person were simultaneously considering where to live and which car to buy (Clemen and Reilly 2001). The individual might prefer a hybrid electric vehicle in a crowded city and a truck in the country. In this case, location and car choice are not preference independent.

With preference independence, the multiplicative equation for two attributes is

$$U = k_1 u_1(x_1) + k_2 u_2(x_2) + (1 - k_1 - k_2) * [u_1(x_1) * u_2(x_2)]$$

If $k_1 + k_2$ sum to less than 1, the multiplicative coefficient is positive, and there is destructive interaction between measures. If $k_1 + k_2$ sum to more than 1, the multiplicative coefficient is negative, and there is constructive interaction between measures.

Let us explore the example of craftsmanship. A single utility function was created for both gaps and misalignment. For this example, we set the two weights to be equal

$$k_1 = k_2 = 0.2 \text{ and therefore } 1 - k_1 - k_2 = 0.6$$

$$U = 0.2u_1(x_1) + 0.2u_2(x_2) + 0.6[u_1(x_1) * u_2(x_2)]$$

Table 5.22 presents the utility scale for gaps. An identical scale applies to misalignment.

Table 5.23 explores pairs of values for two attributes of craftsmanship and calculates total utility with the above equation. With Alternative 1, there is no discernible gap but misalignment is poor. The total utility score is 0.2. The misalignment destroys the total craftsmanship score. With poor misalignment, the maximum attainable craftsmanship score is 0.2. If misalignment improves slightly to 0.1 as in alternative 2, the total score is still only 0.25. In alternative 3 both are very good; the total score is 0.85, which is less than the individual utilities.

TABLE 5.22: Single utility function for gaps.

	Utility Score	Description
Poor	0	Gap is prominent
Mediocre	0.25	Gap noticeable by casual observation
OK	0.5	Gap discernible by careful observation of non-expert
Good	0.75	Gap discernible by careful observation of expert
Very good	0.9	Gap sometimes discernible by careful observation of expert
Excellent	1	No discernible gap

TABLE 5.23: Pairs of values for two craftsmanship measures.

	Weights			
	0.2	0.2	0.6	
Alternative	Gap	Misalignment	Product	Total
1 Excellent and poor	1	0	0	0.2
2 Very good and weak	0.9	0.1	0.09	0.25
3 Both very good	0.9	0.9	0.81	0.85
4 Both good	0.75	0.75	0.56	0.64
5 Both OK	0.5	0.5	0.25	0.35

TABLE 5.24: Single utility function for competitiveness.

	Utility Score	Description
Poor	0	Major competitive disadvantage
Mediocre	0.25	Minor competitive disadvantage
OK	0.5	Competitive
Good	0.75	Minor competitive advantage
Very good	0.9	Moderate competitive advantage
Excellent	1	Major competitive advantage

TABLE 5.25: Pairs of values for two competitiveness measures.

	Weights			
	0.8	0.8	−0.6	
Alternative	Price	Styling	Product	Total
1 Excellent and poor	1	0	0	0.8
2 Very good and weak	0.9	0.1	0.09	0.75
3 Both very good	0.9	0.9	0.81	0.95
4 Both good	0.75	0.75	0.56	0.86
5 Both OK	0.5	0.5	0.25	0.5

Let us explore the example of competitiveness for a product in which style and prices are equally important factors. Table 5.24 presents the utility scale for competitiveness on each attribute.

For this example, we set $k_1 = k_2 = 0.8$ and therefore $1 - k_1 - k_2 = -0.6$

$$U = 0.8u_1(x_1) + 0.8u_2(x_2) - 0.6\left[u_1(x_1) * u_2(x_2)\right]$$

Table 5.25 explores pairs of values for two attributes of competitiveness. With Alternative 1, the product has a major competitive advantage with price but is not competitive on styling. The total utility score is 0.8. The excellent price value can sell the product and poor styling does not undermine the total score. Even with poor styling, the maximum attainable competitive score is 0.8. If styling improves slightly to 0.1 as in alternative 2 and price competitiveness declines slightly, the total score is 0.75. In alternative 3, both are very good; the total score is 0.95, which is more than the individual utilities.

In the two attribute examples, there is one additional coefficient that requires the decision maker make one additional trade-off to assess its value. However, as the number of attributes increase to three or more there is dramatic increase in the number of decision maker trade-offs required to specify the equation. Keeney and Raiffa (1993) provide an extensive discussion of complex utility functions and their specification. They also provide a detailed discussion of different forms of independence.

5.15 Research Issues with Weight Elicitation

The weight-elicitation methodology presented earlier involves assigning points to a rank-ordered list of attributes while explicitly noting the range for each attribute. Edwards and Hutton (1994) term this approach SMARTS, SMART with *S*wing weights. We have found that students routinely find dramatic changes in their weighting if they ignore ranges at first and later redo the weights while considering the ranges. They recognize that considering the ranges better reflects their true judgment. When ranges are considered, a critical concern is how much decision makers adjust their weights to reflect the range. Nitzsch and Weber (1993) found that decision makers do not adequately increase their point assignments as the range grows.

The SMARTS method requires assigning specific points to each attribute as the decision maker moves down the rank-ordered list. This assignment of points is obviously more complex than just rank ordering; it also may seem arbitrary. Barron and Barrett (1996) presented an alternative method, rank order centroid (ROC), that uses just the rank ordering to determine weights. They found that the overall utility scores when using ROC for their alternatives were within 2% of the scores obtained with SMARTS. (See also Bottomley and Doyle (2001) and Joydeep et al. (1995)). Roberts and Goodwin (2002) demonstrated another alternative, rank-order distribution (ROD), which they claim provides an even better approximation to SMARTS weights.

There are several other concerns when assigning weights in an objective hierarchy. One involves the number of attributes used to describe an objective. As more and more attributes are used, there is an observed tendency for the total weight of the objective to increase (Weber and Borcherding 1993). Pöyhönen et al. (2001) carried out detailed studies of individual differences with regard to this bias. Hämäläinen and Alaja (2008) demonstrated the splitting bias in an actual environmental decision. One suggestion for addressing this bias is to review the totals for each objective and verify whether the totals are consistent with the decision maker's preferences. If necessary, the major objective can be rescaled and the attribute weights adjusted proportionately.

Another concern involves the difference between a hierarchical and a nonhierarchical approach to assigning weights (Stillwell et al. 1987). For large objective structures, it is more efficient to assign weights first to the highest level of objective and then subdivide the weights to each sub-objective and ultimately down to each attribute or measure. The global weight for the lowest-level attribute is the product of the values across the different levels. A hierarchical approach to weight elicitation produces significantly more variability among attribute weights than directly eliciting weights of each attribute.

Exercises

Complete Chapter Activities

5.1 How much weight would you assign to each measure when selecting the primary bulb for the lamps in your home?

Measure	Weight
Performance	____
Cost	____
Total	1

5.2 Specify your preference for 10 light bulbs: Performance vs. Cost

Measure	Least Preferred	Most Preferred	Rank	Points	Final Weight
Performance	60	100	____	____	____
Cost	350	150	____	____	____
			Total	____	

5.3   Hotel Manager Preferences for 1000 Light Bulbs: Performance vs. Cost

Measure	Least Preferred	Most Preferred	Rank	Points	Final Weight
Performance	60	100	____	____	____
Cost	35,000	15,000	____	____	____
			Total	____	

5.4   Significant and insignificant price ranges for automobiles

    a. Imagine that a friend is considering purchasing an automobile. Describe under what circumstances you might recommend assigning a relatively large weight to variations in the purchase price.

    b. Under what circumstances, might you recommend assigning relatively little weight to the purchase price?

5.5   Which measures would you scale with a nonlinear utility function?

    a. Home choice: three, four, or five bedrooms

    b. Car acceleration time 0–60 mph: range 6.5–9 seconds

    c. Waiting time on phone with customer service: range 0–12 minutes

    d. Gas mileage: range 20–30 mpg on a highway

    e. Suggest a measure in your work environment that would have a nonlinear utility function

5.6   Concave or convex utility measures

    a. Describe a decision context with a *concave* measure

    b. Describe a decision context with a *convex* measure

5.7   Direct utility assessment of four bedrooms

Directly assess the utility of four bedrooms in a house purchase decision. The possible values are three, four, or five bedrooms. More bedrooms are preferred. Since five bedrooms are best, it has a utility of 1. Three bedrooms are least desirable and have a utility of 0.

- Which change produces a greater value improvement?
    - Change 1—Improve from three bedrooms to four
        - $SUF(4) > 0.5$
    - Change 2 – Improve from four bedrooms to five
        - $SUF(4) < 0.5$
- Which increase is of greater value to you?
- Specify $SUF(4) =$

Cases

5.8   Cell phone plans

    a. Identify a list of measures to be used to compare cell phone plans from more than one service provider.

    b. Gather data for three different cell phone plans. Include at least one nonnumeric measure.

    c. Assign weights to the various measures.

    d. Assign a utility function to the nonnumeric measure.

5.9 Women's retailer—Sales force (use Logical Decisions)

Harry Target is the owner a small sized mid-level retailer of women's clothing. Annual per store sales is $4.5 million. He is thinking about the quality of the people who will service customers. If he hires more experienced sales people, the buying experience should be better. However, because each sales rep spends more time per customer, the average waiting time to be serviced will increase. The relevant data are presented below (Table 5.26).

    a. Create a goal hierarchy.

    b. Create a data matrix.

    c. Assign weights to the measures.

5.10 Used car

Set up the used car example from Section 5.7 in Logical Decisions

    a. Create an objectives hierarchy and measures.

    b. Input the data matrix.

    c. Assess your own weights for the measures—use SMART (swing weight) method.

    d. Identify at least one numeric measure that you believe should have a nonlinear utility scale and assess a nonlinear utility scale.

    e. Assess at least one label measure as nonlinear.

5.11 Kitchen remodeler

Set up the kitchen remodeler example from Section 5.11 in Logical Decisions

    a. Create an objectives hierarchy and measures.

    b. Input the data matrix.

    c. Assess your own weights for the measures—use SMART (swing weight) method.

    d. Identify at least one numeric measure that you believe should have a nonlinear utility scale and assess a nonlinear utility scale.

    e. Assess at least one label measure as nonlinear.

5.12 Blower motor replacement

Your company is considering a change to a standard blower motor that is widely used. There is no specific vehicle program deadline approaching, but the sooner the change is made, the earlier it can be incorporated into future vehicle programs. Currently, there are 43 Things Gone Wrong (TGW) per 1000, and the variable cost is $38. Only one of the following options can be pursued (Table 5.27).

TABLE 5.26: Women's retailer data.

Alternative	Experience	Annual Cost ($)	Average Waiting Time (min)
1	3–5 years experience	$256,000	3
2	6–10 years	$312,000	5
3	More than 10 years	$379,000	8

TABLE 5.27: Blower motor data.

Option	TGW	Variable Cost	Timing
A	40	\$35	1 month
B	30	\$40	10 months
C	15 with $p=0.30$	\$50	12 months with $p=0.25$
	10 with $p=0.40$		15 months with $p=0.25$
	5 with $p=0.30$		18 months with $p=0.50$

Interview a colleague to determine the following:

a. Assess the weights using the SMART method (Swing weights).

b. Create a nonlinear utility function for TGW using the MLS method.

Produce output from Logical Decisions that demonstrates your ability to do all the following and include a limited discussion.

c. Goals hierarchy with the weights

d. Utility function for TGW

5.13 Plant location site selection (case described below)

Interview two people using swing weights. At least one utility scale should be nonlinear. The results of the interviews should be keyed into and analyzed with Logical Decisions software. Use one file and maintain the respective weights in separate preference sets. The report *must* contain a discussion of the differences in objective functions weights.

Background Information

An autosupply company is evaluating potential plant location sites in five countries from diverse regions and with diverse economies: Mexico, the Czech Republic, Poland, South Korea, and South Africa (Canbolat et al. 2007). These five countries were initially selected because they offer low manufacturing costs and also meet minimum selection requirements for achieving the corporate quality standard as well as their concerns about protecting intellectual property. The new plant will manufacture auto brake components and systems. The company assumes that it will initially invest between 210 and 250 million dollars and hire 400 employees for the brake plant. The facility is to serve America, Europe, and Asia-Pacific, in addition to the local market. The goal is to select a site for a new facility in order to maximize its total value to the company.

The ultimate or overall goal of the company is to maximize total value, as decomposed into four subobjectives: minimize total cost, maximize product quality, maximize country stability, and maximize the geographical and demographical location. Table 5.28 presents each measure and its corresponding scale.

Required data were collected from the U.S. Commercial Service, the U.S. Department of State, Central Intelligence Agency, Czech Republic Investment, NAFTA, Czech Statistical Office, Standard and Poor's, Transparency International, International Monetary Fund, and Asia-Pacific Economic Cooperation. Table 5.29 presents 2006 data for alternative countries with respect to each measure.

The risk profile for labor cost in tabular format is presented in Table 5.30 and should be used as input in the multiattribute analysis.

TABLE 5.28:   Country measure levels.

Country Measures	Best 1	2	3	4	Worst 5
High school enrollment	80%–100%	65%–80%	50%–65%	Less than 50%	
Attitude of unions	Co-operative: never work stoppage	Work stoppages less than once a year	Work stoppages more than once a year—infrequent short strike during contract negotiations	Highly organized and aggressive union, frequent work stoppages and strikes at negotiations	
Supplier reliability	High quality, on time delivery	High quality, late delivery	Moderate quality, on time delivery	Low quality, late delivery	
Political stability	No impact of government changes on business No threats for long term stability	Possible impact of government changes on business Some threats for long term stability	Likely impact of government changes on business Likely risk of insurgency Long-term threats to stability		
Membership of free trade agreement	Member of NAFTA, APEC	Member of EU	Member of APEC	South Africa -EU FTA	
People speaking English (%)	More than 40% (very high)	30%–40% (high)	20%–30% (medium)	10%–20% (low)	Less than 10% (very low)
Currency issuer credit ratings (S&P)	A–	BBB+	BBB–		
Unemployment level	More than 15% (very high)	10%–15% (high)	5%–10% (medium)	0%–5% (low)	
Infrastructure quality	Excellent transportation, communication, energy services readily available	Normally good services but specific shortcomings	Widespread shortcomings but basically adequate		

TABLE 5.29: Country data for plant location selection case.

Measure	Mexico	Czech Rep.	Poland	S. Korea	S. Africa
Investment cost ($ M)	230	210	210	250	230
Labor cost ($ M)	18.05	14.25	8.64	26.25	9.97
High school enrollment (%)	50–65	80–100	65–80	80–100	Less than 50
Attitude of unions	Demanding	Co-operative	Co-operative	Aggressive	Rational in Demands
Supplier reliability	Medium	Very high	High	Very high	Low
People speaking English (%)	Low	High	Medium	Very high	Very Low
Political stability	Stable	Highly stable	Highly stable	Very stable	Stable
GDP real growth rate (%) (average of last 5 years)	4.3	0.48	4.76	3.94	2.1
Currency issuer credit ratings (S&P)	BBB–	A–	BBB+	BBB+	BBB–
Inflation rate	12	5.98	10.22	3.62	6.1
Corruption perceptions index	3.7	3.9	4.1	4.2	4.8
Unemployment level (%)	0–5	5–10	10–15	5–10	More than 30
Membership of free trade agreement	NAFTA, APEC	EU	EU	APEC	SA-EU Agreement
Infrastructure quality	Developed	Developed	Moderate	Highly developed	Developed
Regional vehicle production (thousands)	17,870	18,500	18,500	17,497	407

TABLE 5.30: Cumulative risk profile for labor cost.

	Cumulative Probability					
Country	**0**	**0.2**	**0.4**	**0.6**	**0.8**	**1**
Mexico	7.7	10.5	14.0	17.5	22.6	29.1
Czech Rep.	7.4	8.5	11.1	14.1	17.9	20.2
Poland	4.6	5.1	7.7	8.5	11.3	12.3
S. Korea	18.1	20.1	26.0	30.7	38.9	42.3
S. Africa	4.9	5.3	8.3	9.2	13.7	14.4

References

Angus, A. J., Hodge, I. D., McNally, S., and Sutton, M. A. (2003). The setting of standards for agricultural nitrogen emissions: A case study of the Delphi technique. *Journal of Environmental Management*, 69, 323–337.

Arrow, K. J. (1951). *Social Choice and Individual Values*. New York: Wiley.

Barron, F. H. and Barrett, B. E. (1996). Decision quality using ranked attribute weights. *Management Science*, 42, 1515–1523.

Bottomley, P. A. and Doyle, J. R. (2001). A comparison of three weight elicitation methods: Good, better, and best. *Omega*, 29, 553–560.

Canbolat, Y. B., Chelst, K., and Garg, N. (2007). Combining decision tree and MAUT for selecting a country for a global manufacturing facility. *Omega*, 35, 312–325.

Clemen, R. T. and Reilly, T. (2001). *Making Hard Decision with Decision Tools*. Mason, OH: Southwestern Cengage Learning.

Edwards, W. (1977). How to use multiattribute utility measurement for social decision making. *IEEE Transactions on Systems, Man, and Cybernetics, SMC-7*, 326–320.

Edwards, W. and Hutton, B. F. (1994). SMARTS and SMARTER: Improved simple methods for multiattribute utility improvement. *Organizational Behavior and Human Decision Processes*, 60, 306–325.

Ferrell, W. R. (1985). Combining individual judgments. In: *Behavioral Decision Making*, Wright, G., ed. New York: Plenum Press, pp. 111–145.

Hämäläinen, R. P. and Alaja, S. (2008). The threat of weighting biases in environmental decision analysis. *Ecological Economics*, 68, 556–569.

Hamalainen, R. P., Lindstedt, R. K., and Sinkko, K. (2000). Multiattribute risk analysis in nuclear emergency management. *Risk Analysis*, 20, 455–467.

Hasson, F., Keeney, S., and McKenna, H. (2000). Research guidelines for the Delphi survey technique. *Journal of Advanced Nursing*, 32, 1008–1015.

Joydeep, S., Terry, C., and Roy, B. L. (1995). Do ranks suffice? A comparison of alternative weighting approaches in value elicitation. *Organizational Behavior and Human Decision Processes*, 63, 112–116.

Keeney, R. L. and Raiffa, H. (1993). *Decisions with Multiple Objectives: Preferences and Value Tradeoffs*. Cambridge, MA: Cambridge University Press.

Linstone, H. A. and Turoff, M., eds. (1975). *Delphi Method: Techniques and Applications*. Boston: Addison-Wesley Educational Publishers Inc.

Nitzsch, R. V. and Weber, M. (1993). The effect of attribute ranges on weights in multiattribute utility measurements. *Management Science*, 39, 937–943.

Powell, C. (2003). The Delphi technique: Myths and realities. *Journal of Advanced Nursing*, 41, 376–382.

Pöyhönen, M., Vrolijk, H., and Hämäläinen, R. P. (2001). Behavioral and procedural consequences of structural variation in value trees. *European Journal of Operational Research*, 134, 216–227.

Roberts, R. and Goodwin, P. (2002). Weight approximations in multi-attribute decision models. *Journal of Multi-Criteria Decision Analysis*, 11, 291–303.

Smith, G. (2007). *Logical Decisions for Windows: User's Manual*. Fairfax, VA: Logical Decisions.

Stillwell, W. G., Winterfeldt, D. V., and John, R. S. (1987). Comparing hierarchical and nonhierarchical weighting methods for eliciting multiattribute value models. *Management Science*, 33, 442–450.

Weber, M. and Borcherding, K. (1993). Behavioral influences on weight judgments in multiattribute decision-making. *European Journal of Operational Research*, 67, 1–12.

Chapter 6

Value and Risk Management for Multi-Objective Decisions

Taka Bolt is looking for a new warehouse to consolidate all sorting and packaging operations within the warehouse and to improve warehouse material flow. They identified four warehouse sites—FedCo Properties, Center Drive, Wheeling Park, and Deerfield Business Center—and used Multi-Attribute Utility Theory (MAUT) to evaluate them. FedCo Properties ranked first with a utility score of 0.60, followed by Center Drive (0.576), Prospect Park (0.42), and Northbrook Business Center (0.38). FedCo Properties is very strong on a number of truck docks measures, but extremely weak on lease and maintenance cost. Center Drive is strong on lease and maintenance cost, office, and lab space measures, but significantly weak on one highly weighted measure: number of truck docks. What alternative should Taka Bolt select? Is it possible to improve the FedCo Properties' alternative in terms of cost or the Center Drive alternative in terms of the number of truck docks?

Global Electronic decided to install a new conformal coating process. It identified three viable alternatives—Selective Spray Coating, Sil-Gel Potting, and Conformal Coat and Extract—and used MAUT to evaluate them. Selective Spray was ranked first with a utility score of 0.70, followed by Sil-Gel Potting (0.65), and Coat and Extract (0.60). Selective Spray, although it ranked first, involves significantly higher uncertainty and its utility score ranges between 0.62 and 0.72, while Sil-Gel Potting's range is between 0.64 and 0.66. Thus, there is the possibility that the alternative ranked second on average could outperform the highest ranked alternative. In addition, Selective Spray is extremely weak with regard to coating control, which is one of the most important criteria. Which alternative should Global Electronic select? Is it possible to improve the Selective Spray alternative in terms of coating control and reduce uncertainty?

6.1 Goal and Overview

In this chapter, we demonstrate creating added value by a thorough analysis of strengths and weaknesses of the best alternatives and then creating a hybrid solution that is even better.

In Chapter 4, we described a process for structuring multiple objectives and identifying alternatives. Chapter 5 described the process of creating individual utility scores, assigning weights, and ranking alternatives. Ranking alternatives is the first step in MAUT analysis (see Figure 6.1). We do not recommend making a final decision based only on ranking results. Consider the coating selection problem faced by Global Electronic in Chapter 5. Selective Spray, though ranked first, involves significantly higher risk; in a worst case scenario, it would be ranked second. On the other hand,

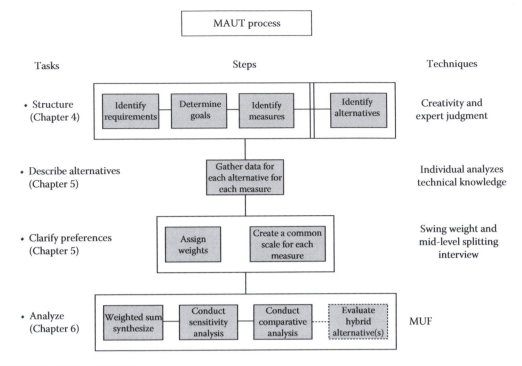

FIGURE 6.1: MAUT process.

Sil-Gel Potting involves less uncertainty but its expected utility is, on average, lower than that of Selective Spray. Is it possible to reduce the uncertainty that Selective Spray poses? If so, is it worth investing to improve the Selective Spray alternative by reducing its risk? Alternatively, can Global Electronic improve or refine the Sil-Gel Potting alternative and increase its overall utility so that it would be ranked first?

In general, a mathematical model should not be used as a shortcut, a means through which managers can make easy decisions, simply by rank ordering alternatives and choosing the one with the highest score. Every mathematical model is an abstraction of reality and, as such, cannot take into account several potentially important issues that are hard to quantify. In addition, many technical decisions involve expert judgment and subjective weights. One cannot reduce the process of making complex decisions by jumping directly to the "bottom line" and implementing the highest scoring alternative. The goal in developing and using a model is to provide decision makers with insight into the strengths and weaknesses of the top alternatives. They should not view the model as providing the answer and, therefore, allow it to usurp their authority to make decisions. Instead, the modeling team should support the decision makers as they go through a six-step process, outlined in the following, that leads to understanding, consensus, and a more informed decision. The primary goal of this chapter is to explore these steps through a series of examples.

1. *Synthesize weighted sum*

    a. Calculate score and rank order alternatives

    b. Identify any potential surprises

    c. Check for validity and mistakes in data input and scaling

    d. Identify and clarify sources of strength for each alternative

This first step of analysis begins with calculating the overall utility scores for all alternatives. At this stage, it is also important to check whether the results make sense. MAUT requires a very structured process for defining different types of measures and their scales as well as for inputting data. It is easy to make a mistake somewhere along the way without noticing. If, for example, the decision maker takes a first look at the results and the rank ordering seems to be badly skewed, this may be due to a scaling error that has gone unnoticed. Imagine what would happen to the rank orderings if the analyst mistakenly forgot to specify that lower cost was better and the software package assumed instead that higher was better. We describe procedures and several charts that are useful for identifying potential mistakes. Ultimately, the only real check on the model is for the decision makers to deeply understand and have confidence in the final rank ordering.

These scores then need to be dissected so that the team can readily understand the reasons for the ranking and examine the relative strengths and weakness of the various alternatives. Logical Decisions for Windows (LDW) software provides charts, tables, stacked bar graphs, and ranking results graphs that help decision makers perceive the strengths and weakness of the alternatives. The ultimate goal is to "update the intuition of the decision maker" who must fully understand the results of the formal analysis (Little 1970).

2. *Comparative analysis: pairs of alternatives*

The first step provided an overview of the strengths and weaknesses of all the alternatives. In this next step, the decision maker focuses on any interesting pair of highly ranked alternatives. Most often, he will compare the top alternative with one of the other highly ranked ones. LDW provides charts and tables that facilitate side-by-side comparisons of pairs of alternatives. This step helps set the stage for brainstorming a hybrid alternative that might outperform all the originally listed alternatives.

3. *Sensitivity analysis*

The purpose of sensitivity analysis is to examine how robust the best alternative is in response to small changes in weights (or data). This step is important for an individual and even more important for a decision-making team whose members' weight preferences could vary significantly. If changes to the weights within a reasonable range do not result in a change in the ranking of the best alternative, then the preferred alternative is considered robust. LDW generates graphs that help clarify weight sensitivity.

4. *Value enhancement*

In this step, decision makers take a close look at the weakness of the best alternatives with regard to those measures and objectives that are highly weighted. They seek an answer to the question "Is it possible to increase the alternative's performance on a specific highly weighted measure?" They will need to brainstorm answers to this question as well as to define how the solution might affect other measures, such as increasing the cost of the alternative in question. The brainstorming might result in creating a hybrid alternative that combines features of the best existing alternatives.

5. *Risk management (not always applicable)*

If the best alternative(s) includes uncertainty, risk analysis and management are applied to reduce the downside risk. The downside risk is a product of the minimum value the alternative might experience and the probability that this would happen. The decision maker's primary interest is only in those measures for which the assigned weights are high and the potential range is wide. As in the previous step, risk management is a clarification and brainstorming process designed to uncover the potential causes of low performance on a measure. The goal is to create countermeasures that will reduce the risk of this happening and minimize the magnitude of any concurrent negative impact.

6. *Cutoff values (not always applicable)*

In many decisions, an alternative may be unacceptable if one of its measures is too low or too high. These are called cutoff values in LDW. They are not eliminated from the analysis, but the bar chart that presents the utility score for each alternative identifies those that fail one or more cutoff values. If an alternative has a high utility score but violates a cutoff, it may be worthwhile to brainstorm a modification that would address its one extremely poor value while maintaining the alternative's other strengths. This process is outlined in Step 4, Value Enhancement.

6.2   Synthesize Weighted Sum

6.2.1   Results Ranking

Ranking the results is the starting point for evaluating the alternatives. LDW sorts the alternatives according to their utilities, which range from the best to the worst. Here, we present the overall ranking of the alternatives for the lighting system for a high-ceiling spacious kitchen area. The fixtures planned for this room would utilize either spot lights or flood lights, which require more expensive bulbs. Mr. Opticast identified incandescent, fluorescent, and halogen bulbs as possible primary bulbs. He had a number of concerns unique to his decision context. Because the ceilings were high, he did not want to have to change bulbs in the 20 fixtures too frequently. He was also concerned with regard to the quality of the light, that is, its ability to accurately represent true colors. In that regard, the halogen lights were best and incandescent lights a close second best. The amount of light was measured in lumens, which enabled a more accurate comparison than watts.

Mr. Opticast identified objectives and assigned the weights as depicted in Figure 6.2. The overall light quality captured more than 50% of the total weight. The amount of light in lumens was more

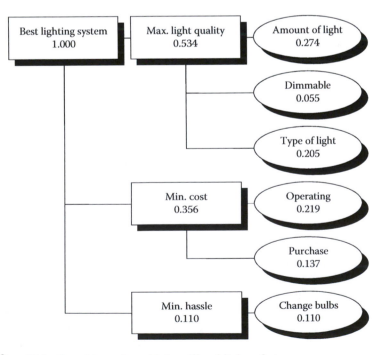

FIGURE 6.2:   Objectives hierarchy—high ceiling kitchen fixtures.

TABLE 6.1: Measures and relevant data for lighting system.

Bulb	Amount of Light	Replace Bulbs	Dimmable	Operating	Purchase	Type of Light
65 W basic	620	10 or more	No	270	50	Incandescent
75 W basic	900	6–9	No	315	76	Incandescent
65 Fluorescent	750	5 or fewer	No	70	90	Fluorescent
75 Fluorescent dim	900	5 or fewer	Yes	80	160	Fluorescent
75 Halogen dim	1020	6–9	Yes	315	150	Halogen

TABLE 6.2: Utility score for category variables.

Replace Bulbs		Type of Light	
Category	Utility	Category	Utility
10 or more	0	Fluorescent	0
6–9	0.5	Incandescent	0.75
5 or fewer	1	Halogen	1

important than the type of light. Although he liked the option of being able to dim the lights, he did not assign too much weight to that measure.

Total cost was assigned one-third of the overall weight total; this factor was split approximately 60–40 between operating costs and purchase costs. A little more than 10% of the overall weight was given to the hassle of replacing bulbs. The relevant data for the five alternatives he considered is presented in Table 6.1.

The data setup included the direct assessment of utility functions for the two measures: bulb replacement (per annum) and type of light. In general, fluorescent light has the lowest scores on the Color Rendering Index, with halogens the best rated. Incandescent light is generally almost as good as halogen. We chose a nonlinear utility scale for the Type of Light (Table 6.2).

6.2.2 Data Check

When using any type of computer model, there are numerous opportunities to make a mistake. It is important to go through some simple verification procedures, which we loosely call the "laugh test." If the first time a decision maker looks at the results he laughs, something is wrong with the data model. The most extreme symptom would be if the perceived worst alternative initially appears as the highest ranked. For example, the decision maker may notice in the stacked bar chart that alternative X draws strength from measure B; yet he knows that, in fact, alternative X is weak on this measure. This is a symptom that the data were input incorrectly or, possibly, that the scale was reversed. LDW does not have data error prevention. The most blatant symptom of bad data is an alternative's utility score that is greater than one or less than zero.

Logical decisions provide a graph as displayed in Figure 6.3 that is useful in tracking down potential modeling or data mistakes. The ranking results graph presents the single utility for every alternative on each measure as well as the overall score. The data are plotted against a utility score of 0–1. Figure 6.3 illustrates the utility of each alternative on each measure for the lighting system decision. The most common symptom would be a value that is either above 1 or below 0. This happens when a data point is either above the most preferred value or below the least preferred. This

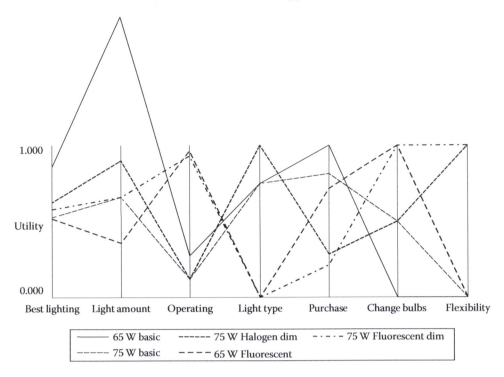

FIGURE 6.3: Ranking results graph for lighting system example—data input error.

can easily happen if an extra 0 is typed in or omitted when inputting data. It could also arise if the range on the measure has been specified too narrowly. Alternatively, you may have forgotten to change the original default range of 0–1. If, for example, you had mistakenly kept the default range for lumens, then a value of 900 would be 900 times as large as the upper limit on the default range. In the following example below, the number of lumens for the 65 W basic incandescent bulb was mistakenly input as 6200 instead of 620. It literally appears off the chart (Figure 6.3) with a value far above 1 for the amount of light.

Another symptom is if the ranking of all of the alternatives on a specific measure seems inverted. There may be an alternative whose utility score on a measure is reported as closest to 1 but the decision maker feels this particular alternative is actually the poorest performer. Conversely, there may be an alternative whose utility score on a measure is displayed close to 0, but the decision maker feels that this alternative is actually the best performer on that measure. Remember that a measure can be either monotonically increasing or decreasing with regard to its utility score. If the model builder made a mistake in specifying the direction of the most and least preferred, the entire measure will look topsy-turvy, as in Figure 6.4. Note that the 75 W basic bulb and 75 W halogen consume the most electricity and, therefore, are extremely weak on operating cost. Yet, in Figure 6.4, they appear extremely strong on this measure. Upon examination, one finds that the measure range has been inappropriately scaled as $350 most preferred and $50 least preferred.

This type of chart is valuable not only for error checking; it can also be used to show aggregate scores for major objectives. Figure 6.5 illustrates the utility of each alternative on each major objective. The halogen bulb is ranked highest. It appears as the best alternative with regard to light quality, but does poorly with regard to cost and hassle. This suggests that there may be an opportunity to improve on the best alternative.

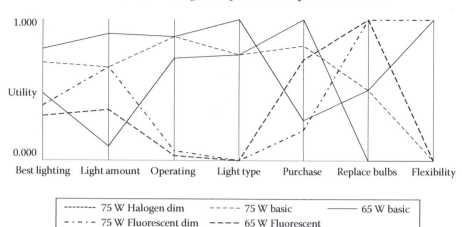

FIGURE 6.4: Ranking results graph for lighting system example: wrong direction on operating cost measure.

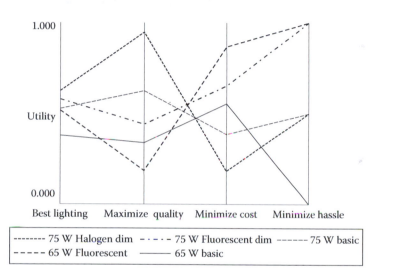

FIGURE 6.5: Results graph for lighting system example.

6.2.3 Understanding the Strengths and Weaknesses of Alternatives

After the software ranks the alternatives, there is a need to examine the performances of the alternatives on the individual objectives or measures, in order to understand the reasons for the ranking. In other words, the decision maker should examine the relative strengths and weakness of the alternatives. This section introduces two types of graph from LDW to investigate the strengths and weaknesses of alternatives: stacked bar graph and comparison of alternatives graph.

Figure 6.6 provides a visual representation of the performance (utility) of each alternative on each major goal. The length of each component is the product of the weight assigned to that objective and the utility scores for each alternative on that objective. The halogen option is clearly the strongest performer on light quality and weakest on cost. The 75 W fluorescent is a close second because it scores well when it comes to minimizing cost and hassle.

Figure 6.7 provides more details. Each segment of the bar for each alternative is the product of the single utility score of that alternative on that measure, multiplied by the weight assigned to that

Ranking for best lighting system goal

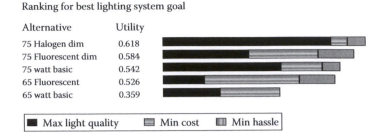

Alternative	Utility
75 Halogen dim	0.618
75 Fluorescent dim	0.584
75 watt basic	0.542
65 Fluorescent	0.526
65 watt basic	0.359

■ Max light quality   ▨ Min cost   ▦ Min hassle

FIGURE 6.6: Objective stacked bar results for lighting system example.

Ranking for best lighting system goal

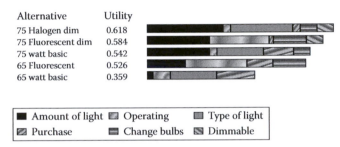

Alternative	Utility
75 Halogen dim	0.618
75 Fluorescent dim	0.584
75 watt basic	0.542
65 Fluorescent	0.526
65 watt basic	0.359

■ Amount of light ▨ Operating   ▦ Type of light
▧ Purchase   ▨ Change bulbs   ◩ Dimmable

FIGURE 6.7: Measure stacked bar results for lighting system.

measure. The halogen draws significant strength from amount of light and type of light. It is weakest with regard to operating cost, meaning that it is a relatively expensive option.

6.3   Comparison of Two Alternatives

Comparisons are more effective when two specific alternatives are placed side by side. This helps a decision maker understand why a particular alternative has a higher utility score than another. In this LDW feature illustrated in Figure 6.7, the length of the bar for each measure represents the percentage of the difference in utilities caused by the measure. Bars on the right of the graph show the measures that provide an advantage for the alternative with the higher overall utility, while the bars on the left identify the measures that favor the alternative with the lower overall score.

Figure 6.8 compares two 75 W equivalent bulbs: halogen and fluorescent. The halogen draws its strengths primarily from type of light and, to a lesser extent, from amount of light. The fluorescent draws its comparative advantage from operating cost and number of bulbs replaced per year. However, these advantages do not compensate for the fluorescent bulb's significant disadvantages on other measures.

6.4   Robustness of a Decision Using Sensitivity Analysis

After the alternatives are ranked and their strengths and weaknesses explored, the robustness of the alternative rankings is tested. Sensitivity analysis investigates the impact of a moderate change

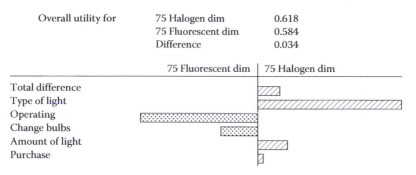

Overall utility for	75 Halogen dim	0.618
	75 Fluorescent dim	0.584
	Difference	0.034

FIGURE 6.8: Comparison between the 75 W halogen and fluorescent.

in the weights (e.g., plus or minus 10% or more) on the overall ranking. If the utility scores of the alternatives are close to each other, the rank ordering will be sensitive even to minor changes in the weights of a number of measures. If the rankings, especially of the best alternative, do not change in response to moderate changes in weights, then the decision is considered insensitive or robust. If any of these steps in the sensitivity analysis result in changes to the ranking, the decision maker should take a closer look at the weights, possibly reevaluating them. This step helps to ensure that the assigned weights truly reflect the decision maker's preferences. Sensitivity analysis is especially important in a group decision-making context, in which there will naturally be some disagreement as to the weights.

Figure 6.9 illustrates the sensitivity analysis graph created by LDW for the weight placed on amount of light. The horizontal axis of this graph corresponds to the weight on amount of light while the vertical axis represents the utility score. The single perpendicular line corresponds to current weight (0.27). The halogen line will be the highest at this point since it is currently ranked first. The ranking of the alternatives is not sensitive to an increase in weight allotted to amount of light, since an increase makes the halogen look even better. As the weight assigned to amount of light decreases, the overall utility of the halogen light decreases even faster than that of its 75 W fluorescent equivalent. If this measure's relative importance were to decline by more than half, the rank orderings would change. The reader should note, though, that a decrease of this magnitude in relative importance for this particular measure is unlikely in this context.

A number of sensitivity analyses were conducted to see whether ranking is sensitive to the weights of other measures. The rankings were found to be most sensitive to the weights assigned to type of light and operating cost. Figure 6.10 shows that the ranking would change if the

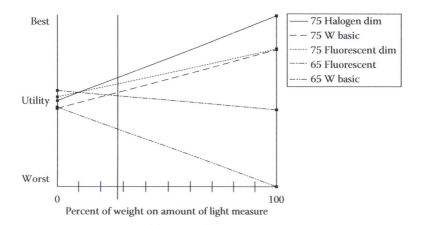

FIGURE 6.9: Sensitivity analysis for weight placed on amount of light.

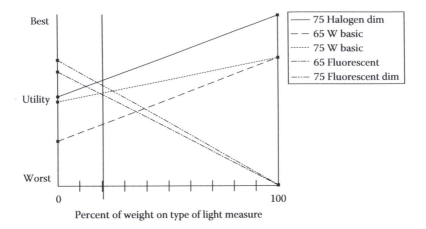

FIGURE 6.10: Sensitivity analysis for weight placed on type of light.

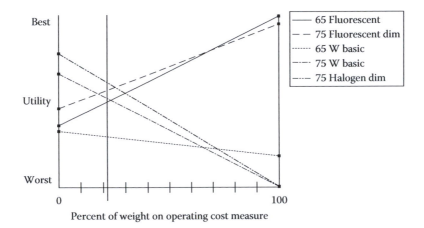

FIGURE 6.11: Sensitivity analysis for weight placed on operating cost.

assigned weight were reduced a little below 0.20. In that case, the 75 W fluorescent would ascend to the top spot.

Figure 6.10 focused on a measure for which the halogen was strong. In Figure 6.11, we look at the weight assigned to operating cost, a measure on which the halogen fares poorly. This is the second highest ranked measure with a weight of 0.22. In the case of cost, a modest increase in the weight to above 0.25 would drop the halogen to second place. In summary, the rank ordering of halogen and fluorescent is highly sensitive to the relative weights assigned to type of light and operating cost. These weights should be carefully reviewed by the decision maker. However, in the following section, we explore a hybrid alternative that makes this point moot.

6.5 Value Enhancement with Hybrid: Lighting Example

After isolating the weaknesses of the best alternative, the decision makers may find that by correcting these weaknesses, they are able to create an even better alternative. Also, weaknesses that appear on a low-weighted measure can be ignored. The lighting example presented earlier provides an opportunity to create a better hybrid. Mr. Opticast had placed a high weight on type of light

TABLE 6.3:   Measures and relevant data for hybrid lighting system.

Bulb	Amount of Light	Change Bulbs	Dimmable	Operating	Purchase	Type of Light
75 Fluorescent dim	900	5 or fewer	Yes	80	160	Fluorescent
75 Halogen dim	1020	6–9	Yes	315	150	Halogen
Hybrid—50–50 split	960	6–9	Yes	200	160	Halogen—Fl

Ranking for best lighting system goal

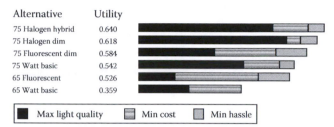

Alternative	Utility
75 Halogen hybrid	0.640
75 Halogen dim	0.618
75 Fluorescent dim	0.584
75 Watt basic	0.542
65 Fluorescent	0.526
65 Watt basic	0.359

Max light quality   Min cost   Min hassle

FIGURE 6.12:   Stacked bar ranking—hybrid lighting system.

for the sake of food preparation and cooking function that was a significant part of kitchen usage. However, upon further reflection, he realized that accurate color rendering was not necessarily crucial in all parts of the kitchen.

Mr. Opticast's hybrid alternative involved using halogen bulbs in half the kitchen and 75 W fluorescents in the other half. The hybrid alternative reduced the average amount of lumens per bulb to 960. He assigned 0.85 as the direct assessment for the type of light since the true color rendering with halogen light was made available where he actually needed it. He set the purchase price at the higher of the two values since he would be buying smaller numbers of bulbs of each type. He approximately split the difference in the operating cost, rounding it up to $200 (Table 6.3). The new rank ordering is displayed in Figure 6.12. The hybrid outperforms the other alternatives.

6.6   Better Alternative through Value Enhancement: Kitchen Remodeling

The Lyons evaluated three different potential builders (Quality Rite, Quality Build, and Cost Conscious) to remodel their kitchen. In the process of obtaining bids and talking to friends, they came up with three broad categories of goals: minimize cost, minimize hassle of construction and any follow-up issues, and maximize quality of the kitchen. Ultimately, they ended up with 15 measures. Table 6.4 presents data on each measure for each contractor.

The result of the LDW software stacked bar chart for the kitchen remodeling example is presented in Figure 6.13. Build Rite is ranked first with a utility score of 0.65, followed by Quality Build with a score of 0.63. Build Rite is the strongest on quality and the second best on hassle, but it is extremely weak on cost. Quality Build is best in terms of hassle and second best in quality and cost. Cost Conscious is ranked third, with a utility score of 0.46.

TABLE 6.4:   Measures and relevant data for kitchen remodeling.

Measure	Build Rite	Quality Build	Cost Conscious
Total labor cost	$34,000	$26,000	25,000
Total material cost	$20,000	$12,000	10,000
Cost overrun history	0% (p=0.33), 2% (p=0.34), 7% (p=0.33)	2% (p=0.33), 5% (p=0.34), 9% (p=0.33)	6% (p=0.33), 9% (p=0.34), 15% (p=0.33)
Kitchen unavailable	13 weeks	10 weeks	9 weeks
Weeks of delay	On time (p=0.33), 1 week (p=0.34), 2 weeks (p=0.33)	1 week (p=0.33), 2 weeks (p=0.34), 3 weeks (p=0.33)	2 weeks (p=0.33), 3 weeks (p=0.34), 4 weeks (p=0.33)
Cleanliness scale	Clean	Messy	Dirty
Follow-up and resolution	Responsive	Highly responsive	Responsive
Creativity scale	Highly creative	Creative	Mundane
Brand and store reputation	Top of line	Moderate price	Moderate price
Percent use of subcontractors	25%	40%	65%
Fit and finish scale	Excellent	Good	Good
Years in business, grouped in ranges	OK	Good	Excellent
Quality of references	Excellent	Good	OK

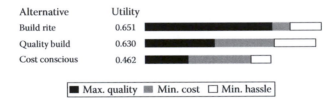

Alternative	Utility	
Build rite	0.651	
Quality build	0.630	
Cost conscious	0.462	

■ Max. quality   ▨ Min. cost   □ Min. hassle

FIGURE 6.13:   Stacked bar ranking for kitchen remodeling example.

A detailed comparison of the best two alternatives is presented in Figure 6.14. The highly weighted measures on which Build Rite is weak are labor cost, material cost, kitchen unavailability, and follow-up and resolution measures. The weights of the respective measures are given inside the parentheses: labor cost (0.10), material cost (0.10), duration of kitchen unavailability (0.08), and follow-up and resolution (0.06). The second best alternative, Quality Build, is weak on creative scale (0.13), fit and finish scale (0.12), brand and store reputation (0.12), and quality of reference (0.09) measures.

Creatively identify a way to improve the alternative's measure level: The process for enhancing an alternative begins with identifying a highly weighted but weak measure level in the best alternative. The decision maker then creatively identifies a way to improve the alternative's measure level and specifies associated changes in other measure levels, such as added cost. To improve the best alternative, Built Rite, the Lyons focused their efforts on the weakness noted earlier: labor cost, material cost, duration of kitchen unavailability, and follow-up and resolution measures. While reducing the labor cost and material cost would yield a significant benefit, Built Rite stated that they would not accept a reduction in price. The contractor also indicated that their creative design requires a longer time for remodeling the kitchen and that they could not guarantee an improvement in the follow-up and resolution measure due to the long commute from their offices to the Lyons'

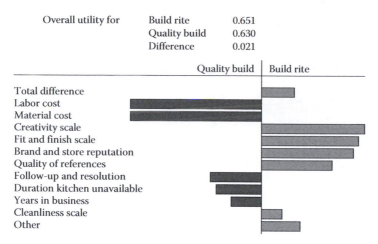

Overall utility for	Build rite	0.651
	Quality build	0.630
	Difference	0.021

FIGURE 6.14:　Comparison of top two kitchen remodelers.

home. In short, they were uncooperative in exploring enhancements, probably because they knew that they were the best.

The Lyons took a closer look at the second best alternative, Quality Build. They realized that two measures on which Quality Build fell short could not be improved: creativity and references. However, the Lyons discussed with the contractors how they might enhance the fit and finish. Quality Build acknowledged that they could improve fit and finish scale from good to excellent by reducing percent use of subcontractors from 40% to 20% and assigning their own experienced employees to this project. Reducing percent use of subcontractors would also improve cleanliness measure from messy to clean. Quality Build asked for additional $3000 for increased labor cost due to the higher cost of experienced employees. Thus, by working together, a hybrid alternative was created.

We named the new alternative "Quality Build + Value Enhancement." Table 6.5 presents data for Quality Build and "Quality Build + Value Enhancement." In the new alternative, fit and finish scale is excellent, cleanliness measure is clean, and percent use of subcontractors is 20%.

TABLE 6.5:　Data for improved kitchen remodeler.

Measure	Quality Build	Quality Build + Value Enhancement
Total labor cost	**$26,000**	**$29,000**
Total material cost	$12,000	$12,000
Cost overrun history	2% ($p=0.33$), 5% ($p=0.34$), 9% ($p=0.33$)	2% ($p=0.33$), 5% ($p=0.34$), 9% ($p=0.33$)
Kitchen unavailable	10 weeks	10 weeks
Weeks of delay	1 week late ($p=0.33$), 2 weeks late ($p=0.34$), 3 weeks late ($p=0.33$)	1 week late ($p=0.33$), 2 weeks late ($p=0.34$), 3 weeks late ($p=0.33$)
Cleanliness created scale	**Messy**	**Clean**
Follow-up and resolution scale	Highly responsive	Highly responsive
Creativity scale	Creative	Creative
Brand and store reputation scale	Moderate price	Moderate price
Percent use of subcontractors	**40%**	**20%**
Fit and finish scale	**Good**	**Excellent**
Years in business but grouped	Good	Good
Quality of references scale	Good	Good

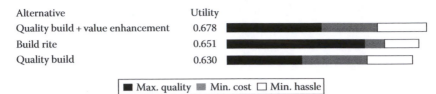

Alternative	Utility	
Quality build + value enhancement	0.678	
Build rite	0.651	
Quality build	0.630	

■ Max. quality ▨ Min. cost ☐ Min. hassle

FIGURE 6.15:   Stacked bar results after value enhancement: kitchen remodeler.

However, the Lyons must pay $3,000 more, which increases total labor cost to $29,000. Note that both alternatives have the same value for all other measures. (Bold values in Table 6.5 correspond to the changes.)

Figure 6.15 presents the utility score for the alternatives. "Quality Build + Value Enhancement" has a utility score of 0.678 and outperforms Build Rite.

6.7   Value Enhancement: Warehouse Selection

Taka Bolt, headquartered in Buffalo Grove, IL, manufactures screws and nuts. In addition to its headquarters, Taka Bolt has a manufacturing and distribution warehouse in Buffalo Grove, IL; a second manufacturing facility in Niles, IL; and a heat treatment facility in Waukegan, IL. Management is looking for a new warehouse within which to consolidate all sorting and packaging operations, thereby improving warehouse material flow. Relocating the warehouse operation will also enable Taka Bolt to expand its Buffalo Grove manufacturing operations. Taka Bolt identified four warehouse sites that meet their requirements: FedCo Properties, Center Drive, Wheeling Park, and Deerfield Business Center.

Taka Bolt evaluated the four alternative sites using MAUT. They identified 4 major objectives, 2 subobjectives, and 11 measures, as shown in Figure 6.16. *Truck traffic handling* and *appearance* measures have constructed scales while other measures have natural metrics. Since as many as 100 trucks may access the site in a day, the site and surrounding area should provide ample room to allow semi-trucks to maneuver as well as room for staging trucks waiting for dock access. *Appearance* takes into account the appearance of the building as well as the appearance of the site. Both *truck traffic handling* and *appearance* measures have three level scales (average, good, very good).

Taka Bolt developed a nonlinear utility function for *number of parking spaces* measure, as depicted in Figure 6.17. One hundred parking spaces would be adequate in servicing employees and visitors to the site. But Taka Bolt management thinks that, if necessary, 75 parking spaces would satisfy their need. The increase from 75 to 100 parking spaces would increase the utility score significantly. The utility score is 0 for 75 parking spaces and 0.8 for 100 parking spaces. However, an additional increase from 100 to 150 parking spaces, although attractive, is of modest value to the management. They utilized linear utility functions for all other natural metrics.

Table 6.6 presents data for each measure for each alternative. Some data, such as number of parking spaces, office and warehouse floor space, and number of truck docks were provided by Taka Bolt's realtor. Taka Bolt easily assessed some measures, such as distance to key facilities and cost. Other measures, including appearance, percent loss of employees, and truck traffic handling, required subjective expert evaluations of each facility.

The weights were developed through discussions with managers of the Warehouse, Materials, and Finance departments. Taka Bolt used the swing weight method with the results summarized in Table 6.7. The number of truck docks received 22% of the total weight, as improving it from 4 to 8 was the most attractive measure for management. Lease and maintenance cost measure accounted

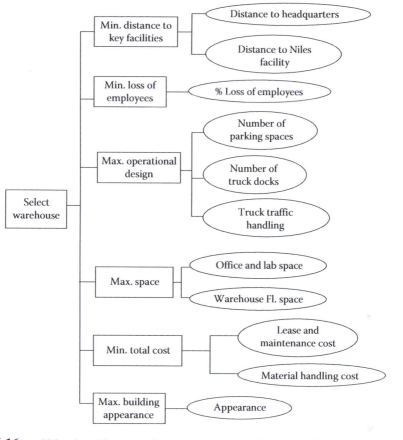

FIGURE 6.16:   Objectives hierarchy for warehouse selection example.

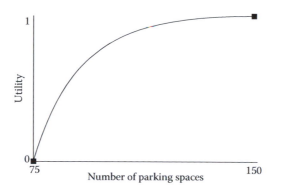

FIGURE 6.17:   Nonlinear utility functions for parking spaces.

for 17% of the total weight. This was followed closely by truck traffic handling measure at 15%. The management was concerned about losing key employees by setting up this new warehouse far from headquarters and gave this measure 13% of the total weight. They were less concerned about material handling cost, distance to key facilities, and appearance measures. Hence, those measures received only a small portion of the total weight.

After assigning weights to specific measures, the decision makers added the totals for the various measures within each objective. They were comfortable with assigning almost half the weight, 0.46, to operational design. This attribute represents more than double the weight assigned to cost.

TABLE 6.6: Data for warehouse site selection example.

Measures	FedCo Properties	Center Drive	Prospect Park	Northbrook Business Center
Number of parking spaces	120	150	100	75
Number of truck docks	7	4	5	5
Truck traffic handling	Average	Good	Very good	Average
Office and lab space (ft²)	4,500	7,500	4,500	5,500
Warehouse floor space (ft²)	62,000	60,000	56,000	64,000
Distance to headquarters (miles)	8	8	40	30
Distance to Niles facility (miles)	7	3	30	20
% Loss of employees	0	0	30	20
Lease and maintenance cost ($)	695,000	630,000	610,000	585,000
Material handling cost ($)	605,000	600,000	620,000	620,000
Appearance	Good	Very good	Good	Average

TABLE 6.7: Weight for each measure for warehouse site selection.

Objective	Measures	Least Preferred	Most Preferred	Swing Weight	Final Weight	Weights of Objectives
Operational design	Parking spaces	75	150	30	0.09	0.46
	Truck docks	4	8	100	0.22	
	Truck traffic handling	Average	Very good	70	0.15	
Space	Office and lab space (sq. ft.)	8,000	4,000	30	0.06	0.15
	Warehouse floor space (sq. ft.)	55,000	65,000	40	0.09	
Distance to key facilities	Distance to headquarters (miles)	40	5	10	0.02	0.03
	Distance to Niles facility (miles)	30	3	5	0.01	
% Loss of employees	% Loss of employees	30	0	60	0.13	0.13
Total cost	Lease and maintenance cost ($)	700,000	580,000	75	0.17	0.20
	Material handling cost ($)	625,000	600,000	15	0.03	
Appearance	Appearance	Average	Very good	10	0.02	0.02

As depicted in Figure 6.18, FedCo Properties ranked first with a utility score of 0.60, closely followed by Center Drive (0.576). These two alternatives have significantly higher scores than Prospect Park (0.42) and Northbrook Business Center (0.38).

FedCo Properties gained its strength from the *operational design* and *loss of current employees* objectives. Center Drive, on the other hand, is weak on the *operational design* objective, which is highly weighted, but strong on the *total cost*, *space*, and *loss of current employees* objectives. Prospect Park is strong on the *operational design* and *total cost* objectives, but extremely weak on the others.

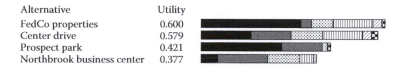

Alternative	Utility	
FedCo properties	0.600	
Center drive	0.579	
Prospect park	0.421	
Northbrook business center	0.377	

■ Max. operational design   ▨ Min. total cost   ▩ Max. space
▥ Min. loss of current employees   ▨ Min. distance to key facilities   ▨ Max. building appearance

FIGURE 6.18: Stacked bar ranking for warehouse site selection example.

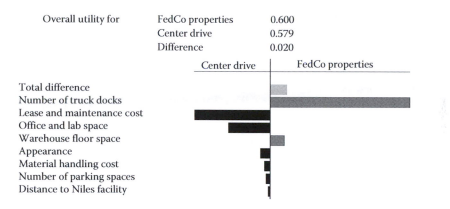

Overall utility for	FedCo properties	0.600
	Center drive	0.579
	Difference	0.020

Center drive | FedCo properties

Total difference
Number of truck docks
Lease and maintenance cost
Office and lab space
Warehouse floor space
Appearance
Material handling cost
Number of parking spaces
Distance to Niles facility

FIGURE 6.19: Comparison of Center Drive and FedCo properties facilities.

Figure 6.19 shows a detailed comparison of the best two alternatives: FedCo Properties and Center Drive facilities. FedCo Properties is extremely strong on the number of truck docks, which is a highly weighted measure, and slightly stronger than Center Drive on the warehouse floor space measure. Center Drive is better than FedCo on six measures, including lease and maintenance cost, office, and lab space. At the same time, Center Drive is extremely weak on the highly weighted measure—number of truck docks; while FedCo Properties has seven truck docks, Center Drive has only four. However, if the number of truck docks in Center Drive increases, it could overtake the FedCo properties in terms of total score.

Identify ways to improve performance on an alternative's highly weighted measure: After this analysis, Taka Bolt management decided to eliminate Prospect Park and Northbrook Business Center and conduct a more detailed analysis of FedCo Properties and Center Drive. The management observed that the FedCo Properties would be more attractive if it could improve the lease and maintenance cost, which was significantly higher than that of all the alternatives. However, the owner of this warehouse was reluctant to reduce lease price. Therefore, Taka Bolt began discussions with Center Drive.

Center Drive's warehouse floor space is 2000 ft^2 smaller than that of FedCo Properties. However, Taka Bolt was not interested in increasing the warehouse floor space at Center Drive; its 60,000 ft^2 area was adequate for their needs. Management examined instead, the possibility of adding new truck docks to the Center Drive site. There was room to add two truck docks. Making this modification would increase the annual lease expense by \$20,000 and reduce warehouse floor space by 2,000 ft^2. Table 6.8 presents data for Center Drive and "Center Drive-Improved" alternatives. (Bold values correspond to the changes.)

Figure 6.20 presents the utility scores after the value enhancement. Modifications to the site would increase the overall score of "Enhanced Center Drive" from 0.579 to 0.644. The increased

TABLE 6.8:　Data for center drive and hybrid alternative.

Measures	Center Drive	Center Drive—Improved
Number of parking spaces	150	150
Number of truck docks	**4**	**6**
Truck traffic handling	Good	Good
Office and lab space (ft^2)	7,500	7,500
Warehouse floor space (ft^2)	**60,000**	**58,000**
Distance to headquarter (miles)	8	8
Distance to Niles facility (miles)	3	3
% Loss of employees	0	0
Lease and maintenance cost ($)	**630,000**	**650,000**

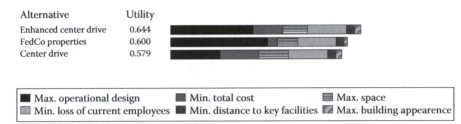

Alternative	Utility
Enhanced center drive	0.644
FedCo properties	0.600
Center drive	0.579

■ Max. operational design　　■ Min. total cost　　▤ Max. space
▨ Min. loss of current employees　　■ Min. distance to key facilities　　▨ Max. building appearence

FIGURE 6.20:　Stacked bar ranking after value management for warehouse site selection.

utility from the addition of two truck docks more than offsets the reduction in the utility score resulting from a modest decrease in floor space and increased annual lease cost. The graph shows that new alternative (Enhanced Center Drive) is better than or equal to FedCo Properties in all objectives except operational design, in which FedCo Properties is slightly better.

6.8　Value Enhancement and Risk Management: Process Selection

Global Electronic will install a new conformal coating process because the upcoming Powertrain Control Module (PCM) design requirements are incompatible with the existing coating process at the plant. These coatings are applied to the printed wiring boards, after the installation of all surface mount devices but before final assembly of the module, to protect circuitry from environmental exposure. Ideally, the process should be capable of selectively applying the coating to various areas of the circuit board, covering some areas while avoiding others. Prescreening of a wide variety of available processes has reduced the number of viable candidates to three: Selective Spray Coating, Sil-Gel Potting, and Conformal Coat and Extract.

The relevant data are presented in Table 6.9. The table also indicates the least and most preferred levels of the measures. A predicted production volume of 250,000 units per year is used in the calculation of costs for the three processes. Only development time involves uncertainty. For example, the development time of Sil-Gel Potting may be 16, 20, or 24 weeks with probabilities of 0.40, 0.50, and 0.10, respectively.

Identify a highly weighted measure with weak performance: The result of the LDW software stacked bar chart for the coating process selection example is presented in Figure 6.21. Selective Spray is ranked first with a utility score of 0.70, followed by Sil-Gel Potting with a score of 0.65. Selective Spray is extremely strong on foreign material, flexibility, material cost, and weight while extremely weak on coating control, facilities and tooling cost, labor cost, and scrap cost. Sil-Gel

TABLE 6.9: Data for coating processes.

	Alternatives			Range	
Measure	**Selective Spray**	**Sil-Gel Potting**	**Coat and Extract**	**Least Preferred**	**Most Preferred**
Flexibility	High	Medium	Low	Low	High
Weight (g)	6	230	20	250	5
Coating control	0	1	1	0	1
Foreign material	Superior	Excellent	Good	Good	Superior
Facilities and tooling cost ($)	300,000	25,000	110,000	350,000	20,000
Labor cost ($)	40,000	10,000	20,000	45,000	9,000
Material cost ($)	17,000	615,000	63,000	650,000	15,000
Scrap cost ($)	95,000	0	11,000	100,000	0
Development time (weeks)	28 ($p=0.15$), 32 ($p=0.45$), 36 ($p=0.35$), 48 ($p=0.05$)	16 ($p=0.40$), 20 ($p=0.50$), 24 ($p=0.10$)	28 ($p=0.10$), 30 ($p=0.20$), 34 ($p=0.60$), 40 ($p=0.10$)	50	15

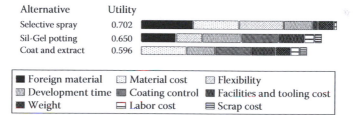

FIGURE 6.21: Stacked bar ranking for the coating processes.

potting draws its strength from coating control, development time, and facilities and tooling cost. On the other hand, Sil-Gel is weak on material cost and weight. Coat and Extract is ranked third with a utility score of 0.60.

A detailed comparison of the best two alternatives is presented in Figure 6.22. The highly weighted measures on which Selective Spray is weak are coating control (0.130), facilities and tooling cost (0.093), and development time (0.148).

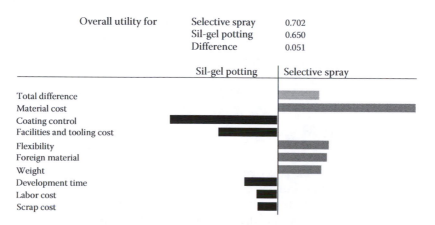

FIGURE 6.22: Comparison of selective spray and sil-gel potting.

TABLE 6.10: Data for selective spray and selective spray + value enhancement alternatives.

Measure	Selective Spray	Selective Spray + Value Enhancement
Coating control	0	**1**
Facilities and tool	300,000	**360,000**
Scrap cost	95,000	**10,000**

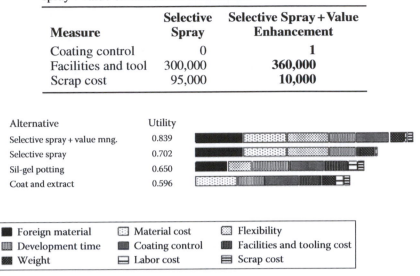

Alternative	Utility
Selective spray + value mng.	0.839
Selective spray	0.702
Sil-gel potting	0.650
Coat and extract	0.596

■ Foreign material   ▦ Material cost   ▦ Flexibility
▦ Development time   ▦ Coating control   ▦ Facilities and tooling cost
▦ Weight   ▱ Labor cost   ▤ Scrap cost

FIGURE 6.23: Stacked bar results after value management.

Identify improvements: To improve the best alternative, Selective Spray, the team focused their efforts on the weaknesses noted earlier—coating control, facilities and tooling cost, and development time. While reducing the facilities and tooling cost for Selective Spray would yield a significant benefit, this attribute is almost impossible to improve.

Development time includes uncertainty, and, therefore, we deal with this measure in the risk management section. The team found that upgrading the coating application nozzles can improve coating control and significantly reduce scrap cost. The supplier of the Selective Spray told Global Electronic that it would have to charge an additional $60,000 to upgrade nozzles and armatures; this would improve coating control from 0 to 1. A value of "0" means that problem areas may affect function, whereas "1" indicates that problem areas will not affect function. This change had the added benefit of reducing the projected scrap cost from $95,000 to $10,000 per year.

Evaluate enhanced alternative: We named the new alternative "Selective Spray + Value Enhancement." Table 6.10 presents data for Selective Spray and "Selective Spray + Value Enhancement." In the new alternative, coating control is 1 and scrap cost is $10,000. However, Global Electronic must pay $60,000, thereby raising the facilities and tooling cost to $360,000.

Figure 6.23 presents the utility score for the alternatives. "Selective Spray + Value Enhancement" has a utility score of 0.84, and significantly outperforms the original Selective Spray alternative.

6.9   Risk Analysis and Management

Risk management is a process of planning and strategizing to reduce the risk of less favorable outcomes associated with the preferred strategy. The key in any risk management strategy is the realization that the originally estimated probabilities of various uncertainties and their associated costs or payoffs can be modified through concerted management oversight and control. For instance, if the probability of a specific supplier meeting a delivery deadline is low, management can reduce this risk by closely overseeing its interactions with this supplier. Procedures may be put

TABLE 6.11: Development time for selective spray if global electronic pays $40,000 more.

Development Time (Weeks)	Probability
28	0.40
30	0.40
32	0.20

in place to remove some of the "usual" excuses for delays, such as late design changes, that are used to justify the failure to meet deadlines. The steps for enhancement of the best alternative(s) through risk analysis and management are summarized as follows:

- Identify a highly weighted measure with significant uncertainty in the best alternative(s) and assess the impact on the overall score of reducing downside risk on that measure.

- Develop a strategy for reducing the downside risk even if this changes other measure levels.

- Evaluate new strategy

Identify a highly weighted measure with significant uncertainty: In the coating example, there is significant uncertainty associated with the projected development time for each of the two best alternatives, Selective Spray and Sil-Gel, as shown in Table 6.9. Development time involves significantly more uncertainty for Selective Spray than for Sil-Gel. There is a 35% chance that development could take as long as 36 weeks. In that case, Sil-Gel's utility score would drop to 0.619 and become less attractive overall than the second ranked alternative. This is due to the high relative importance of development time, the assigned weight of which is 0.148.

Develop a strategy for reducing the downside risk even if it changes other measure levels: Global Electronic contacted the supplier of Selective Spray to explore the possibility of reducing the uncertainty in development time. The supplier asked for $40,000 to cover tooling premiums in order to work overtime; this would reduce development time to a range of 28–32 weeks, as depicted in Table 6.11.

Evaluate newly formed alternatives: We created a new alternative that incorporates a reduction in uncertainty but also increases the facilities and tooling costs. This new alternative has a utility score of 0.705 as compared to the original value of 0.702, as illustrated in Figure 6.24. The more important impact is on the range of the utility score. The lowest score for Selective Spray now exceeds the highest score for Sil-Gel Potting (Table 6.12).

Value enhancement and risk management are complementary activities. In Table 6.13, we present the newest alternative that improves coating control by investing in better equipment and pays for overtime to reduce product development time. The data changes and differences for "Selective Spray" and "Selective Spray with Value Enhancement and Risk Management" are presented in

Ranking for select the best coating process goal

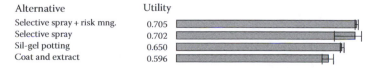

Alternative	Utility
Selective spray + risk mng.	0.705
Selective spray	0.702
Sil-gel potting	0.650
Coat and extract	0.596

FIGURE 6.24: Ranking the results after risk management.

TABLE 6.12: Data changes for selective spray and selective spray + risk management alternatives.

Measure	Selective Spray	Selective Spray + Risk Management
Facilities and tool	300,000	**340,000**
Development time: Expected value	34.19	**29.66**

TABLE 6.13: Data changes for selective spray and selective spray + value enhancement + risk management alternatives.

Measure	Selective Spray	Selective Spray + Value Enhancement + Risk Management
Coating control	0	**1**
Facilities and tool	300,000	**400,000**
Scrap cost	95,000	**10,000**
Development time	34.19	**29.66**

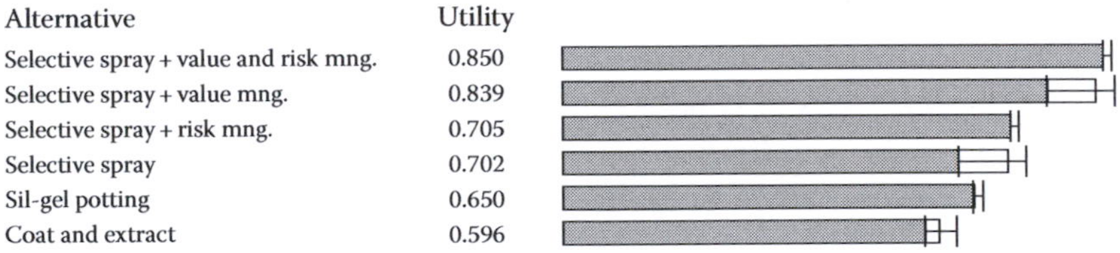

Alternative	Utility
Selective spray + value and risk mng.	0.850
Selective spray + value mng.	0.839
Selective spray + risk mng.	0.705
Selective spray	0.702
Sil-gel potting	0.650
Coat and extract	0.596

FIGURE 6.25: Ranking alternatives after value and risk management.

Table 6.13. The combined enhancements improve the overall average utility score to 0.850, as depicted in Figure 6.25, and also dramatically reduce the risk of lower performance.

6.9.1 Cutoff Values

In many decision contexts, an alternative must meet at least a minimum performance measure on many, if not all, measures. One option is to remove from consideration any alternative that does not meet every cutoff value. There is, however, another option that involves including an alternative that seems to be attractive on a broad array of critical measures even if it is extremely weak on one specific measure. LDW offers this option by highlighting alternatives that fail to meet a cutoff value. We demonstrate the cutoff value by charting the lighting system selection example.

Assume that the maximum purchase budget, or cutoff value, for the lighting is $155. The LDW software uses white bars for rejected alternatives, as shown in Figure 6.26. The utility score of the 75 W fluorescent bulb (0.584) has not changed. However, its bar is highlighted because its cost is $160, which exceeds the cutoff value for cost. Table 6.14 provides the details of the failure.

If the alternative that exceeded a cutoff value is otherwise the highest ranked, the decision maker has a range of options. The simplest is to reassess whether or not the cutoff value was truly critical.

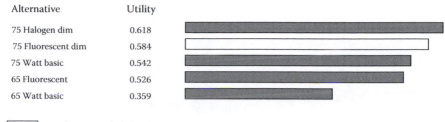

Alternative	Utility	
75 Halogen dim	0.618	
75 Fluorescent dim	0.584	
75 Watt basic	0.542	
65 Fluorescent	0.526	
65 Watt basic	0.359	

☐ Alternative failed at least one cutoff

FIGURE 6.26: Ranking results including cutoff value.

TABLE 6.14: LDW cutoff report.

Alternatives failing one or more cutoffs
75 Fluorescent dim failed 1 cutoff
$160 is above the upper cutoff of 155 for purchase

If he comes to the conclusion that this particular value is not critical, he may relax the cutoff value, thereby removing the stigma of failure. Alternatively, he can apply the process of value enhancement, described earlier, to address the specific cutoff violation.

The issue of cutoff values can arise when dealing with supplier selection decisions in a large organization. Often, the technical staff is asked to provide minimum values for a number of critical performance measures that each and every potential supplier must meet. Purchasing then proceeds to apply these cutoffs as a criterion by which to whittle down the list of candidates. It next focuses on just one or, at most, two measures related to cost for the suppliers who exceeded all of the cutoff values. Finally, the staff may simply pick the lowest cost supplier that meets all of the cutoff values.

In supplier selection problems, one common objective is to maximize quality. Assume that one measure for the quality objective is defective PPM (parts per million). A customer usually defines a maximum PPM (upper cutoff value) above which it is not acceptable. A supplier whose PPM is above the cutoff value is initially rejected irrespective of its other measures. However, many companies have supplier technical assistance programs that can help a particular supplier improve its quality performance. The purchasing department can budget for extra assistance to cover the cost of supplier improvement; this would include supplier training and placement of specially trained quality experts at the supplier's manufacturing facility. In the MAUT analysis, the modeler can create an enhanced alternative that decreases the PPM although this alternative increases investment and operating cost; the program then recalculates the utility score.

Another common objective is to maximize supplier delivery performance. The delivery performance of a supplier is measured using their "on time" delivery percentage. This is a crucial measure if the manufacturer uses a just-in-time or lean production system. Thus, a supplier whose "on time" delivery historical percentage is below a predefined lower cutoff value will be rejected. However, the Original Equipment Manufacturer (OEM) can consider increasing inventory levels to compensate for the relative weakness of the supplier in the area of delivery performance if the supplier's other performance measures are among the best.

In summary, LDW's handling of cutoff values offers decision makers the option of not dismissing an alternative solely on the basis of one or more weak measures. Keeping an attractive, but flawed, option open allows for creative problem solving that can eventually produce superior alternatives.

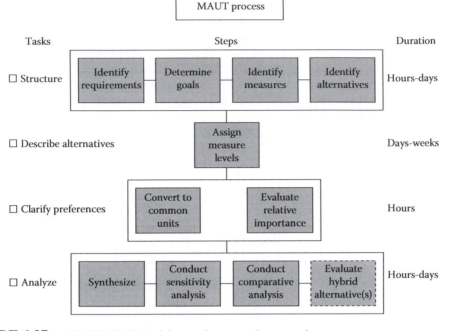

FIGURE 6.27:   MAUT SME/decision makers meeting agenda.

6.10   MAUT and Subject Matter Experts: Process

The MAUT methodology can be used to improve semi-routine decisions made by individual managers or to help a government agency decide on totally new systems. It might take as little as a day or two for a manager to address a local decision over which he has complete authority and personal technical knowledge. Larger decision problems, on the other hand, may spread out over weeks—and sometimes months—if a broad segment of the organization is impacted by the decision. This is true in the design of many military systems that cost billions of dollars to develop and whose core design will last a decade or more. General Dynamics Land Systems, for example, is a major user of this process (see Figure 6.27).

The first step of structuring a problem can be done in less than a week if all the critical players can be quickly gathered to obtain their input and their agreement to buy into the decision-making process. The process might extend to several weeks, however, for a global decision that requires time just to get on everyone's calendar so as to gain input on the decision's structure. The most labor-intensive task is gathering expert opinion on each of the alternatives with respect to each of its measures. Try to envision the number of technical, finance, and marketing experts that will need to be consulted on a problem involving five major objectives with 15 measures while considering 6 or more alternatives. Although the same finance or marketing expert might be able to offer an opinion on several of the alternatives, most likely the technical issues will need to be addressed by specialists in each area.

Clarification of preferences on a local decision can sometimes be accomplished in a matter of hours. Alternatively, the organization may need one or two offsite meetings in which key stakeholders are present and working with the guidance of a trained facilitator. An organization that embarks on its first attempt at providing a multi-objective structure for continually rank ordering new technology or product development projects will need even more time and help. This task of understanding the preferences of multiple constituencies can take weeks if not months.

The actual analysis is conceptually the easiest task. The relative simplicity of the mathematics and the ease of use of the associated software enable decision makers to complete this task in a matter of hours. Extended time is required for organizing, interpreting, and distributing the results to the respective interested parties. They will likely have questions about the results that will require further analyses and extensive communication that could continue over weeks of discussion and debate.

6.10.1 Revisiting the Decision

As every manager knows, decisions often linger and may not be implemented for weeks or even months. During that time span, many things can change. Competitive action and other changing market conditions could force a rethinking of product development priorities. Changes in the supplier community, such as consolidation or bankruptcies, would require reassessment of sourcing decisions. All too often, changes in the internal organization structure of a company will require a review of decisions that have already been made. Newly hired top managers may come with different expertise as well as biases toward or away from a particular technology or supplier based on their experiences. Lastly, many decisions include elements of uncertainty. As uncertainties are resolved and more information is obtained, the decision may need to be revisited or reevaluated.

It is relatively easy to update or revisit a decision that has been analyzed with Logical Decisions. The objectives and measures are unlikely to change, but their relative importance might shift. All that is required to accommodate this new perspective is to introduce a new set of weights and determine whether the decision rank order has changed. As a decision matures, new data may become available for specific measures relevant to certain alternatives. The data matrix can be easily modified to capture the new information.

6.10.2 Role of Logical Decisions Software

The primary role of the various software programs that support multi-objective decisions is to assist in structuring the information and displaying the results in an easy-to-interpret manner. Although the mathematical assumptions may vary between Logical Decisions and Expert Choice, their mathematical complexity and sophistication is about the same. In every case, multi-objective software needs to note and display the overall goals hierarchy and store information about the respective alternatives. The programs also assist in gathering decision maker preferences with regard to the weights assigned to various objectives or measures. The results are then generated, almost instantaneously, so computation time is not a concern. Once the results are in, the analysts can select any of a series of options in order to a conduct sensitivity analysis. Overall, the role of software is outlined as follows.

1. Bookkeeping of information
2. Structure and simplify assessments
   a. Goals hierarchy with measures
   b. Weights: multi-measure utility function (MUF)
   c. Common units: single-measure utility function (SUF—nonlinear)
   d. Cutoff values
3. Calculate overall scores
   a. Keep track of score breakdowns
   b. Simplify comparisons
   c. Keep track of separate preferences (groups)

4. Simulate uncertainty—randomness

5. Sensitivity analysis on weights

6. Ease of updating (new data or alternative)

6.11   Applications

6.11.1   Disposition of Plutonium and Hybrid Strategy

Dyer et al. (1998) present a multi-attribute utility analysis of alternatives for the disposition of surplus weapons-grade plutonium. This is a "real world" example that refers to processes with which the reader may be unfamiliar, but the meaning of these terms is of little consequence to our studies. Rather, the example is brought to illustrate how the process can be used to resolve very complex, politically sensitive problems.

As a result of the strategic arms reduction negotiations between the Untied States and Russia, the United States identified 50 metric tons of weapons-grade plutonium as surplus. The Department of Energy (DOE) was charged with selecting and developing technologies for the disposition of this surplus plutonium that would transform it into forms that are more difficult to use in weapons. Dyer et al. applied MAUT to support the selection of a technology for the disposition of surplus weapon-grade plutonium. The analysis consisted of two phases. Phase I screened 37 candidate alternatives. Phase II focused on the 13 alternatives that withstood the screening process.

As presented in Table 6.15, the alternatives were grouped in three categories: reactor alternatives, immobilization alternatives, and direct disposal alternatives. Reactor alternatives would use surplus plutonium for nuclear reactors that generate electric power. Immobilization alternatives would require the immobilization of the surplus plutonium materials in borosilicate glass. Direct disposal alternatives would involve placement of the plutonium in a borehole. One of these alternatives requires immobilization in an inert matrix, and the other utilizes direct emplacement

A hierarchy of objectives, subobjectives, and measures was developed to evaluate the various alternatives and communicate the results of the analysis to the decision makers and other stakeholders. Figure 6.28 displays a high-level objectives hierarchy. There are three major objectives: nonproliferation, operational effectiveness, and environment, safety, and health (ES&H). The nonproliferation objective consists of five subobjectives while the ES&H objective is made up of three subobjectives.

TABLE 6.15:   Disposition alternatives.

Reactor Alternatives	Immobilization Alternatives	Direct Disposal Alternatives
Existing light water reactors, existing facilities	Vitrification greenfield	Deep borehole (immobilization)
Existing light water reactors, greenfield facilities	Vitrification can-in-canister	Deep borehole (direct emplacement)
Partially completed light water reactors	Vitrification adjunct melter	
Evolutionary light water reactors	Ceramic greenfield	
CANDU reactors	Ceramic can-in-canister	
	Electrometallurgical treatment	

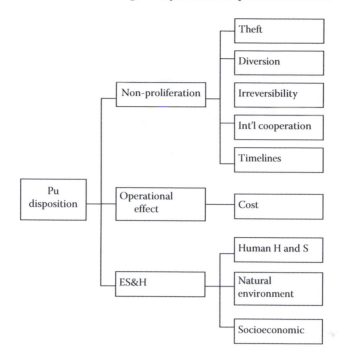

FIGURE 6.28: High-level objectives hierarchy for disposition of surplus weapons-grade plutonium.

The team assessed the utility functions, collected data, and carried out interviews to assign the weights. They then calculated the utility scores for the alternatives and ranked the alternatives, as illustrated in Figure 6.29. Ceramic can-in-can, vitrification can-in-can, and existing reactor and existing facility had higher scores than other alternatives and ranked first, second, and third with the utility scores of 0.6907, 0.6905, and 0.6676, respectively.

After the base-case analysis was completed, the robustness of the ranking was tested by varying the weights and the assumptions. One-way sensitivity analysis was conducted to explore the

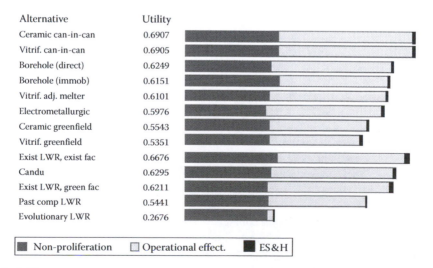

FIGURE 6.29: Overall ranking for disposition of surplus weapons-grade plutonium.

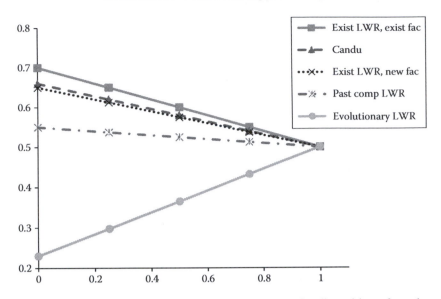

FIGURE 6.30:   Sensitivity analysis for reactor alternatives for disposition of surplus weapons-grade plutonium.

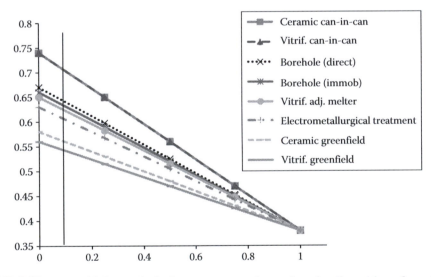

FIGURE 6.31:   Sensitivity analysis for non-reactor alternatives for disposition of surplus weapons-grade plutonium.

impact of varying the weights on the individual measures and objectives. Figure 6.30 presents sensitivity analysis of the weight assigned to Russian cooperation for all of the reactor alternatives. Figure 6.31 presents the sensitivity analysis for the nonreactor alternatives. As the importance of Russian cooperation increases, the reactor alternatives will grow in importance relative to the nonreactor alternatives.

After one-way sensitivity analysis was conducted, two-way sensitivity analysis was performed by varying weights on two major objectives, simultaneously. They also explored the results of changing all the weights simultaneously, using simulation. The weights were randomly selected using a computer simulation program. The results of the sensitivity analysis based on randomly simulated sets of weights showed that the ceramic can-in-can alternative was superior according to its mean and mode. The vitrification can-in-can, existing reactor, and existing facilities alternatives were ranked closely behind. Only two other alternatives, the borehole (direct) and borehole (immobilized) alternatives, were top ranked in all the simulations.

The sensitivity analysis results led to the conclusion that the base-case ranking of the alternatives is relatively insensitive to changes in the base case assumptions over reasonable ranges. Among the reactor alternatives, the existing light water reactors (LWR), existing facilities, and Canada Deuterium Uranium alternatives are typically rated among the top two or three. Among the immobilization alternatives, the vitrification and ceramic can-in-can alternatives dominate the other alternatives.

The sensitivity analysis also indicated that reactor alternatives become relatively more attractive if the proportional importance of Russian cooperation measure increases. One of the primary objectives of the plutonium disposition effort was to prevent the proliferation of nuclear weapons. Thus, Russian cooperation was a key issue and could become an extremely important consideration in the final choice of a U.S. disposition strategy. The analysis, on the other hand, did not fully capture the risk of failure should a single technology be pursued.

A hybrid strategy (parallel development of several technologies) was created and evaluated to better address the ability to influence the Russians to pursue a reciprocal disposition path. The hybrid also addressed the risk of technology failure. If Russian policy is to insist on isotopic degradation, Russia may not approve of a U.S. disposition effort featuring immobilization. In addition, the Russians may be more likely to join a reactor-based disposition program. However, if Russia does not require degradation of fissile material, then the US's most cost-effective and timely course of action would be to pursue immobilization of surplus plutonium in ceramic or glass material. DOE could pursue a joint development approach featuring one reactor technology and one immobilization technology. This parallel development strategy would require higher initial investment costs, but it would also provide additional flexibility in light of the uncertainties.

To evaluate this strategy, the Russian cooperation measure was replaced with a probability distribution over the event that Russia requires an isotopic degradation of fissile material. The weights for the other measures were rescaled while maintaining the original ratios among the weights, so that the sum of the rescaled weights remained one. In this phase of the analysis, three strategies were evaluated: existing reactor, immobilization, and a hybrid of the two.

If the United States chose to utilize an existing reactor in an existing facility, the reactor would not conflict with Russian policy. Therefore, there would be no uncertainty associated with this alternative. The utility score for this alternative is 0.7001.

Alternatively, if the United States selects an immobilization alternative, the Russian policy would be an important factor. The Russians would not begin to dispose of their stockpiles due to dissatisfaction with U.S. proliferation assurance, if the Russians require degradation. In this case, the United States would begin to use the alternative to deploy the existing reactor in an existing facility, which would incur additional cost. This would also mean that the schedule would be delayed. The utility score of this scenario was 0.6665. However, if the Russians do not require degradation, the immobilization program would proceed with no additional cost overruns or schedule delay. The score for this scenario was 0.7474. The expected utility score of the immobilization alternative was 0.7070 using 50–50 probability.

The hybrid deployment strategy requires simultaneous investment in R&D and licensing activities for both the immobilization and reactor technologies. This strategy leads to a higher initial investment cost, but precludes any schedule delays in the future. The expected utility score for the hybrid deployment strategy was 0.7078.

6.11.2   Applications

Two major surveys of the operations research literature contain a wide array of applications of decision analysis (Corner and Kirkwood 1991; Keefer et al. 2004). These surveys cover both decision trees and multi-attribute utility theory. The references for this chapter include selected applications of multi-attribute utility cited in the second survey. These include numerous military, environmental, public utility, and public policy decisions. Only a few of the articles address private company concerns (Kidd and Prabhu 1990; Keeney 1999).

Exercises

6.1   You plan to purchase 20 bulbs to be used in multiple lamps and fixtures throughout your home. The relevant data are provided in Table 6.16. Use the SMART method to assign weights to each measure.

a. Determine the optimal decision using Logical Decisions software.

b. Determine the sensitivity of the optimal decision to the weight assigned to wattage.

6.2   You plan to purchase 1000 bulbs to be used in multiple lamps and fixtures throughout the economy motels you manage. The relevant data are provided in Table 6.17. Use the SMART method to assign weights to each measure.

a. Determine the optimal decision using Logical Decisions software.

b. Determine the sensitivity of the optimal decision to the weight assigned to wattage.

6.3   Cell phone plans: (see Exercise 5.8) Create objectives and measures to be used in evaluating cell phone plans from multiple providers. Gather data for various cell phone plans. Determine the range for each measure. Use the SMART method to assign weights to each measure.

a. Determine the optimal decision using Logical Decisions software.

b. Determine the sensitivity of the optimal decision to weights assigned to the two highest weighted measures.

TABLE 6.16:   Incandescent bulb alternatives for home.

Bulb (W)	Total Annual Operating Cost ($)	Total Bulb Purchasing Cost ($)	Total Annual Cost ($)
60	180	9	189
75	225	10	235
100	300	15	315

TABLE 6.17:   Incandescent bulb alternatives for hotels.

Bulb (W)	Total Annual Operating Cost ($)	Total Bulb Purchasing Cost ($)	Total Annual Cost ($)
60	18,000	900	18,900
75	22,500	1000	23,500
100	30,000	1500	31,500

    c. Create a stacked bar chart and describe the strengths and weaknesses of the various alternatives

    d. Create a pair-wise comparison chart of the two best alternatives.

    e. Create a hybrid solution that improves performance on one or more highly weighted measures of the best or second best alternatives. Increase the cost of that alternative to reflect the improved performance. How does this new alternative compare with the best cell phone on the original list?

6.4 through 6.8

Carry out a comprehensive MAUT analysis for the following exercises. Write a report that includes all of the issues listed below (Each chart should be accompanied by a 50 to 100 word discussion.):

    a. Determine the optimal decision using Logical Decisions software.

    b. Determine the sensitivity of the optimal decision to weights assigned to the two highest weighted measures.

    c. Create a stacked bar chart and describe the strengths and weaknesses of the various alternatives.

    d. If there is uncertainty, can the uncertainty affect the rank ordering of the alternatives?

    e. Create a pair-wise comparison chart of the two best alternatives.

    f. Create a hybrid solution that improves performance on one or more highly weighted measures of the best or second best alternatives. Increase the cost of that alternative to reflect the improved performance. How does this new alternative compare with the best alternative on the original list?

6.4 Refer to Exercise 5.9—the women's retailer exercise. Use the weights and single utility functions you previously assigned, or create new ones.

6.5 Refer to Exercise 5.10—the used-car exercise. Use the weights and single utility functions you previously assigned, or create new ones.

6.6 Refer to Exercise 5.11—the kitchen remodeler. Use the weights and single utility functions you previously assigned, or create new ones.

6.7 Refer to Exercise 5.12—the blower motor. Use the weights and single utility functions you previously assigned, or create new ones.

6.8 Refer to Exercise 5.13—plant location. Use the weights and single utility functions you previously assigned, or create new ones.

References

Baker, S. F., Green, S. G., Lowe, J. K., and Francis, V. E. (2000). A value focused approach for laboratory equipment purchases. *Military Operations Research*, 5(4), 43–56.

Bana E Costa, C. A. (2001). The uses of multi-criteria decision analysis to support the search for less conflicting policy options in a multi-actor context: Case study. *Journal of Multi-Criteria Decision Analysis*, 10, 111–125.

Bresnick, T. A., Buede, D. M., Pisani, A. A., Smith, L. L., and Wood, B. B. (1997). Airborne and space-borne reconnaissance force mixes: A decision analysis approach. *Military Operations Research*, 3(4), 65–78.

Buede, D. M. and Bresnick, T. A. (1992). Applications of decision analysis to the military systems acquisition process. *Interfaces*, 22(6), 110–125.

Burk, R. C. and Parnell, G. (1997). Evaluating future military space technologies. *Interfaces*, 27(3), 60–73.

Chien, C.-F. and Sainfort, F. (1998). Evaluating the desirability of meals: An illustrative multiattribute decision analysis procedure to assess portfolios with interdependent items. *Journal of Multi-Criteria Decision Analysis*, 7, 230–238.

Corner, J. L. and Kirkwood, C. W. (1991). Decision analysis applications in the operations research literature, 1970–1989. *Operations Research*, 39, 206–219.

Davis, C. C., Deckro, R. F., and Jackson, J. A. (2000). A value focused model for a c4 network. *Journal of Multi-Criteria Decision Analysis*, 9, 138–162.

Dyer, J. S., Edmunds, T., Butler, J. C., and Jia, J. (1998). A multiattribute utility analysis of alternatives for the disposition of surplus weapons-grade plutonium. *Operations Research*, 46, 749–762.

French, S. (1996). Multi-attribute decision support in the event of a nuclear accident. *Journal of Multi-Criteria Decision Analysis*, 5, 39–57.

Hall, N. G., Hershey, J. C., Kessler, L. G., and Stotts, R. C. (1992). A model for making project funding decisions at the national cancer institute. *Operations Research*, 40, 1040–1052.

Hämäläinen, R. P., Lindstedt, M. R. K., and Sinkko, K. (2000). Multiattribute risk analysis in nuclear emergency management. *Risk Analysis*, 20, 455–467.

Jackson, J. A., Parnell, G. S., Jones, B. L., Lehmkuhl, L. J., Conley, H. W., and Andrew, J. M. (1997). Air force 2025 operational analysis. *Military Operations Research*, 3(4), 5–21.

Jones, M., Hope, C., and Hughes, R. (1990). A multi-attribute value model for the study of UK energy policy. *Journal of the Operational Research Society*, 41, 919–929.

Keefer, D. L., Kirkwood, C. W., and Corner, J. L. (2004). Perspective on decision analysis applications, 1990–2001. *Decision Analysis*, 1, 4–22.

Keeney, R. L. (1999). Developing a foundation for strategy at Seagate software. *Interfaces*, 29(6), 4–15.

Keeney, R. L. (2001). Modeling values for telecommunications management. *IEEE Transactions on Engineering Management*, 48, 370–379.

Keeney, R. L. and McDaniels, T. L. (1992). Value-focused thinking about strategic decisions at BC hydro. *Interfaces*, 22(6), 94–109.

Keeney, R. L. and McDaniels, T. L. (1999). Identifying and structuring values to guide integrated resource planning at BC gas. *Operations Research*, 47, 651–662.

Keeney, R. L., McDaniels, T. L., and Swoveland, C. (1995). Evaluating improvements in electric utility reliability at British Columbia Hydro. *Operations Research*, 43, 933–947.

Keeney, R. L. and von Winterfeldt, D. (1994). Managing nuclear waste from power plants. *Risk Analysis*, 14, 107–130.

Kidd, J. B. and Prabhu, S. P. (1990). A practical example of a multiattribute decision aiding technique. *Omega*, 18, 139–149.

Little, J. D. C. (1970). Models and managers: The concept of a decision calculus. *Management Science*, 16(8), B-466–485.

McDaniels, T. L. (1995). Using judgment in resource management: A multiple objective analysis of a fisheries management decision. *Operations Research*, 43, 415–426.

Merrick, J. R. W., Parnell, G. S., Barnett, J., and Garcia, M. (2005). A multiple-objective decision analysis of stakeholder values to identify watershed improvement needs. *Decision Analysis*, 2, 44–57.

Noonan, F. and Vidich, C. A. (1992). Decision analysis for utilizing hazardous waste site assessments in real estate acquisition. *Risk Analysis*, 12, 245–251.

Rios Insua, D. and Salewicz, K. A. (1995). The operation of Lake Kariba: A multiobjective decision analysis. *Journal of Multi-Criteria Decision Analysis*, 4, 203–222.

Stafira, S. Jr., Parnell, G. S., and Moore, J. T. (1997). A methodology for evaluating military systems in a counter proliferation role. *Management Science*, 43, 1420–1430.

Toland, R. J., Kloeber, J. M. Jr., and Jackson, J. A. (1998). A comparative analysis of hazardous waste remediation alternatives. *Interfaces*, 28(5), 70–85.

Von Winterfeldt, D. and Schweitzer, E. (1998). An assessment of tritium supply alternatives in support of the U.S. nuclear weapons stockpile. *Interfaces*, 28(1), 92–112.

Chapter 7

Multiple Objective Decisions with Limited Data: Analytical Hierarchy Process

Mike Smith is looking to buy a snow blower to clear his 50 ft long and 12 ft wide, slightly sloped driveway and sidewalk. He also would like to use it when he occasionally travels to his rustic cottage up north, where the driveway is unpaved and very long. Mike has a limited budget and has narrowed down the possible options to three: Monda, Tara, and Zraft. Monda is a single-stage snow blower. It is light and small, a fact that makes it easy to use. It is also the least expensive. Monda, however, has poorer performance because it only clears an 18 in. swath and cannot be used effectively on gravel. Tara is also a single-stage snow blower. It has a 21 in. swath and is more expensive than Monda. Zraft, is a two-stage, heavier snow blower that is more expensive than Monda and Tara, but it cuts a much wider clearing path. Moreover, it can be used on gravel. Mike is evaluating these three snow blowers by considering cost, performance, and ease of use objectives. How should Mike decide which snow blower to select?

MedPhar, a pharmaceutical company, has decided to buy R&D [Research and Diagnostic] project management software. The software decision team consists of experts from project management, purchasing, and information system. They have identified three software alternatives that satisfy their requirements: OG3, GT, and PRS. The team has identified four major objectives: total cost, implementation time, flexibility and reliability, and technical capability and service quality. None of the alternatives dominates the others in terms of all objectives. For example, OG3 is best in terms of flexibility and reliability, but it is the most expensive. Similarly, PRS is the most attractive alternative in terms of total cost, but it is weak on technical capability. GT is very strong on technical capability and implementation time, but weak on other objectives. How should MedPhar decide which project management software to select?

7.1 Goal and Overview

In this chapter, we introduce a second analytical methodology to deal with multiple-objective problems—the Analytical Hierarchy Process (AHP). AHP, when compared to the Multi-Attribute Utility Theory (MAUT), is less data intensive and structured with regard to its measures and scaling. It is less rigid in its data requirement and is built on a natural decision process of pair-wise comparisons. MAUT has a disadvantage in that the process of creating measures and scales can be complex and unnatural. However, MAUT has an advantage in that it can explicitly incorporate the quantification of uncertainty. More importantly, with MAUT, the score of each alternative has absolute meaning, and this facilitates the development of value-added alternatives. In contrast, AHP scores are strictly relative and may be harder to use as a springboard for brainstorming better alternatives. The chapter closes with a detailed discussion of the relative advantages and disadvantages of the two techniques.

AHP was developed by Thomas Saaty (1980) as a comprehensive, logical, and structural framework, to facilitate the understanding of complex decisions. It decomposes the problem into a hierarchical structure, as presented in Chapter 4. The application of AHP approach explicitly recognizes and incorporates the knowledge and expertise of the participants by using their subjective judgments at every step of the process. This is a particularly important feature for decisions that involve relatively limited data.

AHP involves comparisons of objectives and alternatives in a natural, pair-wise manner. It then takes these comparisons and converts them into ratio-scale weights that are combined into linear additive weights for associated alternatives. These resultant scores are used to rank alternatives, thereby assisting the decision maker in making a choice.

AHP has been applied to a wide range of problem situations. See http://www.expertchoice.com for references to over 1000 articles and almost 100 dissertations on the subject. For example, Xerox Corporation has used AHP in over 50 major decisions. These include R&D decisions on portfolio management, technology implementation, market segment prioritization, and engineering design selection. The British Columbia Ferry Corporation used AHP in the selection of products, suppliers, and consultants (Forman and Gass 2001). Lai et al. (2002) used AHP for software selection in a group decision-making setting. Korea Telecommunication Authority used AHP for prioritization, forecasting, and resource allocation (Suh et al. 2004). Tam and Tummala (2001) have used AHP in vendor selection of a telecommunication system.

Rochester General Hospital used AHP to develop and disseminate medical practice guidelines. These practice guidelines are directed at a large class of patients; they have a different focus than clinical practice, which is directed at one patient at a time. AHP was undertaken to reconcile different viewpoints and to develop improved and effective medical care practice guidelines (Dolan and Bordley 1992).

The power and simplicity of AHP, along with a strong marketing effort, have led to its widespread use. It is supported by a number of software products, such as Expert Choice and Logical Decisions. Forman and Gass (2001) and Vaidya and Kumar (2006) provide a broad overview of AHP applications. For the reader interested only in AHP, Expert Choice provides the user with tools that facilitate the framing and development of the hierarchical objectives structure. Logical Decisions, on the other hand, has an advantage in that it enables the modeler to use either AHP or MAUT or a combination of the two (Figure 7.1).

AHP includes the same basic four tasks as MAUT, but the process involved in completing these tasks differs significantly. The four tasks are as follows:

- *Structure*: The process starts by creating goals and specific objectives. However, AHP does not call for detailed scaled measures to define each objective.
- *Describe alternatives*: With MAUT, specific numeric values or scores must be assigned to each measure on each alternative. In contrast, AHP permits a mix of quantitative and non-quantitative descriptions for each alternative.
- *Clarify preferences for measures*: Part I: Identify relative importance of the various objectives/measures.
  a. Develop a pair-wise comparison matrix for the objectives/measures
  b. Check the consistency of the pair-wise comparisons
  c. Calculate weights for the objectives/measures using Eigen values for the matrix of pair-wise comparisons

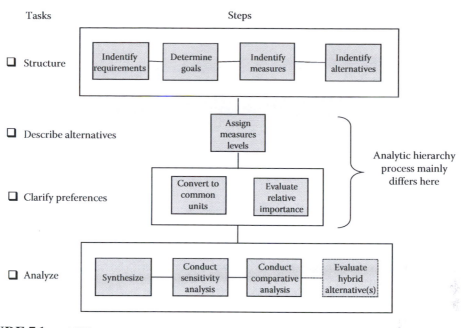

FIGURE 7.1: AHP process.

- *Clarify preferences for alternatives*: Part II: Develop a relative score for each alternative on each measure.
  a. Develop a pair-wise comparison matrix for alternatives on each measure
  b. Check the consistency of the pair-wise comparisons
  c. Calculate a relative score for each alternative on each measure by finding the matrix eigen values

- *Analyze:* Synthesis
  a. Compute an overall score and rank the alternatives. The scores for the alternatives for AHP are strictly relative and must sum to 1.
  b. Analyze strengths and weaknesses.

7.2 AHP Procedure Details and Snow Blower Example

The starting point for AHP is the same as for MAUT. The process begins with laying out an objectives hierarchy. The first example we consider is the choice of a snow blower, as depicted in Figure 7.2. Mike Smith is looking to buy a snow blower to clear his slightly sloped, 50 ft long and 12 ft wide driveway and sidewalk. He also would like to use it when he occasionally travels to his rustic cottage up north, where the driveway is unpaved and very long. Unlike an MAUT objectives hierarchy, AHP, as illustrated in Figure 7.2, includes the alternatives at the lowest level of the hierarchical tree.

1. *Structure the problem:* AHP initially breaks down a complex multicriteria problem into a hierarchy of interrelated elements (objectives, subobjectives, measures, and alternatives.) A hierarchy must have at least three levels: overall objective at the top; major objectives, that define how to accomplish the overall objective, in the middle; and alternatives, not measures, at the bottom. The middle level can have multiple sublevels of objectives or measures. In the

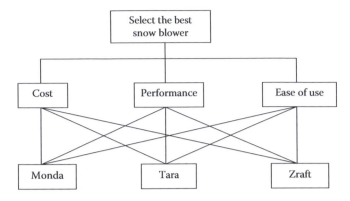

FIGURE 7.2: Snow blower objectives and alternatives.

snow blower decision, for example, Mike specified three objectives at the second level: cost, performance, and ease of use. The lowest level in the pictured hierarchy lists the alternative snow blowers.

2. *Describe alternatives with regard to objectives:* In Table 7.1, we describe the three alternatives. The description of performance is multifaceted.

3. *Identify relative importance of the objectives/measures:* Once the hierarchy has been constructed, we proceed to determine the relative importance of each objective or measure. We recommend, as a first step, to rank order the objectives from most important to least important. This will facilitate maintaining rank order consistency during pair-wise comparisons. We will then use the scale in Table 7.2 as we compare the highest ranked objective to the objective ranked second. The highest ranked objective is next compared to the objective ranked third, and then to the objective ranked fourth, etc. Next we compare the second highest ranked objective to each of the lower ranked objectives.

TABLE 7.1: Description of snow blower alternatives.

Alternatives	Cost	Performance	Ease of Use
Monda	$390	Single stage, cuts 18 in. swath, but picks up gravel mainly used on flat surfaces.	Lighter weight (54 lb)
Tara	$450	Single stage, cuts 21 in. swath, but picks up some gravel. Can handle sloped driveways.	Medium weight (73 lb)
Zraft	$620	Two staged, cuts a larger than 26 in. swath, can handle sloped driveways, and does NOT pick up gravel. It is highly versatile.	Heavy (160 lb)

TABLE 7.2: Interpretation of importance in pair-wise comparison.

Scale	Interpretation
1	Objectives i and j are *equally* important
3	Objective i is *moderately* more important than objective j
5	Objective i is *strongly* more important than objective j
7	Objective i is *very strongly* more important than objective j
9	Objective i is *extremely* more important than objective j
2, 4, 6, 8	Intermediate values. For example, a value of 4 means that objective i is midway between moderately and strongly more important than objective j

The pair-wise comparison of objectives involves asking how much more important objective i is than objective j. Importance is to be described on an integer-valued 1–9 scale, with each number having an interpretation illustrated in Table 7.2. A value of 1 means that the two objectives are *equally* important to the decision maker. At the other end of the scale, 9 means that one objective is *extremely* more important than the other objective. A value of 5 corresponds to *strongly* more important. The link between the descriptive word and the numeric scale is based on the experiences of the developers of AHP.

The pair-wise comparison responses are then captured in matrix form. Suppose we have N objectives/measures. We write down $N \times N$ pair-wise comparison matrix as depicted in Figure 7.3. The entry in row i and column j (call it a_{ij}) indicates how much more important objective i is than objective j. For all i, it is necessary that $a_{ii}=1$ because any objective compared against itself must be equally preferred with a preference value of 1. For consistency, it is necessary that $a_{ij}=1/a_{ji}$.

Now consider the snow blower selection problem. When Mike goes to buy a snow blower, he considers cost, performance, and ease of use as they pertain to each snow blower. Let us compare cost and performance using pair-wise comparison. For the sake of this example, we will assume that cost is more important to Mike than performance. Table 7.3 demonstrates the pair-wise comparison of cost and performance. For example, scale "3" indicates that cost is *moderately* more important than performance.

Table 7.4 shows the mathematical correspondence between the pair-wise comparison scales and the relative weights assigned. For example, scale 3 reflects a weight assigned to cost that is three times as large as the weight assigned to performance. If there were only two objectives, this multiplier of 3 would be exact. However, with multiple comparisons that are not perfectly

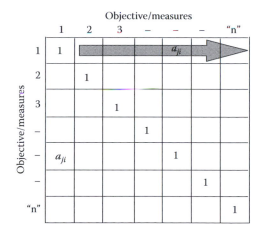

FIGURE 7.3: Pair-wise comparison matrix.

TABLE 7.3: Pair-wise comparison of objectives for snow blower selection example.

Objective	Cost	Performance	Ease of Use
Cost	1	3	6
Performance	1/3	1	3
Ease of use	1/6	1/3	1
Sum	**9/6**	**13/3**	**10**

TABLE 7.4: Pair-Wise comparison of cost and performance.

Scale	Wording and Meaning	Weight Equivalence
1	Cost and performance are equally important	Weight (*Cost*) = 1 × Weight (*Performance*)
3	Cost is moderately more important than performance	Weight (*Cost*) = 3 × Weight (*Performance*)
5	Cost is strongly more important than performance	Weight (*Cost*) = 5 × Weight (*Performance*)
7	Cost is very strongly more important than performance	Weight (*Cost*) = 7 × Weight (*Performance*)
9	Cost is extremely more important than performance	Weight (*Cost*) = 9 × Weight (*Performance*)
2, 4, 6, 8	Intermediate values	

consistent, the calculation of the relative weight is more complicated, as described as follows. In studying AHP, our students have been surprised to learn that the term "moderately" more important results in a weight three times as large. They expected a smaller magnitude difference. However, experienced AHP analysts are comfortable with this correspondence between terms and numeric weights.

7.2.1 Calculate Weights for the Objectives/Measures

The relative importance of each objective/measure is calculated by using the normalized eigenvectors of the comparison matrices. AHP software such as Logical Decisions for Windows (LDW) and Expert Choice automatically calculate weights for measures once pair-wise comparisons are completed. Here, we introduce a simple procedure to calculate weights without a computer.

Normalize the comparison matrix column by column: Once all the relevant pair-wise comparisons have been made, the matrix is normalized in two steps. In the first step, the numbers in each column are added up. In the second step, each value in the column is then divided by the column sum to yield its normalized score. After normalization, the sum of the each column will equal 1.

Average the values row by row: For each row, we determine its average value for the normalized matrix created earlier. The average for each row represents the weight assigned to that objective.

Table 7.3 presents Mike's pair-wise comparison matrix of the three objectives associated with the snow blower decision. *Cost* was ranked first, *performance* second, and *ease of use* third. Each column was then summed. Cost was rated as moderately more important than *performance*. *Cost* was scored between strongly and very strongly more important than *ease of use*. When Mike compared *performance* and *ease of use*, he rated *performance* as moderately more important. The values in Table 7.5 are calculated by dividing each value in Table 7.3 by the sum of its corresponding column.

TABLE 7.5: Normalized matrix for snow blower selection example with row averages.

Objective	Cost	Performance	Ease of Use	Row Average
Cost	6/9 = 1/(9/6) = 0.67	0.69	0.6	**0.655**
Performance	2/9 = (1/3)/(9/6) = 0.22	0.23	0.3	**0.250**
Ease of use	1/9 = (1/6)/(9/6) = 0.11	0.08	0.1	**0.095**
Sum	**1**	**1**	**1**	**1**

For example, the values under *Cost* in Table 7.5 were found by dividing the values in Table 7.3 by (9/6). The columns in Table 7.5 each sum to 1. To determine the weights, the average across each row is then calculated. *Cost* has a weight of 0.655, *performance* a weight of 0.25, and *ease of use* has a weight of 0.095.

7.2.2 Consistency Measurement

There is diverse literature documenting the problems of consistency in decision making, including a number of decision paradoxes. MAUT avoids this problem when calculating its weights by asking exactly enough questions to generate the equations needed to determine the weights. Saaty, in developing AHP, chose instead to address inconsistency head-on, as decision makers are not totally consistent at all levels. AHP accepts a modest level of inconsistent pair-wise comparisons but redirects a review of the weightings when the inconsistent responses are beyond a specified bound. In Appendix A, we present the consistency ratio (CR) calculations.

In this first example, the decision maker has specified that *cost* is moderately, or three times, as important as *performance*. The decision maker has also specified that cost is somewhere between strongly and very strongly more (a scale of 6) important than *ease of use*. This also indicates that to the decision maker, *performance* is more important than *ease of use*. To be absolutely consistent, the decision maker would have to specify that *performance* is twice as important as *ease of use*. Were the decision maker to specify that *performance* is moderately more important than *ease of use*, a scale of 3 would be somewhat inconsistent. This inconsistency is reflected in the normalized matrix by the fact that the values in each row in Table 7.5 diverge somewhat from the overall average. For example, in the first row, the values range from 0.6 to 0.69 and the average is 0.655. (You can confirm that if *performance* were scaled 2 relative to *ease of use*, each value in a row would be identical. The three averages would then be 0.67, 0.22, and 0.11.)

We do not want a decision to be based on judgments that have so low a rate of consistency that the comparisons appear to be almost random. Saaty developed a procedure to help decision makers check consistency and stay within an acceptable range. All of the various software packages automatically calculate this CR. Saaty's guideline is that the value of the CR should be 0.1 or less. If the value is greater than 0.1, the pair-wise comparisons should be reviewed and revised. Software packages also help identify inconsistent comparisons to be reviewed. For the snow blower matrix, the CR was 0.018, well below the 0.1 threshold.

Identify score for each alternative on each measure: We also use pair-wise comparisons to determine how well each alternative "scores" or "satisfies" each objective/measure relative to the other alternatives. To determine these scores, we construct for each objective a pair-wise comparison matrix, in which rows and columns are alternatives. In order to facilitate maintaining consistency in pair-wise comparisons, the decision maker should first rank order the alternatives with respect to each objective from best to least preferred. The decision maker then compares each pair of alternatives with respect to an objective and compares the two alternatives' relative preference, using the same scale presented in Table 7.2. A minor difference is that the term "preferred" should be substituted for the phrase "more important." Thus, alternative A might be "strongly preferred" to alternative B with regard to cost.

In this step, three comparison matrices are created to represent Mike's relative preference for snow blowers with respect to each of three measures. The raw data on cost are presented in Table 7.1. At this point, Mike needs to capture his preferences with regard to these data. The rank ordering is obvious, from lowest to highest cost. Mike's pair-wise comparison rating for each of three snow blowers for the cost measure is summarized in Table 7.6. This pair-wise comparison matrix shows that Monda's cost of $390 is only slightly preferred over Tara's $450, and very strongly preferred over Zraft's $620. Tara's cost is strongly preferred over Zraft's.

TABLE 7.6: Pair-wise comparison matrix for
alternatives for *cost* objective.

		Cost Objective		
		Comparison of Alternatives		
Rank Order		**Monda**	**Tara**	**Zraft**
1	Monda	1	2	7
2	Tara	1/2	1	5
3	Zraft	1/7	1/5	1
	Sum	23/14	16/5	13

TABLE 7.7: Normalized matrix for alternatives on
cost objective for snow blower.

	Cost Objective			
	Comparison of Alternatives			
	Monda	**Tara**	**Zraft**	**Row Average Score**
Monda	14/23	10/16	7/13	0.59
Tara	7/23	5/16	5/13	0.33
Zraft	2/23	1/16	1/13	0.08
Sum	1	1	1	1

Table 7.7 shows a normalized matrix for the three alternatives on the *cost* measure. Monda scored the highest, 0.59 on cost, followed by Tara with a score of 0.33. Zraft is ranked third on cost, with a score of 0.08.

With regard to performance, Zraft was rated best because of its wide swath and its handling of gravel paths. Because gravel was only an infrequent concern, Mike rated Zraft a 3 compared to Tara, which had the same width but was not as effective on a gravel driveway. Zraft was very strongly preferred to Monda, both because of its width and because of its handling of gravel. Tara's width, handling of sloped driveways, and slightly better performance on gravel, led Mike to scale it 5, that is, strongly preferred as compared to Monda (Table 7.8). The CR score for this matrix of performance comparisons is 0.003.

In considering ease of use, Mike preferred a lighter snow blower, but was concerned that it not be so light that it would bounce off snow that had become compacted. He, therefore, rated Tara the best in this category, with Monda a close second best. Tara was strongly preferred over Zraft because of the weight issue. Monda was also moderately preferred over Zraft, because of its lighter weight. The CR score for this matrix delineating ease of use comparisons is 0.004 (Table 7.9).

Synthesis: An overall score for each alternative is calculated by multiplying the alternative's score on each objective by that objective's corresponding weight. The total score for each alternative is calculated as follows and is summarized in Table 7.10.

$$\text{Monda score} = 0.655 * 0.59 + 0.25 * 0.07 + 0.095 * 0.31 = 0.435$$

$$\text{Tara score} = 0.655 * 0.33 + 0.25 * 0.28 + 0.095 * 0.36 = 0.343$$

$$\text{Zraft score} = 0.655 * 0.08 + 0.25 * 0.65 + 0.095 * 0.11 = 0.222$$

TABLE 7.8: Pair-wise comparison matrix for alternatives on performance.

| Rank Order | | Performance Objective | | | |
| | | Comparison of Alternatives | | | |
		Monda	Tara	Zraft	Score
3	Monda	1	1/5	1/7	0.07
2	Tara	5	1	1/3	0.28
1	Zraft	7	3	1	0.65
	Sum	13	4.2	1.476	

TABLE 7.9: Pair-wise comparison matrix for alternatives for ease of use.

| Rank Order | | Ease of Use Objective | | | |
| | | Comparison of Alternatives | | | |
		Monda	Tara	Zraft	Score
2	Monda	1	1/2	3	0.31
1	Tara	2	1	5	0.58
3	Zraft	1/3	1/5	1	0.11
	Sum	23/14	16/5	13	

TABLE 7.10: Scores for each alternative on each measure.

| | Objectives | | | |
	Cost	Performance	Ease of Use	
Weights	0.655	0.25	0.095	
Snow Blower	**Scores on Measure**			**Total Score**
Monda	0.59	0.07	0.31	0.435
Tara	0.33	0.28	0.36	0.343
Zraft	0.08	0.65	0.11	*0.222*

Figure 7.4 shows LDW ranking of snow blower alternatives. Monda is ranked first, with a utility score of 0.435; followed by Tara, with a utility score of 0.343; Zraft is ranked third, with a utility score of 0.223. Monda is extremely strong on cost and it is the second best alternative on ease of use. It is, however, extremely weak on performance measure. Notice that the final scores sum to 1, as required by AHP.

We explored the impact of changing the relative weight assigned to the various objectives. Figures 7.5 and 7.6 display the impact of changes in the weights assigned to cost or performance. The current weight on cost is 0.65 as indicated by the perpendicular line. This would have to be

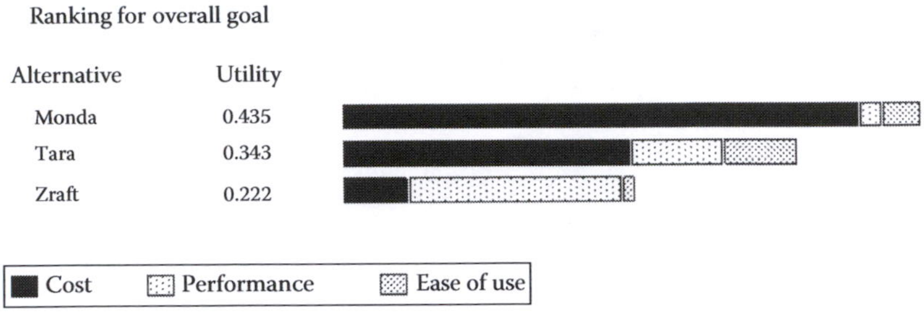

Ranking for overall goal
Alternative
Utility
Monda
0.435
Tara
0.343
Zraft
0.222
Cost
Performance
Ease of use

FIGURE 7.4: Ranking of snow blowers.

Monda
Tara
Zraft
Best
Utility
Worst
0
100
Weight on cost objective (%)

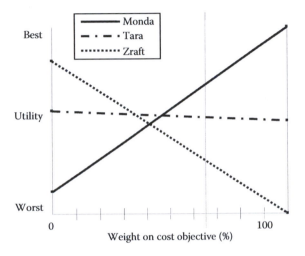

FIGURE 7.5: Sensitivity analysis of weight on cost.

Zraft
Tara
Monda
Best
Utility
Worst
0
100
Weight on performance objective (%)

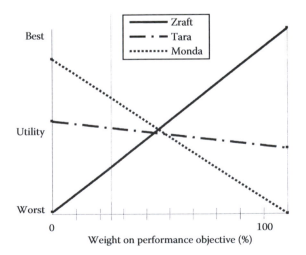

FIGURE 7.6: Sensitivity analysis of weight on performance.

reduced to below 0.5 before the alternative rankings would change. Conversely, the weight on performance would have to increase from 0.25 to 0.45 before the rank orderings change. A weight change of this magnitude would make Zraft the optimal choice. In essence, as long as cost is greatly more important to Mike than performance, Monda remains the preferred alternative. In Figure 7.6, Tara is never the preferred alternative.

7.3 Commercial Snow Throwers Selection

We began demonstrating AHP with a simple example, the selection of a snow blower to clear Mike's driveway and sidewalk. Mike considered three objectives: cost, performance, and ease of use. Now we will consider a decision that involves multiple levels of objectives. Our primary goal is to demonstrate how AHP efficiently handles a multiple objectives hierarchy that has several levels. We will begin by discussing the basis for the assessments articulated by Don Snowden, CEO of Snow Remove.

Snow Remove offers snow clearance and removal to citizens in a local community. It is planning to buy 10 new snow throwers. Snow Remove has identified three two-stage snow throwers: Craft23, Zara MK, and ZSC. Craft23 is excellent for clearing dense, compacted, and/or deep snow. It has an easily adjustable chute deflector and power-driven wheels, which can disengage independently to assist in turning. A 7-horsepower motor supports its 24 in. clearing path. However, it does not throw the snow quite as far as Zara and ZSC. Craft23 is cheaper, but it has some quality problems.

Features-wise, the Zara MK is virtually identical to the Craft23, but the Zara throws farther. The Zara MK is designed to handle long sloping driveways and snowfalls of more than 8 in. Zara MK is the best for gravel drives, since the auger does not touch the surface of the driveway. Zara is known as the highest quality snow thrower, and requires less maintenance, but it is also the most expensive of the three.

Finally, ZSC is designed to handle most wet, heavy snowfalls with ease as well as deeper snowfalls up to 12 in. ZSC has a 28 in. clearing path with a 9-horsepower engine to tackle even the heaviest snow conditions. In terms of quality, ZSC is better than Craft23, but not as reliable as Zara MK. It is cheaper than Zara MK, but more expensive than Craft23. The descriptions of relative performance are summarized in Table 7.11. Table 7.12 lists the relevant cost information. The purchase price for each snow thrower is fixed, but the operating costs, including worker wages, are expressed as ranges. Don cannot be sure how much usage each machine will get each snow season. The main differences in operating costs relate to his estimation of the relative efficiency of the machines, a value that directly impacts worker payroll. In this regard, the wider 28″ path cut by ZSC saves time.

TABLE 7.11: Snow thrower performance description.

	Performance		
Alternatives	**Snow Clearance Capability**	**Ease of Use**	**Reliability**
Craft23	Densely compacted and deep snow. Adjustable chute, clears 24 in. path. Snow throw is smallest	Power wheels for turning	Fair
Zara MK	Densely compacted and deep snow. Adjustable chute, clears 24 in. path. Throws farther than Craft23	Power wheels for turning. Good on sloping driveways and gravel	Excellent
ZSC	Good for wet and very heavy snowfalls, clears 28 in. path and throws farthest	Heavier than the others but usable on all surfaces	Good

TABLE 7.12: Snow thrower cost information per unit.

	Cost	
Alternatives	**Purchase**	**Yearly Operating Cost Range Including Labor**
Craft23	$1100	$1900–$2500
Zara MK	$1450	$1600–$2100
ZSC	$1325	$1500–$1900

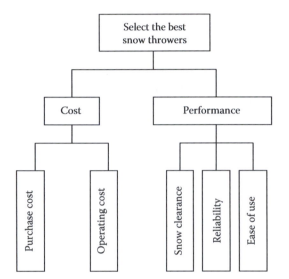

FIGURE 7.7: Goals hierarchy for selecting commercial snow throwers.

7.3.1 Create Hierarchy for Snow Thrower

Figure 7.7 shows the decision hierarchy for selecting the commercial snow thrower. The overall objective of Snow Remove is to select the best snow thrower. The second level of the hierarchy shows that Snow Remove has two major objectives: minimize cost and maximize performance. The last level of the hierarchy represents a set of measures for each major objective. Cost consists of purchase cost and annual operating cost, including maintenance. The performance objective is disaggregated into snow clearance, reliability, and ease of use.

7.3.2 Identify Relative Importance of Objectives/Measures

In this example, the hierarchy has both objectives and measures levels. Here, Snow Remove management compares the importance of major objectives, "cost" and "performance" first. Then they compare the cost measures with each other. Finally, the three measures of performance are compared with each other.

Table 7.13 presents the pair-wise comparison matrix for the two major objectives. From management's perspective, cost is twice as important as performance. Cost accounts for 67% of the total weight, whereas the quality and performance objective accounts for the remaining 33% of the weight. Note that as Snow Remove management compares only two objectives, there can be no inconsistency in this pair-wise comparison matrix.

TABLE 7.13: Pair-wise comparison of major objectives for snow throwers selection.

	Cost	Performance	Weight
Cost	1	2	0.67
Performance	1/2	1	0.33

TABLE 7.14: Pair-wise comparison of measures for cost objective for snow thrower.

	Purchase Cost	Operating Cost	Weight
Purchase cost	1	2	0.67
Operating cost	1/2	1	0.33

TABLE 7.15: Pair-wise comparison of measures under performance objective for snow thrower.

	Snow Clearance	Reliability	Ease of Use	Weight
Snow clearance	1	2	5	0.58
Reliability	1/2	1	3	0.31
Ease of use	1/5	1/3	1	0.11

This pair-wise comparison matrix for measures under the cost objective is shown in Table 7.14. Purchase cost is slightly (less than moderately) more important than operating cost, with a weight of 0.67. This cost is a sure thing that is paid upfront. Operating costs depend on the amount of snowfall and worker productivity.

Table 7.15 illustrates the pair-wise comparison matrix for measures under the performance objective. Snow Clearance is the highest priority measure followed by reliability. Snow Clearance collects 58% of the performance objective weights, reliability garners 31%, and ease of use accounts for only 11%. The CR for this pair-wise comparisons matrix is 0.003.

7.3.3 Identify Score for Each Alternative on Each Measure

Table 7.16 shows pair-wise comparisons matrices of alternatives on each measure and the resultant scores. The rankings are summarized in Figure 7.8. Craft23 is ranked first only on purchase cost. Zara MK is ranked first on reliability and ease of use measures, and second on operating cost and performance. ZSC has the highest scores on operating cost and snow removal, and the second highest scores on purchase cost and reliability. The overall score for each alternative is presented in Figure 7.9. ZSC is clearly ranked the highest, ahead of Craft23, primarily because of its overall advantage with regard to cost. Figure 7.10 presents a more detailed pair-wise comparison of the top two alternatives. ZSC's advantages over Craft23 with regard to snow clearance and operating cost more than compensate for its disadvantage with regard to purchase price. The overall weight currently assigned to purchase price is 0.44. This weight would have to exceed 0.55 before the ranking would change (Figure 7.11). In Table 7.13 and Table 7.14, the relative importance of cost over performance was set as 2 and the relative importance of purchase cost to operating cost was also set at 2. Both values would have to increase to 3 before the ranking would change.

TABLE 7.16: Pair-wise comparison of alternatives on each measure for selection of snow throwers.

	Purchase cost					Operating cost		
	Craft	Zara MK	ZSC	**Score**		Craft	Zara MK	ZSC	**Score**
Craft	1	5	3	**0.64**	Craft	1	0.33	0.14	**0.09**
Zara MK	0.2	1	0.33	**0.11**	Zara	3	1	0.5	**0.29**
ZSC	0.33	3	1	**0.26**	ZSC	7	2	1	**0.62**

	Reliability					Snow removal		
	Craft	Zara MK	ZSC	**Score**		Craft	Zara MK	ZSC	**Score**
Craft23	1	0.17	0.33	**0.10**	Craft	1	0.33	0.17	**0.09**
Zara MK	6	1	3	**0.66**	Zara	3	1	0.25	**0.22**
ZSC	3	0.33	1	**0.25**	ZSC	6	4	1	**0.69**

	Ease of use			
	Craft	Zara MK	ZSC	**Score**
Craft23	1	0.33	0.5	**0.24**
Zara MK	3	1	4	**0.63**
ZSC	0.5	0.25	1	**0.14**

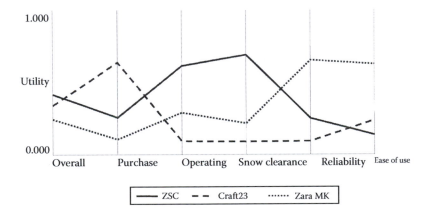

FIGURE 7.8: Utility score for snow thrower on each measure.

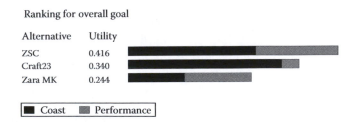

FIGURE 7.9: Overall score for each snow thrower.

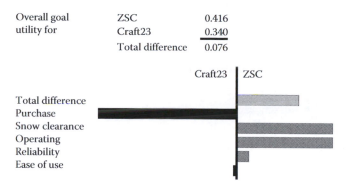

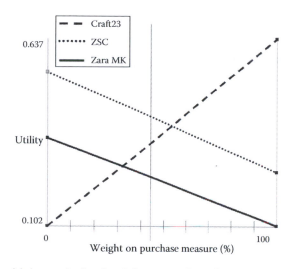

FIGURE 7.10: Comparison of best two snow throwers on each measure.

FIGURE 7.11: Sensitivity analysis of weight on purchase for snow throwers.

7.4 Select a Job

In the snow blower example, there was some concrete data for each of the objectives for every alternative. It was possible to create measures for each objective, score each alternative on each measure, and then apply MAUT. Cost has a natural measure. Performance would have been split into two measures: width and ability to handle gravel. Width would again be a natural measure, but a scale would have been needed to characterize the ability to handle gravel. Finally, weight of the snow blower could have been used as a measure of ease of use. In the next example, job selection, the primary objectives are difficult to quantify. In addition, there is ambiguity as to how well each alternative performs on each objective, especially with regard to the long-term. As a result, AHP is a more natural multi-objective modeling tool than MAUT.

Mary Gill has been working for a medium size chemical company for a long time. She has recently received two offers from (a) a different part of the same organization, and (b) a new organization. We have deliberately not described her current job. This allows each reader to place himself or herself in the position of deciding on a job change relative to his/her current job or status quo.

She has three alternatives as summarized as follows:

1. Her current job
2. A lateral move to different part of same organization
   a. $5000 raise
   b. Assignment—A little more responsibility, with low visibility and little pressure
   c. New Boss—Easy to work for, retiring in a year, committed to his people
   d. Extra 20 minute commute each way
3. Promotion in new organization
   a. $10,000 raise
   b. Assignment—Significantly more responsibility, high visibility, and a good deal of pressure
   c. New Boss—Very difficult to work for, not retiring soon
   d. Relocate family for at least 2 years

Mary must determine which offer to accept. To evaluate the alternatives, she identified three objectives: salary, career path, and work-life balance. From Mary's point of view, job selection is a tough decision and requires trade-offs. For example, job offer 3 has the highest salary, but it requires relocation. Moreover, it seems to offer better career opportunities than her current job because its greater responsibility and higher visibility would enhance her resume. However, the immediate supervisor is not a pleasant person to work with and is not nearing retirement. Thus, the work environment is likely to be less than pleasant and any promotions open to Mary would have to be elsewhere in the company or require another job change.

7.4.1 Create Hierarchy for Job Selection

Figure 7.12 shows Mary's hierarchy. Her overall objective, to select the best job, is at the top of the hierarchy. The second level of the hierarchy shows how each of her three objectives (salary, career path, and work-life balance) contributes to achieving the overall objective. The bottom of the hierarchy demonstrates how each of the three alternatives (current job, job offer 2, and job offer 3) contributes to each of the three objectives.

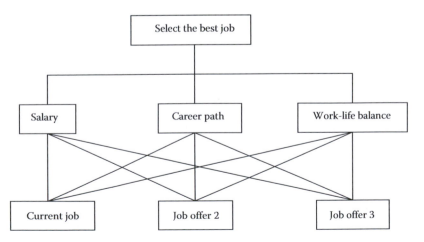

FIGURE 7.12: Hierarchy for selecting a job.

7.4.2 Identify Relative Importance of Objectives/Measures

In job selection, Mary considers salary the most important criterion, followed by career path. Work-life balance is the least important criterion. Of course, this is a subjective judgment and reflects her preference. Mary completed a pair-wise comparison matrix for her three objectives as depicted in Table 7.17. From her perspective, salary is only slightly (but not moderately) more important than career path. The comparison score is a 2. She considers salary moderately more important than work-life balance. She also believes that career is strongly more important than work-life balance.

The CR of her pair-wise comparison is 0.16 (calculated by LDW software), which is greater than 0.1. This means that Mary's pair-wise comparisons are not consistent. When she compared career to work-life balance, she assigned a score of 5. This is a higher value than she assigned when comparing salary to work-life balance. Yet, she rated salary more important than career. Consistency would suggest that career is preferred to salary, which is not how she rated the two relative to each other. Now, assume that she revised her pair-wise comparison as shown in Table 7.18. She reduced the relative preference for career over work-life balance to only 2. The new CR is 0.009. Even if she had scored it a 3 and remained somewhat inconsistent, the ratio would have declined to 0.05, well below the inconsistency cutoff. The relative weights for salary, career, and work-life balance are 0.54, 0.30, and 0.16, respectively.

7.4.3 Identify Score for Each Alternative on Each Objective

In this step, Mary compares each pair of jobs with respect to each objective. To determine these scores, she constructs a pair-wise comparison matrix in which the rows and columns are job offers (see Table 7.19). She found the salary comparisons the easiest to perform. She knew exactly how

TABLE 7.17: Pair-wise comparison for objectives.

	Salary	Career	Work-Life Balance
Salary	1	2	3
Career	1/2	1	5
Work-Life B	1/3	1/5	1

TABLE 7.18: Pair-wise comparison for objectives-revised.

	Salary	Career	Work-Life Balance	Weights
Salary	1	2	3	0.54
Career	1/2	1	2	0.30
Work-life B	1/3	1/2	1	0.16

TABLE 7.19: Pair-wise comparison of jobs on salary.

Rank Order	Job	J1	J2	J3	Score
3	J1	1	1/3	1/5	0.105
2	J2	3	1	1/3	0.258
1	J3	5	3	1	0.637

each job would affect her immediate salary. At this point, she was not concerned about the long-term potential for pay raises in the various jobs. She also believed that the objective, maximize career, captured a great deal of the uncertainty regarding her future pay. With regard to salary, job 3 offers the highest salary. She, therefore, started filling in the matrix by working on row 3. When comparing the salaries for jobs 3 and 1, she felt the $10,000 higher salary was strongly preferred. It would enable her to pay off her credit card debt and begin saving for the future. When comparing the salaries of jobs 3 and 2, she preferred the extra $5,000 with job 3 moderately more than the salary for job 2. The salary for job 2 was moderately preferred over that of job 1. This would enable her to erase her credit card debt but she would not be able to begin accumulating any significant savings. The CR for this pair-wise matrix is 0.004.

Mary then began thinking about the work-life balance objective for each of the three jobs. She was very comfortable with her present home and, on most days, the commute to work was 25 minutes. In addition, she knew her job well and got along with her current boss. She, therefore, preferred her current job to the other two with regard to the work-life-balance objective. However, Mary was struggling with the increased workload, resulting from a series of cutbacks within her organization. Job 2's main drawback was that it added 20 minutes to her commute. This might grow tiresome over the long-term, but she did not have any pressing responsibilities at home after work that would be affected by this longer commute. In addition, she felt that the boss at job 2 was no worse than the one she had now and was perhaps even less demanding. She rated job 1 as moderately preferred over job 2 on this objective. However, she rated the work-life balance of job 1 as extremely preferred over job 3 since the latter involved a major move plus a hard-driving boss. When comparing jobs 2 and 3, she strongly preferred job 2's work-life balance to that of job 3. (CR of 0.028). Her matrix of work-life balance comparisons are summarized in Table 7.20. As a result, job 1 scored 0.672, with job 2 a distant second at 0.265.

Mary found the career comparisons the most difficult to make, but it was her second highest ranked objective and warranted a good deal of thought. She tried to think both in terms of the next year or two as well as the long term. She felt her current job offered little opportunity for career development. She decided that job 2 would offer the best immediate career opportunities. The current supervisor was nearing retirement and Mary considered herself his likely replacement should she join the group. Job 3, with its high visibility, was attractive from a career perspective. However, she was uneasy with the personality of the boss in job 3 and felt he might hinder her career; moreover, his position was not likely to be vacated any time soon. As a result, Mary rated job 2 strongly preferred over job 1 and moderately preferred over job 2. when she compared job 3 to her current job, however, she was very much influenced by the lack of opportunity in her current job. As a result, she initially scored job 3 as strongly preferred to job 1, a score of 5. However, this assessment was inconsistent with her prior statement that job 2 was moderately preferred over job 3. (If job 2 and job 3 were both 5s compared to job 1, then they should be equal when compared to one another.) The CR was 0.13 and above the cutoff. She decided to change her comparison of jobs 3 and 1 to a score of 4, thereby improving the CR to 0.082, within the acceptable consistency standard (Table 7.21).

TABLE 7.20: Pair-wise comparison of jobs on work-life balance.

Rank Order	Job	J1	J2	J3	Score
1	J1	1	3	9	0.672
2	J2	1/3	1	5	0.265
3	J3	1/9	1/5	1	0.063

TABLE 7.21: Pair-wise comparison of jobs on careers.

Rank Order	Job	J1	J2	J3	Score
3	J1	1	1/3	1/3	0.094
1	J2	5	1	3	0.627
2	J3	4	1/5	1	0.280

7.4.4 Synthesis

We compute the overall score for each alternative by synthesizing the objective weights with the scores for each job on each objective as follows:

$$\text{Job 1 Score} = 0.54*(0.11)+0.30*(0.09)+0.16*(0.67) = 0.19$$

$$\text{Job 2 Score} = 0.54*(0.26)+0.30*(0.63)+0.16*(0.27) = 0.37$$

$$\text{Job 3 Score} = 0.54*(0.64)+0.30*(0.028)+0.16*(0.06) = 0.44$$

Table 7.22 summarizes the results that are pictorially represented in Figure 7.13. Job 3 with the new company is ranked higher than the other two jobs. Mary's current job is ranked a distant third. Although it is best with regard to work-life balance, that objective was weighted relatively low. Job 3 was ranked the highest, primarily because of the significant salary raise; Mary had assigned more than half of the weight to salary. Figure 7.14 shows that her optimal choice is somewhat sensitive to this weight. If she were to reduce the weight to 0.44, her choice would change and job 2 would be preferred. However, after reviewing her assignment of weights, she decided she was comfortable with the large weight placed on salary.

TABLE 7.22: Evaluation of jobs.

	Objectives			
	Salary	**Career**	**Work-Life**	
Weights	0.54	0.30	0.16	
	Job scores for each objective			Total score
Job 1 Current	0.105	0.094	0.672	0.19
Job 2 New organization	0.258	0.627	0.265	0.37
Job 3 New company	0.637	0.280	*0.063*	*0.44*

Ranking for overall goal

Alternative	Utility
J3	0.437
J2	0.369
J1	0.194

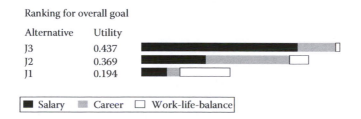

■ Salary   ▨ Career   ☐ Work-life-balance

FIGURE 7.13: Ranking of job selection alternatives.

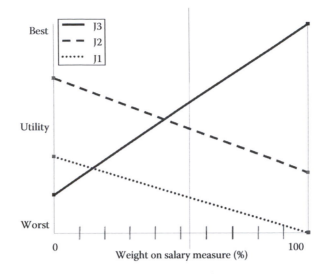

FIGURE 7.14:   Sensitivity analysis of weight assigned to salary.

7.5   Software Selection

This section demonstrates AHP using a more complex example adapted from the literature (Wei et al. 2005). MedPhar, a pharmaceutical company, decided to buy R&D project management software. The project management software team, consisting of experts from project management, purchasing, and information systems, identified three software alternatives that satisfied their requirements: OG3, GT, and PMS.

The ultimate objective is to select the best project management software, as shown in the first level of the hierarchy (see Figure 7.15). The second level of the hierarchy depicts the major objectives: total cost; implementation time; flexibility and reliability; and technical capability and service

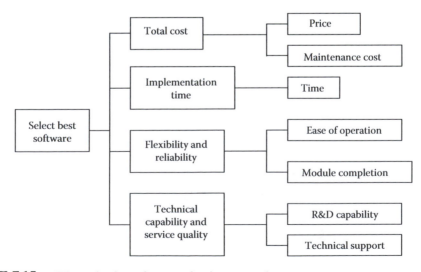

FIGURE 7.15:   Hierarchy for software selection example.

quality. The third level of the hierarchy illustrates the measures. While implementation time has only one measure, the others have two measures. For example, measures under cost are purchasing price and maintenance cost. However, the hierarchy does not show the alternative software packages for the sake of simplicity.

7.5.1   Assess Relative Importance of Measures

After developing the hierarchy, MedPhar management determined the relative importance of seven measures that are used to evaluate the three project management software packages. They ranked the measures from most important to least important. The management considered *module completion measure* the most important criterion. This measure refers to the number and types of modules currently available within the software package. This criterion is followed by *ease of operation, R&D capability* of the software vendor, software purchasing *price*, implementation *time*, *technical support* capability of the vendor, and *maintenance cost*. Table 7.23 shows the pair-wise comparison for software selection, according to MedPhar management's analysis. For example, module completion is only slightly more important than ease of operation and extremely more important than maintenance cost.

The calculated weights are given in the last column of Table 7.23. Module completion has the highest weight (0.39) followed by ease of operation (0.21). R&D capability and price have the same weight (0.13). Time, technical support, and maintenance cost have similar weights. The objective weights can be found by summing their respective measure weights (Table 7.24).

7.5.2   Identify Score for Each Alternative on Each Objective

To calculate a score for each alternative on each measure, MedPhar management used pair-wise comparison matrices as given in Table 7.25. The last columns show the score for each alternative on each measure. For example, OG3 ranked first with a score of 0.55 in module completion, followed by GT (0.24) and PMS (0.21). In terms of ease of operation, PMS has the highest score (0.65).

7.5.3   Synthesis

Figure 7.16 shows a total utility score for each alternative and ranking of the alternatives. OG3 is ranked first with a total score of 0.364. PMS is a very close second best, with a total score of 0.358. GT is the third alternative, with a total score of 0.278. OG3 gains its strength from flexibility and technical capability. It is, however, weak on total cost and implementation time.

7.6   Growth of AHP Pair-Wise Comparison Effort

AHP requires pair-wise comparison to specify the weights and scores for the alternatives. Pair-wise comparison is effective when the number of objectives, measures, and alternatives is small. When N factors are being compared, $N * (N - 1)/2$ questions are necessary to fill in the matrix. Therefore, the number of individual pair-wise comparisons for objectives, measures, and alternatives grows by the power of N^2. The total effort across measures and alternatives grows by three times the power of two. The exponential growth in the number of pair-wise comparison leads to problems of inconsistency.

The use of two or more levels of hierarchies can reduce the number of pair-wise comparisons and help maintain consistency. However, it does little to reduce the number of pair-wise

TABLE 7.23: Pair-wise comparison for objectives.

	Module Completion	Ease of Operation	R&D Capability	Price	Time	Technical Support	Maintenance Cost	Weight
Module completion	1	2	3	3	7	8	9	0.39
Ease of operation	1/2	1	2	2	3	4	4	0.21
R&D capability	1/3	1/2	1	1	2	3	3	0.12
Price	1/3	1/2	1	1	2	3	3	0.12
Time	1/7	1/3	1/2	1/2	1	1	1	0.06
Technical support	1/8	1/4	1/3	1/3	1	1	1	0.05
Maintenance cost	1/9	1/4	1/3	1/3	1	1	1	0.05

TABLE 7.24: Weights for software objectives.

Objectives		Measures	
Name	Weight	Name	Weight
Total cost	0.18	Price	0.13
		Maintenance	0.05
Implementation time	0.06	Time	0.06
Flexibility and reliability	0.60	Ease of operation	0.21
		Module completion	0.39
Technical capability and service quality	0.18	R&D capability	0.13
		Technical support	0.05

TABLE 7.25: Pair-wise comparison of software alternatives on each objective.

	OG3	GT	PMS	Score
OG3	1	2	3	0.55
GT	1/2	1	1	0.24
PMS	1/3	1	1	0.21

(a) Module completion

	PMS	OG3	GT	Score
PMS	1	3	5	0.65
OG3	1/3	1	2	0.23
GT	1/5	1/2	1	0.12

(b) Ease of operation

	GT	OG3	PMS	Score
GT	1	2	3	0.55
OG3	1/2	1	1	0.24
PMS	1/3	1	1	0.21

(c) R&D capability

	PMS	GT	OG3	Score
PMS	1	3	4	0.63
GT	1/3	1	1	0.19
OG3	1/4	1	1	0.17

(d) Price

	GT	OG3	PMS	Score
GT	1	3	3	0.60
OG3	1/3	1	1	0.20
PMS	1/3	1	1	0.20

(e) Time

	OG3	GT	PMS	Score
OG3	1	2	5	0.60
GT	1/2	1	2	0.28
PMS	1/5	1/2	1	0.13

(f) Technical support

	PMS	GT	OG3	Score
PMS	1	1	2	0.40
GT	1	1	2	0.40
OG3	1/2	1/2	1	0.20

(g) Maintenance cost

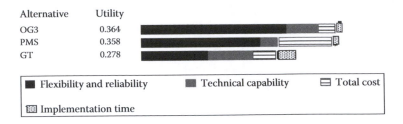

FIGURE 7.16: Ranking of project management software packages.

comparisons involving every possible pair of alternatives for each and every measure. For example, a decision with just one overall objective and nine measures would require 36 pair-wise comparisons to assign weights to the measures. Imagine, instead, that the problem was structured with three intermediate objectives, having three measures each. There would be, therefore, three pair-wise comparisons for the three objectives. There would also be three more pair-wise comparisons for each set of three measures within each objective. As such, there would be a total of only 12 pair-wise comparisons with this hierarchical structure instead of the previously noted 36 pair-wise comparisons with no hierarchy. However, if there were 4 alternatives, the decision maker would have to make 6 comparisons of alternatives for each of 9 measures, a total of 54 pair-wise comparisons.

LDW software can help to improve consistency for large problems and also reduce the total effort required by the growth in the number of objectives, measures, and alternatives. LDW has two options that can facilitate the use of AHP: *Identify Outliers* and *Estimate Ratios*. With *Identify Outliers,* LDW identifies where inconsistencies are the greatest. LDW highlights these cells in red. The *Estimate Ratios* option estimates any unassessed pair-wise comparison based on those assessed so far. This option assumes that there is *no inconsistency*. LDW marks cells with a user-entered preference in blue while estimated cells remain in white. *Estimate Ratios* option makes the evaluation as efficient as MAUT in that only the minimum number of $(N-1)$ comparisons is completed and all the rest are estimated. This is not, however, in line with the spirit of AHP. AHP was specifically designed to ask mathematically redundant questions in order to get at the heart of inconsistent responses and compel decision makers to think clearly about their preferences.

AHP creators also offer a strategy for dealing with a large number of alternatives and measures. They suggest creating a ratio scale within each measure for all possible values. Once that scale is created, each alternative's value within that measure is simply scored on this ratio scale. This eliminates the need for pair-wise comparisons of alternatives. It also simplifies the process of adding a new alternative into the analysis.

7.7　Comparison of AHP vs. MAUT

AHP and MAUT are widely used to deal with multiple objective problems. Both methods have advantages and disadvantages. MAUT and its companion, the multiple-attribute value theory (MAVT) have the stronger base of support in the academic community. The practitioner community, however, gives at least as much credence to AHP. Table 7.26 provides an overview of comparison of the two methodologies in terms of score meaning, hybrid creation, theoretical concerns, adding an alternative, interview process, role of data, and uncertainty. In this table, *bold type* is used

TABLE 7.26: Comparison of AHP and MAUT: An overview.

Issue	MAUT	AHP
Score meaning and hybrid creation	**Absolute relative to theoretical best** Results guide search for improved alternative Alternative scores do not sum to 1	Alternative scores sum to 1 Relative score *Hard to quantify improvement opportunities*
Add alternative	**Gather data on alternative and calculate score of new alternative.**	Perform more alternative comparisons.
Theoretical concerns	Decision makers are not as consistent in assigning weights as MAUT implies	*Rank Reversal*—New alternative can change rank ordering of previous alternatives
Interview and analysis process	Strictly quantitative Abstract assignment of weights No measure of consistency Grows linearly with problem size Requires a SUF for each measure	*Can be qualitative* *Pair-wise comparisons* *Measure of consistency* Can grow exponentially with problem size
Role of data Uncertainty	Explicit Probabilistic estimates can be used or separate risk measure	Implicit Implicit in alternative comparisons or separate risk measure

to distinguish the areas in which one method has an advantage over the other. *Italics* are used to show the greatest areas of concern with a particular method.

1. *Score meaning*: MAUT and AHP rank alternatives in terms of utilities. In MAUT, the actual utility value is meaningful; it represents the percentage of a theoretical alternative that is the best on every measure. For example, a utility score of 0.85 for an alternative shows that this alternative is 85% as good as an alternative that scores the maximum on each measure. On the other hand, AHP score has no absolute meaning; it only reflects a relative ranking. Total AHP scores of all alternatives sum to 1. Therefore, if an alternative is added or removed, the scores of the rest of the alternatives change.

2. *Hybrid creation:* In some contexts a better alternative can be created by combining the best features of two other alternatives. MAUT enables a decision maker to see the value of improving the performance of an alternative on a specific measure. It is easy to interpret investing dollars to improve another measure (see Chapter 6 and Canbolat et al. 2007). With regard to AHP, because scores are relative, it is harder to perceive and harder to anticipate the value of improving the performance of an alternative on a specific measure. However, even within AHP, it is easy to diagnose areas of weakness of an alternative and work to improve the alternative. The University of Chile, for example, used AHP to prioritize research proposal submission and then focused on addressing weaknesses within the highest ranked proposals to great success.

3. *Add alternative:* Adding a new alternative in MAUT involves collecting and inputting data for the new alternative. The new alternative's score is calculated by applying the weights and a single utility function (SUF) to that alternative. However, a problem arises if one or more measures of the new alternative are outside the original ranges. In that case, the weights will need to be reassessed since they are sensitive to ranges. In addition, a new SUF would need to be created for the broader range. In AHP, adding a new alternative

requires the decision maker to compare this alternative with every prior alternative on each measure. In modified AHP, a formal ratio scale is created within each measure. Thus, the new alternative can simply be added to the analysis by specifying how it performs on each measure's ratio scale.

4. *AHP is easier to apply:* AHP easily handles problems with limited data and nonquantifiable measures. The pair-wise questioning process is natural and can be presented verbally, numerically, or pictorially. AHP provides a measure of consistency that helps the decision makers rethink their comparisons. MAUT requires more abstract thinking, especially with regard to the SUF. However, a formal assessment of a measure's SUF is only needed if the curve is nonlinear.

5. *Problem size and effort:* As the number of alternatives and/or measures increase, the number of pair-wise comparisons with AHP can grow exponentially. This problem size challenge can be addressed by creating multiple levels of hierarchy. Creating a ratio scale within a measure reduces the need for more numerous alternative pair-wise comparisons. MAUT, on the other hand, simply grows linearly with the size of the objectives hierarchy or the number of alternatives.

6. *Rank reversal:* Rank reversal is the most serious concern that decision theoreticians have with AHP. Rank reversal describes a situation in which the relative preference between alternatives A and B depends on the existence of a third alternative, C. A is preferred to B when C is not present but B is preferred to A if C is included in the list of alternatives. To a significant segment of the decision analysis community, rank reversal is in violation of the axioms of a "rational" decision maker. For this reason, they consider AHP an inappropriate and arbitrary decision aid. The phenomenon of rank reversal was first explored by Belton and Gear (1983) and expanded upon by Dyer (1990) and Wijnmalen and Wedley (2008a,b) Forman and Gass (2001) and Gass (2005) responded to this challenge by arguing that the presence or absence of rank reversal should not serve as a litmus test to judge a decision model as representative of "rational" decision making. They also point out that, based on simulation and a review of existing studies, rank reversal is a rare phenomenon.

7. *Data:* MAUT is a data intensive methodology. Each alternative is to be clearly described with numeric scales supplied for each and every measure. The scales can be raw measures or categories, as described in Chapter 4. In contrast, AHP can make do with more limited information, as long as the decision maker has enough understanding of the alternatives to make pair-wise comparisons of relative preference with regard to each measure.

8. *Uncertainty:* From the very beginning, MAUT was developed and designed to include probabilities associated with random events. Thus, it readily incorporates probability distributions for individual measures. As a result, utility scores can be expressed as a range. This facilitates the assessment of alternative risk management strategies. Researchers have also explored how to handle multiple uncertainties on measures that interact. In contrast, AHP cannot explicitly represent probabilistic data. The decision maker can implicitly factor uncertainty with regard to a measure into the pair-wise comparisons of alternatives. Alternatively, the hierarchy can include an objective to minimize risk on which all alternatives would be compared.

7.8   Application Capsule: Compare AHP with MAUT: A Case Study

In an effort to reduce risk and boost the productivity of material handling crews, the U.S. Army investigated the use of robotics to perform many of the dangerous and labor-intensive functions normally undertaken by enlisted personnel (Bard 1992). The system that was in place utilized three different-sized rough terrain forklifts, capable of handling 4,000, 6,000, and 10,000 lb each. These could reach speeds of 20 mph, which meant that they were not self-deployable (i.e., a forklift could not keep pace with a convoy on most surfaces). A second problem relates to the safety of the crew.

A heavy-duty cargo-handling forklift was needed to overcome these deficiencies as well as to improve crew productivity. This equipment had to be capable of operating in rough terrain in extreme conditions and travelling over paved roads at speeds in excess of 40 mph. To satisfy the system objectives, either an improvement in the existing system, a modification of a commercial system, or an adaptation of available technology had to be made.

Taking into account mission objectives, three alternatives, as defined as follows, were identified and ranked using MAUT and AHP:

1. Baseline: The existing system, with a new 6000 lb variable reach vehicle.

2. Upgraded system: The baseline upgraded to be self-deployable.

3. USDCH: A tele-operable, robotic-assisted universal self-deployable cargo-handler with micro-cooling for the protective gear, and the potential for full autonomy.

The intent was to determine the strengths and weaknesses of each alternative, and to characterize the conditions under which one might be more appropriate than another. The evaluation team consisted of five program managers and engineers. In evaluating the 3 alternatives, an objective hierarchy (see Figure 7.17) containing 4 objectives and 12 measures was created. The overall goal was to

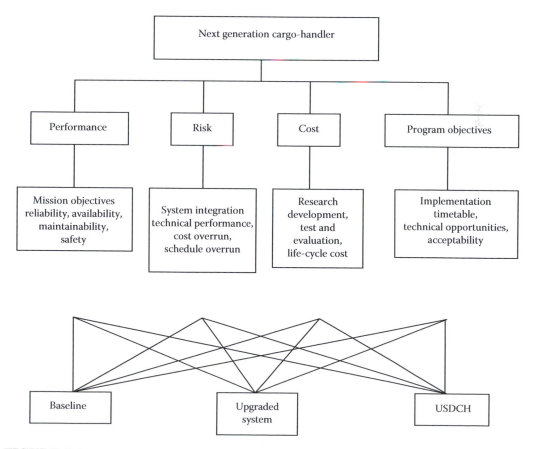

FIGURE 7.17: Objectives hierarchy for cargo-handler problem.

select a "next generation" cargo-handler. The assessment began with a series of questions designed to probe each decision maker's risk attitude over the range of possible outcomes.

The results of the analysis indicated that the group had a modest preference for the USDCH over the baseline. It was recommended that work continue on the development of the basic USDCH technologies.

Appendix 7. A: Consistency Ratio

7.A.1　Consistency Measurement

We present a four-step procedure to check for the consistency of the decision maker's comparisons (Winston and Albright 2009). A denotes the pair-wise comparison matrix and w represents a column vector of the resultant decision maker's weights.

Step 1: Compute Aw. For snow blower example, we obtain

$$Aw = \begin{pmatrix} 1 & 3 & 6 \\ 1/3 & 1 & 3 \\ 1/6 & 1/3 & 1 \end{pmatrix} \begin{pmatrix} 0.65 \\ 0.25 \\ 0.10 \end{pmatrix} = \begin{pmatrix} 2.00 \\ 0.77 \\ 0.29 \end{pmatrix}$$

Step 2: Compute λ_{max}

$$\lambda_{max} = \frac{1}{n} \sum_{i=1}^{i=n} \frac{ith\_entry\_in\_Aw}{ith\_entry\_in\_w}$$

$$\frac{1}{3}\left(\frac{2.00}{0.65} + \frac{0.77}{0.25} + \frac{0.29}{0.10}\right) = \frac{1}{3}(3.077 + 3.067 + 2.90)$$

$$\lambda_{max} = 3.02$$

Step 3: Compute the Consistency Index (CI) as follows:

$$CI = \frac{(\lambda_{max}) - n}{n-1} = \frac{(3.02) - 3}{3 - 1}$$

$$CI = 0.010$$

Step 4: In this step, we select the Random Index (RI) for the appropriate value of n using Table 7.27. We then calculate CR as follows:

$$CR = \frac{CI}{RI}$$

For a perfectly consistent decision maker, the ith entry in $Aw^T = n *$ (ith entry of w^T). This implies that a perfectly consistent decision maker has CI=0. The values of RI in Table 7.27 give the average value of CI if the entries in A were chosen at random, subject to the constraint that all diagonal entries must equal 1 and

TABLE 7.27: Values of RI.

n	2	3	4	5	6	7	8	9	10
RI	0	0.58	0.90	1.12	1.24	1.32	1.41	1.45	1.51

$$a_{ij} = \frac{1}{a_{ji}}$$

If CI is sufficiently small, the decision maker's comparisons are consistent enough to give useful estimates of the weights for his or her objective function. If CI/RI < 0.10, the degree of consistency is satisfactory, otherwise, serious inconsistencies may exist, and AHP may not yield meaningful results. In snow blower selection example n is 3 and RI is 0.58; CI/RI = 0.010/0.58 = 0.017. Thus, Mike's pair-wise comparison matrix does not exhibit any serious inconsistencies.

Exercises

7.1 You plan to purchase 20 bulbs to be used in multiple lamps and fixtures throughout your home. The relevant data are provided in Table 7.28.

a. Use AHP to assign weights to each measure.

b. Are the weights you assigned with AHP significantly different from the weights you used in Exercise 6.1?

c. Use AHP to compare alternatives on each measure. What role do the data play when using AHP?

d. Determine the optimal decision using Logical Decisions software. Were the rankings of the alternatives the same as in Exercise 6.1? Explain.

e. Determine the sensitivity of the optimal decision to the weight assigned to wattage.

7.2 You plan to purchase 1000 bulbs to be used in multiple lamps and fixtures throughout the economy motels you manage. The relevant data are provided in Table 7.29.

a. Use AHP to assign weights to each measure.

b. Are the weights you assigned with AHP significantly different from the weights you used in Exercise 6.2?

c. Use AHP to compare alternatives on each measure. What role do the data play when using AHP?

TABLE 7.28: Incandescent bulb alternatives for home.

Bulb Watts	Total Annual Operating Cost ($)	Total Bulb Purchasing Cost ($)	Total Annual Cost ($)
60	180	9	189
75	225	10	235
100	300	15	315

TABLE 7.29:   Incandescent bulb alternatives for hotel.

Bulb Watts	Total Annual Operating Cost ($)	Total Bulb Purchasing Cost ($)	Total Annual Cost ($)
60	18,000	900	18,900
75	22,500	1,000	23,500
100	30,000	1,500	31,500

    d. Determine the optimal decision using Logical Decisions software. Were the rankings of the alternatives the same as in Exercise 6.2? Explain.

    e. Determine the sensitivity of the optimal decision to the weight assigned to wattage.

7.3 Cell phone plans: (See Exercises 5.8 and 6.3.) Create objectives and measures to be used in evaluating cell phone plans from multiple providers. Gather information for various cell phone plans. Determine the range for each measure.

    a. Use AHP to assign weights to each measure.

    b. Are the weights you assigned with AHP significantly different from the weights you used in Exercise 5.3?

    c. Use AHP to compare alternatives on each measure.

    d. Determine the optimal decision using Logical Decisions software. Were the rankings of the alternatives the same as in Exercise 6.2? Explain.

    e. Determine the sensitivity of the optimal decision to weights assigned to the two highest weighted measures.

    f. Create a stacked bar chart and describe the strengths and weaknesses of the various alternatives.

    g. Create a pair-wise comparison chart of the two best alternatives.

7.4 through 7.8

Carry out a comprehensive AHP analysis for the exercises in the following. Write a report as described at the end of this section. Carry out the following tasks.

    a. Use AHP to assign weights to each measure.

    b. Are the weights you assigned with AHP significantly different from the weights you used in the corresponding exercise in Chapter 6?

    c. Using the data that were originally input, did AHP derived weights change the rank orderings?

    d. Now use AHP for the entire process and do not use the actual data. Use AHP to compare alternatives on each measure.

    e. Determine the optimal decision using Logical Decisions software. Were the rankings of the alternatives the same as in the corresponding exercise in Chapter 6? Explain.

    f. Create a stacked bar chart and describe the strengths and weaknesses of the various alternatives.

    g. Create a pair-wise comparison chart of the two best alternatives.

    h. Describe how you would go about creating a hybrid solution. What steps in AHP would now need to be done?

7.4 Refer Exercises 5.9 and 6.4, the women's retailer exercise.

7.5 Refer Exercises 5.10 and 6.5, the used car exercise.

7.6 Refer Exercises 5.11 and 6.6, kitchen remodeler.

7.7 Refer Exercises 5.12 and 6.7, blower motor.

7.8 Refer Exercise 5.13, plant location and 6.8.

References

Bard, J. F. (1992). A comparison of the analytic hierarchy process with multi-attribute utility theory: A case study. *IIE Transactions*, 24, 111–121.

Belton, V. and Gear, T. (1983). On a short-coming of Saaty's method of analytic hierarchies. *Omega*, 11, 228–230.

Canbolat, Y. B., Chelst, K., and Garg, N. (2007). Combining decision tree and MAUT for selecting a country for a Global Manufacturing Facility. *Omega*, 35, 312–325.

Dolan, J. G. and Bordley, D. R. (1992). Using the analytical hierarchy process (AHP) to develop and disseminate guidelines. *QRB Journal*, 18, 440–447.

Dyer, J. S. (1990). Remarks on the analytic hierarchy process. *Management Science*, 36, 249–258.

Forman, E. H. and Gass, S. I. (2001). The analytical hierarchy process—An exposition. *Operations Research*, 49, 469–486.

Gass, S. I. (2005). Model world: The great debate—MAUT versus AHP. *Interfaces*, 35(4), 308–312.

Hamalainen, R. P. (2004). Reversing the perspective on the applications of decision analysis. *Decision Analysis*, 1, 26–34.

Lai, V., Wong, B. K., and Cheung, W. (2002). Group decision making in a multiple criteria environment: A case study using AHP in the software selection. *European Journal of Operational Research*, 137, 134–144.

Saaty, T. L. (1980). *The Analytical Hierarchy Process*. New York: McGraw-Hill.

Suh, C., Suh, E., and Back, K. (2004). Prioritizing telecommunications technologies for long range R&D planning to the year 2006. *IEEE Transactions on Engineering Management*, 41, 264–274.

Tam, M. C. Y. and Tummala, V. M. R. (2001). An application of AHP in vendor selection of a telecommunications system. *Omega*, 29, 171–182.

Vaidya, O. S. and Kumar, S. (2006). Analytical hierarchy process: An overview of applications. *European Journal of Operational Research*, 169, 1–29.

Wei, C., Chien, C., and Wang, M. (2005). An AHP-based approach to ERP system selection. *International Journal of Production Economics*, 96, 47–62.

Wijnmalen, D. J. D. and Wedley, W. C. (2008a). Correcting illegitimate rank reversals: Proper adjustment of criteria weights prevent alleged AHP intransitivity. *Journal of Multi-Criteria Decision Analysis*, 15(5/6), 135–141.

Wijnmalen, D. J. D. and Wedley, W. C. (2008b). Non-discriminating criteria in AHP: Removal and rank reversal. *Journal of Multi-Criteria Decision Analysis*, 15(5/6), 143–149.

Winston, W. L. and Albright, S. C. (2009). *Practical Management Science*, rev. 3rd edn. Mason, OH: South-Western Cengage.

Part III

Decisions and Management under Uncertainty

Chapter 8

Value-Added Risk Management Framework and Strategies

> The CEO of a major automotive manufacturer has announced plans to source $1 billion in components to distant emerging markets in order to save on manufacturing costs. Middle-level technical managers have identified dozens of supply-chain risks associated with sourcing components of different levels of complexity. These managers thus face the challenge of identifying which components to source, those that will save enough money to more than compensate for the new risks. In addition, they are searching for cost-effective ways to mitigate the risks for those components that will be sourced.

> Petroleum refineries, terminals, pumping stations, and chemical plants store huge amounts of hazardous material in large tanks. Hundreds of accidents have occurred over the past 40 years. Operators and managers are challenged to reduce the likelihood and severity of these accidents.

8.1 Goal and Overview

This chapter provides an introduction to concepts of risk management in a variety of contexts. It explores a wide range of approaches to reducing risk. These range from risk avoidance to mitigating the impact of a negative outcome.

In this chapter, we use the term *risk* as perceived by managers (March and Shapira 1987). For managers, risks are negative consequences of uncertain events. They are usually more concerned about the magnitude of the downside risk than its likelihood. Except for sophisticated financial engineers, these managers do not apply advanced mathematical models to quantify all aspects of the risk. Nevertheless, they are interested in exploring ways to reduce risk and possibly investing money to mitigate the risks or insure against them.

Difficult decisions involve uncertainty, and that uncertainty is tied to risk. Risks can come from uncertainty in financial markets, project failures, legal liabilities, credit risk, accidents, natural causes, and disasters, as well as deliberate attacks from an adversary. Successfully managing these risks is therefore critical to achieving one's objectives.

Uncertainty is an inherent characteristic of many decisions. Manufacturing organizations face internal uncertainties as well those related to their supply chain, external market, and competitive factors. Table 8.1 presents some of the sources of uncertainty within each of these three categories. For example, the product development process involves uncertainty regarding the date of product readiness and the resources required to bring the product to market. There is also uncertainty as to whether or not a new design can meet its targets for performance and cost. The supply chain, especially if it stretches around the world, contributes to further uncertainty regarding its ability

TABLE 8.1:   Sources of randomness in a manufacturing company.

Category	Sources of Uncertainty
Internal	Product development Manufacturing Purchasing
Supply chain	Design Manufacturing Logistics
External—Market	Aggregate demand and pricing dwarfs all other industry concerns Disaggregate demand by product Competitor actions

to handle new designs in a timely fashion, deliver a high-quality product, and manage around-the-world logistics. Global supply chains also face uncertainty around currency fluctuations and inflation in almost any emerging country. Aggregate demand for a new product is always a major uncertainty.

Service industries such as health care face an equally bewildering array of sources of randomness. Imagine a health care system considering building a clinic in a fast-growing county. They have to decide where to locate the clinic and what services to provide. They face significant uncertainty with regard to revenue as the federal government and states continually change the reimbursement formula. The revenue is linked to demand for services, which are uncertain, because the county's population is changing. Last, there could be uncertainty regarding the cost of hiring and keeping critical personnel as well as continuing changes in drug costs (Table 8.2).

In the case of a personal health care decision such as where to obtain treatment for cancer, there are both external and internal uncertainties. The external uncertainties relate to the treatment providers. The internal ones are literally internal; they relate to the uncertain nature of any cancer and the manner in which your specific body responds to the treatment (Table 8.3).

The primary goal of this chapter is to raise the awareness of decision makers as to the need to incorporate risk management strategies into their decisions. It should be obvious that a decision maker cannot manage risks if he does not first recognize their existence. Nor can he determine how much money, time, and management energy to invest in risk reduction if he has not quantified these risks.

TABLE 8.2:   Sources of randomness in health care system decision.

Category	Sources of Uncertainty
Revenue	Federal reimbursement (Medicare and Medicaid) State reimbursement Private health insurance reimbursement
Costs	Drugs Medical personnel Staff
Market	Need for services Changing local demographics Competitor actions

TABLE 8.3: Sources of randomness in personal health decision.

Category	Sources of Uncertainty
External	Doctor competency
	Hospital competency
Internal	Nature of cancer
	Body's response

The chapter begins by describing a risk management process. The first step is risk identification, which we discuss by exploring common risks and their causes. Next, we describe approaches to quantifying the risks. The chapter continues with a discussion of common risk mitigation strategies. We briefly introduce the classical tools used in each stage along with several references to the methodology and its application.

8.2 Overview of the Risk Management Process

Figure 8.1 sets out the risk management process along with the tools that are commonly used to identify, analyze, and manage risk. The first step in risk management is to identify all sources of risk and uncertainty. Once we have characterized risk, we quantify it by assessing the likelihood and impact of each risk factor. The risk analysis phase helps us understand both the overall risk and the high-priority risk factors that should receive more attention in risk management (Morgan 1993; Bell and Schleifer 1996).

Risk identification, quantification, and analysis prepare us to develop risk mitigation and contingency plans. These plans are intended to reduce the likelihood or impact of key risk factors and

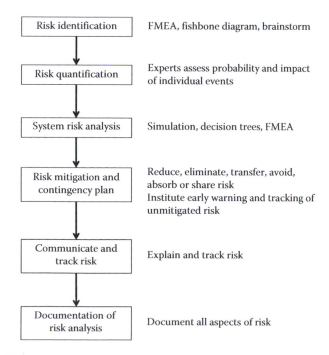

FIGURE 8.1: Risk management process.

help managers choose the strategy best suited for an uncertain environment. Once a plan has been developed, it is critical to raise the organization's awareness level by communicating the nature of the risk and the plans for minimizing it. If the risk is in the public domain, it is important to educate the public about legitimate concerns and reasonable actions to take. It is equally important to avoid hysteria and overreaction. The final step includes the documentation of risk assessment and management plans.

8.3   Risk Identification

We cannot manage or mitigate risks unless we can characterize them, know what they are, know how likely they are, and know what their impact might be. The first step is simply to identify them. The starting point involves identifying all possible drivers of risk and uncertainty that are associated with each of the objectives of a project or decision. This is then coupled with a description of the impact of each uncertainty or risk on a goal or system. In the design process, risk identification is an integral part of FMEA (failure modes effect analysis). In this context, the experts brainstorm all the different ways that a component, piece of equipment, or entire system such as a power plant might fail. In many instances, experts do not need to start from scratch, as the published literature may already articulate the various types of failures and their impact. An experienced company might maintain a comprehensive database of component, machine, and system failures. In addition, the corporation's knowledgeable design engineers might carry in their heads a list of failures based on their experience.

A manufacturing company could begin defining their supply-chain risks by reviewing the applicability of the long list from Chopra and Sodhi (2004), which we have modified slightly (see Table 8.4). First are the broad categories of risks: disruptions, delays, support system breakdowns, forecast problems, ownership of intellectual property, and so on.

Within the broad category of disruptions are natural and man-made disasters. For example, a fire in the main Philips plant in 2000 halted Ericsson's production of cellular phones. This supply disruption resulted in an estimated $400 million in lost revenue for Ericsson and eventually led the company to cease the manufacture of cellular phones (Rice and Caniato 2003). After the 9/11 terrorist attacks, U.S. border crossings with Canada and Mexico were closed, and, as a result, a U.S. automotive company shut down five of its domestic plants because components could not be delivered via trucks or ships. Another major source of disruption in the U.S. automotive supply chain has been supplier bankruptcy.

Delays, and a wide range of their causes, are common in global supply chains. Among the most prominent concerns for global product developers and manufacturers is the potential for communication failures with designers and component and system manufacturers around the world. This was a risk that Boeing underestimated as it developed its new generation of planes, the Dreamliner.

Bonabeau (2007) argues that increasing levels of complexity within organizations and across networks of organizations are making it more difficult even to identify potential risks. He describes environments in which catastrophes do not have a single cause but rather result "from interconnected risk factors and cascading failures." As an example, he refers to the massive power blackout in the northeastern United States on August 14, 2003. Bonabeau makes the case for using sophisticated modeling tools and approaches to simulate and explore these complex environments in order to identify especially rare risks. One tool is *agent-based modeling*. This tool is used to model contexts that involve a number of actors working independently. The actors can be buyers and sellers on a trading floor, workers and customers using different pieces of a corporate IT system, or companies connected to a power grid. Another approach is *diversity-based testing*. This involves using a wide range of individuals to test the reliability of a complex system and search for weaknesses. The open source software movement encourages this type of testing for complex computer programs.

TABLE 8.4: Supply-chain risks and their drivers.

Risk Category	Risk Drivers
Disruptions	Natural disaster; labor dispute; supplier bankruptcy War and terrorism Dependency on single source of supply as well as the availability of alternative suppliers
Delays	High capacity utilization; inflexibility of supply source Poor quality of yield at supply source Excessive handling due to border crossing or change in transportation modes Communication problems with suppliers especially with those from a different culture and language
System problems	Information infrastructure breakdown System integration or extensive systems networking E-commerce
Forecast errors	Inaccurate forecasts due to long lead times, seasonality, product variety, short life cycles, and small customer base
Ownership and protection of intellectual property	Vertical integration of supply chain Global outsourcing and markets
Procurement challenges	Exchange rate risk Percentage of a key component or raw material procured from a single source Industry-wide capacity utilization Long-term versus short-term contracts
Receivables	Number of customers; financial strength of customers
Inventory	Rate of product obsolescence; inventory holding cost Product value; demand and supply uncertainty
Capacity	Cost of capacity Capacity flexibility

A fishbone diagram is a pictorial tool used to group and display the various contributing factors to negative outcomes, often related to quality. It was first developed by Kaoru Ishikawa as a tool for improving quality management in Japanese shipyards. Fishbone diagrams are used to explore opportunities for loss prevention, as in Figure 8.2. Chang and Lin (2005) studied 242 storage tank accidents that occurred over a 40 year period around the world. They grouped the risks into eight categories. Two are man-made errors: operational and maintenance. Three are physical problems involving equipment, such as tanks or pipes. Two relate to the environment: lightning and static electricity. The last category comprises miscellaneous problems.

Chang and Lin also developed six broad categories of risk avoidance and risk management: storage tank design, maintenance, equipment that controls the tanks, workplace (environment), operations, and miscellaneous. Controlling static electricity was one critical risk avoidance strategy in the workplace category. Among the strategies within maintenance were risk-based inspection and ventilation.

The health care industry has embraced fishbone diagrams as part of its push for increased quality and risk management (Frankel et al. 2005; Taner et al. 2007). Figure 8.3 is an example, a diagram for researchers and physicians to clarify factors that can contribute to the failure to diagnose a collapsed lung (pneumothorax) in car crash victims. The risk factors are grouped into six categories: procedures, policies, equipment, people, environment, and other (White et al. 2004).

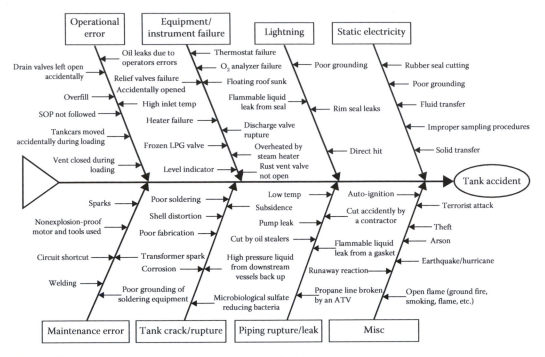

FIGURE 8.2:   Fishbone diagram of storage tank risk.

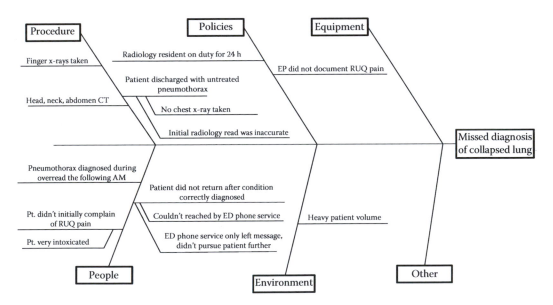

FIGURE 8.3:   Fishbone diagram of causes for delayed diagnosis of collapsed lung in car accident victim.

1. *Activity*: Assume that you are assessing your *career* risk. Identify risk factors that are specific to your career and their potential impact on your career.

Risk factors   Impact

_____   _____

_____   _____

2. *Activity*: Assume that you are assessing your *health* risk. Identify risk factors and impact that are specific to your health and lifestyle.

Risk factors   Impact

_____   _____

_____   _____

8.4   Risk Quantification

In this step, risk is quantified in terms of its likelihood and magnitude of impact. This assessment is first needed to prioritize the risks to be managed or mitigated. It is then used to determine the amount of managerial and financial resources that should be invested in the risk management effort. Although data sometimes are available to estimate the likelihood and magnitude, more often expert judgment is used to make the assessment. There are three methods of quantification that are listed in the following in order of increasing level of specificity (Blanchard 2008; Kossiakoff and Sweet 2003).

- Likelihood impact map
- FMEA
- Expert interviews to determine probabilities and dollar impact

Because the first goal is simply to prioritize risks, managers may use a simple tool such as an impact-likelihood map to place and display the various risks. Table 8.5 is a 3×3 map, although other dimensions may be used. Experts place each of the identified risks into a box. After seeing the range of risks and their location, management focuses first on those risks in the lower-right corner that have been identified as both high likelihood and high impact. Time and energy should also be invested to manage risks that are high impact and medium likelihood and those that are high likelihood and medium impact. Management could then ignore risks at the other end of the spectrum, those of low likelihood and low impact. They might even ignore all low-impact risks regardless of their likelihood if there are enough high priority risks that demand their attention. An oft-debated issue is what to do about high impact, low likelihood risks. Ultimately, the management team must decide where to draw the line on how far to go at this end of the spectrum.

British Telecom is representative of companies that adapt the principles of risk management to their environment (Colwill et al. 2001). BT's risk concerns include a possible terrorist attack. The suggested impact scale goes beyond minor and major to include enterprise risk and even national risk.

The Committee of Sponsoring Organizations (COSO 2004a,b) of the Treadway Commission was originally organized to address financial corruption and fraud. It later developed plans and strategies for managing enterprise risk in financial organizations. Figure 8.4 presents a variety of financial services risks and places each along nonscaled axes of frequency and severity. For example, credit card fraud is a frequent occurrence, but from an organizational perspective, it has

TABLE 8.5:   Likelihood-impact map (3×3).

		Relative Impact		
		Low	Medium	High
Relative Likelihood	Low			
	Medium			
	High			

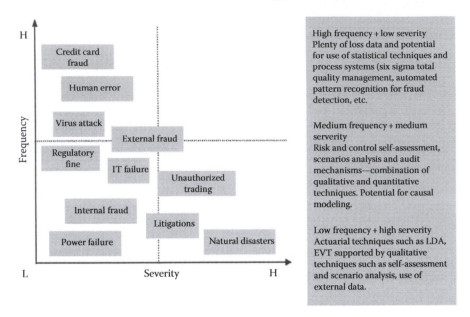

FIGURE 8.4: Likelihood-impact: financial organization risk.

limited impact. Natural disasters such as Hurricane Katrina or the 2011 Japanese earthquake and tsunami are relatively infrequent but have severe impact. Power failures were rated low on both scales. Unauthorized trading was a more substantial risk, both in frequency and severity. In 1995, the British bank Barings was dissolved because one trader engaged in tens of billions of dollars of unauthorized trades that left the company with $1 billion in losses.

When thinking about likelihood, it is important that management understand the concept of aggregation of random events. For example, the likelihood that a specific supplier will go bankrupt might be low, but the chances that any one of a manufacturer's dozens of critical suppliers might go bankrupt may be rated as medium.

3. *Activity*: Classify the risks in activities 1 and 2 mentioned earlier as to their likelihood and potential impact.

Likelihood Level   _____

Impact Level   _____

FMEA involves a more detailed analysis of risks and impact severity. Each risk is scored on a common scale, often 1–10, with regard to both likelihood and impact. (Some FMEA models incorporate a third variable, likelihood of detection.) The two scores are multiplied, and their product is used to rank-order risks. It is preferable that the point scale be grounded with an explanation that would enable an expert to assess each risk. One example of a rating scheme used for system design is presented in Tables 8.6 and 8.7 (Jackson 2010). The impact severity of equipment failure and misalignment ranges across specifications, from needing repairs that require a certain amount of time to a catastrophic failure that injures operators. A failure of this last type is a 9 if there is at least a warning of imminent failure and a 10 if there is no warning. The likelihood table uses the average time between failures as its measure. It compares that average with how long the piece of equipment is typically needed on a continuous basis. The best score of 1 corresponds to the machine continuously operating much longer than it takes to do a job. At the other extreme, a score of 10 corresponds to completing only 10% of a job until the next failure.

TABLE 8.6: Rating for severity of equipment failure.

Severity	Description	Explanation
1	None	Variation within performance limits
2	Very minor	Variation correctible during production
3	Minor	Reparable within 10 minutes
4	Very low	Reparable within 30 minutes
5	Low	Reparable within 1 hour
6	Moderate	Reparable within 4 hours
7	High	Not worth repair; degraded functionality
8	Very high	Not worth repair; mission failure
9	Hazardous—With warning	Affects safety of operator and others but with advance warning
10	Hazardous—Without warning	Affects safety of operator and others but with no advance warning

TABLE 8.7: Rating for likelihood of occurrence of equipment failure.

Occurrence	Description
1	Mean time to failure (MTTF) is 50 times greater than the user's required time
2	MTTF is 20 times greater than the user's required time
3	MTTF is 10 times greater than the user's required time
4	MTTF is 6 times greater than the user's required time
5	MTTF is 4 times greater than the user's required time
6	MTTF is 2 times greater than the user's required time
7	MTTF is equal to user's required time
8	MTTF is 60% of user's required time
9	MTTF is 30% of user's required time
10	MTTF is 10% of user's required time

The product of the FMEA scales provides a means for rank-ordering risks, but the product does not have an absolute meaning. It is analogous to the calculation used for expected values, a value multiplied by likelihood. However, the likelihood measure is not a probability, and the impact measure is not a real value. Consequently, it is not possible to use the product to determine how much money to invest to reduce the expected value of the risk. In order to make such a judgment, the decision maker will need estimates of the probability of each random event or random variable. In Chapter 13, we describe an expert interview process used to estimate probabilities and the distribution of critical random variables.

Canbolat et al. (2008) illustrate how the basic FMEA framework can be expanded and adapted to incorporate actual probabilities and values. This approach was used by Ford to evaluate cost-effective strategies for managing the risk of sourcing to emerging markets. As part of the analysis, Ford developed a list of potential failure modes and their impacts (Table 8.8). It also identified possible causes and the organization that would be responsible for addressing the risk. For example, premium freight is a commonly added cost in automotive supply chains and can be caused by a late purchasing order, defective parts, or a break anywhere along the supply chain. The internal organizations responsible for managing the risk include purchasing, supplier technical assistance (STA), or manufacturing (MP&L). Other risks involve product development (PD) and design and release (D&R) engineers.

A simulation model was developed to determine which components should be sourced to emerging markets (EM) with appropriate risk management strategies and which should continue to be sourced to existing local suppliers. Table 8.9 is a list of cases in broad terms (to protect

TABLE 8.8: Failure modes, causes, and organizational responsibility.

Failure Mode	Potential Effects of Failure	Cause of Failure	Responsible Organization
Prototype	Prototypes not on time	Tight demand for prototype parts	PD
		Late order	Purchasing
Premium freight	Increased air freight cost	Late purchasing order	Purchasing
		Defective parts	STA
		Break in pipeline	MP&L
Component delay	Parts not on time	Shipping delay	MP&L
		Custom problems	MP&L
		National disasters	D&R
		Supplier's production interruptions	STA
		International trade problems	MP&L
Communication problems	Parts not on time	Cultures and language differences	STA
		Engineering specification problems	STA
Warranty	Warranty and recall cost	Defective parts	PD
Currency	Higher cost	Currency fluctuation	Finance
Supplier management	Higher cost	Supplier technical assistance	STA
Inventory management	Higher cost	Inventory holding	MP&L

TABLE 8.9: Case studies for sourcing to emerging markets.

Case Study	Complexity of Component	Potential Saving	EM Supplier Risk after Risk Management	Decision
1	Simple	Medium	Low	EM supplier
2	Simple	Low	Low	Current supplier
3	Simple	Medium	Low	EM supplier
4	Relatively complex	High	Low	EM supplier
5	Simple	Low	Medium	Current supplier
6	Relatively complex	Medium	Medium	Current supplier
7	Complex	Medium	Medium	Current supplier
8	Relatively complex	High	Medium	EM supplier
9	Complex	High	Medium	Dual supplier
10	Complex	High	High	Current supplier

confidentiality). In 5 of the 10 cases, the decision was to avoid the risk completely by sticking with the current supplier. For example, in case 6, the potential savings were rated as medium but so was the risk—even with a risk management strategy. Thus, the potential savings did not warrant the added risk of sourcing to emerging markets. In case study 9, the part was considered complex, but the potential savings was high. The best strategy in that case was dual sourcing.

8.5 Systems Risk Analysis

The systems risk analysis phase usually combines the various identified and characterized risk elements into a single-quantitative risk estimate. In complex situations, the decision maker will need to develop a model of the entire system. This could be just a decision tree with multiple random events or a stochastic simulation model that captures many random elements and their interaction. These modeling tools for capturing the effect of multiple risks are illustrated in Chapters 9 and 10.

8.6   Risk Mitigation Framework

The ultimate goal of risk identification and risk analysis is to prepare for risk mitigation. Risk mitigation includes reduction of probability of a risk event occuring and/or reduction of impact of a risk event if it occurs. We will present two different frameworks for exploring a wide range of risk mitigation strategies.

Morgan (1981) suggested the conceptual approaches to controlling risk, as shown in Figure 8.5. To illustrate these risk management strategies, we discuss three applications as presented in Table 8.10: getting shot, feeling the side effects of a drug, and losing a job. Modifying the environment is a risk management strategy intended to avoid the risk completely. It usually involves a grand scale that is often beyond the influence of the risk manager. In the shooting example, it entails creating a society less prone to violence. In the drug example, it might mean reducing the overall prevalence of the disease or not developing the drug at all. However, the pharmaceutical manufacturer's decision maker would have little interest in these options since the company makes its profits from new drugs. The job loss example provides the best opportunity to actually control one's environment, by seeking jobs that rarely encounter layoffs such as those in the federal government or the military.

The second approach focuses on modifying or avoiding the exposure process. People can avoid exposure to the risk of being shot by avoiding people and locations where guns are prevalent. Society can reduce exposure by banning guns and imposing stricter penalties for crime. In the drug context, a company could work to design optimum clinical trials that more accurately predict the efficacy and safety of the drug. They could also develop technology that predicts the safety and efficacy of drug. In the job loss example, the worker can suck up to his boss and make him look good so as to reduce the worker's chance of being fired.

The third approach, modifying or avoiding effects processes, means that if one is shot at, the impact on the victim is minimized. This can be accomplished with a bullet-proof vest or by learning to duck quickly. In the drug example, it includes identifying early symptoms of the drug's unacceptable safety hazards experienced by patients. It would also involve developing solutions to counter the possible side effects. With regard to job loss, a working spouse with a successful career will attenuate the effects of the worker's job loss, as will having a second job.

The final strategy is mitigating or compensating for effects once they have occurred. When shot, it is useful to have a high-quality emergency medical system close by that is highly skilled

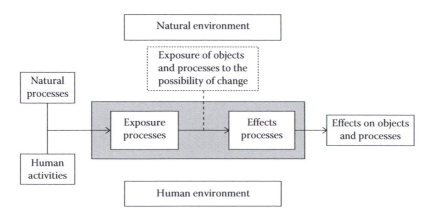

FIGURE 8.5:   Controlling risk.

TABLE 8.10: Four categories of risk abatement strategies.

Risk	Modify Environment	Avoid or Modify Exposure Processes	Avoid or Modify Effects Processes	Mitigate or Compensate for Effects
Risk of getting shot by handgun	Eliminate poverty and other causes of crime Teach anger management	Ban handguns Harsh penalties for gun usage Avoid high crime areas Live with people who do not get angry	Wear bullet-proof clothing Duck	Good emergency trauma centers Carry health insurance Sue the shooter
Drug causing negative side effects	Do not develop medicine Prevent the disease	Design clinical trials that more accurately predict safety and efficacy of drug Identify right population for drug demonstration	Identify early signs of symptoms of the drug's safety problems Ask doctors to monitor patients closely	Carry adequate insurance Ask patients to stop using drug at earliest signs of side effects
Job loss	Work for federal government or military Work for company in growth area Work for company with tradition of no layoffs Build your own company Live in area with low unemployment	Increase your value to the organization Develop relationships with key organizational decision makers on jobs Help boss look good	Have second job Have a spouse with complementary career	Low debt Significant savings Low-cost lifestyle Large social network Updated list of other potential jobs

in treating gunshot wounds. With drugs, it could include asking patients to stop using drugs and visit their doctors or hospitals as soon as symptoms appear. The drug manufacturers will also typically carry liability insurance to cover any extreme costs. For the worker who has lost his job, it is important to have a broad social network to facilitate finding a new job. In addition, a low-cost lifestyle and high savings would enable the worker to avoid temporarily the catastrophic effects of loss of income.

4. *Activity*: Describe risk management strategies that you could use to reduce your career risk.

5. *Activity*: Describe risk management strategies that you could use to reduce your health risk.

8.7 Risk Communication, Perception, and Awareness

Ideally, the philosophy and practice of risk management should be part of the culture of an organization in much the same way safety concerns should be part of hazardous work settings. All members of an organization should be aware of critical uncertainties and be attuned to tracking and identifying early warning signs of both trouble and opportunity. This applies not only to equipment or system failures in a manufacturing plant but also to marketing or investment risks and opportunities. Unfortunately, years of research have shown that people generally do not accurately assess risks on their own (Slovic et al. 1979). They are unable, for example, to rank order the greatest risks they face on a daily basis. This poor perception is compounded by the internet, which relishes highlighting the rare occurrence as newsworthy. Thus, a key element in development of an appropriate risk management culture involves a communication strategy that identifies the most important risks and their relative magnitude. Hospitals are prime examples of organizations that strive to inculcate a culture of risk management with regard to universal precautions, such as regular hand cleansing to reduce the spread of infection (Rutala and Weber 2007).

The development of risk consciousness requires unremitting attention by top-level managers and demonstration of management's commitment to this issue at every opportunity. One might argue that, considering the constant barrage of safety messages, nothing more remains to be said; the object, however, is not to say something new but to keep repeating the same message over and over until it becomes part of the culture. People on construction sites may initially assume that safety is the responsibility of the safety engineer, but the consistent message has to be that it is everyone's responsibility. Safety consciousness is achieved when all personnel understand that they cannot ignore any unsafe condition. Similarly, risk consciousness is achieved when everyone knows that he cannot ignore any potentially risky condition.

Sometimes, public awareness campaigns can be misdirected, however. There have been widespread campaigns in which children are told to be aware of strangers. Yet most risks that children face are the result of actions of friends, family members, and acquaintances. In some instances, corporate efforts toward risk communication are more about protection from potential litigation than actually reducing likely risk. The pharmaceutical industry, for instance, bombards us with information about the numerous side effects of every drug we take. It shows up in small print along with every prescription. It overwhelms television ads for drugs. Similarly, auto manufacturers put stickers on sun visors warning of risks from airbags for children riding in the front seat.

The same challenge arises in the public policy domain. Lawmakers attempt to manage societal risk by restricting certain types of risky behavior, such as smoking or drunk driving (Morgan et al. 2002). Alternatively, they impose costs, as when the automotive industry was required to reduce the likelihood of rollover accidents by making design changes. As a result, in 2009, the number of traffic deaths declined to its lowest level in 60 years, and the rate of fatalities per mile driven continues to go lower.

When considering risks to the public, it is critical that decision makers take a total systems perspective. There was once a debate about whether to require air travelers to purchase a seat and use a car seat for any infant traveling with them. Studies, however, indicated that the net impact would be an increase in infant deaths. The car seat proposal might save an infant in an extremely rare airplane crash, but the added cost would lead to more parents choosing to drive rather than fly between distant locales. Driving has a much higher fatality rate per mile compared to flying.

8.8   Alternative Risk Mitigation and Elimination Strategies

Earlier, we used Morgan's (1981) diagram to frame the risk management environment. Here, we review a list of different ways of managing risk.

8.8.1   Reduce or Eliminate Risk

Totally eliminating risks is usually either not possible or too expensive. A decision maker should thus investigate how to reduce the probability or impact of risks. Figure 8.6 identifies a number of supply-chain risk mitigation strategies (Chopra and Sodhi 2004). There are generally no silver bullets, and all strategies come with costs. For example, adding capacity will reduce the likelihood of delays, reduce inventory, and probably be the most expensive risk mitigation strategy. In contrast, increasing manufacturing ability to respond rapidly to changes in demand seems to greatly reduce risk in multiple areas with no negative consequences. This is the strategy that Dell has used with its supply chain in order to cope with the short lifecycle of its products.

Pooling demand involves combining the demand from multiple sources. These sources can be different locations, customers, and product types. This can be accomplished by centralizing manufacturing or inventory. It may require designing a product portfolio so that it can all be manufactured in one location. In Figure 8.6, there is no increased risk associated with this strategy. However,

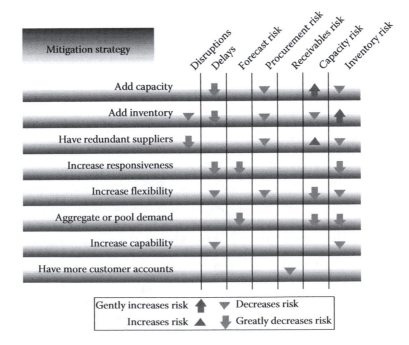

FIGURE 8.6:   Risk mitigation for supply chain risks.

pooling demand to be serviced from one location may increase logistics costs. Pooling demand in one location can increase the company's exposure to currency fluctuations. Toyota, a model of global manufacturing flexibility, is struggling with too much capacity in Japan as the yen sores to record heights. This contrasts sharply with Honda, which has distributed its capacity more evenly around the world.

There is an ongoing debate regarding the relative risk management benefits of single-source and multiple-source suppliers. Motorola, for instance, buys many of its handset components from multiple vendors to lower the probability of disruption. In the well-known case of a catastrophic fire at a chip manufacturing plant, Nokia reacted quickly and found alternatives, while Emerson's slow response led to a loss of market share.

Japanese and U.S. automotive manufacturers have adopted divergent strategies with regard to sourcing to suppliers. The Japanese have focused on building long-term relationships with individual suppliers around the globe. When a critical source supplier plant went down, Toyota's supply-chain network worked together to develop alternatives quickly and also to bring the affected plant back to operation in a timely fashion. For multiple reasons, U.S. automanufacturers have traditionally favored dual sourcing. One reason is the concern that a single-supply source would be able to hold companies hostage to price increases year after year. Another is that unions would be able to strike a single supplier in order to halt vehicle production. In addition, a single-quality problem stemming from one source would affect a large volume of cars.

Over the last few years, however, Ford has embraced a partnership approach similar to that of the Japanese with the goal of improving quality. By focusing on a single supplier, Ford is better able to detect and address quality concerns at an early stage. This change in strategy has been critical to Ford's rise to the top in U.S. quality rating surveys. In addition, Ford hopes to copy Toyota's aggressive pursuit of yearly price reductions by working with their sole source suppliers to take out cost through coordinated efforts of improved design for manufacturability. At the same time, it is not considered likely that an aggressive union at a supplier would hold a U.S. car company hostage, because such an action could drive the supplier into bankruptcy and would thus be suicide for the local union.

A number of product development risks and their drivers are listed in Table 8.11. The first concern for any new product is the initial demand. Lower than expected demand may be the result of either inaccurate forecasts or the product missing the market. A disciplined QFD (quality function deployment) can help align a product with customer wants and needs. In addition, volume forecasters need to be careful not to be overly optimistic regarding their new product's prospects in the marketplace.

One risk that is often underestimated is using suppliers whose design process is unproven but who offer a significant savings in manufacturing. The use of such suppliers may lead to late design changes, which, along with communication barriers, pose a special risk in new relationships, particularly with overseas suppliers. Furthermore, the supplier may not have the necessary design experience to consistently meet the main manufacturer's needs, especially under time pressure.

8.8.2 Risk Transfer and Contracting

One effective risk management strategy is to allocate risks to the parties best able to manage them. Contracts and insurance are the principal way risks are transferred. Contractors may generally take risk only in exchange for adequate rewards. Insurance companies take risks for a payment (premium) linked to the probability of occurrence and the size of the hazard associated with the risk. Insurance coverage can range from straight insurance for high-impact risks with low probability (e.g., fire and tornado) to sophisticated financial derivatives such as hedge contracts to avoid unanticipated losses in foreign exchange markets. Unfortunately, for the global economy, the financial industry and especially AIG did not fully comprehend the nature of the complex risk transfers they contracted. Customers who purchase extended warranties are essentially buying a form of insurance.

TABLE 8.11: Product development risk and risk mitigation.

Risk	Risk Driver	Risk Mitigation
Initial demand is low	Product not aligned with customers' wants Inaccurate forecasts Product not competitive	Use QFD and extensive customer clinics and surveys Assess competitive products Validate the forecasting model assumptions
Changing demand totals and mix	Competitive actions Technological advances	Modular design to facilitate upgrades R&D integrated into product development Flexible manufacturing systems
Product does not meet performance goals	Mismanagement of product development process Too many advances in technology designed into one new product	Focus on performance targets at each product gate review Incremental strategy for integrating new technology
Suppliers fail to deliver systems in timely fashion	Late design changes Communication problems Supplier's limited technical expertise	Disciplined approach to design changes that are clearly communicated to suppliers Use proven suppliers with long-standing relationship
Product quality problems	Late design changes Too aggressive cost-cutting targets Poor integration of design and manufacturing	Disciplined approach to design changes Realistic balance between manufacturing cost and design Experienced teams of product and manufacturing engineers Require product designers to gain manufacturing experience

In theory, the major U.S. automotive manufacturers hold their suppliers liable for recalls involving the supplier's parts. In practice, a supplier whose business relationship is measured in millions of dollars could not possibly cover the costs of a major recall that could reach $100 million. Thus, this form of risk transfer holds its own risks.

8.8.3 Avoid Risks

The most obvious way to avoid risk is to use an alternative without the risk. For example, sourcing to emerging markets for a complex and critical system can be too risky. The risk can be avoided by using a well-established supplier for that system instead. Similarly, companies may hesitate to set up businesses in countries with large swings in their currency exchange rate. In the realm of product development, a company may emphasize incremental improvements in its lineup rather than launch new products with unproven technology.

8.8.4 Buffer Risk

Risk buffering is the establishment of some reserve or buffer that can absorb the affects of many risks. Holding a safety inventory is one example of a buffer that reduces the production disruption risk. Buffering can also include the allocation of additional time, manpower, machines, or other resources to support a project. With regard to the launch of a radically new product, a company may also continue to manufacture a current product until it can determine the new product's acceptance level.

8.8.5 Absorb or Pool Risk

If risks cannot be eliminated, reduced, transferred, or avoided, they must be absorbed. There should be sufficient financial margin to cover the risk should it occur. However, one party alone may not bear all the absorbed risks. Risks can be reduced by pooling them through consortium, joint ventures, and partnerships. The oil industry routinely forms joint ventures when the search for oil involves large investments with much uncertainty.

8.8.6 Risk Control and Contingency Plans

Risk control means assuming a risk and developing contingency plans to reduce it. It should begin with establishing a process for tracking key variables so as to have adequate warning when things go awry. Disaster planning falls within this category of risk management. The tsunami of 2004, for example, led to a global early warning system for oncoming massive waves. Often, construction projects include contingency plans in case of bad weather. The military routinely develops contingencies when facing the uncertainties of war. By contrast, NASA included only minimal contingency plans in its space shuttle program in the event of damage upon liftoff. That changed after the second shuttle disaster. Banks are required to keep contingency funds in reserve to cover the potential cost of a failure to repay loans. Sometimes, the contingency plan is in the form of a written contract, as with an athlete whose salary is contingent on performance. Similarly, a prenuptial agreement is a contingency plan should a marriage fail.

8.8.7 Flexibility and Delayed Decisions

One cost-effective risk management strategy involves developing the organizational flexibility to respond quickly to changing conditions. This could entail a flexible manufacturing system that can respond to changes in the product demand mix. It could involve a flexible workforce that can cover random shortages of skilled labor. Flexibility could even be integrated into design, enabling substitution of critical components to adjust to variations in product performance required in different parts of the globe. One clothing manufacturer developed a process that enabled it to delay coloring its garments until late in the manufacturing process so as to better match products to the hottest colors in that fashion cycle.

An interesting example of facility and service flexibility relates to the design of movie theater complexes. Gone are many of the massive theaters that can only run one or two movies at a time for several days and seat hundreds of customers in one showing. They have been replaced by groups of smaller theaters that can show a wide array of films simultaneously. The number of theaters carrying each movie can be adjusted easily on a daily basis according to demand, time of day, and day of the week.

8.8.8 Assume Risk

The last option is simply to accept the risk as the cost of doing business. This implies that the risks associated with going ahead are less than the risks of not going forward. If risk assumption is chosen, it should be clearly defined, understood, and communicated to all stakeholders.

8.8.9 Managerial Focus and Actions to Reduce Risk

Management can use brainstorming sessions to identify a wide variety of approaches both internally and externally and thus manage risk associated with different variables. Table 8.12 catalogues a broad array of management activities that can reduce risk. Internally, the company can establish strong cost controls that include setting specific milestones and establishing a quick response if

TABLE 8.12:   Management actions reduce risk.

Internal Management of Firm	External Arrangements
Cost controls	Financial payment controls
Setting milestone	Reduce accounts receivable
Monitoring outflows	Increase accounts payable
Quick response	Delivery times
Productivity increases	Supplier cooperation
Incentive systems	Contract arrangements
Labor coordination	Take or pay clauses
Technological innovation	Penalty clause or warranty
Computer simulation of performance	Incentive clause
Extensive prototype testing	Performance-based contingent claim
Use of proven designs	Match exposure to interests
Pilot plant	Length of contract commitment
Product improvements	Reliability requirements
Marketing studies	Termination option
Field tests	Variable usage option
Shared development	Financial markets
Tried and true fallback systems	Hedges
Manufacturing capacity	Options
Flexible machines	Derivatives
Commonality of product design	Shared ownership (risk sharing, alliance or
Agile workforce	joint venture)
Globally integrated planning	Insurance against contingencies
Spare capacity (machines, parts)	Targeted marketing

financial milestones are missed. Management can develop incentives and develop better labor coordination to reduce any production risks. In the area of technology innovation, they can use extensive computer simulations to model performance as well as carefully test prototypes.

From an external perspective, the company can carefully review all accounts receivable in a timely fashion and work cooperatively to ensure suppliers are meeting deadlines. Contracts offer a diverse range of risk management concepts that can include penalty and incentive clauses, length of contract, provisions for premature cancellation, and measurable performance objectives. In the area of financial markets, a company can share risk in partnerships or hedge its bets.

8.8.10   Documentation

Risk assessment and risk management plans should be part of the documentation of every critical decision point. Risk mitigation plans should be documented, independently reviewed, critiqued, and reworked as needed so that management may overview the riskiness of the problem.

Exercises

8.1   Assume that you are assessing your *career* risk. Identify risk factors that are specific to your career and their potential impact on your career.

    Risk factors   Impact

    _____   _____

8.2 Assume that you are assessing your *health* risk. Identify risk factors and impact that are specific to your health and lifestyle.

Risk factors   Impact

_____   _____

8.3 Classify the risks in 1 and 2 mentioned earlier as to their likelihood and potential impact in a 3×3 likelihood-impact table.

		Relative Impact		
		Low	Medium	High
Relative likelihood	Low			
	Medium			
	High			

8.4 Describe risk management strategies that you could use to reduce your career risk.

8.5 Describe risk management strategies that you could use to reduce your health risk.

8.6 Assume that you are assessing the risks your *organization* faces in the near term. Identify risk factors that are specific to your organization and its potential impact on it. If you are a non-working student, your organization is the university.

Risk factors   Impact

_____   _____

_____   _____

8.7 Classify the risks in your organization as to their likelihood and potential impact in a likelihood-impact table.

8.8 Describe risk management strategies your organization could use to reduce its risk.

8.9 What risk messages does your organization routinely communicate, and how does it do so?

8.10 What risk messages does your organization occasionally communicate, and how does it do so?

References

Bell, D. E. and Schleifer, A. Jr. (1996). *Decisions under Uncertainty*, 2nd edn. Cambridge, MA: Course Technology Inc. (CTI).

Blanchard, B. B. (2008). *Systems Engineering Management*, 4th edn. Hoboken, NJ: John Wiley & Sons.

Bonabeau, E. (2007). Understanding and managing complexity risk. *MIT Sloan Review*, 48(4), 62–68.

Canbolat, Y. B., Gupta, G., Matera, S., and Chelst, K. (2008). Analysing risk in sourcing design and manufacture of components and sub-systems to emerging markets. *International Journal of Production Research*, 46, 5145–5164.

Chang, J. I. and Lin, C. (2005). A study of storage tank accidents. *Journal of Loss Prevention in the Process Industries*, 19, 51–59.

Chopra, S. and Sodhi, M. (2004). Managing risk to avoid supply chain breakdown. *MIT Sloan Management Review*, 46(1), 53–61.

Colwill, C. J., Todd, M. C., Fielder, G. P., and Natanson, C. (2001). Information assurance. *BT Technology Journal*, 19(3), 107–114.

Committee of Sponsoring Organizations of the Treadway Commission (2004a). *Enterprise Risk Management—Integrated Framework: Executive Summary and Framework.*

Committee of Sponsoring Organizations of the Treadway Commission (2004b). *Enterprise Risk Management—Integrated Framework: Technical Applications.*

Frankel, H. L., Crede, W. B., Topal, J. E., Roumanis, S. A., Devlin, M. W., and Foley, A. B. (2005). Use of corporate Six Sigma performance-improvement strategies to reduce incidence of catheter-related bloodstream infections in a surgical ICU. *Journal of the American College of Surgeons*, 201, 349–358.

Jackson, P. L. (2010). *Getting Design Right, A Systems Approach*. Boca Raton, FL: CRC Press.

Kossiakoff, A. and Sweet, W. N. (2003). *Systems Engineering Principles and Practice*. Hoboken, NJ: John Wiley & Sons.

March, J. G. and Shapira, Z. (1987). Managerial perspectives on risk and risk taking. *Management Science*, 33, 1404–1418.

Morgan, M. G. (1981). Choosing and managing technology-induced risk. *IEEE Spectrum*, 18, 53–60.

Morgan, M. G. (1993). Risk analysis and management. *Scientific American*, 269(1), 32–41.

Morgan, M. G., Fischoff, B., Bostrom, A., and Atman, C. J. (2002). *Risk Communication: A Mental Models Approach*. Cambridge, U.K.: Cambridge University Press.

Rice, J. B. and Caniato, F. (2003). Building a secure and resilient supply network. *Supply Chain Management Review*, 7, 22–30.

Rutala, W. A. and Weber, D. J. (2007). How to assess risk of disease transmission to patients when there is a failure to follow recommended disinfection and sterilization guidelines. *Infection Control and Hospital Epidemiology*, 28(2), 146–155.

Slovic, P., Fischhoff, B., and Lichtenstein, S. (1979). Rating the risks. *Environment*, 21(3), 14–21, 36–39.

Taner, M. T., Sezen, B., and Antony, J. (2007). An overview of six sigma applications in healthcare industry. *International Journal of Health Care Quality Assurance*, 20, 329–340.

White, A. A., Wright, S. W., Blanco, R., Lemonds, B., Sisco, J., Bledsoe, S., Irwin, C., Isenhour, J., and Pichert, J. W. (2004). Cause-and-effect analysis of risk management files to assess patient care in the emergency department. *Academic Emergency Medicine*, 11, 1035–1041.

Chapter 9

Spreadsheet Simulation for Decisions with Uncertainty

Goldgate Corporation has been given a government contract to develop a new product. This project consists of three sequential major activities: (1) system design, (2) component design, and (3) testing. The engineers believe that each of these three activities includes significant uncertainty, and, therefore, total project time is uncertain. Engineers are tasked to estimate the time uncertainty associated with each activity. The contract includes a penalty clause, a charge of $100,000, if the project is not finished in 130 days. What is the probability that Goldgate will finish this project within 130 days? Is there a cost-effective strategy that will facilitate timely completion and avoid or reduce the probability of a penalty?

Sonica, located in Alabasca, New York, manufactures televisions. Sonica wants to reduce purchasing cost by 5% within the next year. The company is planning to source more components from low-cost countries such as India and China. Mary Layton, Vice President of Purchasing, thinks that they can reduce piece price up to 30% for some components and parts by sourcing to emerging markets. However, she knows that these cost-saving opportunities come with significant risks. These include international logistics, cultural and language differences, foreign exchange rate fluctuation, duty/custom regulations, quality problems, and political and economical stability. Would Sonica save money by sourcing to emerging markets after taking into account the risks?

9.1 Goal and Overview

This chapter is a basic introduction to stochastic simulation that is used to model a collection of uncertainties. It is a descriptive tool that produces a risk profile but unlike decision trees does not identify the optimal decision. The software used is @Risk an Excel add-on that is included with the book.

Stochastic simulation is a general purpose modeling tool that can be used to study and analyze a wide array of uncertain situations and related decisions. It captures the elements of uncertainty in the problem by representing each uncertain variable or event with a probability distribution. The stochastic simulation model replicates the uncertainty by generating random numbers according to the assumed probabilistic pattern for each uncertain variable. The simulation model links the various uncertainties according to the specific problem description. For example, in the Goldgate situation, there are three random time variables. Management is interested in the uncertainty related to the sum of the three variables.

Once a stochastic simulation model is developed, it is run multiple times, or *iterations*. The results of hundreds of iterations are tabulated and used to estimate the mean and variability of key outcome measures such as total time to complete a project. The simulation model enables decision makers to better understand the current system. More importantly, they can experiment with changes to the system and simulate the impact on key outcome measures.

There are a wide array of software packages used to develop simulation models. These range from Excel add-ins (@Risk, and Crystal ball) to specialty languages (Simul8, Siman, and Arena). In this chapter, we use @Risk. Its main value is that it easily transforms any spreadsheet cell into a random variable and converts output into easy-to-read charts. One does not need to be a statistics expert to perform the simulations and analyze risk.

More complex situations such as a production line or hospital emergency room are modeled using specialty languages and advanced simulation tools. These have sophisticated graphics and a long list of functions to facilitate development of the model.

In this chapter, we present how to structure and analyze a basic problem including uncertainty through stochastic simulation. The key output will be the risk profile for the project. The chapter begins with the Goldgate product development example to demonstrate the basics of building and running a simulation model. Management is very concerned about the risk that the project may not be completed in 130 days. What is the probability that the project will not be completed within 130 days? They also wish to assess the impact of alternative strategies on development time. For example, what happens if component design time is reduced by employing more engineers? Simulation will answer these types of questions in a quick and cost-effective manner. We then grow the problem size through two more complicated examples: profit forecasting for drug development and global sourcing risk analysis.

9.2   Using @Risk Spreadsheet Simulation

@Risk uses Monte Carlo simulation to generate multiple random values according to user-specified probability distribution functions. After running the simulation many times and tabulating results, the modeler obtains a range of estimates for the possible outcomes. These outcomes can be averaged as well as charted as a cumulative probability distribution function. The steps in developing a simulation model proceed as follows:

Identify the variables and define the probability distribution functions: In this step, the variables that will affect the outcomes are identified, and the probability distribution functions for random input variables are defined. Both discrete and continuous probability distributions can be used to represent random variables. The probability distributions can be specified by subject matter experts or developed from data.

Construct the model and perform the simulation: The model is formulated by building the relationships between the input variables and the output variables. In the Goldgate example, the relationship is just the sum of the variables. With each iteration of the model, a new set of possible values are sampled from each input distribution, and output results are generated. The software package @Risk automatically keeps track of these output values. A distribution of possible outcomes is generated by running all possible scenario outcomes. The output probability distributions then give a decision maker a complete, realistic picture of the range of possible outcomes.

Compare the simulation results for the alternatives: Typically, the simulation model is run for multiple strategies. The results for each of these strategies are compared and contrasted both in terms of expected values and risk profiles. Each strategy comes with its own cost to implement that strategy. The decision maker must then decide on the preferred balance of cost and predicted performance.

9.3   Project Acceleration Investment

Goldgate Cooperation launched a project to develop a new product. This product development project consists of three sequential major activities: (1) system design, (2) component design, and

(3) testing. The engineers believe that each of these three activities include significant uncertainty, and, therefore, total project time is uncertain. Goldgate wants to finish the project within 130 days.

9.3.1 Identify the Variables and Define the Probability Distribution Functions

Engineers predict that all three activities involve uncertainty. For the sake of simplicity, assume for now that there is no decision to make, and Goldgate management wants simply to analyze uncertainty. Figure 9.1 shows an influence diagram for this example. The objective of Goldgate management is to analyze uncertainty in the total project time, which is influenced by the three random variables: (1) system design, (2) component design, and (3) testing.

Engineers are tasked to estimate the time for each activity. Table 9.1 summarizes the activities and the probability distribution functions for the time for each task. The development of system design is estimated to be uniformly distributed between 30 and 40 days. Component design time is also uniformly distributed between 40 and 60 days. Testing time is normally distributed with a mean of 40 days and a standard deviation of 15 days. The marketing department predicts that if the project cannot be finished in 130 days, the company will incur a loss of $100,000 as a result of lost sales.

9.3.2 Construct the Model and Perform the Simulation

Total project time is the sum time of all the activities, which is calculated by the @Risk simulation model. Figure 9.2 presents a cumulative risk profile, showing that project completion time is highly variable. The total project time varies between 69 and 179 days with a mean of 125 days. There is 38% chance that the total project time will exceed 130 days, after which Goldgate will incur a loss of $100,000. Figure 9.2 also lists the 5th and 95th percentiles. There is only a 5% chance that total project time will be within 98 days. The 95th percentile is 152 days. This means that there is a 5% chance that total project time will be more than 152 days.

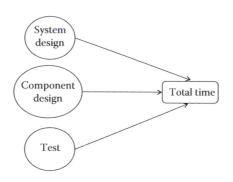

FIGURE 9.1: Influence diagram for Goldgate project management.

TABLE 9.1: Activity time for the product development example.

Activity	Development Time (Days)
System design	Uniform (30, 40)
Component design	Uniform (40, 60)
Test	Normal (40, 15)

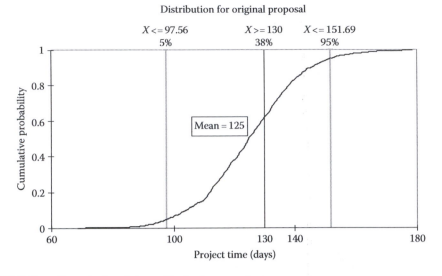

FIGURE 9.2: Cumulative risk profile for base project.

TABLE 9.2: Activity development time for new strategies.

	Development Time (Days)		
Activity	**Base**	**More Engineers**	**Test Technology**
System design	Uniform (30, 40)	Uniform (30, 40)	Uniform (30, 40)
Component design	Uniform (40, 60)	Uniform (30, 40)	Uniform (40, 60)
Testing	Normal (40, 15)	Normal (40, 15)	Uniform (20, 30)

The current plan faces significant risk in not meeting the deadline. The engineering department has proposed two alternatives to reduce risk. Alternative A involves hiring three more engineers to conduct component design. Engineers believe that this will reduce the component design time from a Uniform (40, 60) to a Uniform (30, 40). The cost of hiring three more engineers for this project is $25,000. Alternative B involves using a new testing technology that can reduce test time from a Normal (40, 15) to a Uniform (20, 30). Using new technology for this project will increase the project cost by $50,000. Table 9.2 presents data for all three options.

Alternative A reduces the average total project time from 125 to 110 days. Its cumulative risk profile is illustrated in Figure 9.3a. Total project time varies between 56 and 158 days, and there is still a 10% chance that the total project time will exceed 130 days. Alternative B also reduces the average project time to 110 days. The range of total project time narrows with Alternative B as the total project time fluctuates between 91 and 129 days (see Figure 9.3b). Alternative B eliminates the risk of paying penalty since the maximum total project time is lower than 130 days.

9.3.3   Compare the Results for the Alternatives

The software @Risk can place the risk profiles side-by-side. Figure 9.4 depicts the cumulative risk profile for the three options. In this example, since the least time is preferred, the curve further to the left side is better. Alternative A is always probabilistically better than the base project. It is said to stochastically dominate the base case. For any given time to completion such as 100 days, Alternative A has a higher probability than the base case of meeting that value.

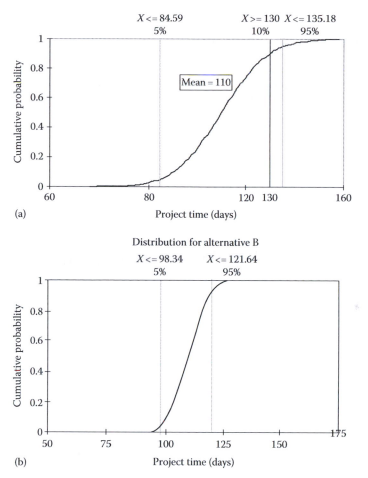

Distribution for alternative A

Distribution for alternative B

FIGURE 9.3: Cumulative risk profile for Alternative A and B.

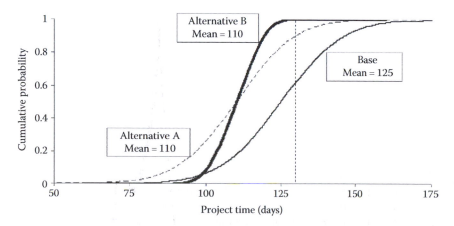

Distribution for total project time

FIGURE 9.4: Cumulative risk profiles for product development time for 3 alternatives.

TABLE 9.3:   Comparison of base project and alternatives A and B.

Alternative	Cost ($)	Expected Duration (Days)	Probability of Penalty	Expected Cost ($)
Base	0	125	0.38	38,000
Alternative A	25,000	110	0.10	35,000
Alternative B	50,000	110	0	50,000

Similarly, there is a 90% probability that it will meet the 130 day deadline as compared to 62% for the base case.

The risk profile for Alternative B curve crosses over the other curves. With B, the sum of the minimum values for each task is higher than the minimum of other alternatives. At the other extreme, its maximum total time is lower than others' total time. The mean values for A and B are the same, but the range of outcomes is much broader for A than B.

Table 9.3 compares expected time, risk, and costs of the base project and alternatives. The expected time of the base project is 125 days, with no additional cost. There is a 38% chance that Goldgate will incur a penalty, and, therefore, the expected cost of the base project is $38,000. On the other hand, if the company follows Alternative A, it will incur an additional cost of $25,000, but the probability of penalty will be reduced to 10%. The expected cost of Alternative A is $35,000. This is found by adding the upfront cost, $25,000 to the expected penalty cost, $100,000 * 0.10. Alternative B eliminates the likelihood of late penalty; however, its additional cost is $50,000. In terms of expected cost, Alternative A is the most attractive option. Yet, it is not obvious that all decision makers would be willing to spend $25,000 upfront to reduce, although not eliminate, the risk of meeting the deadline. They may worry about spending money upfront but still not solving the problem or eliminating the risk of penalties if the project is late. Other decision makers might be willing to spend the $50,000 upfront to avoid the cost of missing the deadline and to appear proactive. In summary, probabilistic analysis can clarify the risks, but hard decisions remain to trade off cost and uncertainty.

9.4   Profit Forecasting for Drug Development

BSG, a global pharmaceutical company, is interested in purchasing the right to develop and market the drug GEN-257 in the United States from a Japanese biotechnology company (GenBio). GEN-257 has just successfully completed global Phase 3 trials and will file for approval by the U.S. Food and Drug Administration (FDA) for marketing in the United States. GenBio is asking for $700 million to sell GEN-257's marketing rights in the United States. GEN-257's patent will expire within 7 years.

BSG's licensing team has been charged to analyze GEN-257's potential over 7 years to determine whether GEN-257 is worth acquiring. Because of the uncertainty in the product demand and cost, the team decided prior to negotiations to use @Risk to simulate the Net Present Value (NPV) before tax and the Internal Rate of Return (IRR).

BSG uses NPV and IRR for evaluating long-term projects. The NPV is a standard method for including the time value of money to evaluate projects. It indicates how much value an investment or a project adds to a company by calculating the present value of the project's future net cash flow. The IRR on an investment is the annualized effective compounded return rate that is earned through the life of the investment. In BSG's licensing decision, the IRR shows the discount rate that reduces the net present value of a stream of income inflows and outflows. If the IRR of the project is higher than the desired rate of return, the project will be approved.

9.4.1 Identify the Variables and Define the Probability Distribution Functions

BSG's commercial team was asked to forecast sales and operating margins for GEN-257. The team employed a normal distribution to describe the uncertainty surrounding forecasted sales. Table 9.4 shows the team's projections for price per unit and units sold. The team projects that the unit price in the 1st year would be $700 and would increase slightly over the forecasting period. First year sales are forecasted to be 300,000 units with a standard deviation of 15,000 units. Sales are projected to increase each year through year 5 and decline in year 6. GEN-257's patent will expire within 7 years. Since GEN-257 is expected to launch next year, year 1, it will be in the market for 6 years before its patent expires.

BSG calculates Cost of Goods Sold (COGS) and operating cost as a percent of revenue using historical revenue and cost data. COGS includes the direct costs such as material and labor, which are attributable to the production of goods. The team used a triangular distribution (30%, 35%, and 40%) to project COGS as a percentage of revenue. They projected 30% as the minimum, 35% most likely, and 40% as the maximum. Operating costs include day-to-day expenses incurred in running a business, such as sales and administration. For the operating cost, a normal distribution was utilized with a mean of 10% and a standard deviation of 1% as percent of revenue.

BSG's licensing team estimates there is 90% chance that FDA will approve GEN-257 for marketing in the United States. The team also predicts that the filing will cost $10 million and take 1 year for the launch. BSG employs a 10% discount rate when calculating the NPV.

9.4.2 Construct the Model and Perform the Simulation

The commercial team used a cash flow spreadsheet analysis to calculate expected NPV. Table 9.5 shows expected values of cash flow analysis (gross income, costs, and net income). Gross income in a given year is calculated by multiplying the price per unit by the number of units sold in that respective year. It is then multiplied by 0.9, the probability of FDA approval. The expected value of gross income in the 1st year is $189,000,000 (= $700 * 300,000 * 0.9). COGS is the product of gross income and COGS percentage. In year 1, expected COGS would be $66,150,000 (=189,000,000 * 35%). Operating cost is calculated by multiplying gross income by the operating cost percentage. In year 1, the expected value of operating cost is $18,900,000 (=189,000,000 * 10%). Net income is estimated by subtracting all costs from gross income. In year 0 (prelaunch), net income is −$710 million (licensing and registration cost). After the launch of the product, net income would be positive. Expected net income in the 1st year would be $103,950,000 (=189,000,000−66,150,000−18,900,000).

The NPV was calculated using Excel functions, the negative and positive cash flows, and initial investment. The expected NPV is $302 million with an IRR of 21%.

The licensing team is concerned about the uncertainty surrounding some important variables and projections (units sold, COGS percentage, and operating cost percentage). The cumulative risk profile of the NPV helps the team see possible outcomes as depicted in Figure 9.5. The NPV varies between a loss of $710 million and a profit $597 million with a mean of $302 million. The most negative outcome is associated with the 10% chance that the FDA does not approve the marketing of the drug in the United States. However, if the company wins FDA's approval, the NPV would be $240 million or more. This means that in the worst case scenario (lowest sales and highest costs), the NPV would be $240 million. The 90th percentile is $480 million, which means that 90% of the

TABLE 9.4: Price and units sold projections for profit forecasting example.

Year in Market	1	2	3	4	5	6
Price/unit ($)	700	710	730	740	760	770
Units sold (000)	N(300, 10)	N(500, 25)	N(700, 35)	N(800, 40)	N(900, 45)	N(800, 40)

TABLE 9.5: Cash flow spreadsheet analysis for profit forecasting example.

Year in Market	0	1	2	3	4	5	6
Price/unit ($)		700	710	730	740	760	770
Units sold (000)		300	500	700	800	900	800
Gross income (000) (price * units * probability)		189,000	319,500	459,900	532,900	615,600	554,400
Gross income * % COGS		66,150	111,825	160,965	186,480	215,460	194,040
Operating cost (000) (gross income * % operating cost)		18,900	31,950	45,990	53,280	61,560	55,440
Initial cost (000) (licensing and registration cost)	710,000						
Net income (gross income–cost of revenue–operating cost–initial cost)	−710,000	103,950	175,725	252,945	293,040	338,580	304,920

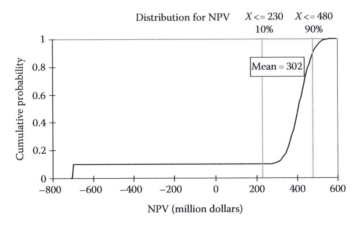

FIGURE 9.5: Cumulative risk profile for forecasted profit.

time, the NPV is at or below $480 million and that there is a 10% chance that the NPV will be more than $480 million.

The NPV value is positive, and the IRR exceeds BSG's hurdle rate (defined here as the discount rate of 10%). However, the BSG senior management is concerned about the huge initial investment of $710 million. The senior management asked the licensing team to develop a profit sharing option to reduce initial investment cost and offer royalty payment to GenBio. They named the original alternative "buy option" and called the new option "profit sharing option."

The licensing team worked on several alternatives and decided to offer the following option to GenBio. BSG would pay $350 million dollars instead of $700 million and pay GenBio a 22% royalty on all GEN-257 sales. Figure 9.6 illustrates the cumulative risk profile for the NPV for this profit sharing option. The NPV varies between −$360 and $492 million with a mean of $247 million. The mean IRR of the profit sharing option is 27%. The 10th percentile is $150 million, and the 90th percentile is $377 million.

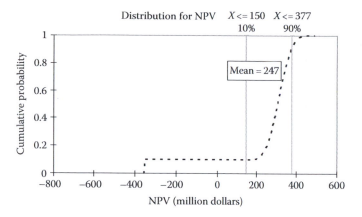

FIGURE 9.6: Cumulative risk profile of forecasted profit with profit sharing option.

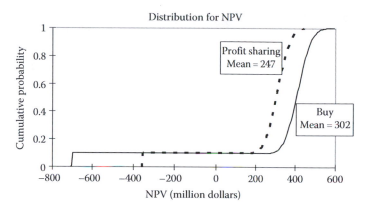

FIGURE 9.7: Cumulative risk profile of BSG profits for both options.

9.4.3 Compare the Simulation Results for the Alternatives

Figure 9.7 shows cumulative risk profiles for both options. The profit sharing option reduces the initial investment from $710 to $360 million. However, the NPV of this option is $55 million less than the buy option. There is a NPV of $247 million for the profit sharing option vs. $302 million for the buy option. Because the upfront investment has been cut in half, the profit sharing option has a much larger IRR of 27% compared to buy option's IRR of 21%.

BSG has decided to offer the profit sharing option to GenBio. From GenBio's point of view, they would receive $350 million as upfront payment in year 0. If GEN-257 receives marketing approval from the FDA, the NPV of the royalty payment would be up to $483 million. On the other hand, if the FDA does not approve the drug, they would not get any royalty payment. GenBio's NPV for the profit sharing option varies between $350 and $833 million (see Figure 9.8). Expected total NPV for GenBio would be $755 million, which is $55 million more than the buy option.

9.5 Global Sourcing Risk Analysis

Sonica, located in Alabasca, New York, manufactures televisions. Sonica wants to reduce purchasing cost by 5% within the next year. The company is planning to source more components to

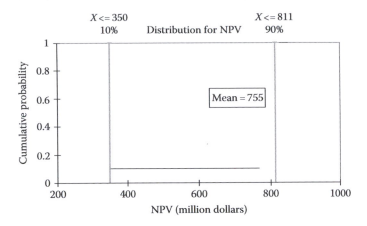

FIGURE 9.8: Cumulative risk profile of GenBio's profit for profit sharing option.

low cost countries such as India and China. Mary Layton, Vice President (VP) of purchasing, thinks that they can reduce piece price up to 30% for some components and parts by sourcing to emerging markets. However, she knows that these saving opportunities come with significant risks that can drive up other costs. These risks include cultural and language differences, foreign exchange rate fluctuation, duty/custom regulations, quality problems, and political and economical stability. International logistics related to inventory management, border-crossing procedures, and transportation delays create challenges that could impact product availability.

VP Layton created a multifunctional team, Global Sourcing Risk Analysis (GSRA), to analyze outsourcing risk. She asked the team to present their findings using a case study approach. GSRA identified a variety of global sourcing risk factors. For the case study, they chose a component that offers great cost savings when sourced to an Indian supplier, Indi Electronics.

9.5.1 Identify the Variables and Define the Probability Distribution Functions

This global sourcing case study of a TV component is adapted from a real-world modeling effort developed to support Ford's outsourcing decisions (Canbolat et al. 2008). Sonica is considering an Indian supplier (Indi Electronics) as a replacement for a local current supplier. The piece price of the part is $25 when bought from the current supplier; while Indi Electronics is asking for $20 per part including transportation and duty costs. The potential saving per part is $5 if risk is not factored into the equation. However, Sonica wants to investigate the risk of sourcing from Indi Electronics. If the incremental risk of sourcing from Indi Electronics is less than potential savings, the company will switch to Indi Electronics.

The GSRA team conducted a comprehensive literature review and examined the existing Sonica sourcing plan to identify risk factors in the global sourcing process. There are a wide variety of sources of risk, and no one person or department within Sonica is aware of all of them. The team interviewed a number of executives who are experienced in global sourcing, supply chain management, and production operations. The goal of the interviews was to identify and quantify global sourcing risk factors.

From the discussions, the team identified risk groups, their potential effects, and risk factors. These are presented in Table 9.6. The first three columns of Table 9.6 show risk groups, their potential effects, and risk factors. One of the main problems in global sourcing is that when components are delayed, production may be curtailed or the company pays a premium in logistics to make up for lost time. The causes of component delay are shipping delay, customs-related issues, national emergency-related failures, supplier's production interruptions, international trade problems, culture and language differences, and engineering specification problems. As can be expected, many

TABLE 9.6: Risk factors and probability distribution functions.

Risk Group	Potential Effects	Risk	Probability	Effect
Premium freight	Parts not on time and increased air freight cost	Defective parts	0.01	Uniform (2, 3)
		Break in pipeline	0.02	Uniform (1, 3)
Component delay	Parts not on time	Shipping delay	0.20	Normal (3, 1)
		Custom problems	0.10	Uniform (0, 3)
		National disasters	0.01	Gamma (7.5, 0.6)
		Supplier's production interruptions	0.01	Normal (4, 1)
		International trade problems	0.01	Normal (3, 1)
		Cultures and language differences	0.10	Normal (4.5, 1.5)
		Engineering specification problems	0.05	Normal (4, 1)
Warranty	Warranty and recall cost ($R/1000$)	Defective parts	0.30	Uniform (0.5, 2)
Currency	Higher cost	Currency fluctuation		Normal (0.002, 0.042)
Supplier management	Higher cost	Supplier technical assistance	0.6	Normal (75,000, 10,000)
Inventory management	Higher cost	Inventory holding		

of the risk factors are related to the complexity of logistics in the long pipeline. Even a highly responsive overseas supplier will often have to resort to expensive air freight to compensate for the longer supply chain. In addition, customs clearance often takes more time since September 11, 2001. The uncertainties associated with each of these can result in added costs that must be assessed when sourcing to a distant low-cost supplier.

Because of the complexity and vulnerability of international logistics and longer pipeline, companies often increase the inventory level to deal with product delay risk. However, excess inventory increases inventory holding cost. For products with short life-cycles, there is the risk of product obsolescence. The longer pipeline also reduces the company's flexibility to react to changing markets. The other failure risk groups include increased warranty issues and problems in the ongoing management of the supplier relationship.

The fourth and fifth columns of Table 9.6 depict the probability of the specific risk and probability distribution function for the effect of risk. For example, the probability of a break in the component delivery pipeline is 5%. This will cause Sonica to use air freight. In addition, a break in the pipeline will delay components 1–3 days (Uniform distribution). Similarly, there is 1% chance that the supplier's production is interrupted. In that case, the impact of component delay is represented by a Normal distribution with a mean of 4 days and a standard deviation of 1 day. With respect to warranty, there is 30% chance that there will be defective parts. If there is a quality problem, Rejects/1000 is estimated to range between 0.5 and 2 according to a Uniform distribution.

Currency fluctuation is a special area of concern in sourcing to emerging markets. If the currency of India, the Rupee, declines in value when compared to the U.S. dollar, Sonica will benefit from that change. If the Indian Rupee increases in relative value, Sonica will pay more in U.S. dollars for the same component. As a result, the projected component cost savings will decrease

TABLE 9.7:   Data for global sourcing case.

Variable	Value
Fixed premium freight cost	$100,000
Number of parts per shipment	20,000
Number of shipments per year	12
Purchasing price per part	$20
Stock out cost per part	$30
Stock out cost per day	$20,000
Cost per repair	$150
Inventory holding cost per part per month	$0.17

or even disappear. For example, in 2004, $1 was equivalent to 45.3 Indian Rupees. In 2005, the Indian Rupee became 4% stronger and $1 was equivalent to 43.6 Indian Rupees. As a result, components become 4% more expensive. The team analyzed fluctuation in Indian currency with respect to the U.S. currency between 1994 and 2009. Based on this analysis, the team modeled percent change in the Indian Rupee using a Normal distribution with a mean of 0.002 and a standard deviation of 0.042.

The GSRA team believes that there is a 60% chance that Sonica will need to provide technical assistance to Indi Electronics to meet Sonica's quality standards. If Indi Electronics needs technical support, Sonica will incur an annual cost that is represented by a normal distribution with a mean of $75,000 and a standard deviation of $10,000. In this case study, there is no risk of obsolescence. The model also includes inventory holding and stock out cost.

Table 9.7 presents additional data needed to determine the cost impact of different risks. For example, if Sonica uses air freight, it will incur a fixed cost of $100,000 per flight.

9.5.2   Construct the Model and Perform the Simulation

The GSRA team converted the risk into dollars. Table 9.8 summarizes how to convert risk into dollars for a period between two regularly scheduled shipments. Component delay involves delays of varying duration that set back the launch date of a product or interrupt the regular sales of the products. In each case, the cost is a lost opportunity cost that equals

$$(\text{product profit margin}) * (\text{daily sales rate}) * (\text{total delay duration}).$$

In some specific situations, Sonica has a policy of paying premium freight to reduce the duration of the delay. The cost of a sales interruption delay is also linked to the company's inventory policy. This cost is moderated by the inventory available to compensate for the short-term delay.

TABLE 9.8:   The impact of risk in terms of dollars.

Risk Group	Risk Calculation/Definition
Premium freight	Fixed cost + variable cost * number of products per day * duration of lost sale
Component delay	Variable cost * number of products per day * duration of lost sale
Warranty	R/1000 * warranty cost per part
Currency	Percent increase or decrease in the ratio of Indian Rupee to the U.S. dollar * purchasing price per part * number of parts per shipment
Supplier management	Fixed annual supplier technical assistance cost
Inventory management	Average inventory * purchasing price * holding cost as a percent of price per period (i.e., time between shipments)

Warranty cost is calculated by multiplying $R/1000$ and repair cost per component. The company can use historical warranty cost data when dealing with existing suppliers. However, the prediction of the $R/1000$ includes higher uncertainty if the buyer has no prior relationship with a supplier. The supplier management annual cost is a Sonica policy decision cost that reflects its commitment of technical resources to manage the relationship and improve the supplier's process and product quality.

The model calculates total risk per shipment by adding up all of the risks. The model also calculates total risk per part. This statistic is easier to use to compare strategies. The GSRA team simulated two different safety stock levels: (1) 10 day and (2) 20 day inventory.

9.5.3 Check the Results

The simulation run assumes that inventory levels are set equal to 20 day safety stock. As illustrated in Figure 9.9, the dollar value of risk varies from approximately −$2 to + $7.03 with a mean of $0.58. Minus risk means a potential further savings opportunity for a part sourced to the Indian supplier. The standard deviation of the total risk is $1.21. The model also lists the 5th and 95th percentiles. There is only a 5% chance that total risk is −$0.95 or less. The 95th percentile is $2.14, which means that 95% of the total risk is at or below $2.14 and that there is a 5% chance that total risk will be more than $2.14. This represents a substantial increase in cost to Sonica that would cut into the projected $5 savings; however, there does not seem to be a significant risk of not benefiting from sourcing to Indi Electronics.

The model ranks the risk groups according to the mean risk in dollars as presented in Table 9.9. In working with new suppliers in an emerging market, it is often not possible to assess the quality and experience of the supplier management team. In this instance, supplier management includes the highest mean risk ($0.31 per part). The supplier management risk varies between $0.18 and $0.45. Premium freight risk is $0.15 per part and includes significant variability, $0–$5 per part. The high inventory level eliminates any direct cost associated with component delays. Because the Indian Rupee is more likely than not to decline in value against the U.S. dollar, Sonica expects to save $0.04 per part due to currency exchange rate fluctuation.

9.5.4 Compare the Simulation Results for the Alternatives If Possible

The team is concerned about the high safety stock level. From the simulation results, they know that a 20 day inventory eliminates the risk of product unavailability. They decided to explore the

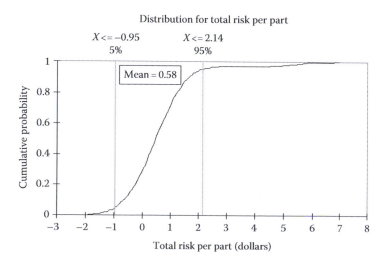

FIGURE 9.9: Cumulative risk profile for global sourcing example.

TABLE 9.9: Risk groups in global sourcing risk analysis example.

Risk Group	Minimum	Mean	Maximum	5%	95%
Supplier management	0.18	0.31	0.45	0.24	0.38
Premium freight	0.00	0.15	5.00	0.00	0.00
Inventory management	0.07	0.10	0.11	0.09	0.11
Warranty	0.00	0.06	0.30	0.00	0.26
Component delay	0.00	0.00	0.00	0.00	0.00
Currency	−2.68	−0.04	2.58	−1.43	1.34
Total	−2.25	0.58	7.03	−0.95	2.14

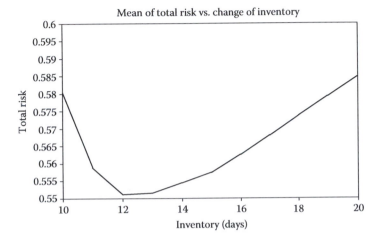

FIGURE 9.10: Sensitivity simulation showing the impact of inventory on risk.

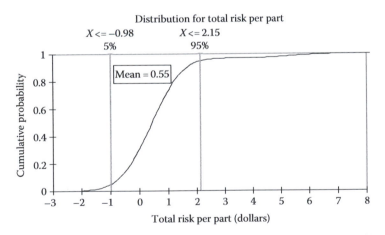

FIGURE 9.11: Cumulative risk profile for global sourcing (13 day safety stock).

impact of reducing inventory level to as low as 10 days. Figure 9.10 illustrates how the mean value of the total risk as measured in dollars per part changes as the inventory level is adjusted. The shape of the sensitivity analysis curve is concave. A 12 or 13 day inventory level minimizes the total risk.

Indi Electronics offers $5 potential savings per part, and the expected value of global sourcing risk is $0.55 if it holds a 13 day inventory (see Figure 9.11). Were Sonica to switch to the Indian supplier for this component, Sonica's expected cost saving will be $4.45 per part after the risk is accounted for. Based on the GSRA team analysis, the team's recommendation to Sonica was to source this component from the Indi Electronics and hold 13 day safety stock.

9.6 Real-World Applications @Risk

All of the cases presented in the following come from summaries on the Palisade corporate website. We often quote directly from the website summaries without modification. These cases are intended to illustrate the value of @Risk. http://www.palisade.com/cases/

9.6.1 Using Simulation to Model Blood Screening Safety

One of the political goals of the European Union (EU) is to achieve a more unified approach to public health and practice. Pereira (2003) examined how improved donor screening for hepatitis B affects health outcomes and medical costs across the EU.

Increased sophistication in blood donor screening has virtually eliminated the risk of post-transfusion HIV and other serious viral infections in many parts of the world. The only significant disease for which transfusions continue to pose a risk is from the hepatitis B virus (HBV). The current risk of acquiring this infection through blood transfusion is below 1:75,000 blood units. The risks for HIV and hepatitis C stand around 1:1 million and 1:300,000 transfused blood units, respectively.

A new technology for detecting HBV, based on nucleic acid testing (NAT), was introduced in the late 1990s. NAT was more sensitive than other detection methods, known as HBsAg assays, though it was considerably more expensive. The primary concern of this study was the issue of allocating resources to NAT at the expense of diverting them from other health care priorities. The decision was complicated by the fact that improved versions of the HBsAg assay were under development and would be licensed soon. They are nearly as sensitive as NAT but substantially cheaper.

Spreadsheet simulation enabled Pereira to take full account of uncertainties in his analysis of cost-effectiveness. For some uncertain variables, he had enough data to describe a probability distribution function. For others, he employed a triangular distribution that used three parameters, minimum, most likely, and maximum values.

The simulation model predicted that 0.97% of EU patients with post-transfusion HBV would die of liver disease. The mean loss of life expectancy was 0.178 years per patient, and the expected value of lifetime costs of treating HBV-related complications was estimated to be $4700 per patient. For the EU population, the projected cost of NAT testing for each life-year gained was $6,519,000. Using the enhanced sensitivity HBsAg method, the cost for each life-year gained was $888,000. The simulation model demonstrated that NAT would provide a small health benefit at a very high cost. Under some circumstances, however, the cost effectiveness of enhanced-sensitivity HBsAg assays would be acceptable for new public health interventions.

9.6.2 Assessing U.S. Agriculture Policy

The U.S. government spent $24 billion on farm programs to support income and reduce risk in 2000. The government programs include direct payments and subsidies on crop insurance premiums. Gray et al. (2004) studied the impacts of U.S. farm programs on farmland risk and return using a spreadsheet simulation model. They examined whether these programs individually and collectively reduce agricultural risk and explored better ways to handle agricultural risk. The researchers modeled the uncertainties associated with crop yield and price. After running base-line simulations, the researchers added the individual farm programs into the model to determine their impacts. The @Risk simulation model demonstrated that a combination of all government programs would raise average farm incomes by almost 45%. Additionally, the programs would reduce the economic risks associated with farming by half. The simulation model allowed researchers to examine how the programs interact with one another to alter the return distribution.

9.6.3 Federal Highway Administration Simulates Life Cycle Cost Analysis

The Federal Highway Administration (FHWA) developed a model and made arrangements with 10 states and 2 pavement associations to develop case studies. These studies illustrated the application of spreadsheet simulation to life-cycle cost analysis in pavement design (Herbold 2000). With the simulation model, state highway agency personnel were able to analyze the possible outcomes, their likelihoods, and consequences. They modeled uncertain variables such as initial cost, future pavement rehabilitation cost, and year of rehabilitation using probability distribution functions. They calculated NPV for each project using the simulation. The case studies demonstrated that with limited training in probabilistic principles and in the application of risk analysis software, state highway agency personnel were able to apply the probabilistic approach to their current life-cycle cost-analysis procedures.

9.6.4 Pension, Insurance Researchers Simulate Project's Key Indices

Ahlgrim et al. (2004) developed a simulation model for calculating requirements for pension funds, life insurance, or long-term care for the Casualty Actuarial Society and the Society of Actuaries. They examined and summarized the relationships among economic variables, particularly relating to interest rates, inflation, and equity returns. Summaries of that information were posted on the societies' web sites.

The model provided an integrated framework for sampling future financial scenarios that represent a reasonable approximation of historical values. It produced output values for interest rates, inflation, stock and real estate returns, dividends, and unemployment. The model proved useful for a variety of actuarial applications, including dynamic financial analysis, dynamic financial condition analysis, pricing embedded options in insurance contracts, solvency testing, and operational planning.

The simulation model was able to correlate (1) the performance of stocks and bonds, (2) the housing market, and (3) natural disasters with interest rates, inflation, and unemployment. Capturing the interplay among these variables created a far more accurate model. The insurance industry benefited since unforeseen events could create havoc with insurance rates. Better modeling tools result in better prepared insurance companies and more consistent pricing for insurance buyers.

Exercises

9.1 *Order quantity*

Jenny Parker, a purchasing manager at a local retailer, needed to decide how many shirts to order for the upcoming season. After reviewing historical data, she noticed that demand for shirts includes uncertainty and is normally-distributed with a mean of 15,000 units and a standard deviation of 2,000 units per season. They buy shirts for $15 and sell them for $30. Any shirts left at end of the sale are sold at a discounted price of $10. Jenny wants to maximize profit from shirts by taking into account sales, purchasing cost, inventory shortage, and excess inventory costs. She would like to evaluate three order quantities: 13,000 units, 15,000 units, and 17,000 units.

a. Construct a simulation model to find optimal inventory policy. Estimate mean profit for three order policies.

b. Compare cumulative risk profiles of profits for three order policies.

c. Do you recommend a better order policy to Jenny? What is the optimal order quantity?

9.2 *Project time*

DallCom has recently won a contract from a government agency to design, develop, and produce a prototype of a special laptop. The company must produce the prototype within 45 weeks. The project team identified major activities, activity sequencing, and time to complete an activity as depicted in Table 9.10. For example, activity B starts after activity A is completed and the uncertainty with respect to the time of activity B is represented using a uniform distribution between 4 and 6 weeks. Activity D starts when both activity B and C are finished.

The project manager developed a Critical Path Method (CPM) network for this project as illustrated in Figure 9.12. Activities are indicated as nodes while arrows show the sequence in which the activities must be completed. The time until task D can be stated as the sum of the time for task A and the maximum of the times for task B and C.

a. What is the probability that the company will complete the project within 45 weeks?

b. Compute the average critical path time and the frequency that each path is critical (the critical path is the longest sequence of connected activities through the network and is defined as the path with zero slack time).

9.3 *Portfolio selection*

Infosol, a small information technology (IT) company, identified six possible projects for the upcoming year. The project teams calculated expected revenues, project investment costs, and staff requirements for the projects as shown in Table 9.11. A normal distribution was employed to describe uncertainty related to investment cost. For example, investment cost of project 1 was estimated to have a mean of $2 million and a standard deviation of $200,000. To characterize uncertainty surrounding staff requirements, a triangular distribution was used. Values were specified for the minimum, most likely, and maximum

TABLE 9.10: Prototype development activity design and time estimates.

Activity	Immediate Predecessors	Time (Weeks)
A	—	N(25, 5)
B	A	U(4, 6)
C	A	U(2, 3)
D	B, C	U(3, 4)
E	D	N(7, 1)
F	E	U(2, 3)

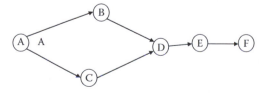

FIGURE 9.12: CPM network for prototype development example.

TABLE 9.11: Data for project selection case.

Project	Project Investment Cost (000)	Staff Requirement	Probability of Success	Expected Revenue
1	N(4,700, 200)	18	0.85	N(8,000, 600)
2	N(8,000, 600)	35	0.70	N(32,000, 3,000)
3	N(2,300, 100)	8	0.95	N(4,300, 100)
4	N(12,000, 800)	45	0.50	N(35,000, 5,000)
5	N(4,700, 200)	15	0.60	N(19,000, 2,000)
6	N(5,000, 400)	8	0.90	N(11,500, 2,500)

TABLE 9.12: Alternative project portfolios.

Portfolio	Projects
1	1, 2, 3, 6
2	2, 4
3	2, 3, 5, 6

staff requirements. The fourth column is the probability of success for each project. Once a project is selected to fund, the company would incur the cost. But, it would make money from the project only if it succeeds. The last column of Table 9.11 shows forecasted revenue using a normal distribution for successful projects.

Infosol has constraints on both investment and labor. The company has a project investment budget of $20 million and 80 staff for the upcoming year. Funding all six projects requires $36.7 million and 139 staff. The company identified three different portfolios as shown in Table 9.12. The expected cost of each portfolio and its staff requirements are within budget and labor capacity. The management will choose one of these three portfolios to maximize NPV and minimize risk.

a. Use basic probability theory to calculate the expected value of the net profit for each project as well as for each portfolio.

b. Structure the simulation model to calculate the NPV for all three portfolios defined earlier.

c. Use the simulation to estimate mean, 5th and 95th percentiles of the NPV for each portfolio. Estimate the probability of negative NPV for each portfolio. Compare your averages to the expected value found in Part A.

d. Compare cumulative risk profiles of the NPVs. Which portfolio is best for Infosol? Why?

References

Ahlgrim, K. C., D'Arcy, S. P., and Gorvett, R. W. (2004). Modeling of economic series coordinated with interest rate scenarios. Available at http://www.soa.org/research/finance/research-modeling-of-economic-series-coordinated-with-interest-rate-scenarios.aspx

Canbolat, Y. B., Gupta, G., Matera, S., and Chelst, K. (2008). Analysing risk in sourcing design and manufacture of components and sub-systems to emerging markets. *International Journal of Production Research*, 46, 5145–5164.

Gray, A. W., Boehlje, M. D., Gloy, B. A., and Slinsky, S. P. (2004). How U.S. farm programs and crop revenue insurance affect returns to farm land. *Review of Agricultural Economics*, 26, 238–253.

Herbold, K. D. (2000). Using Monte Carlo simulation for pavement cost analysis. *Public Roads Magazine*, 64, 2–6.

Pereira, A. (2003). Health and economic impact of post transfusion hepatitis B and cost-effectiveness analysis of expanded HBV testing protocols of blood donors: A study focused on the European Union. *Transfusion*, 43, 192–201.

Chapter 10

Decisions with Uncertainty: Decision Trees

Peter King, the manufacturing supervisor of a supplier to Fired Motors, is faced with a tough decision regarding investment in capacity expansion at a manufacturing plant in Pontiac. His research has narrowed down his investment options to four, but he is having difficulty choosing because of the uncertainty surrounding these alternatives. What is the short-term demand for the product? How long will Fired Motors keep the product in its lineup? Will the OEM maintain him as the sole supplier or hedge its risks by multisourcing? What would be the payback period and return on investment? (ROI)

The product development team at Dial Inc., a manufacturer of PCs, is trying to define its products amid technological uncertainty. The team must decide between two alternatives: (1) a proven technology that is known to be viable or (2) a prospective technology, Lion, that offers superior price-to-performance results but whose viability is uncertain. What should the firm do?

1. Stick with the proven NiHi technology
2. Commit to Lion, the prospective technology, despite its risks
3. Wait for more information and defer commitment until a later time

Note that even sticking with the existing NiHi technology is risky in terms of market share and technology obsolescence. Dial's competitors may introduce a new technology into their PCs, successfully rendering Dial technology obsolete. The waiting strategy is also risky because it may jeopardize the launch date of the latest products (Krishnan and Bhattacharya 2002).

10.1   Goal and Overview

The goal of this chapter is to present decision trees as an analytic approach to making decisions involving uncertainty. It is a simpler, more transparent modeling tool than stochastic simulation discussed in the previous chapter. It is a normative decision-making tool that identifies the optimal decision based on expected value or expected utility. Stochastic simulation, in contrast, is primarily descriptive. Like simulation, the decision tree output includes a risk profile of all the alternatives.

Decisions are first compared through expected value analysis; the optimal decision is the one with the best expected value. Further analysis enables the decision maker to understand and interpret the related strengths and weaknesses of the alternatives with the goal of developing an even better alternative that mitigates some of the risks. The tool is used in situations where the decision maker is faced with a discrete set of limited alternatives, and uncertainty plays a major role in the future outcome of these decisions.

TABLE 10.1: Random elements in decision environment.

Time needed to complete task or reach goal	Resources required
How long does each task take to complete?	Globalization
	Currency fluctuations
Can the project deadline be met?	National and regional politics and economies
Cost	Team dynamics in cross-culture paradigm
Variable costs of production for a totally new product	Is the task doable?
	Will something specific happen?
Uncertainty in learning curves impact on long-term cost variability	Performance
	Unfixed bugs in new software
Breakdowns, personnel training	Health care problem solved
Revenue	NVH, fuel economy, warranty claims for a car
Retail	Effectiveness and toxicity for a drug
Commercial	
Government	
Local and global	
Sales—market demand	
Unanticipated competitive actions	

The first task in developing a decision tree is to identify key quantifiable uncertainties that directly impact the outcome of the decision. As a starting point, Table 10.1 presents some random variables and outcomes that arise in a variety of decision contexts. The key challenge is to explicitly account for the uncertainty upfront, not to drive the uncertainty out of the decision process.

10.2   Early Users of Decision Trees

The oil industry was one of the first to use decision trees, incorporating available information to help determine the risks involved in various decisions. Significant uncertainty is commonplace in the industry, involving billions of dollars in costs and revenues. Uncertain variables include the potential yield of oilfields, the cost of extraction, political instability in various regions, and the fluctuating price of oil. Determining potential yield, for example, involves extensive and expensive data collection before drilling. Decision trees have thus allowed companies such as ExxonMobil, ChevronTexaco, Phillips Petroleum, BP, Amoco, and Shell to analyze their risks and evaluate their decisions for maximum profitability (Walls et al. 1995).

Other industries seeking to maximize net present value (NPV) are major users of decision trees as well. The power and electric industry has used decision trees to select the best alternatives when analyzing, for example, alternative methods for hauling coal, including salvaging a grounded ship, buying a new ship, or subcontracting for delivery (Bell 1984). At the Ohio Edison Company (Borison 1995), decision trees were used to select particulate emission control equipment for three units of Ohio Edison's W. H. Sammis coal-fired plant. Pharmaceutical companies, such as Novartis Pharma AG, Pzifer, Abbott Laboratories, and Bristol-Myers Squibb, routinely invest hundreds of millions of dollars in R&D while facing multiple uncertainties in the drug development pipeline. High-profile manufacturers who have used decision trees include DuPont, Xerox, AT&T, Eastman Kodak, General Motors, and Ford. The Decision Analysis Affinity Group (DAAG, www.daag.net), which comprises about 50 organizations, hosts an annual conference on the state of the field and maintains a Web site with numerous success stories. (For an extensive survey of decision tree applications, see Corner and Kirkwood 1991; Keefer et al. 2004.)

10.3 Concepts

Structuring a problem involves developing and understanding the basic components underlying a decision. An influence diagram, discussed in Chapter 2, is a user-friendly, interactive tool for structuring a problem. The problem-structuring phase helps decision makers reach a clear statement of alternatives, uncertainties, and values. In this chapter, we introduce schematic trees as another aid for structuring the problem.

Schematic trees: A schematic tree contains more information than an influence diagram. Branches in a schematic tree reflect the decision alternatives or the outcome of the random events. It more clearly lays out the sequence of events if outcomes are not symmetric. Schematic trees also form the first step in transition from an influence diagram, which is a communication tool for structuring the problem, to a decision tree used to analyze the decision problem. A schematic tree is the skeleton structure of a decision tree without any of the specific numeric values or detailed descriptions of branches.

Decision (analytic) trees: The four basic components or building blocks of a decision tree are decisions, uncertainties, branches, and the information content. The decision analysis employs a rollback procedure to determine the alternative with the best expected value. A basic tree consists of a single decision node followed by a major uncertainty. Figure 10.1 represents a simple decision between two investments in automation. The product is sold as an optional add-on to a major purchase. The major uncertainty is related to the percentage of customers who will purchase an option on a new product. Simple decision trees can be easily analyzed by hand. However, as trees grow in size, a number of software packages can facilitate the analysis. PrecisionTree, a Palisade Decision Tool, is used to model the decision trees in this book. The software is available as a Microsoft Excel add-in.

Objective function: Decision trees model the effect of uncertainties on a single objective function. These objectives can be to minimize cost, maximize profit, maximize ROI, or minimize time to finish the project.

Information content: Information content in a decision tree includes the description of the specific decision alternatives, the list of possible outcomes of each random event, probability of the occurrence of uncertain events, the costs or revenues associated with a particular decision, and the formula for calculating the objective function.

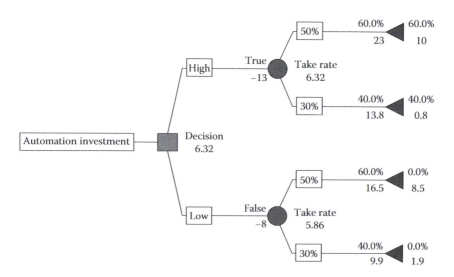

FIGURE 10.1: A basic decision tree with one decision node and one random node.

Calculation by rollback procedure and expected value analysis: A decision tree is solved using the rollback procedure to determine the optimal policy using expected value as the criterion. The expected value is the probability-weighted average of all the possible payoffs in the decision tree. The risk attitude of the decision maker can also be taken into account in the rollback procedure by incorporating a utility function. Decision trees that account for risk attitude are introduced in a subsequent chapter.

Risk profiles: Each decision alternative is associated with a set of outcomes of the objective function values. A risk profile is a graphical representation of the outcomes versus the probability of their occurrence. It allows decision makers to compare and understand the risks involved with a particular decision alternative. A detailed study of the risk profile can help identify actions that could reduce the risks.

Conditional decisions: A conditional decision is one that follows an uncertain event. The preferred decision will be influenced by the outcome of the prior uncertain event. One class of conditional decisions naturally arises when information is gathered. The ultimate decision follows the uncertain outcome of gathering information.

Advanced analysis: This involves studying the strengths and weaknesses of the alternatives and the robustness of the decision-through-sensitivity analysis. These concepts are discussed in the next chapter.

The complexity of the decision tree model varies with the decision context. A decision tree model could be symmetric or asymmetric; it can involve a single decision or a sequence of decisions. In a symmetric tree, the sequence of random events is the same, regardless of the decision path, although the values will be different. In an asymmetric tree, some decision or random events may not arise if the decision maker chooses alternative A instead of alternative B.

Decision trees can be structured as information-seeking models. They evaluate the probabilistic value of gathering more information prior to a decision. The information might be test results from a medical procedure or a test run of a pilot plant. It could be a large-scale marketing experiment to determine market demand or a drug experiment on animals to determine side effects.

10.4   Influence Diagrams and Schematic Trees

We begin with an influence diagram to frame the problem, proceed to schematic tree representation, and follow with decision trees and analysis. Definitions of the basic terms are provided at the end of this chapter.

Influence diagrams are used to frame a decision's basic elements. Their primary role is communication, to obtain agreement on the objectives and critical uncertainties. However, influence diagrams do not provide details as to the specific options and outcomes. Schematic trees and decision trees display more details while using some of the basic building blocks of influence diagrams. Decision nodes with alternatives are represented as rectangles. The uncertain or chance events are represented as circles or ovals. The nodes are connected by the branches, which also represent the flow of information.

10.4.1   Symmetric Decision Trees

Star Electronic, a cellular phone manufacturer, is exploring optimum production capacity for a new phone. The new product requires a new production line and there is uncertainty regarding its yield. Management is focusing on three capacity options. Their competitor's new product may have either marginal or significant impact on the demand for Star's new product, which could be high, medium, or low.

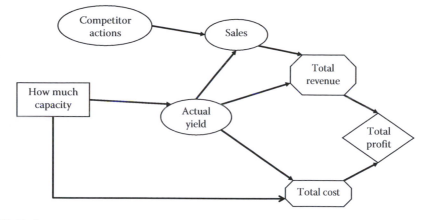

FIGURE 10.2: Influence diagram for capacity planning example.

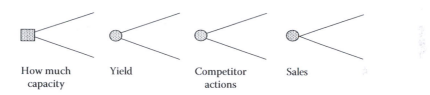

FIGURE 10.3: Schematic tree for capacity planning example.

What are the issues involved in this decision problem? What does the decision maker value the most? How is the outcome of the decision measured? What scenarios would lead to a change in the choice of the alternative? The decision is about how much to invest in manufacturing, even though there is uncertainty regarding demand. An influence diagram and schematic tree model of this case are shown in Figures 10.2 and 10.3, respectively.

First, consider the influence diagram. The overall objective of the problem is to maximize profit, represented by the diamond shape. Profit is influenced by total cost and total revenue. Total revenue is directly influenced by sales volume, while total cost is influenced by capacity decision and the outcomes of yield and sales volume uncertainties. The competitor's action has impact on Star Electronic's sales volume.

The schematic tree representation is read from left to right and represents events in the sequence they may take place. It shows that the problem involves a capacity decision. The outcome of the decision depends on the uncertainties in manufacturing yield, competitor's action, and total demand. All three of these uncertain events will be used to calculate the ultimate net revenue.

When compared to a complete decision tree, the benefits of a schematic tree become apparent as the number of nodes increase. For example, if a decision scenario contains two decision nodes with three alternatives at each node and three random events with three branches at each node, then a full tree would involve $2 * 2 * 3 * 3 * 3 = 108$ branches. Representation of such a tree in the exploded decision tree format is cumbersome.

10.4.2 Asymmetric Decision Tree

In a symmetric decision tree, all paths (scenarios) contain the same sequence of decisions and random events, which is not the case for an asymmetric decision tree. Some random events or subsequent decisions are relevant for only a portion of the overall sequence of decisions and random events.

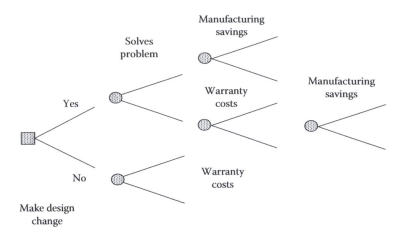

FIGURE 10.4:   Design change schematic tree.

Case study: It is only 6 months before the vehicle launch of the MX36. A sound emanating from the instrument panel has been detected on some test drives. However, engineers are not able to reproduce the sound in a controlled environment. They are fairly certain the problem is from a series of three assembled parts. They have a new single modular design that can be implemented quickly and should solve the problem. There is a potential added benefit from the modular design: reduced manufacturing and assembly cost.

The decision in this case is whether or not to go with the proposed modular design, and the issues are presented in a schematic tree, Figure 10.4. The uncertainty results from several issues. Will the new design solve the problem? What are the likely warranty costs and manufacturing savings? The uncertainty regarding solving the problem as well as the manufacturing savings are only relevant if a new design is introduced. The objective function of the problem can be stated as either maximizing savings or minimizing total costs. Other components of an overall corporate objective are not considered in the model, such as customer satisfaction and lost sales. One interesting aspect is that manufacturing savings may be realized even if the proposed design does not solve the noise problem. A schematic tree representation clarifies asymmetries; this is difficult to do with an influence diagram.

10.4.3 Sequential Decisions

Good Food, a frozen food packing company, will expand its capacity in response to increasing demand. The management has identified three competitive manufacturing systems from three different companies; each has a different forecasted production rate and cost. Good Food will select one kind of machine and determine the number of machines to buy. There is uncertainty regarding the initial throughput of each type of machine. All three machines use new technologies that are unfamiliar to the company's manufacturing engineers. Therefore, management assumes that there is uncertainty associated with the learning curve that will affect the annual improvements in production rate. Good Food's competitor may impact total demand by increasing its production rate and reducing its price. There is significant uncertainty regarding the competitor's response.

In many instances, it is useful to split the decision into a sequence of decisions. For example, consider the case of Good Food deciding on the purchase of a new machine for capacity expansion. One could model this case as a single decision scenario: How many machines of each type should be purchased? On the other hand, it could be modeled as a sequence of decisions where the initial decision is which technology to select and the sequential decision addresses the issue of how many machines of the selected type to purchase.

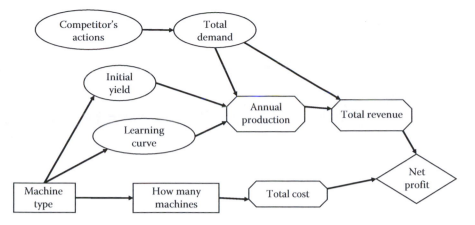

FIGURE 10.5: Influence diagram for machine planning for capacity example.

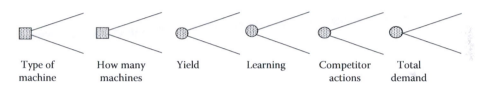

| Type of machine | How many machines | Yield | Learning | Competitor actions | Total demand |

FIGURE 10.6: Schematic tree for machine planning for capacity example.

The important thing to note is that the second decision, how many machines to purchase, may not be the same for all types of machines under consideration given the difference in throughput rates. The influence diagram and schematic tree for Good Food are given in Figures 10.5 and 10.6, respectively.

The schematic tree in Figure 10.6 shows that the capacity planning decision is split into two separate decisions. There could be a third sequence to this capacity planning decision. When buying the equipment, the company also must decide on a contract for maintenance and repair services. What type of maintenance contract should be purchased? There is no right answer as to whether or not to divide and represent the decision as one decision or a sequence of decisions. If the maintenance contracts are relatively standard across all companies, then there is no real need to incorporate a separate decision. Similarly, there is no need for a separate decision node regarding the number of machines if the throughput of the various machines is almost equal.

1. **Activity:** Present an example of a sequence of two or more decisions followed by an uncertainty.

_____

_____

10.4.4 Information-Seeking Trees

There are opportunities to seek additional information before making the primary decision. Usually, there is a cost associated with gathering the additional information. That cost could be direct, such as the cost of a survey or experiment, or indirect, such as the lost opportunity cost resulting from delaying the decision until after the information is gathered and interpreted. These types of decision scenarios represent a major class of applications in decision trees. Oil drilling decisions, new product or process development decisions, and personal health decisions offer the

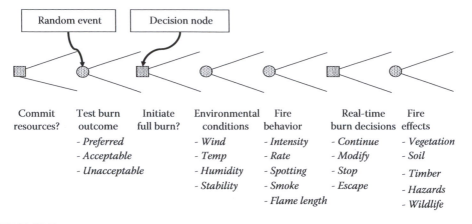

FIGURE 10.7: Schematic tree—controlled forest fire. (From Cohan, D. et al., *Interfaces*, 14(5), 8, 1984.)

opportunity to delay the major decision while seeking additional information by carrying out more tests. These types of models have an uncertain node between the two sequential decisions that represent the unknown outcome of gathering more information. When the first decision is made to gather more information, the range of possible outcomes is known but not the actual results. However, information will be analyzed, and the uncertainty resolved before the second decision is made.

Case study: An information-seeking schematic tree was applied in a U.S. Forest Service study by Cohan et al. (1984) for conducting prescribed forest fires. Prescribed fires are the controlled fires used to reduce major fire hazards created by heavy volumes of logging slash or naturally accumulated forest residues. A prescribed fire is a complex and variable process. Many key factors in the process are uncertain, including environmental conditions and the effects of the fire. Even if the environmental conditions are fully known, fire behavior is difficult to predict. Fire behavior can be described in terms of flame length and height, fire line intensity, rate of spread, and other characteristics. Weather conditions such as temperature, wind speed, and relative humidity influence the outcome of forest fires. A prescribed fire, including the possibility of it spreading out of control, affects forest resources, air and water quality, buildings, and the safety of the burn crew. Decision analysis techniques enable forest managers to explicitly incorporate key uncertainties in the planning and decision-making process.

The forest services' schematic tree is shown in Figure 10.7. The tree includes three decisions and four random events. Overall, there will be $2^7 = 128$ branches in the full-blown decision tree. The decision is divided into three parts with additional information gathered before each of the subsequent decisions. The first decision is to commit resources to the burn, for example, sending personnel and equipment to the site. Then, there is the decision to actually initiate the burn while there are uncertainties regarding the environmental conditions, fire behavior, and ultimate effects of the fire. Information from a test burn may be available before the initiate burn decision is made. Even though manpower and resources have been gathered to carry out a full burn, the results of a test burn might indicate that proceeding is a potentially dangerous decision. Once a full-burn decision has been implemented, the forest service monitors the fire's progress and makes real-time decisions to modify or possibly shut down the burn.

Product development decisions offer an analogous situation. A manager must decide whether resources should be committed to develop a new product or not. There are uncertainties pertaining to technology, manufacturing cost, demand, and economies of scale. Initial tests may be conducted to gain more information on the feasibility of developing the product, but this could delay the product's launch. In addition, the market can be surveyed as the product comes closer to realization.

Once the product is launched, real-time pricing discount decisions must be made based on how the market responds and competitors react.

2. **Activity:** Information gathering and decisions: Think of a decision scenario where decisions are interspersed with random events.

Decision _____

Random event _____

Decision _____

Random event _____

10.5 Constructing and Analyzing a Simple Decision Tree

In this section, we create and analyze a basic decision tree for an automation investment problem that consists of one decision node and one random node. We use this simple example to demonstrate basic steps and follow that with more complex examples in the following sections.

Boss Controls (BC) is gearing up to manufacture an option to be offered on 1 million new cars worldwide. The key uncertainty is the percentage of vehicle buyers who will choose to order the option. This percentage is called the take rate. Initial estimates are that the take rate for the option could be as low as 30% or as high as 50%. Past experience indicates the probability that the take rate will be low is 0.4.* The plan calls for BC to deliver the option to the automotive manufacturers at a price of $60. Timothy O'Leary, VP for Imaginative Products, is considering two alternatives, low investment or high investment in automation. The level of automation directly affects the investment cost and the subsequent variable cost. Relevant data for both alternatives are presented in Table 10.2. Should BC choose a high- or low-investment strategy?

10.5.1 Decision Tree Construction/Layout

The information content in a decision tree consists of four components: layout, probabilities, values, and the formula. Probabilities reflect uncertainty, values correspond to cost or profits, and the formula captures how the values interrelate to determine the overall objective function.

The decision tree contains three kinds of nodes: decision nodes (boxes), chance or random nodes (ovals), and end nodes (triangles). A rectangle is used to represent a decision node, and all branches

TABLE 10.2: Data for automation investment.

Strategy	Investment Dollars ($Million)	Variable Cost ($)	Net Sales Revenue (NSR)	Take Rate (%)	NSR ($Million)
Low	8	27	$(60-27) *$ take_rate $* 10^6$	30	9.9
				50	16.5
High	13	14	$(60-14) *$ take_rate $* 10^6$	30	13.8
				50	23.2

---

* Take rate: If a component is offered as an option in the final product (e.g., a car), then the take rate is the percentage of products in which the component will be assembled. For example, if the take rate of a component is 20%, and the number of cars in which it can be provided as an option is 100,000, then the consumption of the component will be 100,000 * 0.20 = 20,000.

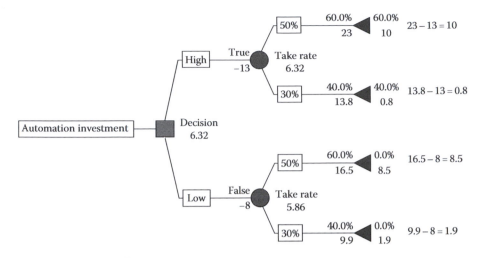

FIGURE 10.8: Boss Controls decision tree—automation investment.

emanating from it reflect alternatives, one of which is chosen by a decision maker. The branches that stem from a chance node are the possible outcomes of the uncertain variable or event and must be mutually exclusive and collectively exhaustive. In other words, only one of the specified outcomes will occur and the sum of the branch probabilities must equal 1. The branches in the decision tree lay out the paths from one node to another and lead to the end of the path. The values associated with the decision and chance nodes are stored on the branches.

The first decision context is which manufacturing investment option to choose. The decision has to be made now, under the influence of uncertainty in the take rate. The structure of the decision tree is shown in Figure 10.8, with estimated payoffs in the millions of dollars. The automation investment decision node has two branches, depicting low-investment and high-investment options. Each decision branch has a cost value associated with it. The low-investment option has a cost of $8 million, which is accorded a negative value, −8. The corresponding value for the high-investment option is $13 million.

Each random branch of the take rate has three kinds of information associated with it: the branch's descriptive name, its probability, and net sales revenue. This is calculated by subtracting the variable cost from the sale price of $60 and then multiplying the take rate and forecasted sales volume. The topmost take-rate branch for the decision path of "high" investment is labeled "50% take." This event has a 60% chance of occurring. For the low investment, the variable cost is $27. Since the sale price is $60, BC nets $33 per unit. If the take rate were only 30% and the sales volume a million, the net sales revenue would be ($33) * (.3) * 10^6 or $9.9 million. The corresponding value for a high take rate would be ($33) * (.5) * 10^6 or $16.5 million. For the high-investment decision and low take rate outcome, the net sales revenue would be ($60–14) * (.3) * 10^6 or $13.8 million. The corresponding value for high investment and high take rate would be $23 million.

Each sequence of branches represents a scenario that reflects a combination of decisions and outcomes of random events. The bottommost path or scenario corresponds to the decision to make a low investment, and the take rate turns out to be only 30%. At the end of the path is a triangle. Alongside, it is a value that, in this case, is $1.9 million. In this tree, we use the default assumption that the end value is calculated by adding up the values along the path (e.g., −8 + 9.9 = 1.9). Each scenario has a probability of occurrence given a specific set of decisions that were made. If there were multiple random events, a scenario's probability is calculated by multiplying all the probabilities along the path. (This is demonstrated in the next case study.) In this instance, there is only one random event. Thus, if the decision is to make a low investment, the probability of the bottommost path is just 0.4. The topmost path corresponds to the decision to make a high investment and the take rate turns out to be 50%. Its value is $10 with a probability of 0.6.

10.5.2 Probabilities on Branches

The probabilities in the decision tree provide the information regarding the occurrence of a particular event. The probability distribution can be either continuous or discrete, depending on the way the model is structured. In most cases, the probabilities are either a discrete distribution or modeled as a discrete approximation of a continuous distribution. Gathering the data on probabilities itself is a complex procedure and may involve one or more interviews with the experts on the specific issue. The interview procedure and potential biases are discussed in detail in Chapter 13. For now, we assume the probabilities are given. In the automation investment case, the marketing expert forecasted that the take rate could be as low as 30% or as high as 50%. She was comfortable with considering just two possibilities. She estimated the probability of a low take rate to be 0.4 and the probability of a high take rate to be 0.6.

10.5.3 Values on Branches

Values represent information regarding cost, profits, or parameters associated with the decision and chance nodes. For example, if a low-investment decision is made, then an investment of $8 million will be necessary. Similarly, an investment of $13 million will be required if a decision of high investment has been made. In decision tree construction, these values are inputs on the respective decision tree branches as shown in Figure 10.8. In this example, the investment required for the alternatives is given and thus the construction of the tree becomes easier. This may not always be the case. Determining these values may require just a simple calculation or it could be as complex as involving a complete spreadsheet model.

There are actually two values associated with each decision branch. In addition to the investment, each investment decision branch sets the variable cost per unit. For low investment, this variable cost is $27 per part. For high investment, the cost is $14 per part. The software package PrecisionTree does not allow more than one value per branch to be specified. However, other software packages do allow for this. The random event outcome branches also have a relevant value, the percentage take rate. The values correspond to the branch names. The 30% take rate corresponds to a value of 30%. Because of the difficulty in handling multiple values on the decision branches, a formula was used to calculate the net revenue for each chance outcome and decision combination. The net sales revenue depends on the sales volume, take rate, selling price, and variable cost of manufacturing. The relation is given by the formula:

$$\text{Net sales revenue} = (\text{sales volume}) * (\text{take rate}) * (\text{selling price} - \text{variable cost})$$

Thus, for high investment and a high take-rate branch, revenue is equal to $23 million ($10^6$ * 0.50) * (60 − 14). These values will be calculated inside the spreadsheet that contains the decision tree.

10.5.4 Objective Formula

An overall objective to be maximized defines the relationship between the values in different branches. The value at the end node is calculated using the formula that relates all the values incurred in that particular path. The simplest formula and the software default assumption is that the values along a path simply add. For example, for the path "High" investment followed by "50% take," the values are "investment" and "revenue." The relation between these values is

$$\text{Profit} = \text{Net sales revenue} - \text{Investment}$$

Therefore, the value related to this end node is $(23 - 13) = \$10$ million.

In the case of symmetric decision trees, the formula usually remains the same for all the end nodes. But for asymmetric decision trees, or other models, the relation or the formula between the values may change.

The above example uses total net profit as its ultimate objective and therefore the values simply sum. However, a reasonable alternative measure is ROI. The formula is

$$\text{ROI} = (\text{Net sales revenue/Investment}) - 1$$

The ROI for "High" investment followed by "50% take" is

$$\text{ROI} = (23/13) - 1 = 0.77 \text{ or } 77\%.$$

The ROI version of this same tree will be addressed when we discuss why the optimal decision turns out to be different from the case in which profit is used as the ultimate objective.

10.5.5 Optimal Alternative

A decision tree model is guided by its objective. The objective function in the problem captures the values of the decision maker and must be defined by a single numeric function (or payoff) to be maximized or minimized. For example, the objective function could be to minimize total costs, maximize the net profit, maximize ROI, minimize time to finish the project, and so forth. A multiple objective function could be used in a decision tree. It would have to convert multiple measures for a decision-outcome path into a single numeric value based on a weighted sum. PrecisionTree does not naturally handle this type of problem without using sophisticated elements of Excel.

The basic decision tree analytic process calculates the expected value for each of the alternative decision sequences. It then recommends choosing the sequence with the optimal expected value. The maximum value path is selected for decision trees that focus on profit, and the minimum is selected when cost is the primary focus. For now, we use the actual calculated values to be optimized. In Chapter 12, we introduce the concept of a decision maker's risk attitude and demonstrate how to incorporate this facet into the decision tree. When risk attitudes are incorporated, the decision maker's utility function is maximized.

The expected value of the decision tree is evaluated using a process called "rolling back the tree" (also referred to as "folding back the tree" in some decision analysis books). As the term suggests, the process starts at the rightmost end of the tree and is rolled back to each preceding node until reaching the first node or root of the tree. There are three basics rules in the rollback procedure:

1. Start at the rightmost end of the tree. Use the values at the end nodes (consequences) when rolling back.
2. At random nodes calculate the expected value: If the end node branches are emanating out of a chance node, calculate the expected value for the set of branches for this random event.
3. At decision nodes select the best value path: If the values are emanating out of a decision node, select the branch that best suits the decision objective. That is, if the objective is to minimize, select the lowest value, and if the objective is to maximize, select the highest.

Starting from the rightmost end of Figure 10.9, notice that the last node in the tree is a chance node (take rate). There are two branches for each of these nodes. The expected value is calculated by

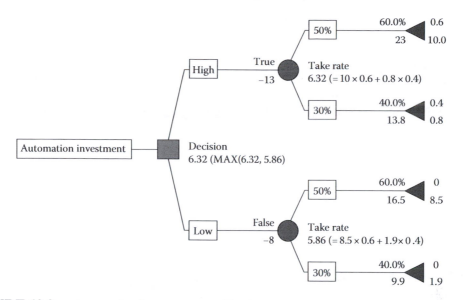

FIGURE 10.9: Automation investment—rollback using expected value analysis.

using the formula $\sum_i p_i x_i$ where i represents the index for the branch emanating from the chance node. The variable p_i is the probability of occurrence of the event associated with the particular branch, and x_i is the end node value at that particular branch.

The rollback calculation for this step is also shown in Figure 10.9.

Expected value at bottom chance node$=0.4 * 1.9+0.6 * 8.5=5.86$

Expected value at top chance node$=0.4 * 0.8+0.6 * 10.0=6.32$

Thus, we obtain two expected values, 5.86 and 6.32.

Rolling back further, note that the next node encountered is a decision node. Thus, apply Rule #3 and take the larger of the two values, 5.86 and 6.32, since the objective of the problem is to maximize profit. The value 6.32 corresponds to the "high-investment" alternative and investing $13 million in automation is the optimal choice.

10.6 Risk Profile/Cumulative Risk Profile

Expected value analysis was used to arrive at the optimal decision. However, expected value has its limitations as a decision-making criteria. It does not account for the risk attitude of the decision maker and assumes that the decision maker is risk-neutral. Further, expected value analysis has the inherent assumption of probabilistic rules that the outcome will converge to this value if the experiment were to be conducted a large number of times. Unfortunately, rarely will the decision maker face the same decision over and over so as to benefit from the law of large numbers. Nevertheless, a company can benefit from the consistency associated with using expected value as the routine criterion for the vast majority of decisions. Large multinationals make multimillion dollar decisions on a weekly basis. Still, it is valuable to consider the overall probability distribution. This enables decision makers to explore strategies of risk management that address concerns of potentially serious negative outcomes even though the expected value appears attractive.

10.6.1 Caution about Expected Values

As a phrase of caution, we recommend "Do NOT expect the Expected Value!" The term "expected value" is misleading since there is no reason to expect to see the expected value when performing a random experiment. Consider a simple example: a typical six-faced die. What is the probability of rolling a 3? What is the expected value of the outcome of a roll of the die? The probability that the up-face is 3 is 1/6 and the same is true for every other value. As a result, the expected value is:

$$\text{Expected value} = \sum p_i x_i = \frac{1}{6} \times 1 + \frac{1}{6} \times 2 + \frac{1}{6} \times 3 + \frac{1}{6} \times 4 + \frac{1}{6} \times 5 + \frac{1}{6} \times 6 = 3.5.$$

Anyone can win a lot of money by betting against observing the expected value on any roll of the die. It is impossible to observe a 3.5 on any roll of the die, yet that is its expected value! What does the term represent, then? It only means that if the die were rolled a large number of times, then the average of all the outcomes of all the rolls of the die will be close to 3.5. This is true even though on any individual roll, the actual value will be any whole number between one and six. How close to 3.5 will the expected value be? It is a function of the variance and the number of rolls. Every introductory statistics book provides formulas to estimate how close the observed average is likely to be as a function of the number of repetitions of the die.

If expected value is not the only value to consider in a decision, how should managers deal with decisions that involve uncertainty? The actual outcome of any specific decision can be significantly different from the expected value. Decision makers should consider the risks as represented in the risk profile. They need to understand the sources of these risks and see if anything can be done to influence those issues. The next chapter explores the concept of risk management and presents a paradigm for including it as a natural element of decision tree analysis.

In our experience, decision makers are comfortable seeing the expected value and risk profile and using both in an ad hoc fashion to balance averages and risk. However, early decision analysis researchers recognized the need for converting the risk profile into a single value. They developed utility theory to capture manager's differing preferences for assuming risk. This theory is discussed in Chapter 12. Decision makers are, however, uncomfortable with the abstract nature of the concept of utility theory. In addition, large companies tend to be risk-neutral for a wide range of decisions (Howard 1988). Even moderate risk aversion might not affect which alternative is optimal. With that in mind, the experienced decision analyst does not assess a utility function to capture management's attitude toward risk unless absolutely necessary. Instead, the analyst presents and discusses the risk profile of the decision.

A risk profile is the density function or cumulative distribution of the objective function for a decision. A risk profile can be constructed through the following three-step approach:

Calculate the probability of occurrence of the end node values. If there is a single chance node, this is the probability of occurrence of that event. If there are multiple chance nodes along a decision-event path, the probabilities along the path are multiplied to determine the probability of that endpoint value.

Group together equal end node values for a particular decision and add their probabilities.

Graph the end node values (*x*-axis) against the probability of their occurrence (*y*-axis). To determine the cumulative distribution function, it is necessary to accumulate the probabilities.

To construct a risk profile for the automation investment case, observe that there are four end node values. The values and their probability of occurrence are organized in Table 10.3. The end node values and the probability of their occurrence are graphed for each alternative to develop a risk profile. The risk profile is shown in Figure 10.10, and the cumulative risk profile is shown in Figure 10.11.

TABLE 10.3: End values and their probability of occurrence.

Decision	End Node Value	Probability of Occurrence	Cumulative Probability
Low investment	1.9	0.40	0.40
	8.5	0.60	1.00
High investment	0.8	0.40	0.40
	10.2	0.60	1.00

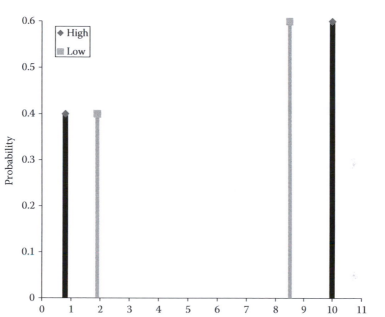

FIGURE 10.10: Automation investment—risk profile.

The above figures show the range of profit for each alternative. The high-investment option has a much wider profit range, from a low of $0.8 million to a high of $10 million. For the low-investment alternative, the maximum profit is $8.5 million and the minimum is $1.9 million.

10.7 Complex Symmetric Decision Tree: Make or Buy

Western Co. manufactures household appliances. The company has a design for a key component, but the engineers are not sure that the current design will be feasible when manufacturing begins. If it is not, substantive redesign of the component will be needed quickly. If Western Co. manufactures the component itself, it forecasts that with the redesign, manufacturing costs will increase by 8%. The decision to make or buy must be made now before there is time to fully validate the design. If Western Co. signs a contract with the supplier for a specific piece price and then a redesign is required, the supplier is likely to increase the price arbitrarily by 15%. The demand for the product is also uncertain. The data is presented in Table 10.4.

The influence diagram and schematic tree for this case are shown in Figure 10.12. The objective function is to minimize total costs. Total costs consist of the fixed costs and the variable

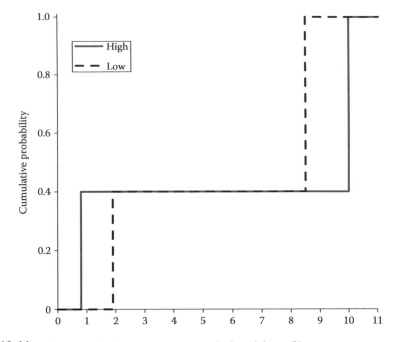

FIGURE 10.11:   Automation investment—cumulative risk profile.

TABLE 10.4:   Data for make/buy decision for Western Co.

Random events and probabilities		
Design feasibility	Probability that current design will work	0.4
	Probability that part will need a major redesign	0.6
Demand	Probability of low demand (1 million)	0.3
	Probability of medium demand (1.25 million)	0.5
	Probability of high demand (1.5 million parts)	0.2
Costs		
Make in-house	Fixed cost: facility investment (millions)	$55
	Variable cost per part if current design works	$100 per part
	Variable cost per part if there is a major redesign	8% increase
		$108 per part
Buy from supplier	Fixed cost (million dollars)	$0
	Variable cost per part if current design works	$140 per part
	Variable cost per part if there is a major redesign	15% increase
		$161 per part

manufacturing costs. Fixed costs depend on the decision; the variable cost depends on both the decision and the uncertainty in design. The total cost is affected by market demand. The formula for calculating the end values is:

$$\text{Total cost} = \text{Fixed cost} + (\text{Variable manufacturing cost/part}) * (\text{Market demand}).$$

The complete decision tree for the make/buy decision is shown in Figure 10.13. The end values in the tree are calculated using the formula given above. A brief description of the tree follows:

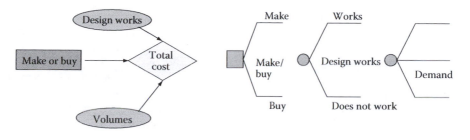

(a) Influence diagram   (b) Schematic tree

FIGURE 10.12: Make/buy structure.

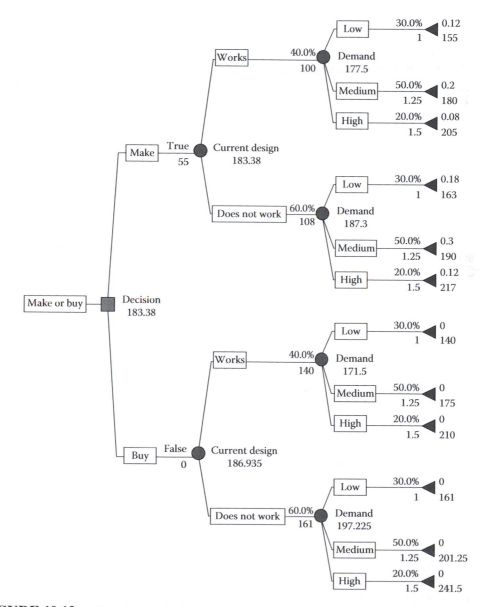

Decision node—make/buy: Western Co. must choose to make the component in house or buy from a supplier. An assumption for any decision node is that the branches in the decision node describe all the possible alternatives and no other significant alternative has been left out in the evaluation. Thus, not making the part is not an option in this case.

Payoff at each alternative: If the "make" alternative is selected, then a fixed cost of $55 million would be incurred and recorded on the corresponding branch. If the "buy" alternative is selected, then Western would incur no fixed cost.

Chance nodes—Design feasibility and demand: This problem includes two uncertainties: design feasibility and demand. Design feasibility has two possible outcomes and therefore two branches: Design works and design does not work. There is a 40% chance that the design will work and a 60% chance that it will not. For the "make" option, the piece price would be $100 per part if the design works or $108 per part if the design does not work. For the "buy" option, the corresponding variable costs per part would be $140 if the design works or $161 if the design does not work. Both the probabilities and the piece price are recorded on the branches for the random event, either that the design works or does not work.

The marketing department of Western Co. estimates three possible outcomes for annual demand: 1 million with a probability of 0.30, 1.25 million with a probability of 0.50, and 1.5 million with a probability of 0.20. These are recorded on each repetition of the demand branches.

End nodes: The value of the outcome is calculated by a formula that relates all the values on the branches in a path. Unlike the previous investment example, it makes no sense to add up the values along the branches to determine the total costs. The values correspond to the fixed costs, the variable cost per unit, and the total volume. In this make/buy case, the formula is

Total cost = fixed cost + (variable manufacturing cost/part)*(market demand).

Consider the path: Decision (Make) → Design works (Yes) → Demand (Low).

The "make" decision incurs a $55 million fixed cost. If the "design works," the variable cost will be $100 per unit. A "low" volume corresponds to selling 1 million components. The total cost for the "make" decision and the random event outcomes would be $155 million. This appears at the end value node.

The probability that the value $155 million occurs is calculated by multiplying the probabilities of the chance events that arise in the path leading to that particular node. The probability that the design works is 0.40, and the probability that demand is low is 0.30. Therefore, the probability that the outcome is $155 million is Probability (Design works = Yes) * Probability (Demand = Low) = 0.30 * 0.40 = 0.12

Similarly, one can calculate the value of the outcome and the probability of occurrence for that outcome for all the paths. Consider one more path.

Consider the path: Decision (Buy) → Design works (No) → Demand (Medium).

The "buy" decision incurs no fixed cost. If the "design does not work," the variable cost will be $161 per unit. A "medium" volume corresponds to selling 1.25 million components. The total cost for the "buy" decision and the random event outcomes would be $201.25 million. The likelihood of this happening is 0.6 * 0.5 or 0.3 if Western uses a supplier. (PrecisionTree reports zero as the relevant probability, because the optimal decision is to make the part in house; therefore, this path cannot occur.)

10.7.1 Dependence/Independence of Random Events

A decision tree can have more than one chance node in succession. The order of presentation of these events on the decision tree depends on the relation between the events represented by these nodes. In general, if the outcome of one event influences the outcome of the other, then the sequence must follow the logic of dependence. If the outcome of one event does not influence the other, that is, if they are independent events, then the sequence of events does not matter. For example, consider the two chance nodes "design works" and "market demand" in Figure 10.13. In this case, the outcome of market demand remains the same, regardless of whether the design works or not. Thus, the order of the two nodes does not matter mathematically, although logically the issue of design effectiveness will be known before the demand.

Now consider the case that if the design does not work, there will be a significant delay in bringing the product to market. This delay could influence the demand for the product due to external factors such as competitors' products. In addition, if reports surface that the company has had a problem with the original design, this too could reduce potential demand. In this instance, the demand probabilities would be conditioned on the outcome of the random event "design works" and sequencing of the nodes would be important. The "design works" node would have to precede the "market demand" node.

10.7.2   Rollback/Expected Values

Consider the "make" half of the tree shown in Figure 10.14. Start from the rightmost end to rollback this part of the tree. There are six end node values for the "make" decision. To calculate the expected value, we take a probabilistically weighted sum of the branch values for branches emanating from the same chance node. For example, the first three end nodes with values 155, 180, and 205 are used to determine one expected value, and end nodes with values 163, 190, and 217 are used to find the other expected value.

$$\text{Expected value at node } \#14 = 0.30*155+0.50*180+0.20*205 = 177.5$$

$$\text{Expected value at node } \#15 = 0.30*163+0.50*190+0.20*217 = 187.3$$

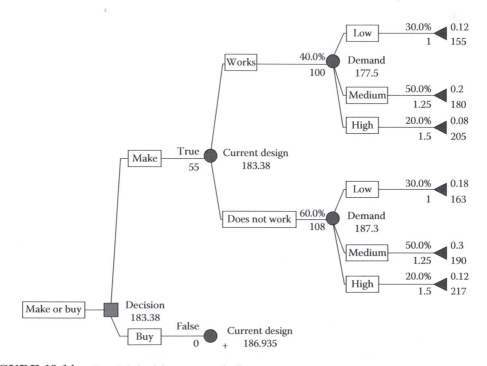

FIGURE 10.14:   Partial decision tree make/buy.

The values have the following meaning. If the decision is to "make" the part in-house and the design works, then the expected value of the total costs is $177.5 million. If the decision is to "make" the part in-house and the design does not work, then the expected value of the total costs is $187.3 million. These are called conditional expected values, conditioned on the outcome of an earlier random event, either current design works or does not work.

To determine the unconditioned expected value for the decision to make the part, continue the rollback process at the chance node "current design," as shown below.

$$\text{Expected value at node} = 0.40*177.5 + 0.60*187.3 = 183.38$$

Therefore, the expected value of the total cost for the "make" decision is $183.38 million. Similarly, if the tree were rolled back for the "buy" decision node, the expected cost for this node is $186.94 million.

First, we would find the conditional expected value for the decision to buy and the outcome showing that the design works:

$$= 0.30*140 + 0.50*175 + 0.20*210 = 171.5$$

Next, we would find the conditional expected value for the decision to buy and the outcome showing that the design does not work:

$$= 0.30*161 + 0.50*201.25 + 0.20*241.5 = 197.225$$

The unconditional expected value for the buy decision is

$$0.4*171.25 + 0.6*197.225 = \$186.935 \text{ million}$$

The next node encountered in the process is the decision node, at which point we apply rule #3 and select the lower cost value. In the PrecisionTree software, TRUE appears on top of the selected decision branch and FALSE appears on top of the other decision branches. The minimum expected cost for the make/buy decision case is $183.38 million, and the optimal decision is to make the part in-house.

10.7.3   Risk Profile

The optimal strategy based on expected value analysis is to make the part in house; the expected cost is $183.38 million. If the "buy" decision is made, the expected cost is $186.9 million. Is a $3.55 million difference in the expected values significant enough to drive the decision on whether to make the item in-house? If it were hundreds of millions, would it matter more? If you cannot expect to see the "expected value," does this difference warrant the decision?

The impact of savings in this case is less than 2%. But should percentages be used to differentiate between such decisions? Now consider the situation at a higher level. Assume a company makes or buys $9 billion worth of components a year. The decision described above might be made 50 times. If in each instance, the company used the expected value as its criterion, the total net difference would be $177.5 million, 50 times $3.55 million. With that many repeated decisions, the law of large numbers comes into play and the expected value will approach the sum total of the realized outcome of the 50 independent decisions. Thus, although a small percentage difference may not seem significant, the impact of such differences can add up to a significant value. Most companies would work hard to consistently reduce their expected costs by 2%.

The outcomes and the respective probabilities for the make and buy decisions are summarized in Table 10.5. These outcomes are arranged in ascending order. Values in the probability column are obtained by multiplying the probabilities along the path to the end point value. The range in outcomes is $62 million for the "make" decision, a minimum cost of $155 million and a maximum cost

TABLE 10.5: Make/buy outcomes and their probabilities.

Make Decision			Buy Decision		
Value ($Million)	Probability	Cumulative Probability	Value ($Million)	Probability	Cumulative Probability
155	0.12	0.12	140	0.12	0.12
163	0.18	0.30	161	0.18	0.30
180	0.20	0.50	175	0.20	0.50
190	0.30	0.80	201.25	0.30	0.80
205	0.08	0.88	210	0.08	0.88
217	0.12	1.00	241.5	0.12	1.00

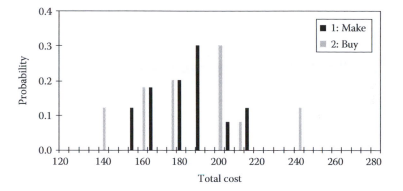

FIGURE 10.15: Risk profile for make/buy.

of $217 million. The range for the buy decision is much wider, $101.5 million. There is a minimum cost of $140 million and a maximum cost of $241.5 million.

The risk profile for make/buy is shown in Figure 10.15. The lowest total cost of $140 million can be incurred if the "buy" decision is made and the probability of this outcome is 0.12. On the other hand, the highest total cost, $240 million, is also associated with this decision with a probability of 0.12. This is associated with a design failure and high demand. For the "make" decision, the total costs can only be as high as $217 million with a probability of 0.12, but the lowest possible cost is $155 million. The decision to buy the product could result in a cost as low as $140 million but could backfire by increasing the costs up to $240 million if the design does not work and demand is high. These differences are reflected in the corresponding variances: 359.88 for the "make" decision and 875.23 for the "buy" decision.

Outcome variability is driven by two factors: design uncertainty and demand uncertainty. The risk profile can also be studied in cumulative form. The cumulative risk profile (Figure 10.16) answers questions such as "What is the probability that the total costs will be less than $200 million if it is decided to make the part in-house?" or "What is the probability that the total costs will be greater than $210 million?" PrecisionTree software generates both risk profiles.

Some interesting observations from this plot can be made:

$$P(X \leq 180) = 0.5 \text{ for both}$$

$$P(X \leq 190) = 0.8 \text{ for make and } 0.5 \text{ for buy}$$

$$P(X \leq 220) = 1.0 \text{ for make and } 0.88 \text{ for buy}$$

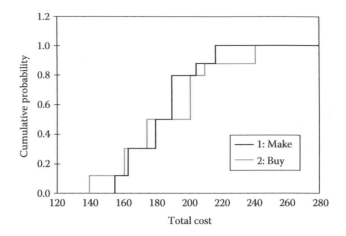

FIGURE 10.16:   Cumulative risk profile for make/buy.

What strategy should be adopted? Should the part be purchased or made in-house? In the present scenario, there are fewer variations in outcomes for the "make" decision than in the "buy" decision. Thus, on average, the "make" decision is not only preferred but is equal in terms of overall risk. The main benefit of the "buy" decision is that it does not involve a large $55 million investment upfront.

10.7.4   Stochastic Dominance

The risk profile offers a decision maker a comprehensive comparison of the risks, but how does a risk profile enable a decision to select one alternative over another? Unfortunately, there is no simple rule for comparing and then selecting one risk profile over another when the risk profiles overlap, as represented by lines crossing in the cumulative profiles. Utility theory converts the respective risk profiles into single numbers that can be compared, called certainty equivalents. In some instances, one risk profile dominates another, thereby enabling the decision maker to toss out the dominated decision. Dominance can be deterministic or stochastic. In this example of cost minimization, lower values are preferred and a risk profile that is more to the left in the graph is preferred. If an alternative is dominated, it means that regardless of the utility function form, the alternative will never be preferred.

Consider a case where Western Co. floats out a bid to purchase the part and two suppliers respond. Cost data for all alternatives are given in Table 10.6. Under the new scenario, the price offered by Supplier B is $143 per part. Western is not sure how much the suppliers will increase their respective prices if the design does not work. It will then have to be redesigned and the supplier will use the revised design as an excuse for a significant price increase. In each case, they believe that there is a 50–50 chance the supplier will charge a premium of either 10% or 20%. The two suppliers are independent of each other and could react differently to the design change. This adds one more random event to the tree (Figure 10.17).

TABLE 10.6:   Updated cost data for make/buy example.

	Make	**Supplier A**		**Supplier B**	
		Price	**Probability**	**Price**	**Probability**
Design works ($/Part)	100	140		143	
Percent premium if design does not work	8%	10%	0.5	10%	0.5
		20%	0.5	20%	0.5

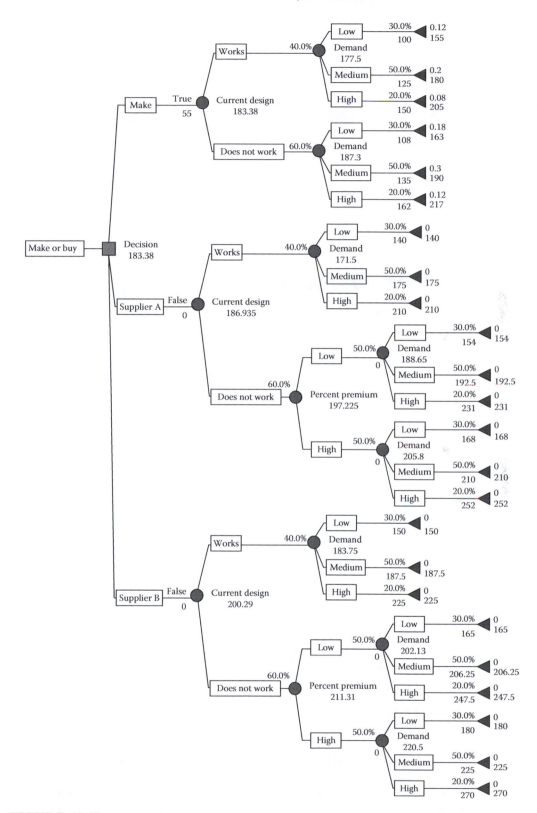

FIGURE 10.17: Updated decision tree for make/buy case.

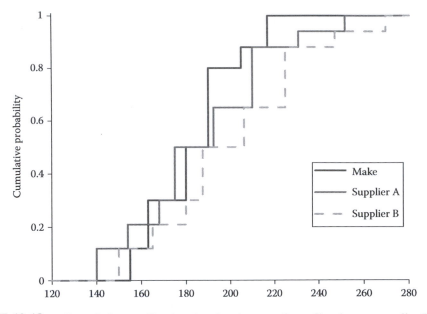

FIGURE 10.18:   Cumulative profile showing dominance of supplier A over supplier B.

The cumulative risk profiles for the three alternatives are shown in Figure 10.18. Supplier A's profile overlaps or lies consistently to the left of supplier B. Supplier A is said to stochastically dominate supplier B. For any specific value of total cost on the x-axis, supplier A has a "higher probability" of charging that amount or less when compared to supplier B. It is possible that supplier B will be cheaper if, for example, it charges a premium of only 10% when supplier A charges a premium of 20%.

10.8   Asymmetric Tree: Design Change

It is only 6 months before the vehicle launch of the MX36. A sound emanating from the instrument panel has been detected on some test drives. Engineers are not able to reproduce the sound in a controlled environment, however. They are fairly certain the problem is from a series of three assembled parts. They have a new single modular design that can be implemented quickly and that should solve the problem. There is a potential added benefit from the modular design, reduced manufacturing, and assembly cost.

Engineers estimate that there is an 80% probability that the problem will be solved by the proposed design. The relevant planning horizon is 1 year, with sales volume forecasted to be 100,000 units and the estimated cost per warranty $50. A 1% warranty problem will cost the company $50,000; a 5% rate will cost $250,000. The manufacturing savings could be nothing or as high as $250,000. It is assumed that if the modular design solves the problem, then there will be no warranty costs due to the noise. The data are given in Table 10.7. The corresponding decision tree is shown in Figure 10.19.

The tree is asymmetric because if the design change fixes the problem, there is no longer any uncertainty regarding warranty costs. Those costs will be zero. The total cost for each path is the sum of the cost of the design change and the cost of the warranty claims minus any manufacturing savings. In Figure 10.19, costs are negative values and savings are positive values. The worst case scenario involves making a design change, but the change does not solve the problem. In addition,

TABLE 10.7: Data for design change.

Warranty		Manufacturing Savings	
(% of Volume)	Probability (%)	Dollars per Unit	Probability (%)
0	50	0	30
1	30	2.50	70
5	20	—	—

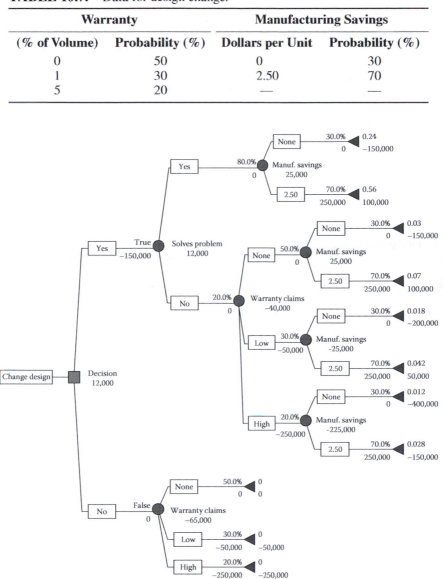

FIGURE 10.19: Decision tree for design change example.

to make matters worse, the warranty is at the highest level and there are no manufacturing savings. The total costs for this scenario is (−$150,000 − 250,000 + 0), which is equal to −$400,000. The best case scenario involves a design change that works and there are manufacturing savings as well. The net cost is (−150,000 − 0 + 250,000), which yields a positive benefit of $100,000. The same maximum value occurs even if the fix does not work but the warranty turns out to be zero anyway and there are manufacturing savings.

The optimal decision is to go with the proposed modular design with an expected value of savings of $12,000. The tree shows that the most favorable outcome is a savings of $100,000 with a probability of 56%. This is much better than the best outcome of the "do nothing" alternative that results in no savings with a probability of 50%. On the other hand, the worst outcome of the optimal alternative is a loss of $400,000 with a probability of 1.2% when compared to a "do nothing" loss

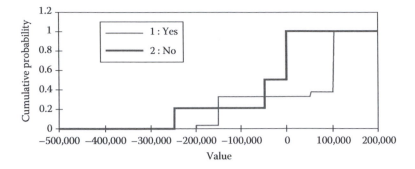

FIGURE 10.20:   Design change—risk profile.

of $250,000 with a probability of 20%. The probability of losing $100,000 or more with the new design is 20%, whereas it is 30% with the current design. Overall, the benefits of manufacturing savings make the modular design more attractive than the current design. The crossing lines in the cumulative risk profile (Figure 10.20) indicate the lack of a dominant alternative.

10.9   Sequential Decisions

Sam and Don (S&D) Company is a leading supplier of critical components in the computer industry. Due to increased demand, S&D is investigating manufacturing capacity expansion for one of its popular products, a large external hard drive. The company's current capacity for the hard drive is 1 million units. The decision to add capacity involves several options and is complicated by uncertainties. The drives sell for $170; the current variable cost of manufacture per unit is $140.

The S&D hard drive department identified three alternatives to expanding their manufacturing capacity: (1) buy additional old technology production lines, (2) buy new technology production lines, or (3) increase productivity on the existing lines. Each old technology line costs $1.5 million and adds 150,000 units of capacity. The management is planning to install three or four additional lines if they buy old technology. S&D's marketing department estimates that demand would be 1.2 million with a probability of 0.3, 1.4 million with a probability of 0.5, and 1.7 million with a probability of 0.2.

New technology reduces the variable unit cost from $140 to $130 due to more automation and less labor. The capacity of each new technology production line is 300,000 units and each line costs $10 million. There is also a $500,000 training cost associated with the new technology. This cost is the same whether one or two lines are installed. S&D is planning to install either one or two new technology lines if it decides to buy the new technology. The throughput of new technology involves significant uncertainty. The engineers predict that there is only a 20% chance that 100% of the new technology capacity can be used, while there is an 80% chance that capacity utilization will be 85%.

A machine rehabilitation project will result in improved efficiency and productivity. Engineers predict that there is a 60% chance that productivity improvement will result in a 25% increase in capacity (i.e., the total capacity will be 1.25 million units), and there is a 40% chance that capacity will increase by 30%. S&D would need to invest $3 million to improve productivity. The rehabilitation project will not only result in increased capacity but also reduce variable costs. The project will result in a $1 million reduction in variable cost.

The first decision S&D management faces involves choosing one of three alternatives to expand capacity. If the old technology or new technology is selected, S&D must specify the number of production lines to buy. The productivity improvement alternative involves no further decision. But this strategy has uncertainty regarding the impact of productivity improvement on capacity. The new technology faces uncertainty regarding throughput. All the alternatives are affected by uncertain demand. (See Figure 10.21 for the complete picture.)

In this case, S&D's objective is to maximize profit (revenue − variable cost − investment cost). The profit calculation is complicated by the fact that it cannot sell more than its capacity. Thus, if demand exceeds capacity, its total profit is based on its capacity. Conversely, if capacity exceeds demand, its profit is calculated using the demand. We can calculate end values of the old technology alternative using the following formula:

$$\text{Profit} = \text{MIN (demand, additional capacity + current capacity)}$$
$$* \ (\text{price} - \text{variable cost}) - \text{investment cost}$$

Table 10.8 shows the calculated value for each end node. For example, consider the impact of using the old technology and adding three lines. The three lines cost $4.5 million and increase the capacity by 450,000 units. If the demand is low, the capacity of 1.45 million exceeds the demand of 1.2 million. The company makes a net profit of $30 per unit of demand. The total net profit is $31.5 million. If demand is low, then

$$\text{Profit} = \text{MIN } (1.2, \ 0.45+1)*(170-140)-4.5 = 1.2*30-4.5 = \$31.5 \text{ million}$$

As another example, consider the impact of using the new technology and adding two lines. The cost for the lines is $20 million plus $500,000 for training. The two lines add potentially 600,000 units of capacity if productivity is high. The variable cost is now only $130 and the profit per drive is $40. There are two uncertain events. Now consider what happens if the lines only achieve 85% of their potential productivity and demand is high, 1.7 million units. Actual capacity is only 1.51 million units, which is all the company can sell. Its net profit would be $39.9 million.

$$\text{Profit} = \text{MIN } (1.7, \ 0.6*0.85+1)*(170-130)-20.5 = \$39.9$$

10.9.1 Roll Back Tree: Expected Value Calculation

The expected value for three old technology lines is $36 million ($0.3 * 31.5+0.5 * 37.5+0.2 * 39$). The expected value for four lines is $35.4 million. If old technology is chosen, the preferred alternative for S&D is to buy three production lines.

With regard to new technology, the expected value calculations proceed in two steps. The effect of random demand is determined first and then the impact of productivity level. Consider just one line. The expected value with high productivity equals ($0.3 * 37.5+0.5 * 41.5+0.2 * 41.5$), which is $40.3 million. The expected value with low productivity equals ($0.3 * 37.5+0.5 * 39.7+0.2 * 39.7$), which is $39.04 million. These are conditioned on the productivity uncertainty. Next, we roll back the tree to account for the 20% probability that productivity will be high:

$$0.8*39.04+0.2*40.3 = \$39.292$$

Thus, the unconditional expected value of one line is $39.29 million. The corresponding value for two new technology lines is $34.12 million. If S&D chooses new technology, it prefers to buy one production line with an expected value of $39.29 million. The expected value of productivity improvement is $35.72 million.

We know the expected values of three alternatives regarding capacity expansion. S&D management selects an alternative with the highest expected value. The preferred alternative is to buy one new technology production line whose expected value is $39.29 million.

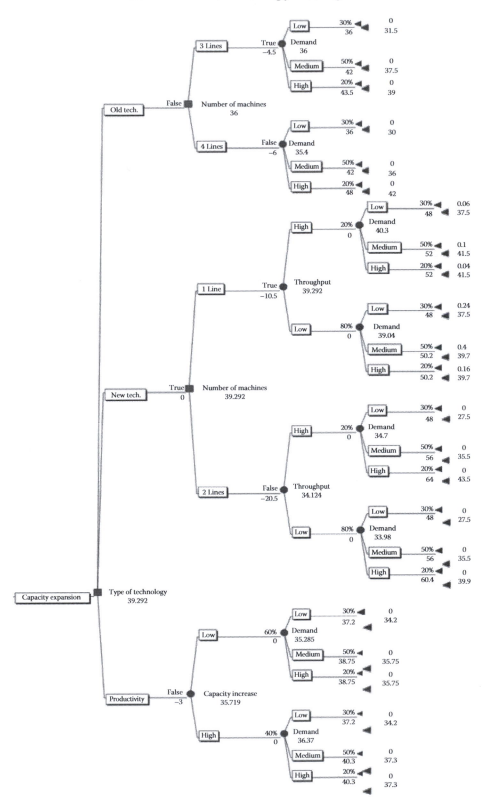

FIGURE 10.21: Decision tree for capacity expansion case.

TABLE 10.8:   End values for capacity expansion example.

Main Alternative	Sub Alternative	Throughput or Productivity	Demand	End Value
Old technology	Three lines		Low	MIN(1.2, 0.45 + 1) * (170 − 140) − 4.5 = $31.5
			Medium	MIN(1.4, 0.45 + 1) * (170 − 140) − 4.5 = $37.5
			High	MIN(1.7, 0.45 + 1) * (170 − 140) − 4.5 = $39
	Four lines		Low	MIN(1.2, 0.6 + 1) * (170 − 140) − 6 = $30
			Medium	MIN(1.4, 0.6 + 1) * (170 − 140) − 6 = $36
			High	MIN(1.7, 0.6 + 1) * (170 − 140) − 6 = $42
New technology	One line	High	Low	MIN(1.2, 0.3 + 1) * (170 − 130) − 10.5 = $37.5
			Medium	MIN(1.4, 0.3 + 1) * (170 − 130) − 10.5 = $41.5
			High	MIN(1.7, 0.3 + 1) * (170 − 130) − 10.5 = $41.5
		Low	Low	MIN(1.2, 0.3*0.85 + 1) * (170 − 130) − 10.5 = $37.5
			Medium	MIN(1.4, 0.3*0.85 + 1) * (170 − 130) − 10.5 = $39.7
			High	MIN(1.7, 0.3 * 0.85+1) * (170 − 130) − 10.5 = $39.7
	Two lines	High	Low	MIN(1.2, 0.6+1) * (170 − 130) − 20.5 = $27.5
			Medium	MIN(1.4, 0.6 + 1) * (170 − 130) − 20.5 = $35.5
			High	MIN(1.7, 0.6 + 1) * (170 − 130) − 20.5 = $43.5
		Low	Low	MIN(1.2, 0.6 * 0.85 + 1) * (170 − 130) − 20.5 = $27.5
			Medium	MIN(1.4, 0.6 * 0.85 + 1) * (170 − 130) − 20.5 = $35.5
			High	MIN(1.7, 0.6 * 0.85 + 1) * (170 − 130) − 20.5 = $39.9
Productivity		Low	Low	MIN(1.2, 0.25 + 1) * (170 − 139) − 3 = $34.2
			Medium	MIN(1.4, 0.25 + 1) * (170 − 139) − 3 = $35.75
			High	MIN(1.7, 0.25 + 1) * (170 − 139) − 3 = $35.75
		High	Low	MIN(1.2, 0.3 + 1) * (170 − 139) − 3 = $34.2
			Medium	MIN(1.4, 0.3 + 1) * (170 − 139) − 3 = $37.3
			High	MIN(1.7, 0.3 + 1) * (170 − 139) − 3 = $37.3

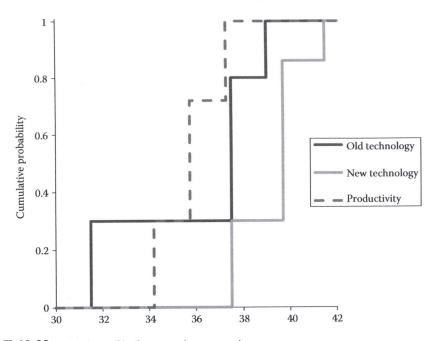

FIGURE 10.22:   Risk profile for capacity expansion case.

10.9.2 Deterministic Dominance

Earlier, we introduced stochastic dominance. Here, we discuss deterministic dominance. Deterministic dominance can be detected in the cumulative risk profiles by comparing the value where one cumulative risk profile reaches 100% with the value where another risk profile begins. In the capacity expansion case, the outcomes from the productivity improvement alternatives range between $34.2 and $37.3 million. On the other hand, the worst outcome of one new technology line is a profit of $37.5 million. This is more than the highest potential profit with the productivity improvement alternative, as shown in Figure 10.22.

In Figure 10.22, further to the right represents more profit. Clearly, one line of new technology stochastically dominates three old technology lines. These two risk profiles never intersect. However, it requires a deeper analysis to notice that the new technology profits are always better than the old technology. The random event demand is exactly the same for each alternative. If the demand is low, with new technology, the profit is $37.5 million compared to $31.5 million for older technology. If demand is high, with old technology the profit rises to $39 million. However, with new technology the profit is at least 39.7 million. If an alternative always yields better outcomes for every possible outcome of uncertainty, this dominance is called deterministic dominance; otherwise, it is stochastic dominance. Productivity improvement and old technology alternatives do not dominate each other since their profiles intersect.

10.10 Robustness of Optimal Solution through Sensitivity Analysis

Consider the automation investment problem in which the expected values of high- and low-investment alternatives are $6.32 and $5.86 million, respectively. The optimal choice, high investment, involves new technology. In the initial analysis, management used point estimates for the capital equipment costs and variable costs, but this actually masked some of their uneasiness. The management is concerned that the capital equipment estimate could be off by ±7%. There is even more concern regarding the variable cost of high investment that could be off by ±10%. They set the take-rate probabilities at 0.4 for the low take rate and 0.6 for the high take rate. However, there is also uncertainty regarding this probability.

Do the ranges in these parameters change the optimal choice? For example, would a change of ±7% in the capital equipment cost of the high-investment alternative change the optimal choice? What happens if the probability of a low take rate is really 0.2 or 0.6?

Decision makers are usually concerned about the impact of possible changes in the estimated parameters on the optimal decision, and they ask similar questions—exactly those that sensitivity analysis helps a decision maker address. We introduce one-way sensitivity analysis here to observe the robustness of the highest ranked alternative to possible changes in numerical values of key parameters, such as probabilities or payoffs. The process involves specifying a parameter range and the number of intermediate values to consider. The Precision Tree software simply calculates the expected value for each intermediate value and graphs the expected value over the range for the various alternatives.

In one-way sensitivity analyses, one variable is changed at a time. Reports generated by this analysis include tornado diagrams and spider graphs, where the results of multiple one-way analyses can be presented and compared together (see Chapter 11). In a later section, we introduce two-way sensitivity analyses in which two variables are changed simultaneously.

10.10.1   One-Way Sensitivity Analysis

One-way sensitivity analysis examines the effect of a single variable on the expected value of a model. This value could be either the payoff related to an event (deterministic sensitivity analysis) or the probability related to a chance occurrence (probabilistic sensitivity analysis).

In this section, we demonstrate one-way sensitivity analysis using the automation investment example. Management is especially concerned about the fixed and variable cost of high investment and the probability associated with the take rate. In addition, the low-investment alternative is well tested, and there is hope that continuous improvement could reduce the variable cost by 5%. Table 10.9 presents base, minimum, and maximum values for these variables.

Running sensitivity analysis in PrecisionTree is straightforward. PrecisionTree quickly evaluates the expected value associated with intermediate values within the specified range. The software user specifies the number of intermediate values. PrecisionTree generates two graphs to display the changes in the expected value, as illustrated in Figures 10.23 and 10.24. These two graphs show that expected profit increases as the fixed cost of high investment decreases. The fixed cost is a negative $13 million. Plus or minus seven percent corresponds to a range of −$13.91 to −$12.09 million. The value of the selected variable, fixed cost of high investment, is plotted on the *x*-axis and the expected value of the profit is plotted on the *y*-axis. In Figure 10.26, an intersection point indicates a change in the optimal decision. Figure 10.27 displays just the optimal value of the expected profit as the fixed cost of high investment varies from $12.09 to $13.91 million. A bend in the graph indicates a change in the optimal decision.

TABLE 10.9:   Possible ranges for some input variables.

Variable Analyzed	Minimum	Base	Maximum
	Values		
Fixed cost: high investment	$12.09M	$13M	$13.91M
Variable cost: high investment	$12.6	$14	$15.4
Variable cost: low investment	$25.65	$27	$27
Probability: low take rate	20%	40%	60%

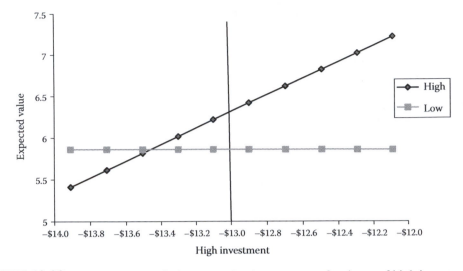

FIGURE 10.23: Sensitivity analysis automation investment—fixed cost of high investment.

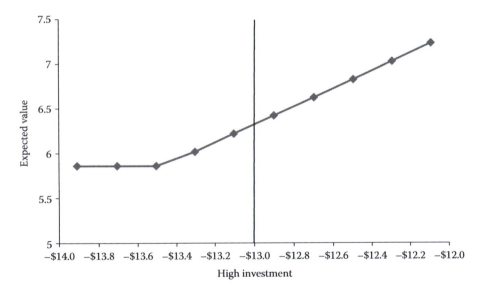

FIGURE 10.24: Sensitivity analysis for optimal alternative for automation investment—fixed cost of high investment.

In Figure 10.23, at the breakpoint of $13.46 million, each alternative yields the same profit, $5.86 million. The high-investment alternative has the higher expected value as long as its investment cost is less than $13.46 million, in the figure to the right of −13.46 million. Above that value, the optimal decision changes to low investment. At that point, the expected value of the model is no longer affected by increases in high investment cost, because the low-investment alternative becomes the optimum decision. The line is parallel to the x-axis.

Figure 10.25 shows the expected profit changing as the probability of a low take rate varies from 0.3 to 0.6. There is no intersection point in Figure 10.25. The expected value increases as the probability declines from 0.6 to 0.2, but the optimal choice is always high investment except when probability approaches 0.6. Therefore, the optimum alternative is relatively robust to changes in this probability.

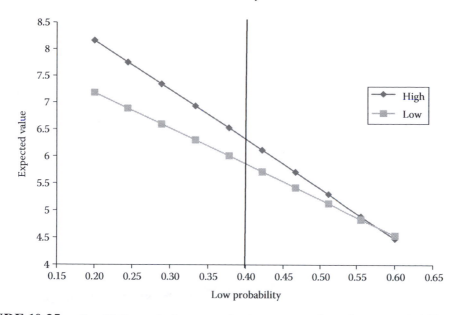

FIGURE 10.25: Sensitivity analysis automation investment—low take rate probability.

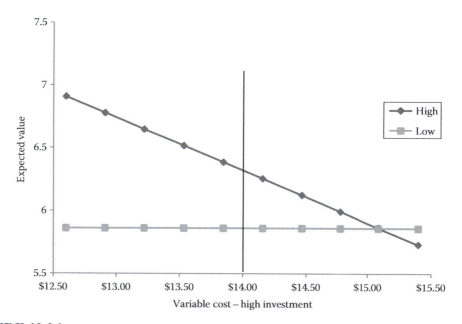

FIGURE 10.26: Sensitivity analysis automation investment—variable cost of high investment.

3. Activity (Figure 10.26): Interpret sensitivity analysis—the variable cost of high investment ±10%.

    a. How sensitive is the optimal solution to changes in this variable?

    _____

    b. What do you notice with regard to the slopes?

    _____

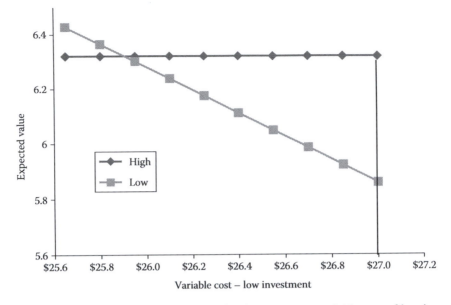

FIGURE 10.27:   Sensitivity analysis automation investment—variable cost of low investment.

4. Activity (Figure 10.27): Interpret sensitivity analysis—the variable cost of low investment −5%.

   a. How sensitive is the optimal solution to changes in this variable?

   _____

   b. What do you notice with regard to the slope?

   _____

10.10.2   Two-Way Sensitivity Analysis

Two-way sensitivity analysis is for a decision maker who wants to examine the joint impact of changes in two critical variables. All possible combination values of the two variables are generated and their associated expected value is calculated. Using one-way sensitivity analysis, we can conclude that both variable cost and fixed cost of high investment are crucial variables with significant impact on the value of the optimum decision. We use two-way sensitivity analysis to investigate the combined impact of changes in these two variables.

The strategy region graph in Figure 10.28 shows which decision is optimal given the changes in two selected variables. A triangle indicates that low investment is preferred and a diamond indicates that the high investment is preferred. In general, the low investment is optimal for paired values in the upper left quadrant. For example, if the fixed investment is $13.35 million, and the variable cost is $14.25 per part, the low-investment alternative is always preferred. As the high-investment fixed cost decreases to $12.8 million or less, then regardless of the variable cost, the high-investment alternative is optimal.

10.11   Real World Applications

10.11.1   Postal Automation (Zip + 4) Technology: A Decision Analysis

During the early and mid-1980s, the U.S. Postal Service (USPS) was faced with the decision of whether to extend postal automation, and if so, by what means. The key new variable was the

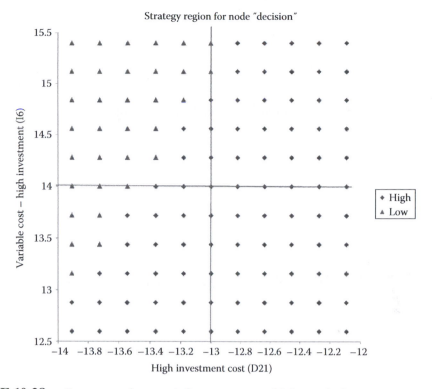

FIGURE 10.28:   Strategy region graph for two-way sensitivity analysis.

introduction of Zip + 4 codes. The office of technology assessment (OTA) was asked to investigate the advisability of proceeding with phase 2 of the postal automation strategy on both technical and economic grounds. The USPS's strategy was judged to be technically feasible, although a technology other than the USPS's choice was deemed a worthy alternative (Ulvila 1987, 1988).

The decision ultimately centered on which technology would perform the best in terms of economic savings (Figure 10.29). The USPS's choice was for single-line optical character readers,

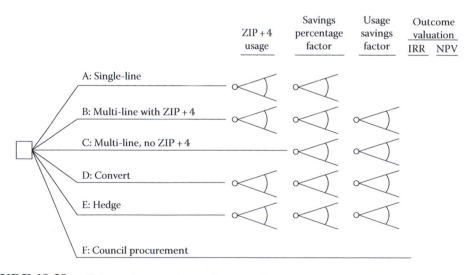

FIGURE 10.29:   Schematic tree of postal automation.

while the proposed alternative was for a multiline reader that could convert a full address that used only the standard five-digit zip code to Zip+4. Both types of reader would then place a bar code on the letter, with the bar code read by an automatic sorter. The advantage of Zip+4 is that the automatic sorter can then sort the mail to the level of a carrier route, rather than a post office or postal zone. This question was further complicated by uncertainty about the use of ZIP+4 by consumers. Historically, the USPS had overestimated consumer use of its innovations.

To address this complex and uncertain situation, Decision Science Consortium was contracted to perform a decision analysis of postal automation alternatives. A complex decision tree with six decision branches was developed. Each decision branch except the one for canceling the automation altogether was subjected to a probabilistic analysis of three factors: rate of Zip+4 usage, savings percentage factor, and usage savings factor. Next, for each path in the decision tree, a detailed cash-flow analysis was developed to compare the outcomes of the various alternatives.

The results of this analysis indicated that the NPV of the five automation alternatives ranged between $900 million and $1.5 billion. On an expected value basis, all these options were preferable to canceling the automation, and the option to convert from single-line to multiline optical character readers was the optimal decision.

Sensitivity analyses were performed to consider the uncertainty in the evaluations. Based on these analyses, the following conclusions were reached:

1. Any continuation of postal automation was better than canceling.
2. Converting to multiline optical character readers was preferred.
3. Uncertainty about the cost of the multiline readers contributed very little to the uncertainty of its NPV.
4. Uncertainty about Zip+4 usage contributed the most to variations in NPV.

Thus, the USPS's main arguments against the use of multiline readers, that their price and performance were uncertain, were found to be insignificant when compared with other factors, particularly the uncertainty of the rate of Zip+4 usage. This analysis formed the basis of the OTA's report and recommendations to Congress, and the decision was made to convert to multiline readers. The savings to the USPS (and taxpayers) were estimated to be $1.5 billion, some $200 million more than would have been saved with the USPS's first choice.

10.11.2 Drug Tests for Student Athletes

In the spring of 1987, the athletic director at Santa Clara University presented a proposal to the university's Athletic Board of Governance to test all student athletes for drug use. Some straightforward techniques of operations research, including a decision tree, were applied to the question of whether to test any single individual for the presence of drugs (Feinstein 1990).

The heart of this analysis was a decision tree: a simple decision node with branches "test" and "do not test" which then progressed to three outcomes: drug user identified, false accusation, and unidentified drug user. Tables were constructed to determine the probability that a person is a drug user given a positive test result. Tests having reliabilities between 75% and 99% and possible proportions of drug users in the general population ranging between 5% and 16.6% were included in the tables. The tables of probabilities were then used to determine the test reliability requirements necessary to reduce the probability of false positive results to an acceptable level.

Based on this analysis and the ensuing discussion, the board voted unanimously to recommend to the university president not to begin drug testing of student athletes. The board had determined that no available test would acceptably reduce the probability of making a false accusation. Ultimately, the president adopted the board's recommendation. The chairman of the board later indicated that the analysis of the decision using a decision tree was the prime factor behind the recommendation.

10.11.3 Fourth and Goal

Imagine a football team trailing by three points with a minute left in the game and facing fourth and goal from less than two yards out. The coach must decide whether to kick a field goal to tie the game or go for the win by attempting to score a touchdown. More often than not, the team will call a timeout to discuss the decision. These discussions usually revolve around whether to go for a touchdown rather than kick a field goal, and, if so, what play to run. Thus, in most cases, the decision to go for a touchdown on fourth down is not made until after the team has decided what plays to run on first, second, and third down.

Hurley (1998) argues that the fourth-and-goal conference should never involve the decision to go for a touchdown, because that decision (assuming a short distance from the end zone on fourth down) should have been made prior to first down. The author, an assistant college football coach as well as an operations researcher, came to this conclusion after using a decision tree to analyze a decision of his own coaching staff. That decision turned out to be wrong, not due to the choice to go for the touchdown, but due to the timing of the decision. The coaching staff all agreed later that, had they already decided prior to first down that they would go for a touchdown on fourth and short, the sequence of plays selected on first, second, and third down would have been quite different. In other words, if you know that you will be playing four-down football before first down, that knowledge affects the plays that will be called.

10.11.4 Oglethorpe Power Corporation Decides

Oglethorpe Power Corporation (OPC), a wholesale power generation and transmission cooperative, provides power to consumer-owned distribution cooperatives in Georgia. In Georgia, OPC produces 20% of the power and Georgia Power Company meets the remaining power demand. Georgia has the ability to produce surplus power, while Florida buys power from outside the state to meet its increasing demand because of its rapidly growing population. As a result, there is a substantial power flow from Georgia and nearby states into Florida.

Late in 1990, OPC management learned that Florida power corporation (FPC) wanted to add a transmission line to Georgia capable of transmitting an additional 1000 MW. The key decision facing OPC was whether to add this additional transmission capacity at a cost of $100 million or more with an annual savings of $20 million or more. Because of the multiple options and uncertainties involved, OPC used a decision tree to address the problem (Borison 1995).

OPC's analysis included a series of decisions combined with uncertainties, as demonstrated in Figure 10.30. Their initial decision was to choose among the three alternatives associated with the line decisions: build a new transmission line in a joint venture (integrated transmission system [ITS]), build it alone (no ITS), or not build a new line (no line). The subsequent decisions include control of new facilities and whether to upgrade existing facilities to satisfy Florida's demand. The uncertainties OPC faced included the cost of building new facilities, competition from other power sellers to Florida, Florida's demand for power, the OPC share of Florida power, and spot price. The decision tree helped the OPC management better understand the decision process as well as their competitive situation with FPC before making a final decision.

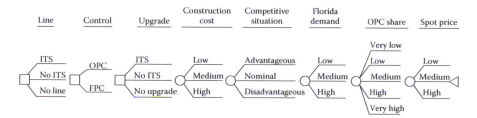

FIGURE 10.30: Decision tree showing for the transmission line problem.

Basic Terminology/Glossary of Terms

Alternatives: Options available to the decision maker when making a decision.

Collectively exhaustive events: Events that together define all the possible occurrences in an uncertain situation.

Conditional probability: A revised probability that arises from changing the estimate of the probability of an event based on additional knowledge that might affect its outcome.

Continuous random variable: A variable that can take an infinite number of values, such as a person's net worth.

Cumulative risk profile: A graphical representation of the random variable outcomes of an event plotted against the cumulative probability of the occurrence of that event. It helps the decision maker determine if one alternative stochastically dominates another.

Discrete approximation: Approximation of a continuous random variable by a discrete random variable.

Discrete random variable: A variable that can assume only a countable number of values, such as the number of correct answers on an exam.

Expected value: The probability-weighted average of all possible values of a random variable. In the decision tree context, it is the probability-weighted average of all possible payoffs that can result from a decision or sequence of decisions.

Independent events: Events A and B are called independent events if the occurrence of event A in no way affects the probability of occurrence of event B and vice versa.

Mutually exclusive events: Events that do not contain any common element among them.

Probability: A number that measures the likelihood that a particular event will occur.

Risk profile: A graphical representation of the random variable outcomes of an event plotted against the probability of occurrence of that event. It helps the decision maker understand the risks associated with the decision.

Uncertain/random/chance event: In this book, these terms are used interchangeably. Such events refer to those occurrences whose outcome is not in control of the decision maker but that influence the outcomes of decisions. A decision maker can only make educated probabilistic estimates about those occurrences.

Values: The attributes of a decision problem that a decision maker considers most important.

Exercises

Complete Chapter Activities

10.1 Sequential decisions: Present an example of a sequence of two or more decisions followed by an uncertainty.

10.2 Information gathering and decisions: Think of a decision scenario where decisions are interspersed with random events.

10.3 Refer to Figure 10.26: Interpret sensitivity analysis—the variable cost of high investment ±10%.

a. How sensitive is the optimal solution to changes in this variable?

b. What do you notice with regard to the slopes?

10.4 Refer to Figure 10.27: Interpret sensitivity analysis—the variable cost of low investment −5%.

a. How sensitive is the optimal solution to changes in this variable?

b. What do you notice with regard to the slope?

Decision Tree Examples

10.5 The owner of the Down Home restaurant is considering two ways to expand operations: open a drive-thru window or serve breakfast. There are increased annual costs with each option and a one-time cost associated with the drive-thru. Labor and marketing costs are annual costs and include new staff and more ads in local media. Layout redesign is a one-time cost. Details are provided in the Table 10.10.

The forecasted increase in income is affected by whether a competitor opens a restaurant down the street or not. The manager currently believes that the competitor is unlikely to open a new restaurant. He estimates the probability to be 0.65. Table 10.11 provides estimates of increased revenue for each scenario.

The owner of the restaurant is focused just on next year. He, therefore, decided to simply add the one-time cost for the redesign to the annual labor and marketing costs.

a. Calculate the net profit of each combination of decision and competitor action.

b. What is the best alternative if no competitor opens nearby? What is the best alternative if a competitor opens nearby?

c. Draw the associated decision tree.

d. What decision should the company follow and what is the expected value?

e. Let p represent the probability that the competitor will open a restaurant down the street. Use PrecisionTree to find the value of p that equalizes the expected values.

f. Recall that the owner treated the layout redesign the same as other annual costs. Would the decision change if he considered only 50% of these redesign costs this year.

10.6 The Red Hen company is launching its new food for sale in supermarkets throughout Michigan. The sales department is convinced that its spicy chicken soup will be a great success. The marketing department is considering an intensive advertising campaign. The advertising campaign will cost $2,000,000 and if successful produce $9,600,000 in added revenue. If the campaign is less successful (25% chance), the added revenue is estimated at only $3,600,000. If no advertising is used, the revenue is estimated at

TABLE 10.10: Down home restaurant costs.

	Costs		
	Annual		**One Time**
Decision	**Labor**	**Marketing**	**Layout Redesign**
Drive-thru window	56,000	20,000	100,000
Breakfast	77,000	10,000	—

TABLE 10.11: Down home revenue estimates.

	Competitor	
Decision	**Open (0.35)**	**Not Open (0.65)**
Drive-thru window	220,000	260,000
Breakfast	120,000	200,000

$7,000,000 with probability 0.7 if customers are receptive and $3,000,000 with probability 0.3 if they are not.

a. Draw the associated decision tree.

b. Should Red Hen invest in an intensive advertising campaign?

c. Write an equation to calculate the expected value for each decision as a function of the probability that the major advertising campaign will be effective (p)?

d. Use PrecisionTree to find the value of p for which Red Hen would be indifferent between the two choices?

10.7 A group of high school students has decided to start a summer business. They plan to design and color T-shirts and sell them to clothing stores in their community. They will need special equipment that they can buy or rent. After negotiating with a company about equipment, they figure out that they have three options to start their business:

- They can buy all the equipment and do the design and printing themselves. In this case, they have to pay for equipment but they can recover part of the money at the end of summer by reselling the equipment. The cost of buying equipment is $8500, and they can resell it at 40% of the original price. The cost of printing will be $1.1 per T-shirt.

- The second option is renting the equipment and returning it at the end of summer. The renting cost is $2000 for the whole summer and a variable cost of $1.60 per print.

- The third option is outsourcing the printing. In this case, they do the designs themselves but send them to a company for printing. The company charges them $2.1 per T-shirt.

  The market demand for colored T-shirts is uncertain. After doing market research, they summarized their estimates in Table 10.12.

a. They believe that they can sell each T-shirt for $5.5. Construct a decision tree to help them make their decision.

b. What is the best option if the demand is 2500 T-shirts?

c. What is the best option if the demand is 5500 T-shirts?

d. What is the best option if the demand is 8500 T-shirts?

e. Which option is the best for them? What is its expected profit?

10.8 The Buy&Wear clothing store uses an interesting strategy to attract customers to return each week. Each week any unsold dress is reduced by 25% of its original price. On each dress is a label with its original price and the date hung on the rack. Customers know that a $50 dress placed out on November 7th will be priced only $37.50 on November 14th if it is not sold before then. It will be reduced by another $12.50 on November 21st if it is still unsold. After two more days, any unsold dress is sent to a local charity. Each week, there is a 0.5 probability that the dress will be sold.

TABLE 10.12: T-Shirt demand estimates.

Demand (Number of T-Shirts)	Probability (%)
2500	15
5500	50
8500	35

a. Laila saw a dress she really liked and knows the almost identical dress is available online for $60. The current store price is $50. Construct a decision tree to determine whether or not she should buy the dress now or gamble and wait to try to buy it next week if it remains unsold. (If she comes back next week and finds the dress has been sold, Laila will buy it online.)

b. Just before finalizing her decision, she found another place online that sells the same dress for $55. Why might a lower price online affect her purchase decision in this store? Should she buy the dress now or gamble and wait to buy it in the second week if available?

c. She just saw a more expensive dress for sale at $80. These more expensive dresses have only a 30% chance of being sold each week and again they tell the customers that every week they reduce the price by 25% of the original price. She checked and found a similar dress for $90 online. Construct a full decision tree for 3 weeks of possible discounts.

10.9 A contestant on the Smart Quiz show must decide whether to stop or try to answer another question. The contestant is first asked a question about U.S. history. If the contestant answers correctly, she earns $1000. Historically three out of four contestants answer the first question correctly. If answered incorrectly, the game is over. If answered correctly, the contestant can leave with $1000 or go on and attempt a question about U.S. Civil War. If answered correctly, the contestant wins an additional $1200. If the answer is incorrect, the contestant loses all previous earnings and is sent home. Historically, two out of three contestants answer this question correctly. The third question is about music. This question is worth $1500, and the same rule applies. The chance of answering this question correctly is 50–50.

a. Draw a decision tree to determine the number of questions to attempt to maximize expected earnings. What is the best decision and what are the expected earnings?

b. Some contestants may feel more or less knowledgeable about the third question category. Let p represent the probability that a contestant will answer the third question correctly. Write an equation to calculate the expected value for attempting the third question as a function of p.

c. Use PrecisionTree to find the cutoff value of p such that a contestant should attempt the third question for any value higher than p?

d. The Smart Quiz show is considering changing the reward for answering the third question correctly. Let m represent the amount of money a contestant will earn for correctly answering the third question. Write an equation to calculate the expected value for the last decision as a function of m (assume a 50–50 chance).

e. Use PrecisionTree to determine the relationship between m and the optimal decision. What does this intersection point represent?

10.10 The U_Gov software company is considering submitting a bid for a state government contract to install their software on 50,000 computers. It will cost U_Gov an estimated $80 per computer to install their software. The government would use their software to oversee the management of tens of thousands of large and small contracts the government signs every year. There is only one other potential bidder for this contract, Simplexo Computers, Inc. Simplexo has a good reputation with this kind of contract. As a result of its lesser experience, the U_Gov bid must be at least $5 less per computer installation than Simplexo's in order to win the contract. Simplexo Computers is certain to bid and is generally more expensive. The company's management believes that it is equally likely that Simplexo will bid $110, $100, or $90 per computer installation.

What are the possible bids that the company should consider?

The company's bidding decision is complicated by the fact that it is currently working on a new process to install software remotely through the internet. If this process works as hoped, it may substantially lower the cost of installations. However, there is some chance that the new process will actually be more expensive than the current installation process. Unfortunately, the company will not be able to determine the cost of the new process without actually using it to install the software. The higher the company's bids the more money it makes if it wins the contract. However, the higher the bid, the less likely it is to win the contract. If the company decides to bid, it will cost $20,000 to prepare all of the relevant documents required to submit the bid. The company will incur this expense regardless of whether it wins or loses the competition. With the proposed new installation process, there is a 0.3 probability that the cost will be $55 per computer and a 0.50 probability that the cost will be $70 per computer. Unfortunately, there is also a 0.2 probability that the cost will be $90 per computer.

a. Construct a decision tree to model this situation.

b. Based on your decision tree, do you recommend they bid, and if so, what should they bid per installation?

c. Under the optimal policy, what is the probability they will win the contract?

d. What is the overall expected profit if they bid on the contract?

e. If they win the contract, what is their expected profit?

10.11 The Mint Free software company released a beta version of a software package. It expects a large number of requests for help from the users dealing with problems from bugs in the software. These problems include crashing, lock up, and incompatibility errors. The company has established a help desk to handle telephone requests. The company trained two groups of software support specialists. Group 1 has just been hired and trained; meanwhile, specialists in Group 2 are senior technicians capable of solving problems with 100% certainty. However, their salaries are much higher than the newly trained specialists. The company pays specialists based on the number of the problems they attempt to solve. Group 1 salaries are $25 per problem and Group 2 salaries are $40 per problem. The software company must decide which specialist to assign a problem in order to minimize the cost of support. If they assign a problem to a Group 1 specialist and he is unable to solve the problem, they reassign it to a senior specialist. In this case both specialists are paid. This costs the company $65 per problem. To address this issue, they developed an automatic system to predict the chance a Group 1 specialist can solve a problem.

a. A crashing problem was just received, and the prediction software forecasts a 75% chance of success for a Group 1 specialist. Draw a decision tree for this problem.

b. Based on the decision tree, what kind of specialist should be assigned the problem?

c. Another problem, a compatibility error, was received and the prediction software forecasts a 55% chance of success for a Group 1 specialist. Draw a decision tree for this decision.

d. Based on the decision tree, what kind of specialist should be assigned the problem?

e. Mint Free wants to determine the conditions under which it should assign a problem to a Group 1 specialist. Use PrecisionTree to find this probability.

f. In the previous question, what was the role of the two salaries in determining the break even value of p? Assume that the salary of Group 1 specialists is x and the salary of Group 2 workers is y. Write an equation to calculate the expected value of the cost for each decision as a function of p, x and y.

10.12 (Continue the previous problem) after finishing the first phase, management figured out that they need a Group 1.5 category of specialists that are more knowledgeable than Group 1 but not necessarily experts. They are to be paid $30 per problem because of their higher success rate than Group 1.

a. A problem was just received and the prediction software forecasts 70% chance of success for a Group 1 specialist and 85% chance of success for a Group 1.5 specialist. Group 2 can solve the problem for sure. Draw a decision tree for this problem. (Assume that if a Group 1.5 specialist fails to fix the problem, the company next assigns it to Group 2.)

b. Based on the decision tree, what kind of specialist should be assigned the problem first?

c. For another problem the prediction software forecasts 50% chance of success for a Group 1 specialist and 75% chance of success for a Group 1.5 specialist. Experts can solve the problem for sure. Based on the decision tree, what kind of specialist should be assigned the problem first?

10.13 An automotive part has to go through two different metal lathes to be shaped properly. Each lathe operation has a cost and associated scrap rate. For example, when Lathe 1 processes a part, the cost is $100 and the risk of being scrapped is 10%. Each part that successfully processed by both lathes is sold for $500. The net profit is equal to the number of parts sold minus the cost of processing all parts. The cost of processing includes both finished and scrapped parts. Table 10.13 shows the cost and scrap rate of each lathe. Scrapped parts are worthless.

a. What is the probability that a part will end up being scrapped? Does the order of processing make a difference?

b. The lathe processes can be done in either order. Draw a decision tree to determine the optimal sequence of processes to maximize the expected value of the net profit per part.

c. Which process should be done first?

d. How sensitive is the optimal strategy to the cost of processing by Lathe 2?

10.14 Continuing the previous problem, suppose there are three required lathe processes. Parts that complete these three processes are sold for $800. Table 10.14 has the associated cost and scrap for each lathe.

a. How many different sequences need to be considered?

b. What is the probability that a part is ruined?

c. Draw a decision tree to determine the optimal sequence of processes that maximizes the expected value of the net profit.

d. In which order should the processes be done?

TABLE 10.13: Two-lathe processing costs and scrap rates.

	Cost ($)	Risk of Scrap (%)
Lathe 1	100	10
Lathe 2	150	20

TABLE 10.14: Three-lathe processing costs and scrap rates.

	Cost ($)	Risk (%)
Lathe 1	100	10
Lathe 2	150	20
Lathe 3	200	25

e. Explain how you can use a pair-wise comparison of processes to find the optimal sequence?

f. If there were four processes, how many pair-wise comparisons would need to be made to find the optimal sequence?

Decision Trees: Cases

10.15 Specialty Brakes—construct and solve tree by hand

Specialty Brakes is a medium-size company that manufactures brake drums and is a supplier to OEMs as well as to the spare part market. Sales manager William Frail reports the results of a market survey upon which he has predicted sales for the next 3 years (Table 10.15). With this information, Michael Bake, vice president of operations, estimates probabilistic demand for each of the next 3 years.

Dex Peditor, the production manager, reports that he has studied the predictions and has been looking at avenues for expanding the capacity of the plant. The required expansion could be achieved by investing in additional production lines. Each additional line costs $3 million and adds 150,000 units of capacity. Addition of lines also results in a corresponding decrease in unit variable cost of manufacturing. The selling price per unit is $50. Table 10.15 shows the data relating to the demand, and Table 10.16 shows the relative costs of adding production lines.

a. Construct a schematic tree

b. Construct and solve tree by hand

TABLE 10.15: Demand data.

Level	Average Annual Demand (Million #)	Probability
Low	0.55	0.3
Medium	0.80	0.4
High	0.90	0.3

TABLE 10.16: Cost data.

Proposal	Investment per Year ($Million)	Average Annual Capacity (Million #)	Variable Cost ($ per Unit)
No change	0	0.60	26.00
Add one line	3	0.75	22.00
Add two line	6	0.90	21.00

c. Present your decision

d. Re-solve this problem using PrecisionTree software

e. Comment on the cumulative risk profile

Notes:

1. If demand exceeds capacity, demand is not met.

2. Profit = (units manufactured) × (selling price/unit − variable cost/unit) − investment

3. For "Add one Line," if demand = 0.90, capacity = 0.75, then units manufactured = 0.75,

   a. Revenue = 0.75 * (50 − 22.00) − 3 = $18.0M

10.16 Sprocket manufacturing

The manager of this production firm is faced with a potential problem: shortages in the supply of a vital gear sprocket. He is considering the development of an in-house capability to make the sprockets, knowing that the firm needs 35,000 such sprockets each year. If the firm elects to make the sprockets internally, then it must select one of the three possible production processes. Each has different costs, as shown in Table 10.17.

Adopting a manual process may lead to union problems. The manager estimates the probability of these problems as 25%, and knows that if they do occur, then the variable cost will increase by $0.50 per sprocket. Further, if the semi-automatic or fully automatic option is selected, then additional training of personnel will be required. The amount and therefore the cost of training are uncertain. For the semi-automatic process, the fixed cost of basic training will add $15,000. However, there is a 20% chance the workers will need extra training at a cost of $5,000. For the fully automatic process, the fixed cost of basic training is $16,000, but there is a 30% chance they will need advanced training that will add $10,000 to the fixed cost.

The firm has the option of purchasing the sprockets from their external supplier at a cost of $2.90 per sprocket; however, the manager estimates that there is a 40% chance that this supplier will fail to develop the process in a timely manner. If he does fail, then one of the three in-house processes must be developed; however, time pressure would increase the fixed cost of each process by 50%.

a. Construct the appropriate decision tree using PrecisionTree software.

b. What is the optimal strategy, and what is the final expected total payoff?

c. If the manager were to develop an in-house facility, which production process should he select and why?

10.17 WeExcel Inc. in danger of missing a deadline

TRAP Inc. is a premier provider of solutions for many of the best-run e-businesses. It is currently developing its third-generation software to help companies collaborate with suppliers on complex engineering projects. WeExcel Inc., consultants in Supply Chain

TABLE 10.17: Cost data for sprocket manufacturing case.

Production Process	Fixed Costs/Year ($)	Variable Cost/Unit ($)
Manual	20,000	2.5
Semiautomatic	50,000	1.5
Fully automatic	80,000	1

Solutions, had won from TRAP a major $12 million contract for developing one of the critical modules in the software. WeExcel has 12 months to deliver the project, with payment to occur upon delivery.

Six months into the project, WeExcel has spent $3 million of the expected $6 million investment on developing the platform on which to write the software. An internal review of the progress now finds a problem. At the current pace, they estimate that there is only a 70% chance of completing the project on time. The firm's experts feel that if the project deadline could be delayed 1 month, they could deliver, with a 95% probability, an outstanding product that would satisfy the needs of TRAP.

Mite Stifid, president of WeExcel, has to decide whether or not to inform TRAP regarding the possible delay in the project and to ask for a 1 month extension. From past experience, he knows that TRAP has been very particular about deliverables. Considering his performance in earlier projects, he feels that there is a 50% chance that the deadline might be extended. But he also knows that there is a 50% chance that the order might be cancelled altogether. In this event, they would lose the $3 million that they had already invested.

Stifid is also considering an option of offering a rebate of $500,000 if TRAP gives the extension. He feels this would improve the probability of receiving a 1 month extension to 70%. He also has an option of investing an extra $2.8 million in the next 6 months in the project (over and above the $6 million originally planned). By doing so, the probability of delivering the module by the 12 month deadline will improve to 90%.

The management estimates a loss of goodwill in the amount of $1 million if they fail to deliver the module. In the event that they get the extension in the deadline and still fail to deliver the software module, the loss of goodwill may amount to $3 million.

a. Construct the schematic tree.

b. Construct the appropriate decision tree using PrecisionTree software.

c. What is Mite Stifid's optimal strategy?

10.18 Colonel Car Company: late design change

The Colonel Car Company has experienced a significant number of warranty claims associated with the driver's side mirror in its luxury car. Almost 10% of the warranty claims in the first 6 months of operation relate to this mirror. The design engineers at Colonel have come up with a new design for the driver's side mirrors. They believe the new design will not only reduce the warranty claims but also cut production costs. The decision is complicated by two factors. (1) Production of next year's model is to begin in exactly 6 months, which means there is insufficient time to carry out a comprehensive production run test of the new design. (2) This model year has been classified as a quiet year to discourage design changes. Colonel executives have to decide whether or not to pursue the design change of the mirrors.

The design change is expected to eliminate a major problem with a key component of the mirror, which accounted for 60% of the $250,000 in mirror warranty claims. The anticipated reduction in warranty costs amounts to $150,000. (Each 10% reduction in warranty claims saves $25,000.) In addition, this simpler design will reduce the variable cost of production with forecasted annual savings of $75,000. The new design will require new tooling that will cost a total of $35,000.

There is a 20% chance that the new design will not perform as expected and will, in fact, make matters worse. If that happens, the company cannot simply revert back to the old design. Production will have to be interrupted while the old tooling is restored on an emergency basis. As a result, Colonel would incur a $250,000 cost in terms of lost production and emergency installation.

If the company is to go ahead with the new design, it should move quickly and complete the changes within the next month. If the changeover is started but not completed by the 1 month deadline, there will be an added penalty of $100,000 for incorporating a new design after the deadline. There is a 70% chance that the design change can be implemented before the deadline.

a. Construct a schematic tree for the Colonel late design change.

b. Determine the optimal decision.

c. Discuss the risk profile.

References

Bell, D. E. (1984). Bidding for the S. S. Kuniang. *Interfaces*, 14(2), 17–23.

Borison, A. (1995). Oglethorpe power corporation decides about investing in a major transmission system. *Interfaces*, 25(2), 25–36.

Cohan, D., Haas, S. M., Radloff, D. L., and Yancik, R. F. (1984). Using fire in forest management: Decision making under uncertainty. *Interfaces*, 14(5), 8–19.

Corner, J. L. and Kirkwood, C. W. (1991). Decision analysis applications in the operations research literature, 1970–1989. *Operations Research*, 39, 206–219.

Feinstein, C. D. (1990). Deciding whether to test student athletes for drug use. *Interfaces*, 20(3), 80–87.

Howard, D. A. (1988). Decision analysis: Practice and promise. *Management Science*, 34, 679–695.

Hurley, W. J. (1998). Optimal sequential decisions and the content of the fourth-and-goal conference. *Interfaces*, 28(6), 19–22.

Keefer, D. L., Kirkwood, C. W., and Corner, J. L. (2004). Perspective on decision analysis applications, 1990–2001. *Decision Analysis*, 1, 4–22.

Krishnan, V. and Bhattacharya, S. (2002). Technology selection and commitment in new product development: The role of uncertainty and design flexibility. *Management Science*, 48, 313–327.

Ulvila, J. W. (1987). Postal automation (Zip+4) technology: A decision analysis. *Interfaces*, 17(2), 1–12.

Ulvila, J. W. (1988). 20/30 Hindsight: The automatic zipper. *Interfaces*, 18(1): 74–77.

Walls, M. R., Morahan, G. T., and Dyer, J. S. (1995). Decision analysis of exploration opportunities in the onshore US at Phillips petroleum company. *Interfaces*, 25(6), 39–56.

Chapter 11

Structured Risk Management and the Value of Information and Delay

> Anco, an oil exploration company, has identified a site under which there may be an oil reservoir. It would cost $5 million to drill for oil there. If the oil exists, the company will spend $150 million to develop the field. Anco has estimated the Net Present Value (NPV) of high and low reserves, excluding drilling, seismic, and development costs, to be $700 and $100 million, respectively. A seismic test would cost $500,000. The test results could be positive, inconclusive, or negative. Should Anco perform a seismic test or make its drilling decision without it?

> BioTech has won approval from the Food and Drug Administration (FDA) to market its new anticancer drug Astena in the United States. It has been negotiating with BSG to license the sale of the drug in the European Union (EU). The negotiations are at a stalemate. Biotech is seeking 20% royalties on forecasted sales of $11 billion in the EU. BSG is offering only 15% based on their sales forecast of $8 billion. How can the two companies reach a negotiated agreement with such divergent sales forecasts?

11.1 Goal and Overview

In the previous chapter, we introduced decision trees as a tool for analyzing decisions in the presence of uncertainty. This chapter presents a structured approach to risk management that is developed around decision trees.

Classical decision analysis refers to random outcomes as states of nature. In the earliest applications, states of nature described outcomes such as the amount of oil in the ground. The very term "state of nature" seems to imply that there is little a decision maker can do to change a random event. This chapter is intended to counter this prevailing notion by exploring all the options a decision maker has to manage the states of nature within the framework of decision trees. Even in the classic example of oil drilling, where decision makers are unable to change the amount of oil in the ground, they can take actions that affect the amount of oil that can be recovered. In this way, they can affect the probability distribution of recoverable oil and, in effect, manage the uncertainty.

The chapter's main goal is to facilitate the search for risk management strategies within the modeling framework of decision trees. We present a structured analytic approach that identifies critical variables to control. One element in this approach is the valuation of perfect control of randomness. Perfect control means a manager can drive the uncertainty toward its ideal value. The expected value of perfect control of one variable is a theoretical bound on the value of risk management of that variable.

Complex problems can include dozens of risk factors. A decision analyst may not have time to analyze all risk factors. Therefore, it is vital to highlight which variables have the most significant impact on the outcome. We present two commonly used tools (tornado diagram and spider plot) to identify high impact variables. We then describe how to evaluate the impact of these strategies on the expected value using a structured risk management procedure (Chelst and Bodily 2000). The steps of this procedure are distinct from standard sensitivity analysis. Sensitivity analysis is generally a process for assessing the robustness of the optimal strategy by determining the range of values for which it remains optimal. However, as Clauss (1997) points out, the practical manager is more interested in how he can get more value out of the optimal solution than how robust the strategy is. Our procedure is designed to increase the overall value of the optimal strategy in the presence of uncertainty. The process moves the analyst and decision maker toward the goal of finding "the best of the best." It is intended to be a catalyst in the search for new strategies with less risk and higher expected values (Rothkopf 1996).

A core risk management strategy involves obtaining information in order to reduce or eliminate uncertainty. Information gathering includes conducting surveys, consulting experts, establishing pilot plants, performing tests, performing analysis, doing research, or reading books or journals. In almost all cases, there is a cost for gathering this information. This may be a direct cost associated with a survey or experiment or the indirect cost of delaying the decision until the new data arrive. A key question in each decision context is, "How much is this information worth in terms of its quantifiable impact on the decision?" The early developers of decision analysis recognized the role of information and built into the decision tree methodology a process for assessing the value of both perfect and imperfect information.

Closely tied to the concept of information is a strategy to delay critical pieces of the decision until some or all of the uncertainty is resolved. We illustrate this concept of delay in the context of contingent contracts. In negotiations, often the two sides of the contract disagree on what the future holds. Rather than attempt to reach a consensus on the future, they structure the agreement to reflect their divergent perspectives. Each side's share of the total value is not decided until the uncertainty is resolved at a later date. Service contracts with sports stars and CEOs often include contingency clauses.

Another strategy of delaying a decision involves the use of real options. Real options are used to manage risk in large investment projects such as research and development (R&D), investment in information technology, capacity expansion, partnership, and merger. An option holder has the right but not the obligation to make an investment. This is similar to a financial call or put option on a common stock. In a real options approach, managers design strategies that provide them the flexibility to make ongoing decisions as uncertainties are clarified. They are able to rapidly terminate projects that are not working out or expand those that are performing well.

11.2   Identify High-Impact Variables

In this section, we illustrate the use of decision tree software to identify and highlight risks that should be the focus of risk management efforts. This approach draws upon the sensitivity analysis capability of decision tree software. In Chapter 10, we introduced sensitivity analysis to explore the effect of a single variable on the expected value of a model. Here, we explore how to review a number of variables simultaneously in order to identify the most significant ones. This section presents two commonly used tools, a tornado diagram and a spider plot.

We use Boss Controls (BC) automation investment decision that was introduced in Chapter 10. BC is gearing up to manufacture an option that will be offered on one million new cars worldwide. Initial estimates are that the take rate for the option could be as low as 30% or as high as 50%. Based on experience, it is also estimated that the probability of a low take rate is 0.4. The plan calls for BC to deliver the option to the automotive company at a price of $60. BC is considering two alternatives that differ significantly in the level of investment in automation and the related variable cost of production. Relevant data for both alternatives are presented in Table 11.1. The decision tree for BC is presented in Figure 11.1.

TABLE 11.1:   Data for BC automation investment.

Investment Decision	Investment Dollars ($Million)	Variable Cost ($)	Net Sales Revenue (NSR)	Take Rate (%)	NSR ($Million)
Low	8	27	$(60-27) *$ take_rate $* 10^6$	30	9.9
				50	16.5
High	13	14	$(60-14) *$ take_rate $* 10^6$	30	13.8
				50	23.0

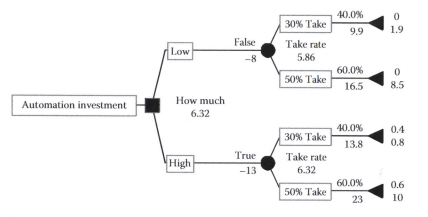

FIGURE 11.1:   Decision tree BC automation investment.

11.2.1   Tornado Diagram

A tornado diagram compares one-way sensitivity analysis for many inputs at once. Unlike a sensitivity analysis line chart, a tornado diagram only considers the two extreme values of a variable and no values in between. It stacks all the variables from widest to narrowest impact, so that their effect on the value of the model can be compared (Eschenbach 1992).

To create a tornado diagram, the decision analyst elicits pessimistic and optimistic values of each uncertain variable from the subject matter experts. Table 11.2 shows the pessimistic and optimistic values of variables for the automation investment. For example, BC's fixed investment cost of the high investment option can vary between ±7% of base value ($13 million).

The decision analysis software calculates the expected value of the decision for the pessimistic and optimistic values of each uncertain variable (Figure 11.2). The variable for which the expected value of the model has the widest range is listed in the top bar. Bars for the other variables are placed in descending order to create the tornado diagram. The variables with the longest bars are the prime

TABLE 11.2:   List of viable ranges for bc automation investment decision.

Variable	Pessimistic Value	Optimistic Value
Fixed investment of high investment	+7% of base	−7% of base
Price	−10% of base	Base
Variable cost of low investment	+10% of base	−10% of base
Variable cost of high investment	5% of base	Base
Probability of low take rate	−0.2 absolute	+0.2 absolute
Low take rate (30%)	−10% absolute	Base
Volume	−15% of base	Base

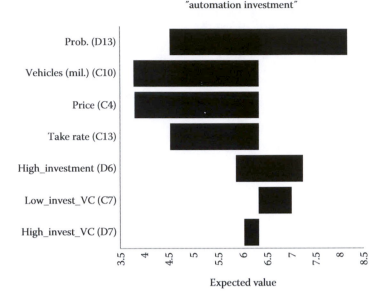

FIGURE 11.2: Tornado diagram for BC automation investment.

candidates for creative efforts in risk management. On the other hand, the variables associated with short bars do not deserve much attention.

Figure 11.2 depicts how profit for the automation investment case varies over the extreme range of values for seven input variables. For instance, the estimated probability of a low take rate is 0.4. When this is changed by ±0.2, there is a huge impact on the expected value. The expected value ranges from $4.54 to $8.16 million. In contrast, a 10% change in the variable cost of low investment has a more modest impact. A 10% increase has no impact because the optimal decision is a high investment and does not incur this cost. A 10% decrease changes the decision, and the optimal decision's profit increases to $7 million.

The estimated sales volume of 1 million cars is the second most significant variable. The range on this variable is only one-sided, because no one believes that the actual sales volume could exceed the manufacturer's projection, but some believe that it could be lower by 15% or only 850,000 units total. This would reduce the expected profit to $3.78 million, a decrease of more than 40%. Similarly, they do not expect a price increase, but BC fears it may be pressured to reduce the price by 10%. This impact would be somewhat less than 40%. The variable cost ranges have the least impact on the total profit.

11.2.2 Spider Plot

A spider plot provides more detailed information than a tornado diagram but in a more complex format. The spider plot is a graph of the change in each variable against the value of the model. It is created by calculating the values of the model not only for the two extremes but also for a specified number of intermediate values. Consequently, a spider plot can demonstrate the slope of the relationship and any nonlinearity as shown in Figure 11.3. Its limitation is that only a few variables can be clearly displayed together in one chart (Eschenbach 1992). Only four variables are included in Figure 11.3. In contrast, there is no practical limit to the number of variables that can be effectively displayed in a tornado diagram.

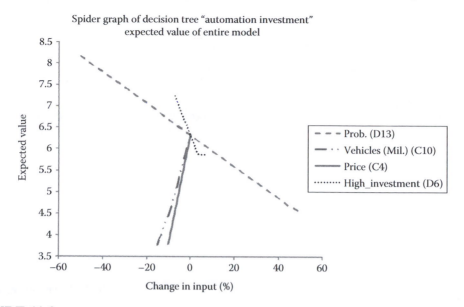

FIGURE 11.3: Spider plot for BC automation investment.

In Figure 11.3, the *x*-axis corresponds to the range of the input variable and is plotted as a percentage change in its value. The *y*-axis represents the expected value of the total profit in this case and is measured in millions of dollars. The variables that contribute to the largest total variations in expected value are the ones whose *y*-axis range is greatest. As before, the probability of a low take rate has the highest impact on the profit, which ranges from less than $5 million to more than $8 million. A 15% decrease in sales volume causes the expected value to drop below $4 million.

The slope of each line depicts the relative change in the outcome for each percent change in the independent variable. The steeper the curve, the more sensitive the expected value is to percent changes in the input variable. The steepest slope is associated with changes in the price. Each percentage change in price has a bigger impact on the expected value than a corresponding percentage change in the probability of the low take rate. As seen in Figure 11.3, the fixed cost of high investment is the factor least critical for profitability.

In summary, both the magnitude and probability of a low take rate are the critical variables with regard to expected total profit. BC should consider investing in joint advertising with the automotive company to reduce the likelihood that the majority of car buyers will not consider spending more to purchase BC's option. The marketing campaign might also increase the magnitude of the low take rate even if the option does not become overwhelmingly popular. Another variable of interest is the volume of vehicle sales. However, this is a variable that BC cannot significantly impact as a component supplier. Finally, it is important for the company to maintain the sales price against any pressure to offer a price reduction. Investing time to reduce variable costs at this juncture is not as critical as the income side of the equation.

11.3 Risk Profiles and Structured Risk Management

The link between decision trees and risk management was established decades ago (Covello 1987). A number of case studies in the literature used a decision tree to evaluate risk management strategies. For example, Balson et al. (1992) employed a decision tree to manage environmental risk, and Engemann and Miller (1992) applied it to manage operations risk at a bank. A 1987 issue of the journal *Risk Analysis* includes a number of articles that discuss the role of decision trees in risk analysis and management. *Risk Management* (Bell and Schleifer 1995) also discusses how decision trees can be used as part of the risk management paradigm.

In all of the aforementioned studies, the decision trees were used to model alternatives that corresponded to specific risk management actions. The approach proposed by Chelst and Bodily (2000) is intended for use at an earlier stage, before specific risk management alternatives have been defined. The steps are intended to trigger the search and definition of risk management alternatives that would then be incorporated into a revised decision tree and evaluated.

An experienced decision analyst might already be including risk management in two stages of the decision analysis paradigm. The first opportunity arises in the interview of a subject matter expert. As part of the assessment of the probability distribution of a model parameter, the interviewer routinely asks for clarification as to the nature and causes of the uncertainty. This interview process could simultaneously uncover opportunities for risk management. Second, once the tree is constructed and evaluated, the decision analyst might vary a specific random variable to ascertain the sensitivity of the optimal decision to the input values and the sensitivity of the total value function to the input value.

When reviewing a discrete random variable in the decision tree, decreasing the specific probability of the most negative outcome or reducing its negative effects reduces downside risk. In the case of a continuous random variable, risk management would improve the risk profile in three ways. An illustrative risk profile for a hypothetical best alternative is shown in Figure 11.4a. A decision maker may take actions that serve to shift the risk profile to the right, thereby adding value for all

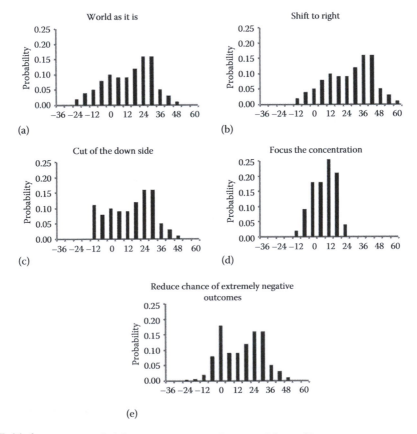

FIGURE 11.4: Impact of risk management actions on risk profile.

possible outcomes. This impact is shown in Figure 11.4b. This shift right might be brought about, for example, by eliminating altogether an operating cost in a project. Alternatively, the analyst may find ways to cut off the downside risk and move those outcomes to some guaranteed level. This would increase the mean and, more importantly, remove the most disastrous possibilities. This is shown in Figure 11.4c. A guarantee for a minimum purchase quantity in a contract might provide such a shift. Insurance is another example of how to cut off the downside risk. However, insurance costs money, and so its expense would generate a leftward shift in the whole risk profile and reduce the overall expected value. Here is an example wherein a manager may think that a downside risk is deleted rather than traded for additional cost; managers should be honest and admit the cost of their perceived changes in risk.

Last, management may be able to centrally concentrate the uncertainty in the risk profile, thereby reducing the risk, even though the mean performance does not change. Figure 11.4d shows a risk profile for such a risk-reducing activity. In fact, this risk profile was created in a way that reflects risk sharing. If we could sell half of a risky opportunity for a price equal to half its expected value and keep the rest of the risky opportunity, we would have the focused concentration, as shown in Figure 11.4d.

Virtually, all management attempts to improve a risky opportunity do so by one of these effects or some combination of them. In each instance, some least desirable scenarios are eliminated. Even if these scenarios are still possible, the risk profile can be improved by reducing the probabilities of extremely negative outcomes. This is reflected in Figure 11.4e. The concept of "magnitude reduction" of either poor outcomes or their probability is consistent with the way managers view risk in terms of worst case scenarios labeled "downside risk" (March and Shapira 1987). However, the literature notes that managers tend to not quantify the probability of a worst-case scenario. Thus, even though managers can appreciate actions that significantly reduce the likelihood of occurrence, they tend to perceive greater value in the risk profile changes characterized in Figures 11.4b through 11.4d.

11.3.1 Adding Value and Reducing Risk in the "Optimal" Strategy

Let us place ourselves at that stage in a decision analysis where the manager has reviewed the risk profile of each alternative and identified the most favorable risk profile. This is referred to as the "optimal" strategy. "Optimal" is in quotation marks to make it clear that while one could claim this is the best alternative of those presented, it may be possible to creatively improve upon it.

At this stage, the manager responsible for presenting it to a decision board and subsequently implementing it should examine the strategy to see whether it has any weak or even unacceptable outcomes. Inevitably, it will have some downside risk, and then the question becomes how to improve the alternative by reducing risk. In decision trees normally presented in textbooks, there are usually at most three discrete random variables. A decision maker could simply review all the end values of the branches and find the worst value(s) *within the optimal strategy*. With this as the first focus, there are a number of specific changes to evaluate in terms of improvement in the expected value of the optimal strategy.

Step 1. Perfect control: Within the optimal strategy, select a *random event* that appears along the path to the worst-case scenario endpoint value. Calculate the expected value of perfect control of the event by assigning a probability of 1 to its most favorable branch and determining the net increase in expected value.

Step 2. Reduce risk by changing a probability: Select the *branch* of the random event that leads down the path to the worst value in the optimal strategy. Reduce the probability of that branch by some easily multiplied increment such as 0.1. Add that probability to the neighboring complementary branch.

Step 3. Reduce negative impact by changing a value of random variable: For the same worst-case scenario branch, change the lowest endpoint value by some easily multiplied increment such as

$1 or $1 million and recalculate the expected value. Repeat the process by improving the value such that it matches the value of the next worst branch or use some other realistic bound on the maximum improvement.

Step 4. Change a given parameter: In many contexts, there are parameters that are part of the calculation of the values. Change the value of a deterministic parameter that is linked to the optimal path in some logical, easily multiplied increment such as $1 or $1 million and recalculate the expected value. (The fact that the parameter was initially a given does not preclude management actions from improving its value.)

Repeat the aforementioned process for another random event and its branches that appear in the optimal strategy.

The previous steps focused on the optimal strategy. However, there may be more cost-effective options for risk management of the second best alternative that could result in it outperforming the original optimal. Thus, repeat the process for the worst path on the *second best strategy*. (It might be worthwhile to repeat this process for more than two alternatives if the expected values of the lower-ranked alternatives are close to the optimal. If, however, there are large differences in expected value, the additional analysis is not likely to be worthwhile.)

The net change in the expected value will provide the manager with insight as to the payoff of seeking risk management strategies for different random events and key parameters. Once the expected incremental value of risk management actions is established, the decision maker can create and evaluate cost-effective strategies whose cost is less than the net change in expected value. If the decision maker is sufficiently risk averse, he may even choose to spend more than the expected value of the change. The steps described earlier enable the decision maker to incorporate his attitude toward risk in a direct fashion. This may be preferable to constructing a formal utility function, which decision makers often find to be abstract.

11.4 Make or Buy Example: Discrete Decision Tree Analysis

In this section, we introduce risk management for a discrete decision tree analysis using the make or buy example introduced in Chapter 10. Western Co. is to decide whether to manufacture a component in-house or to buy it from a supplier. The company has a design for the part but is unsure whether the design will work as is. The data are summarized in Table 11.3. Because of long lead times, it has to make the decision now before the design can be totally validated. Thus, one of

TABLE 11.3: Data for make or buy decision for Western Co.

Random Events and Costs		
Design feasibility	Probability that current design will work	0.4
	Probability that part will need a major redesign	0.6
Demand	Probability of low demand (1 million parts)	0.3
	Probability of medium demand (1.25 million parts)	0.5
	Probability of high demand (1.5 million parts)	0.2
Make in-house	Fixed cost: Facility investment (millions of dollars)	$55
	Variable cost per part	
	If current design works	$100 per part
	If there is a major redesign	8% increase to $108 per part
Buy from supplier	Fixed cost (millions of dollars)	$0
	Variable cost per part if current design works	$140 per part
	Variable cost per part if there is a major redesign	$161 per part

the uncertainties is whether or not the current design will work. If there is a need for a late major design change, it will be difficult to keep cost efficiencies in place, and the variable cost will go up by 8%. This, in turn, will affect the cost of producing the component in-house. If instead a contract is signed now with a supplier and the design has to be changed significantly, the supplier will almost certainly use the late design change to justify a 15% increase in the part price.

Figure 11.5 presents the entire tree. The expected value for the *make in-house* alternative is $183.38 million; for the *buy from a supplier* alternative, the expected value is $186.94 million. Not only is the *make* alternative preferred in terms of its expected value but it also has less risk associated with it. Its total cost cannot exceed $217 million (redesign and high demand). In contrast, the cost of buying from a supplier could range as high as $241.5 million. This would occur if the design turned out to be unfeasible and the demand was high. (See Figure 11.6 for the risk profiles.)

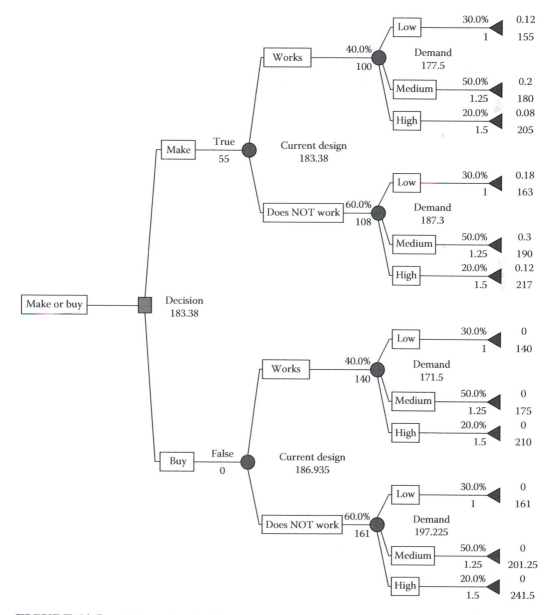

FIGURE 11.5: Make or buy decision tree.

Choice comparison for node 'decision'

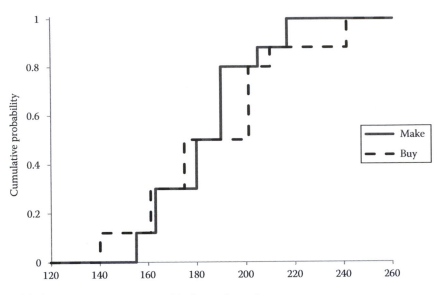

FIGURE 11.6:   Cumulative risk profile for make or buy.

The highest cost in the optimal strategy occurs if the design does not work and the demand is high. As discussed earlier, we identify a random event that appears in the optimal strategy (Figure 11.5). The random event selected is "design failure." What if management could totally control the design process and remove any chance that the design will not work? (step 1 mentioned earlier.) To calculate the expected value of perfect control, we assign zero to the probability that the design will not work. The impact is dramatic. The optimal strategy now is to outsource the product with a net decrease in expected value of $11.88 million. Perfect control represents the ideal. However, if the probability of a design success were to improve by just 0.1, the net savings would be $980,000. (step 2 mentioned earlier.)

In this example, the variable costs per part are linked to the outcome of the random event, "design failure," as well as to the decision of "make or buy." It was estimated that the in-house variable cost would increase by $8–$108 if there is a need for a major redesign What if management could reduce that by $1 and hold the variable cost to no more than $107? (step 3 mentioned earlier.) The expected value decreases to $182.645 million, a reduction of $735,000. If they could reduce the impact of the redesign on manufacturing cost to only $103, the total expected cost would be $179.705 million, a savings of $3.675 million. These numbers suggest that there is value in investing time and energy to hold the line on a variable cost increase if there is need for a major redesign.

The company is facing strong pressure to continue its relationship with its supplier. It is therefore interested in taking a close look at managing the risks associated with going with the supplier. To study this issue, we change the parameter that specifies the percent increase (15%) that the supplier has historically added to the part price as a result of a redesign (step 4 mentioned earlier). A 1% reduction in the adjustment to 14% has a zero dollar impact because the optimal decision remains "buy." However, a supplier commitment to hold the line to an 8% adjustment, equivalent to the "make" adjustment, would have a major impact. It would make the buy option optimal and reduce the expected cost by more than $3.6 million.

Table 11.4 summarizes the results of analysis. The greatest potential for risk management is linked to reducing the probability of design failure with a savings of $11.88 million. This, however, may be technically difficult to accomplish. In contrast, a commitment from the supplier not to

TABLE 11.4:   Summary of risk management alternatives for western Co.

Factor	Change	Cost Savings	Risk Reduction Strategies
Reduce **cost** increase linked to redesign	From $8 to $7 From $8 to $3	$730,000 $3.65 million	If redesign is needed, try to contain added cost of manufacturing.
Reduce **risk** that design will not work	From 0.6 to 0.5	$980,000	Modify design quickly to reduce need for major redesign later
	From 0.6 to 0.3	$4.16 million	New optimal: Use supplier
	From 0.6 to 0.0	$11.9 million	Value of perfect control
Manage uncertainty of demand	Not appropriate		Does not make sense to reduce total demand to lower total cost.
Percentage price increase by supplier if design does not work	From 15% to 14% From 15% to 8%	$0 $3.5 million	Obtain commitment from supplier not to take advantage of redesign to raise prices disproportionately
Supplier price reduction if volumes are high	Up to $8 reduction in price	No impact	Negotiate major price reduction for high volumes
EVPI of design feasibility		$2.4 million	Test feasibility of current design

increase prices unreasonably if the design fails would save $3.65 million. The likelihood of a redesign would need to be cut in half, from 0.6 to 0.3, in order to achieve a similar gain. Interestingly, in that instance, the optimal decision would involve the supplier anyway.

A second uncertainty, "demand," also has a significant impact on the total cost. The total cost is highest when the demand is highest. However, it makes no sense to talk about managing the risk of high demand. Assuming the company makes a profit on every part, it does not want to reduce the magnitude of the high demand or its probability. If this problem had, however, been framed in terms of net profit, then management would want to take a closer look at "low" demand's impact on net revenue. This illustrates the value of focusing on the right overall performance measure when managing risk; it affords a wider range of potential improvements.

In the paragraphs above, we have evaluated the impact of risk management of key variables on the value of the objective function for the optimal strategy and the second best strategy. The types of risk management strategies needed to achieve these gains would likely be diverse. They could involve any one or more of the following activities:

Work on the design to reduce the chances for a major redesign.

Perform testing on the design to clarify whether or not it will work.

Invest effort to ensure that any major redesign does not increase the cost per part by as much as 8% for the in-house option.

Negotiate tighter guidelines on the supplier's right to increase the price if a major redesign is needed.

11.5   Perfect and Imperfect Information

Decision makers often collect information when they face uncertainty in order to reduce or eliminate uncertainty and thus increase the expected value of their decisions. Information gathering takes many forms: market research, experiments and testing, detailed mathematical and statistical analysis, expert interviews, prototyping, pilot plants, and literature review.

Although additional information is desirable, gathering it can be expensive and may delay the project. It is important to weigh the value of new information against its related cost. In general, if the cost of new information is less than its expected value, it is not worth gathering. (In a later chapter, we discuss risk aversion that could justify paying for information to reduce extreme risk even if on average this risk reduction does not cover the cost of the data.) This section presents procedures to determine when it is worth collecting additional information to eliminate or reduce uncertainty. We will first consider the case where the information is perfectly reliable, certain to indicate which outcome will happen. Next, we look at the more common situation where the information is an imperfect predictor of outcome. For example, market surveys do not perfectly predict customer behavior, and medical tests (Ades et al. 2004) produce both false negative and false positive results. Bickel (2008) explores the ratio of value between perfect and imperfect information. Eppel and von Winterfeldt (2008) in a study of nuclear waste estimated the value of sample information as between $0.8 and $17.7 million. The main example for this section is an extension of the BC automation investment decision (Figure 11.1).

11.5.1 Expected Value of Perfect Information: One Uncertain Event

If information is always correct, it is said to be perfect. Perfect information about an uncertain event lets the decision makers know the outcome of the uncertain event before the decision is made. In BC, imagine an expert who, like a clairvoyant, always correctly identifies a situation in which the take rate will be high. If the expert says the take rate will be high, we know that this will be the case. Similarly, if the expert says the take rate will be low, then it will be low. Although this scenario is unrealistic, the concept of perfect information is still useful, because it provides an upper bound on the value of any information.

How does a decision maker decide whether to request the advice of an expert? In the automation investment example, the optimal investment level is "high." If the expert says the take rate will be high, the BC management would still choose the "high" investment option, since its profit is $2 million more than the "low" investment option ($10 million versus $8.5 million). In this case, the information does not change the optimal decision and provides no value. On the other hand, the expert may say that the take rate will be low. In that case, the "low" investment option has a higher value; the low investment yields $1.9 million in profit versus the high investment's $0.8 million. In this case, the information has value since it leads to a different action.

We usually think about information value after the fact. But it would be much more useful to weigh the cost and value of information before we hire an expert or conduct some experiments or surveys. However, because we do not know in advance what the expert will say, we need to think probabilistically about the information and determine its expected value. Only if the expected value of information is more than the cost of obtaining the information would we consult an expert or perform a test to gather information.

We apply the following reasoning to calculate expected value of perfect information (EVPI). Based on experience, we had estimated a 40% chance of a *low take rate*. Thus, before we ask the expert his opinion, we believe there is a 40% chance he will say that he knows for sure the take rate will be low. Conversely, before we ask the expert his opinion, we believe there is a 60% chance he will say that he knows for sure the take rate will be high. This logic is characterized as pre-posterior analysis; to many decision analysis novices, it often seems counterintuitive. We are attempting to imagine the situation after the expert gives his opinion and our subsequent decision before we have even asked his opinion. The probability distribution of his advice is just the original probability estimates of each outcome. This logic is represented in a decision tree format in Figure 11.7 by placing the uncertainty to be explored with the expert in front of the decision.

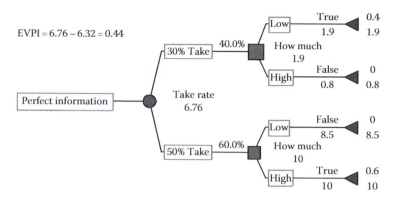

EVPI = 6.76 − 6.32 = 0.44

FIGURE 11.7: EVPI tree for BC.

In constructing the tree, we need to place the calculated end values for each path in the appropriate places. There is a 40% chance the expert will forecast a low take rate; BC will then make a low investment and earn $1.9 million. There is a 60% chance the expert will forecast a high take rate; BC will then make a high investment and earn $10 million. The expected value is now $6.76 million as compared to $6.32 million, a net improvement of $0.44 million. This net improvement is the EVPI.

EVPI = Expected value with perfect information − Expected value without information

11.5.2 Expected Value of Perfect Information: Multiple Uncertain Events

The BC decision includes one decision and only one uncertain event. Next, we illustrate how to calculate EVPI for a more complex symmetric decision tree using the Western Co.'s make or buy decision. Figure 11.8 presents the original decision and data for Western Co. There is one decision and two uncertain events, design feasibility and demand. Thus, there may be value in asking the opinions of experts in design and demand forecasting to help resolve each uncertainty. The expected value without information is $183.38 million.

The management of Western Co. is interested in evaluating the impact of resolving both uncertain events. The process is identical to that applied to the BC' decision tree.

1. Place an uncertain event at the start of the tree along with its associated probabilities on the respective branches.
2. Make sure to place the correct end values along each path. (This is often the most difficult task to carry out correctly.)
3. Calculate the new expected value.
4. Calculate the net change in expected value and this EPVI.

Figure 11.9 presents the decision tree with perfect information on design feasibility. If the design works, the *buy* option will have the lowest expected cost, $171.5 million. On the other hand, if the design does not work, the *make* option will result in the lowest expected cost, $187.3 million. There is an estimated 40% chance that the design will work as is and a 60% chance that there will need to be a redesign. The expected cost with perfect information on design feasibility is $180.98 million, compared to $183.38 million with no information. The EVPI about the design feasibility is $2.4 million.

Design feasibility	Probability	Make costs	Buy costs
Works	0.4	100	140
Does NOT	0.6	108	161
	Premium	8%	15%

Demand	Prob.
1	0.3
1.25	0.5
1.5	0.2
Fixed costs	
Make	55
Buy	0

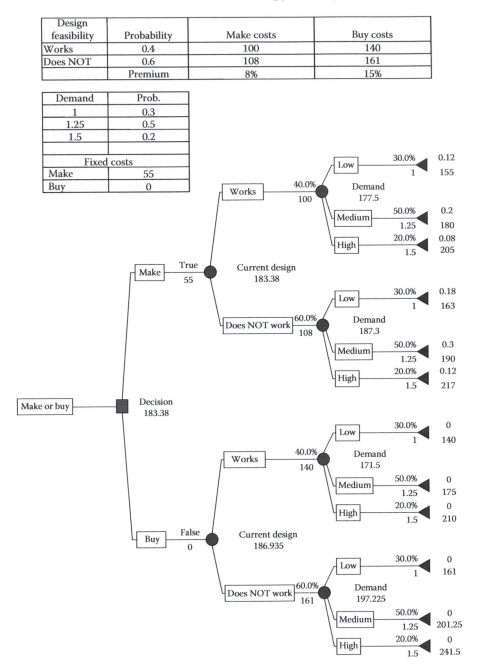

FIGURE 11.8:   Western Co. make or buy decision.

Figure 11.10 presents the decision tree with perfect information on demand. If the company knows that the demand is low, it would select the *buy* option, since its expected cost is lower than the *make* option. The company would choose the *make* option if it knows the demand will be medium or high. The expected cost with perfect information about demand is $181.22 million. Thus, the EVPI for demand is $2.16 million. This is $240,000 less than the EVPI of design feasibility.

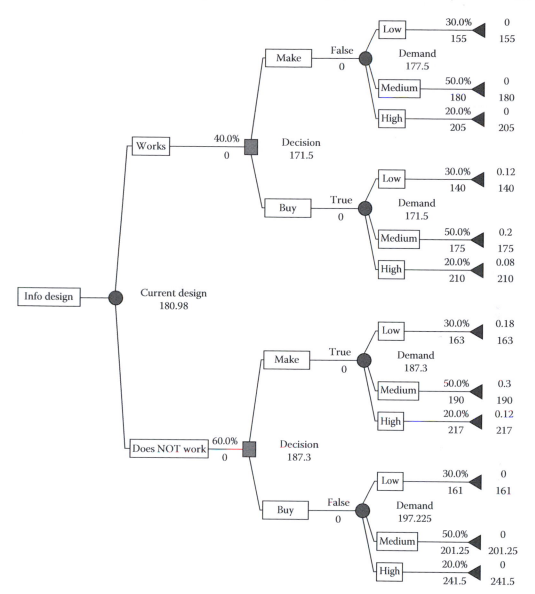

FIGURE 11.9: EVPI of design feasibility for Western Co.'s make or buy decision.

It is possible to obtain expert information about both random events. When we do the calculations, it becomes clear that the EVPI for separate events cannot simply be added to determine the combined EVPI. To determine this EVPI, it is necessary to place both random events before the decision as in Figure 11.11. The placement sequence of these events in the tree does not matter as long as they are both before any decision. There are now six distinct paths prior to any decision. If the design works and demand is either low or medium, the preferred decision is *buy*. If the design works and demand is high, the preferred decision is *make*. If the design does not work and demand is low or medium, the preferred decision is *buy*. If the design works and demand is high, the preferred decision is *make*. The overall expected cost with all uncertainties resolved is $180.22 million, compared to $183.38 million with no information. This EVPI is $3.36 million, which is significantly less than the sum of the individual values.

EVPI for demand = 183.38 – 181.22 = 2.16 million

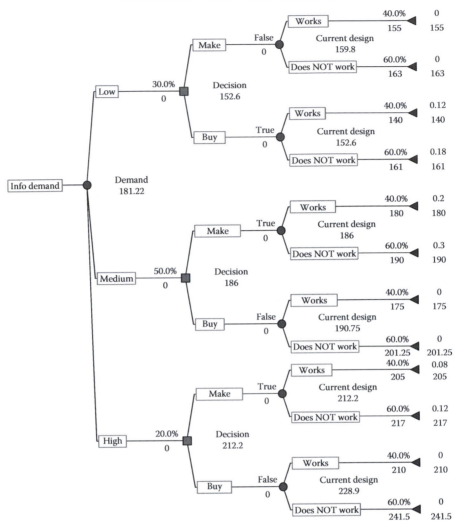

FIGURE 11.10:   EVPI of demand for Western Co. make or buy.

11.6   Imperfect Information: Bayes' Theorem

Decision makers rarely have access to perfectly reliable information. This concern was recognized and modeled in the early application of decision analysis in the oil industry. Oil companies gather extensive seismic information, but there is still uncertainty regarding the petroleum and gas deposits in the study area (Pickering and Bickel 2006). We extend the decision tree analysis to deal with imperfect information often called sample information.

Imperfect information will change a decision maker's estimates about a chance event or random variable, but it does not eliminate uncertainty. For example, market research results are unlikely to be perfectly reliable. The sample size and the design of the questionnaire contribute to a less than perfect prediction. In addition, what people say does not perfectly align with their future behavior. A forecaster's prediction of future stock prices, interest rates, and exchange rates is, needless to say, not perfectly accurate.

Total combined EVPI for both uncertain events = 183.38 − 180.22 = 3.16 million

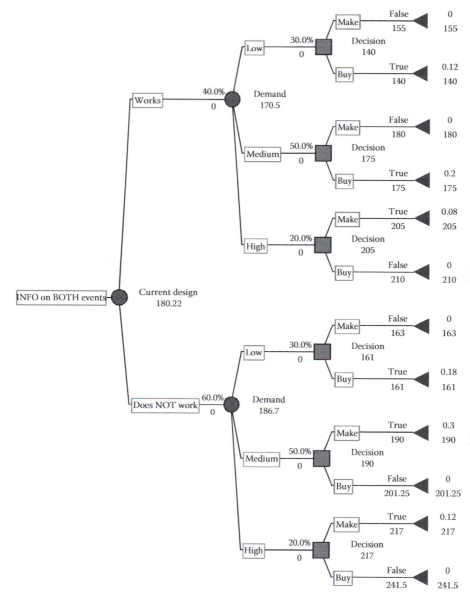

FIGURE 11.11: EVPI of both demand and design feasibility for make or buy.

Calculating expected value of imperfect information (EVII) is more complicated than determining EVPI. In both instances, the decision tree starts with an uncertain event before the decision. However, when we gather imperfect information, we do not eliminate the uncertainty; all we can do is revise initial probability estimates in the light of new information. As a result, the uncertain event appears twice, first in an estimate of what the imperfect information will likely say and later as the same random event but with updated estimates of the probability that are based on the concept of conditional probability.

We present two methods to revise the initial probabilities: Bayes' Theorem and expert opinion. Classically, the decision analyst uses Bayes's Theorem to determine the conditional probabilities. This works in a data-rich environment that often uses the same standard testing procedures over and over. As a result, they can estimate the accuracy of the testing procedure. However, in many onetime decisions, there is little formal data on the accuracy of the imperfect information. The only alternative is to use subjective expert estimates of the conditional probabilities.

Imagine the following situation: a fetus has been tested for a rare genetic disorder that occurs in 1 per 10,000 infants. The testing for the gene is highly accurate. If the genetic defect is present, it will find the defect 99.5% of the time. However, there are also occasionally false positives; even when there is no genetic defect, test results come back as false positives 2% of the time.

1. Activity: The test results just came back positive. What is the likelihood of the fetus having the genetic defect? Record your intuitive estimate of this probability. _____

If you are like many people, your gut instinct says that the probability is above 95%. The research literature on probability misconceptions points out how poor our instincts are in this case. The key problem is that for many people their intuition does not know how to factor the initial prevalence of the disease, which in this case was 1 in 10,000. Most people would have given the same answer whether the prevalence was 1 in a 100, 1 in 10,000, or 1 in a million. Bayes' Theorem adjusts for the prevalence, but let us first develop an intuitive understanding of what the probability should be.

Imagine 10,000 fetuses took this test, and this population included a fetus with the genetic defect. That one fetus will almost certainly produce a positive test result. However, there is a 2% false positive rate for fetuses without the defect. The 9999 fetuses without the defect would generate ~200 false positive, 2% of 9999. The 10,000 population would therefore generate a total of 201 positive test results but only one fetus has the defect. Thus, there is only a 1:201 risk of having the genetic defect even after receiving a positive test result. This is a 0.005 probability.

2. Activity: Apply the same logic for a genetic defect that has an incidence rate of 1 in a 100. What is the likelihood of a fetus with a positive test result having this genetic defect? _____

Bayes' Theorem is credited to Rev. Thomas Bayes, an eighteenth century British clergyman. It is the most common way to revise probabilities when new information becomes available.

Let events D and D^c represent the two possibilities: have the genetic defect (D) and do not have the defect (D^c). Let Pos correspond to a positive test result. Bayes' Theorem states as follows:

$$P(D|Pos) = P(Pos|D)P(D)/[P(Pos|D)P(D) + P(Pos|D^c)P(D^c)]$$

The numerator of the right hand side of the equation represents the joint probability of a positive test and having the disease. The denominator represents the total probability of a positive test result.

Here is another example of Bayes' Theorem applied to a medical test. Suppose 1 in 500 men in a certain age group has one type of cancer. Screening for this type of cancer is 99% accurate; that is, if someone has the cancer, the test will be positive 99% of the time

$$(P(Pos|Disease) = P(Pos|D) = 0.99).$$

Let us also assume that if someone does not have the cancer, the test will yield a false positive 4% of the time.

$$(P(Pos|No Disease) = P(Pos|D^c) = 0.04).$$

Now imagine a friend has taken the test and his doctor somberly intones that he has tested positive. Does this mean he is likely to have the specific cancer? What is the likelihood he has this cancer? We can calculate the probability using Bayes' Theorem.

$$P(D|Pos) = P(Pos|D)P(D)/[P(Pos|D)P(D) + P(Pos|D^c)P(D^c)]$$

TABLE 11.5: Bayesian posterior (after positive result) probabilities.

Initial Event Probability	Test Accuracy			
	0.7	**0.8**	**0.9**	**0.95**
0.1	0.21	0.31	0.50	0.68
0.3	0.50	0.63	0.79	0.89
0.4	0.61	0.73	0.86	0.93
0.45	0.66	0.77	0.88	0.94
0.5	0.70	0.80	0.90	0.95
0.6	0.78	0.86	0.93	0.97
0.7	0.84	0.90	0.95	0.98

$$= 0.99 * (1/500)/[0.99 * (1/500) + (0.04) * (1 - 1/500)]$$

$$= 0.00198/0.0419 = 0.047$$

There is a less than 5% chance that he has this cancer. It is analysis such as this that has caused a great deal of debate recently about the value of routine screening of individuals in age brackets with low incidence of certain cancers. For example, the US Preventive Services Task Force recommended that the starting age for regular mammograms should be 50 years of age for nonhigh risk women instead of the previously recommended 40 years of age. They also recommend testing every other year instead of every year.

In the earlier examples, the underlying disease prevalence was small, less than 1%. After receiving the test results, the chance of having the disease was still small when compared to the overall accuracy of tests. In Table 11.5, we determine the posterior Bayesian probabilities for a wide range of initial probabilities and test accuracies. The initial probabilities range between 0.1 and 0.7. The test accuracy rates are between 0.7 and 0.95. To simplify the presentation, we set the two measures of accuracy to be equal. The likelihood of a true positive equals the likelihood of a true negative.

For low initial probability events such as 0.1 or even 0.3, the test accuracy and final probability are far apart. However, with an initial probability of 0.5, the final or posterior estimates equal the test accuracy. In general, when the initial probability was between 0.4 and 0.6 and test accuracy was at least 0.9, the difference between the posterior probability and the accuracy was 0.03 or less.

11.6.1 Western Co.: Make or Buy with Imperfect Information

Here, we demonstrate the value of imperfect information with the Western Co.'s make or buy decision and with an oil drilling example. Western Co. is trying to decide whether to manufacture a component in-house or to contract with a supplier. Design feasibility is a key concern for management. The EVPI was found to be $2.4 million, which is a bound on the maximum value of any imperfect test. Experts initially estimated that the current design will work with a probability of only 0.4. However, engineers believe that a complex test can be used to almost validate or invalidate the design. In the past, if the designed worked, 98% of the time the test results were good. On the other hand, if the designed failed, 94% of the time the results were bad.

Will conducting this test reduce expected cost? If yes, how much would Western Co. be willing to pay for this complicated test? We answer these questions by calculating EVII. In Bayes' Theorem, an initial probability estimate is known as a "prior" probability. In the make or buy example, the prior probabilities are 0.4 that the design will work and 0.6 that the design will not work. When Bayes' Theorem is employed to modify a prior probability in the light of new information, the result is known as a "posterior" probability. We apply conditional probability and Bayes' Theorem to determine the necessary probabilities for the revised decision tree pictured in Figure 11.12.

The tree starts with the uncertainty as to what the test results will show. This part of the tree is directly analogous to the EVPI tree. There we asked what the probability distribution of the expert's opinion was. Here, we ask, in advance: what is the probability distribution of the test results? This first step requires some calculation using the concept of probabilistic concept of partitioning. The test results can turn out to be good in either one of two ways.

1. The design is good, and the test reflects this fact.

2. The design is bad, and the test provides a false positive result.

Let G = good test results and B = bad test results (complement of G)
   W = design works and W^c = design does not work (complement of W)
   The mathematical equation is

$$P(G) = P(G|W)\,P(W) + P(G|W^c)\,P(W^c)$$

$$= (0.98)\,(0.4) + (1 - 0.94)\,(0.6) = 0.428$$

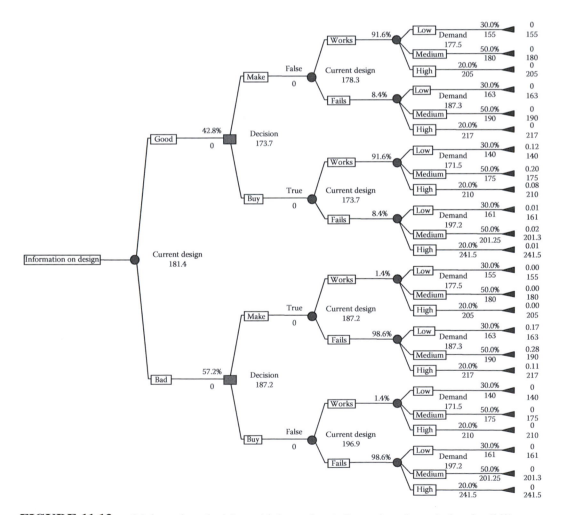

FIGURE 11.12:   Make or buy decision with imperfect information about design feasibility.

We can determine P(B) by using the concept that complementary event probabilities must sum to 1.

$$P(B) = 1 - 0.428 = 0.572$$

These probabilities do not match the initial estimates that the design works or does not work. This indicates that this testing procedure has a slightly negative bias; the positive test results are less likely than the original estimates of design success. These probabilities are placed on the first branches that begin the tree with the uncertain event, namely, test results. Once the test results are in, management of Western must decide whether to make or buy the part. However, there is still uncertainty as to the design, because the tests are not perfect. This uncertainty appears after the decision, but the posterior probabilities need to be updated using Bayes' Theorem. These probability estimates are now different because of the test findings. We need four conditional probabilities to complete the tree:

$$P(W|G) \text{ and its complement } P(W^c|G)$$

$P(W|B)$ and its complement $P(W^c|B)$

Using Bayes' Theorem to find the posterior probabilities, we find

$$P(\text{Design Works}|\text{Test Results Good}) = P(W|G)$$

$$P(W|G) = P(G|W) \ P(W)/[P(G|W) \ P(W) + P(G|F) \ P(F)]$$

$$P(W|G) = 0.98 * 0.40/[0.98 * 0.40 + 0.06 * 0.60] = 0.916$$

$$P(W^c|G) = 1 - 0.916 = 0.084$$

Similarly,

$$P(W|B) = P(B|W) \ P(W)/[P(B|W) \ P(W) + [P(B|W^c) \ P(W^c)]$$

$$= 0.02 * 0.4/[0.02 * 0.4 + 0.94 * 0.6] = 0.014$$

$$P(W^c|B) = 1 - 0.014 = 0.986$$

There has been a dramatic change in the probability estimate as to whether or not the design works based on the test results (see Table 11.6). If the test is positive, the likelihood that the design works has increased from 0.4, our initial estimate, to 0.916. If the test results come back negative, the likelihood that the design works drops to 0.014. The posterior probabilities are included on

TABLE 11.6: Posterior probabilities for the feasibility of design.

Results of Design Test	Design Feasibility Probabilities	
	Work	**Does Not Work**
Good	0.916	0.084
Bad	0.014	0.986

appropriate branches in the completed decision tree (Figure 11.12). Now, the optimum decision depends on the test results. If the test result is good, the optimal decision changes. The *buy* option's expected cost is $173.7 million, which is $4.7 million less than that of the *make* option. On the other hand, the Western Co. management will stay with *make* if the test result is bad. In this case, their internal cost is projected to be $187.2 million compared to $196.9 million for the *buy* alternative. In summary, the optimal alternative is a conditional decision that depends on the outcome of the design test. The expected values for these conditional decisions must be multiplied by the corresponding probability of a good and bad test result.

$$\text{Expected Cost} = 0.428(173.7) + 0.572(187.2) = \$181.42$$

EVII = Expected value with imperfect information − Expected value without information

The expected cost with imperfect information is $181.42 million, compared to $183.38 million with no information. The EVII in this example is $2.04 million. This is almost as good as the EVPI for design, which is $2.4 million. Extensive testing of the design is estimated to cost less than $200,000, and it therefore makes sense to proceed with testing before deciding to make or buy the item.

11.6.2 Anco Oil with Three Outcomes

There are two primary sources of information for oil companies: seismic readings and test drill holes. The seismic readings can help not only predict the presence of oil but also the breadth of the oil reservoir system at different depths underground. There are different levels of seismic measurement sophistication, and accuracy and new technologies are being developed. The cost of a test well in deepwater can be $20 million. Equally important, drilling such a well can delay oil field development by 6 months. These sources of data are used not only to determine whether or not to proceed but also to plan the level of development required and the location of key wells. Even with these sources of information, there will still be significant uncertainty regarding the total amount of oil recovered from the field (Prange et al. 2008). In recent years there has been dramatic improvement in the quality of seismic information. The Anco case is a simplified example of a common oil drilling decision. We have reduced the range of field size to two values and the outcome of the seismic readings to three possibilities.

Anco, an oil exploration company, has identified a site under which there may be an oil reservoir. The probability that oil exists at this location is 10%. It would cost $5,000,000 to drill exploratory wells to determine the presence of oil. If oil is found, the company plans to spend $150 million to develop the field. Even if oil is found, there is uncertainty regarding the amount. The oil reserve would be either high or low, with probabilities of 0.2 and 0.8, respectively. Anco estimated NPV of high and low reserves excluding drilling, seismic, and development costs. The NPV of high and low reserves are estimated at $700 and $100 million, respectively.

A seismic test is available and would cost $500,000. The result of the test would be one of the following: *strong, inconclusive,* or *weak.* Based on data from a wide variety of oil fields, if there is oil, there is a 60% chance of a *strong* reading, a 30% chance of an *inconclusive* reading, and a 10% chance of a *weak* reading. If there is no oil at that location, there is a 5% probability of a strong reading, a 20% probability of an inconclusive reading, and a 75% probability of a weak reading. Anco's decision analysis group wishes to develop a decision tree for this situation and solve it to obtain a recommendation for senior management. The three possible results of the seismic test require a small modification to the Bayes' Theorem presented earlier, but the basic logic is unchanged. In real-world oil exploration, the seismic data are actually continuous. Decision analysts and data analysts working with the oil industry have developed sophisticated models to determine the conditional probabilities of oil presence based on the results of different types of tests.

The tree used to calculate EVII begins with the branches of the possible seismic readings. These are calculated using the concept of partitioning. We consider the different possible conditions that could lead to a seismic reading characterized as *strong, inconclusive*, or *weak*. For example, a strong reading could result from the presence of oil or even if no oil is present.

$$Let\ S = Strong$$

$$I = Inconclusive$$

$$W = Weak$$

$$P(S) = P(S|Oil) * P(Oil) + P(S|No\ oil) * P(No\ oil)$$

$$= 0.6 * 0.1 + 0.05 * 0.9 = 0.105$$

$$P(I) = P(I|Oil) * P(Oil) + P(I|No\ oil) * P(No\ oil)$$

$$= 0.3 * 0.1 + 0.20 * 0.9 = 0.210$$

$$P(W) = P(W|Oil) * P(Oil) + P(W|No\ oil) * P(No\ oil)$$

$$= 0.1 * 0.1 + 0.75 * 0.9 = 0.685$$

We now use Bayes' Theorem to find the posterior probabilities, the updated likelihood of oil after seeing the seismic results (see Table 11.7). If the seismic readings are *strong*, then the updated probabilities are

$$P(Oil|S) = P(S|Oil)\ P(Oil)/[P(S|Oil)\ P(Oil) + P(S|No\ oil)\ P(No\ oil)]$$

Substituting in values for the conditional probabilities and priors,

$$P(Oil|S) = 0.6 * 0.10/(0.6 * 0.10 + 0.05 * 0.90) = 0.06/0.105 = 0.571$$

$$P(No\ Oil|S) = 1 - 0.571 = 0.429$$

If the seismic readings are *inconclusive*, the updated probabilities are

$$P(Oil|I) = P(I|Oil) * P(Oil)/[P(I|Oil) * P(Oil) + P(S|No\ Oil) * P(No\ Oil)]$$

Substituting in values for the conditional probabilities and priors,

$$P(Oil|I) = 0.3 * 0.10/(0.3 * 0.10 + 0.2 * 0.90) = 0.03/0.21$$

$$P(Oil|I) = 0.143$$

TABLE 11.7: Posterior probabilities for presence of oil.

Seismic Test	Presence of Oil	
	Oil	No Oil
Strong	0.571	0.429
Inconclusive	0.143	0.857
Weak	0.015	0.985

Last, if the seismic readings are *weak*, the updated probabilities are

$$P(Oil|W) = P(W|Oil) * P(Oil)/[P(W|Oil) * P(Oil) + P(W|No\ Oil) * P(No\ Oil)]$$

$$P(Oil|W) = 0.1 * 0.10/(0.1 * 0.10 + 0.75 * 0.90) = 0.01/0.685 = 0.015$$

Figure 11.13 presents the decision tree for the oil drilling example. The tree shows the expected value with and without a seismic test. The first decision Anco needs to make is whether to use a seismic test. If they do not use a seismic test, they can employ original probability estimates. The expected NPV for drilling without a seismic test is $2 million.

If they select a seismic test, they use marginal and posterior probabilities for seismic test and oil presence. Anco's decision analysis group included all probabilities and values on appropriate branches in the tree. The expected NPV with a seismic test is $4.23 million after accounting for the cost of the test. The optimal decision is conditioned on test results. If the seismic test result is strong,

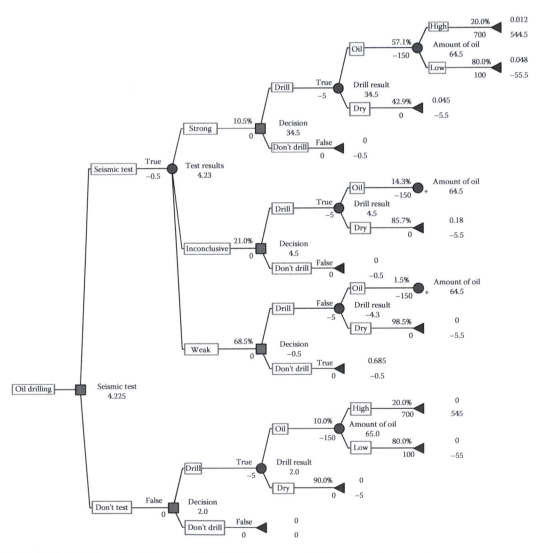

FIGURE 11.13:   Decision tree for oil drilling case with imperfect information.

they drill, and the expected value is $34.5 million. On the other hand, if the test result is inconclusive, they would still drill, but the expected value is only $4.5 million. If the results are weak, the optimum decision is "don't drill." The EVII for this problem is $2.73 million compared to a cost of $0.5 million for the seismic data.

11.6.3   Boss Controls with Expert Judgment

In the two previous cases, the decision maker was using standard test procedures. The test for design feasibility has been validated independent of the likelihood that a specific design will work. The decision maker knew the probability of a positive result, given a workable design, P(Pos|W). For the decision, he wanted to calculate the reverse conditional probability, P(W|Pos). To determine this probability, he needed to use Bayes' Theorem coupled with an initial estimate of the design feasibility. Similarly, the seismic data have been validated independent of the presumed likelihood of the presence of oil. Thus, the decision maker needs Bayes' Theorem to update the estimate of the presence of oil based on the test results. Bayes' Theorem is especially important when the initial estimates of a specific outcome can vary widely, as in some of the medical examples presented earlier.

There may be other decision contexts in which information gathering does not involve a formal test, and the initial probability estimates are not extremely low. Experienced information gatherers may be able to directly forecast the relevant conditional probability of interest. We demonstrate this with the BC example.

In BC's automation investment decision, suppose that management interviews a focus group to forecast the take rate. The focus group reaction to the option is a useful but imperfect predictor of the actual take rate. Focus group facilitators have a wide range of experience with a similar focus group responding to new car options like the one BC is considering. In general, the group's reaction can be described as either Enthusiastic (E) or Good (G). Their reactions can then be used to update the estimate that the take rate will be at the high end or low end of the estimates. Table 11.8 shows focus group's success at predicting demand. The probabilities here are conditioned on the response of the focus group. For example, P(take rate is "high"|focus group was "enthusiastic") equals 0.7. In other words, in the past, if the focus groups were "enthusiastic," the take rate ended up being at the "high" end 70% the time. However, if the focus groups' reactions were just "good," then 80% of the time the take rate was at the "low" end. If the response is only "good," there is only a 20% chance the take rate will be "high." Organizers of focus groups know that the process tends to have an optimistic bias in that groups are enthusiastic 80% of the time.

Figure 11.14 shows a decision tree representation of imperfect information on the take rate for the BC automation investment problem. The first event is the focus group's forecast on the take rate. These probabilities are determined by past experience. After receiving the focus group's reaction, BC management makes its investment level decision. The next event in this example contains conditional probabilities determined by expert judgment and based on past experience, as reported in Table 11.8.

TABLE 11.8: Conditional probabilities describing focus group's performance.

Focus Group	Take Rate Probabilities Conditioned on Focus Group Reaction	
	High	**Low**
Enthusiastic	0.7	0.3
Good	0.2	0.8

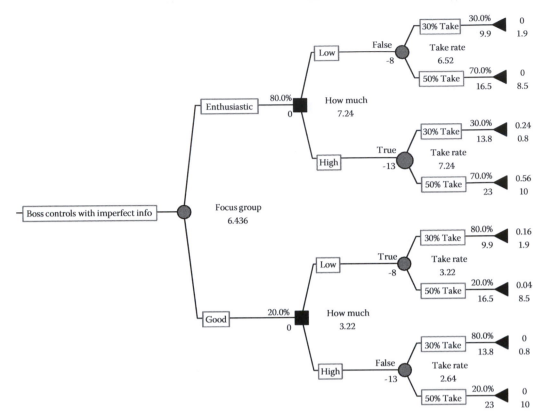

FIGURE 11.14: Decision tree of EVII for BC automation investment.

The conditional probabilities should be consistent with original estimates of the take rate probabilities. In the following, we verify that the probabilities are, in fact, consistent. We can define the following partition formulas:

$$P(\text{low take rate}) = P(\text{low take rate}|G)\ P(G) + P(\text{low take rate}|E)\ P(E)$$

$$= 0.80 * 0.20 + 0.30 * 0.80 = 0.40 \text{ (original estimate)}$$

$$P(\text{high take rate}) = P(\text{high take rate}|G)\ P(G) + P(\text{high take rate}|E)\ P(E)$$

$$= 0.20 * 0.20 + 0.70 * 0.80 = 0.60 \text{ (original estimate)}$$

In the decision tree displayed in Figure 11.14, the expected value when using a focus group is $6.436 million. If the focus group is enthusiastic, BC management's optimum decision would be "high" investment. On the other hand, BC management would choose "low" investment if the focus group reaction is only "good." Recall that expected value of this investment problem without using a focus group is $6.32 million. The EVII is the difference between two expected values, just $0.05 million or $116,000. The BC management should not pay more than $116,000 for the focus group's forecast. This contrasts with the EVPI, which was worth $440,000. Clearly, in this case, the focus group would provide only limited value in part because of its optimistic bias.

11.7 Conditional Decisions and Information Seeking Trees: Flu Virus Detection Technology

Omega Biotech's R&D department is evaluating proposals for a new technology to test for bird flu virus in animals. They prescreened two major technologies: Y and Z. Omega has experience with Technology Z, whereas Y represents a breakthrough but includes significant uncertainty. If it succeeds, the manufacturing department predicts that Omega can save millions of dollars in production costs. If Omega chooses Technology Z, they estimate that they will need to invest $10 million to develop it. The manufacturing department predicts that the total production cost for 5 years will be either $15 or $20 million, with equal likelihood. Technology Y will require $13 million to develop but could ultimately fail completely. Consequently, Omega is considering development of a prototype of Technology Y. The prototypes are very good but imperfect predictors of the viability of the new technology. Based on experience, they have found that 70% of the time they can successfully develop prototypes and 30% of the time the prototype phase fails. In the past, if the prototype results were positive, 90% of the projects were ultimately successful. On the other hand, if the prototype results were negative, only 20% of the projects turned out to be successful. These are the probabilities that are conditioned on the outcome of the prototype development phase.

If Omega successfully develops Technology Y, it estimates that the 5-year production cost will be either $4 million with a probability of 0.7 or $7 million with a probability of 0.3. However, if Omega decides to attempt to develop Technology Y and fails, it still has the option to develop Technology Z.

Figure 11.15 shows schematic for technology development case. The optimal decision path is a *conditional* decision. It is conditioned on the outcome of gathering information. Given that Omega Biotech pursues the development of prototypes of Technology Y, there is an a priori uncertainty about what can be learned from the prototype experience. This uncertainty cannot be resolved until the prototype is developed and results are analyzed. The results of this information gathering phase will determine which path to pursue. Nevertheless, an initial decision must be made even as to which alternative to pursue before all of the information is gathered. The gathering of this information can be expensive, either because of the cost of running the experiment or because of the delays incurred as a result of waiting for information to be gathered. The pharmaceutical industry is constantly faced with these types of costly information-gathering decisions along the path of developing one successful drug.

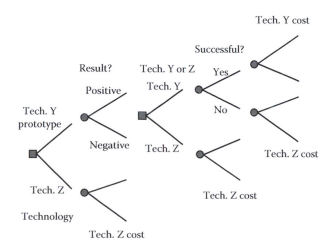

FIGURE 11.15: Schematic tree for technology development example.

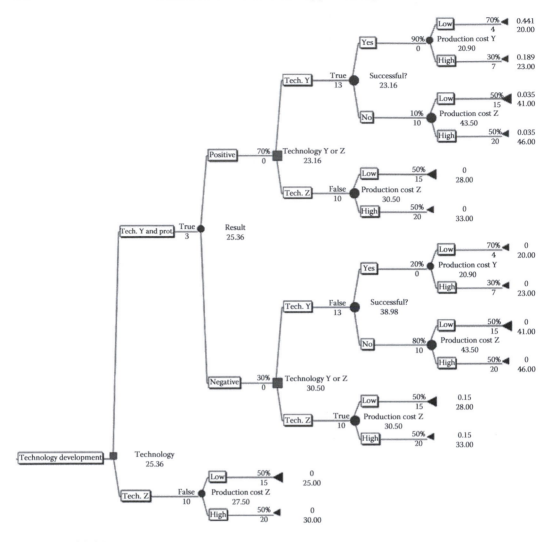

FIGURE 11.16: Decision Tree for Omega case.

Figure 11.16 illustrates the decision tree for the technology development case. To be complete, we include in the tree the possibility of continuing to pursue Technology Y even if the prototype phase is negative. Omega's objective is to minimize total cost of technology development and 5 year production. All the values along the branches relate to costs incurred with that branch. The total cost at each end node is the sum of the respective costs along the way.

If Technology Z is chosen as in the bottom of the tree, there is one uncertain variable regarding production cost. The production costs will be either low, $25 million (=10+15), or high, $30 million (=10+20). The expected cost of Technology Z is $27.5 million (=25 * 0.5 + 30 * 0.5). The calculation of expected value of Technology Z is trivial as compared to the analysis involved with first pursuing Y.

The pursuit of Y involves uncertainty regarding the success of the prototype phase, uncertainty regarding the information gathering, and final uncertainty regarding production costs.

Let us consider several of the longest paths. Prototypes are developed and the results are positive. Omega Biotech proceeds with the development of technology Y. Assume again it is successful and the production costs turn out to be low. The costs are $3 million for the prototypes, $13 million for the development, and $4 million for production. The total cost is $20 million. If production costs turn out to be high, the total cost is $23 million.

If production cost is low, the total cost $= 3 + 13 + 4 = \$20$ million

If production cost is high, the total cost $= 3 + 13 + 7 = \$23$ million

The expected value of successful development is $20.9 million ($20 * 0.7 + 23 * 0.3$).

If they pursue but do not successfully develop Technology Y, Omega can still develop Technology Z. If they succeed in the prototype phase but fail with development, the costs still include the prototype phase, Technology Y development, Technology Z development, and final production costs.

If Technology Z's production cost is low, the total cost $= 3 + 13 + 10 + 15 = \$41$ million

If its production cost is high, the total cost $= 3 + 13 + 10 + 20 = \$46$ million

If Omega attempts but fails to develop Technology Y, the expected cost will be $43.5 ($= 41 * 0.5 + 46 * 0.5$) million.

Based on the expected value, the initial optimal decision is to develop prototypes of Technology Y with an expected value of $25.36 million. However, if the prototype phase fails, it should drop Y and work on Z. The difference between the expected values of these two technologies is $2.14 million. The outcome of the decision to initially pursue Y ranges between $20 and $46 million. These two alternatives do not dominate each other as in Figure 11.17. While the best outcome of Technology Y is $20 million, the best outcome of Technology Z is $25 million. On the other hand, the worst outcome associated with starting on Technology Y is $16 million more than with Technology Z. There is significantly more uncertainty associated with the optimal decision path than with simply developing technology Z. In Chapter 12, we introduce utility theory to incorporate risk attitude into the decision process.

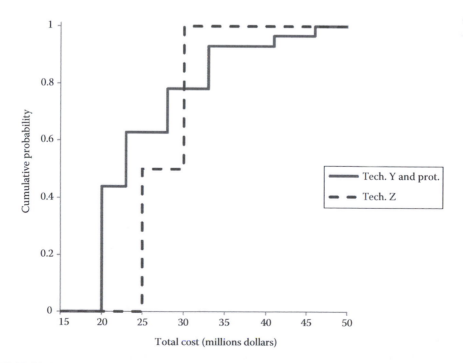

FIGURE 11.17:   Cumulative risk profile for technology development case.

11.8   Contingent Contracts Reduce Risk

A contingent contract is a contract to do or not to do something depending on the future event. For example, a buyer can make an offer to purchase a house contingent upon his obtaining a loan to cover 80% or more of the purchase price. If he cannot obtain the loan, he is not required to proceed with the purchase of the house. This type of contingency clause protects just one side of the relationship from getting stuck if some untoward event occurs.

Two sides of a negotiated business agreement may have different expectations about the future and be very suspicious of the other side's prediction. As the negotiations progress, the forecast differences between two parties dominate the discussions and can create deadlocks that are hard to break. Contingent contracts can help to avoid these kinds of impasses and enable two parties to reach an agreement despite their different predictions (Bazerman and Gillespie 1999).

Contingency contracts can be useful in a wide array of business arrangements when the two sides have different perspectives on the future value of the relationship. For example, contingency contracts play an important role in personal service contracts for both athletes and senior executives. The typical pay package includes a base salary plus bonuses based on future performance that is undetermined at the time the contract is signed.

In the pharmaceutical industry, contingent contracts are common. Pharmaceutical companies often use contingent contracts when they negotiate licensing a drug candidate or developing a partnership to co-develop a drug. During the negotiations, both parties often have different estimates for the probability of successfully developing and commercializing a medicine as well as different forecasts of annual sales. It is extremely difficult to bridge these differences. Therefore, the contract might have the licensee make milestone payments upon successfully completing certain phases. Royalty payments to a licensor would be paid later and be based on actual sales.

Contingent contracts also reduce risk by sharing it among two or more parties. For example, when a retailer agrees to buy a large volume of products from a vendor, it faces the risk that demand for the products will not meet the expectations. If that happens, it will be left with a pile of unsold goods. The retailer can temper its risk by offering the vendor a contingent contract. If the products' sales exceed expectations, it will share some extra profits to the vendor. However, if sales fall short of expectations, the vendor will provide a rebate on the unsold units. The sharing of upside gains and downside losses significantly reduces the risk associated with excess goods.

11.8.1   BioTech and BSG Negotiations

Consider a negotiation between a small biotechnology company (BioTech) and a global pharmaceutical company (BSG) on the marketing rights to BioTech's anticancer medicine, Astena, in the EU. BioTech has successfully developed Astena and won the approval to market it in the United States. However, the company does not have a presence in the EU. BioTech is negotiating with BSG to sell Astena's marketing rights in the EU. Biotech has asked for a total payment of $2.4 billion over the next 10 years. On the other hand, BSG has offered a total payment of only $1.1 billion.

Both companies were eager to close the deal, but after several heated debates they were unable to do so. They noticed that they had different expectations about the probability of winning drug approval in the EU and sales in that market. BioTech is very confident that Astena will win the approval in the EU, since it satisfies an unmet medical need and has already been approved by the US FDA. However, BSG thinks that winning approval from the FDA does not guarantee approval from the corresponding EU agency. They predict that there is a 90% chance that Astena will be approved in the EU.

Their sales projections also differ. BioTech projects total sales of $10–$12 billion during Astena's patent life since Astena provides better benefit risk profiles than current therapies. Though BSG accepts the claim of Astena's being better than current medicines, they think Astena's price per

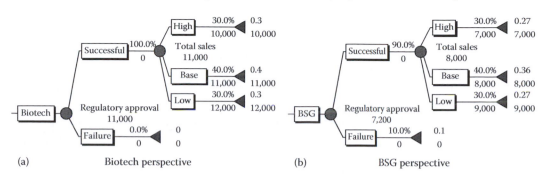

FIGURE 11.18: Total sales from perspectives of Biotech and BSG.

therapy will decline as new medicines enter the market over the next 10 years. BSG's cumulative sales forecast for Astena is between $7 and $9 billion.

The companies agreed to use a contingent contract to avoid deadlock and close the deal. Figure 11.18a shows Biotech's perspective on the probability of regulatory approval and total sales; Figure 11.18b illustrates BSG's perspective. From BioTech's perspective, there is no uncertainty regarding EU approval of the drug. It forecasts total sales to be $10, $11, or $12 billion with a probability of 0.3, 0.4, and 0.3, respectively. The excepted value is $11 billion and BioTech is asking for a royalty that is 20% of total sales. BSG believes approval is not certain and places a 0.9 probability on approval by the EU. BSG estimates that total sales could be $7, $8, or $9 billion with a probability of 0.3, 0.4, and 0.3, respectively. The expected value of sales is $8 billion, but that is conditioned on approval. Thus, the overall expected value of total sales is $7.2 billion. As they project a lower expected value, they are willing to offer BioTech only a 15% royalty on sales.

With a contingent contract, BioTech and BSG do not have to bridge their difference of opinion about future events. The differences become the core of the agreement. BioTech and BSG agreed that if sales are lower, the royalty percentage that BioTech receives would be lower, and conversely, it would be higher if sales are higher. They agreed that BSG will pay BioTech $100 million upon closing and another $50 million upon regulatory approval. BSG will make additional royalty payments on sales ranging from 12% to 22% according to a step function as listed in Table 11.9.

For example, let us consider what happens if sales are $8 billion. BSG would make a royalty payment of $1.08 billion. There is a payment of $7 billion * 12% + ($8–$7) billion * 24%. In this case, BioTech will receive a total payment of $1.23 billion, including $100 million upfront and a $50 million milestone payment. This $1.23 billion is 15.3% of the total $8 billion in sales. If sales are $11 billion, BSG would pay a royalty payment of $1.98 billion. This is based on the following calculation:

$$\$7 \text{ billion} * 15\% + \$1 \text{ billion} * 24\% + \$1 \text{ billion} * 27\% + \$1 \text{ billion} * 30\% = \$1.98$$

BioTech will receive a total of $2.13 billion ($2.35 billion + $0.15 billion). This is equal to 19.4%. For $12 billion in sales, BioTech would receive $2.49 billion, which is 20.8% of total sales. If Astena does not receive regulatory approval in EU, BSG will lose only $100 million, and BioTech will receive only $100 million from this deal.

11.9 Real Options

The real options approach coined by Myers (1977) extends from its application in finance to a wide range of real-life decisions made under uncertain circumstances. A financial option gives its holder the right to buy or sell a stock at a set price, X, on or before its expiration date. A *buy* option

TABLE 11.9: Royalty payments to biotech.

Total Net Sales	Incremental Sales Royalty Rate (%)
Up to $7 billion	12
Between $7 and $8 billion	24
Between $8 and $9 billion	27
Between $9 and $10 billion	30
Between $10 and $11 billion	33
Between $10 and $12 billion	36

turns favorable when the share price goes above X but does not obligate the owner of the option to buy the stock if the price stays below X. A *put* option works in reverse, giving the option holder the right to sell at a fixed price on or before the option expiration date. The key aspect of options is that a current dollar value can be placed on the flexibility provided by an option. Financial options are routinely traded in financial markets.

Real options models focus on an underlying source of uncertainty, such as the outcome of a research project, exchange rate, or market conditions. The outcome of the underlying uncertainty will be revealed over time. A real option is the right, but not an obligation, to undertake some business decisions such as to invest, defer, grow, contract, or abandon a capital investment contingent upon the arrival of new information. A real option deals with investments with option-like characteristics, but they are not traded as securities in financial markets. They share the characteristic of a financial option in that they provide a methodology for placing a dollar value on the flexibility the real option provides. This is critical to obtaining the support of corporate finance for decisions that do not directly generate revenue but rather reduce the risk of losses.

Real options have been suggested as a capital budgeting and strategic decision tool, because they place a dollar value on future flexibility (Trigeorgis 1996; Amram and Kulatilaka 1999). They can be applied to plans to bring a new technology to market. A technology option enables a company to bring the new technology to the market if the market is attractive, but does not obligate the company to do so if the market is unfavorable. Fund managers who invest in a manufacturing plant may also buy an option to expand the factory or alternatively sell the factory depending on market conditions.

Table 11.10 presents common real option categories, their descriptions, and application areas. A defer option refers to the possibility of waiting until more information becomes available. For example, Paddock et al. (1988) argued that there is significant value associated with the flexibility on the date to initiate a project. For example, this would be valuable for the purchaser of an offshore oil lease. He can adjust project start-up as the price of oil fluctuates over time.

Options can be used when dividing a project into stages. Staging offers the possibility to make investment in stages; based on new information, the decision maker can decide whether to proceed further or stop. An "alter operating scale" option represents the possibility to adjust the scale of the investment depending on whether market conditions turn out favorably or not. For example, a guaranteed work agreement with a union may include the option to shut down the facility for a certain period until demand increases. A switching option allows management to change the mode of operating assets depending on the price of a key resource. For example, a real option enables a power company to invest in a plant's flexibility to switch the energy source from oil to natural gas or coal depending upon their changing relative prices.

A strategic options perspective provides a proactive assessment of future opportunities under uncertainty (Bowman and Moskowitz 2001). A real options' framework is a road map that optimizes decision making by enabling managers to take multiple contingencies into account, develop responses as the uncertainty is revealed, and phase the investment accordingly. There is growing interest in real options to guide both capital budgeting and strategic decisions in dynamic environments. Many firms in finance, biotechnology, manufacturing, natural resources, R&D, and

TABLE 11.10:   Common real option types.

Option	Description	Relevant Application Industries
Defer	Project that can be postponed allows learning more about project outcomes before making a commitment	Real estate development, farming, paper products, offshore oil lease
Stage	A multistage project whose construction involves a series of cost outlays could be delayed or killed in a midstream	R&D intensive industry such as pharmaceuticals or other long development capital intensive projects
Alter operating scale	A project whose operating scale can be expanded or contracted according to market conditions	Mining, facilities planning, fashion apparel, consumer goods
Abandon	Project can be abandoned permanently when market conditions worsen severely and project resources could be sold or put to other more valuable uses	Capital intensive industries (airline, railroad), new product introduction, financial services
Switch	The project permits changing its output mix or producing the same outputs using different inputs in response to changes in the price of inputs and outputs	Any good sought in small batches or subject to volatile demand (e.g., consumer electronics, toys, machine parts)
Explore	Start with a pilot or prototype project and follow-up with a full-scale project if the pilot or prototype succeeds.	High production cost areas

information technology have adopted a real options perspective. For example, in 1990, Roche cut a market-based real options deal with Genentech in which Roche paid an option premium to buy Genentech publicly traded stock at certain dates in the future for a certain price (Lavoie and Sheldon 2000). Roche was able to mitigate any technical risk it would have assumed by doing research itself and turned it into a market-based risk handled with the use of option price. Oil companies use real options as a gauge to value unexplored oil fields. For example, Anadarko Petroleum (Coy 1999) used a real option method to outbid competitors for a tract of land in the Gulf of Mexico.

In the early-1990s, Merck wanted to enter a new line of business that required the acquisition of technology from a small biotechnology company code-named Gamma (Luehrman 1994; Sender 1994; Thackray 1995). This proposed relationship was called Project Gamma. Gamma had patented its technology but not developed commercial applications. Merck proposed licensing this technology with the goal of developing a new product. Merck projected that it would take another 2 years of R&D activities after licensing the technology. The major uncertainties included whether Merck could develop a product from this technology and the commercial possibilities if a product were developed. After completion of R&D, Merck would assess the new product's commercial viability. If the market forecast justified going ahead, Merck would need to construct a plant and make the associated marketing, working capital, and other start-up expenditures. These activities would take another year to complete. Merck employed the real options to evaluate this business relationship opportunity with Gamma (see Figure 11.19 for Merck's options and major uncertainties in this project).

Under the terms of the proposed agreement, Merck would pay Gamma a $2 million license fee over a 3 year period. Merck would also pay royalties if the product came to market. Merck had the option to terminate the agreement at any time if it was dissatisfied with the progress of the research.

Merck estimated the stock price assuming that the technology was successful and the plant was built. The stock price here refers to the value of the project and is calculated using present value of the cash flow. The stock price excluded the cash flows for building the plant and the associated start-up cost, and the upfront licensing and development costs. The exercise price was the cost of building

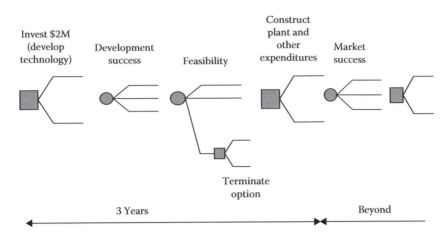

FIGURE 11.19: Merck's options and major uncertainties in Project Gama.

the plant and the associated start-up cost that would be incurred if Merck decided to commercialize technology. The cost of the option included the upfront licensing and development costs.

Merck examined the value of the option for 15 different cases. The analysis demonstrated that the value of the option exceeded the cost of the option in 13 of 15 cases. Based on the analysis, Merck agreed to license the technology and begin working on its commercial development.

Exercises

11.1 Refer to Exercise 10.5, Down-Home Restaurant, and use Table 11.11 to:

    a. Generate Tornado and Spider plots for the variables

    b. Based on the tornado plot, which variable range has the large impact? Which one has the smallest impact?

    c. Based on the spider plot, which variable has the highest rate of change?

    d. Calculate the EVPI of the competitor decision.

11.2 Refer to Exercise 10.6, the Red Hen Company, and use Table 11.12 to:

    a. Generate Tornado and Spider plots for the variables

    b. Based on the tornado plot which variable range has the highest impact?

    c. Which one has the lowest impact?

    d. Based on the spider plot, which variable has the highest rate of change?

11.3 Refer to Exercise 10.7, T-shirt business, and use Table 11.13 to:

    a. Generate Tornado and Spider plots for the variables

    b. Based on the tornado plot which variable has the largest impact? Which one has the smallest impact?

TABLE 11.11: Variable ranges for down-home restaurant.

Variable	Pessimistic Value	Optimistic Value
Layout redesign cost	+7% of base	−7% of base
Competitor open chance	+10% of base	Base
Increase in income (launch and competitor open)	−10% of base	+10% of base
Drive thru labor work	+15% of base	Base

TABLE 11.12: Variable ranges for red hen company.

Variable	Pessimistic Value	Optimistic Value
Unsuccessful campaign chance	+14% of base	−14% of base
Ad cost	+10% of base	Base
Profit if ad is successful	−15% of base	+15% of base

TABLE 11.13: Variable ranges for T-shirt business.

Variable	Pessimistic Value	Optimistic Value
Rental cost	+10% of base	−10% of base
Probability of 5500 T-shirt demand	−7% of base	Base
Sales price	−10% of base	Base
Buying equipment cost	+14% of base	−14% of base

c. Based on the spider plot, which variable has the highest rate of change?

d. Calculate the EVPI of the demand.

e. MarQuery is a marketing company that evaluates markets and predicts demand based on surveys. The result of their evaluation would be one of the following: *strong*, *ok*, or *weak*. Based on MarQuery experiments, if the demand is high, there is a 70% chance of a *strong* result, a 25% chance of an *ok* result, and a 5% chance of a *weak* result. If the demand is medium, there is a 35% probability of a strong result, a 50% probability of an ok result, and a 15% probability of a weak result. If the demand is low, there is a 10% probability of a strong result, a 30% probability of an ok result, and a 60% probability of a weak result. Find EVII of market demand based MarQuery service.

11.4 Refer to Exercise 10.8, Buy & Wear, and use Table 11.14 to:

a. Generate Tornado and Spider plots for the variables

b. Based on the tornado plot which variable has the highest impact? Which one has the lowest impact?

c. Based on the spider plot, which variable has the highest rate of change?

11.5 Refer to Exercise 10.10, U_Gov software, and use Table 11.15 to:

a. Generate Tornado plot for the variables.

b. Based on the tornado plot which variable has the highest impact? Which one has the lowest impact?

c. Calculate the EVPI of the new process cost.

11.6 Refer to Exercise 10.11, automotive part, and use Table 11.16 to:

a. Generate Tornado and spider plots for the variables.

b. Based on the tornado plot which variable has the highest impact? Which one has the lowest impact?

TABLE 11.14: Variable ranges for buy and wear.

Variable	Pessimistic Value	Optimistic Value
Online price	+15% of base	−15% of base
Selling probability	+10% of base	−10% of base
Weekly discount percentage	−7% of base	Base

TABLE 11.15: Variable ranges for U_gov software.

Variable	Pessimistic Value	Optimistic Value
Simplexo highest bid	−10% of base	Base
Probability of new process cost is $90	+10% of base	−10% of base
Simplexo lowest bid	−10% of base	+10% of base
Fixed cost	+15% of base	Base
Number of installations	−7% of base	+7% of base

TABLE 11.16: Variable ranges for automotive parts.

Variable	Pessimistic Value	Optimistic Value
Lathe 1 cost	+10% of base	−10% of base
Lathe 2 cost	+10% of base	−10% of base
Risk of lathe 1	+7% of base	Base
Risk of lathe 2	+7% of base	Base

c. Based on the spider plot, which variable has the highest rate of change?

d. Calculate the EVPI of the Lathe 1 risk.

e. Calculate the EVPI of the Lathe 1 risk.

11.7 Refer to Exercise 10.15, Specialty Brake.

a. Michael Bake is concerned about the accuracy of the estimates associated with the variable cost for both the one line and two line options. Assume the estimates could be off by 5%. What are the ramifications?

b. Of all the potential parameters, which should Michael Bake be most concerned regarding a 5% (deleterious) deviation from its base value? What 5% improvement represents the greatest opportunity?

c. Generate spider plot for the four parameters with largest impacts.

d. Based on the spider plot, which variable has the highest rate of change?

e. Calculate the EVPI of the average annual demand.

f. MarketSurvey is a marketing company that evaluates markets and predicts demand probabilities. The result of their evaluation would be one of the following: *strong*, *ok*, or *weak*. Based on MarketSurvey experiments, if the demand is high, there is a 75% chance of a *strong* result, a 15% chance of an *ok* result, and a 10% chance of a *weak* result. If the demand is medium, there is a 40% probability of a strong result, a 50% probability of an ok result, and a 10% probability of a weak result. If the demand is low, there is a 0% probability of a strong result, a 35% probability of an ok result, and a 65% probability of a weak result. Find EVII of the market demand based on the MarketSurvey service.

11.8 Refer to Exercise 10.18, Colonel Car Company case study.

a. Determine the EVPI regarding:

i. New design failure

ii. Meeting the 1 month design implementation deadline

b. If the company decides to perform the speedy life cycle test, they cannot meet the 1-month deadline. However, they will have a better understanding and more reliable information on the impact of new design before making the change decision. The life cycle test costs $60,000. If the design was good, the test will predict so with 99% reliability. However, if the design is bad, the test will find the problem with 95% certainty. Remember, initially, the design manager felt that there was only a 20% chance that the design would not perform well.

    i. Evaluate whether or not to use the speedy life-cycle test. The test currently costs $60,000. What is the maximum value you would be willing to spend on this test? Give an intuitive explanation for your results.

    ii. Construct a Tornado diagram involving the variables of your choosing for the tree. Which variable has the highest impact?

References

Ades, A. E., Lu, G., and Claxton, K. (2004). Expected value of sample information calculations in medical decision modeling. *Medical Decision Making*, 24, 207–227.

Amram, M. and Kulatilaka, N. (1999). *Real Options: Managing Strategic Investment in an Uncertain World*. Boston, MA: Harvard Business School Press.

Balson, W. E., Welsh, J. L., and Wilson, D. S. (1992). Using decision analysis and risk analysis to manage utility environmental risk. *Interfaces*, 22(6), 126–139.

Bazerman, M. and Gillespie, J. (1999). Betting on the future: The virtues of contingent contracts. *Harvard Business Review*, 77(5), 155–160.

Bell, D. E. and Schleifer, Jr. A. (1995). *Risk Management*. Cincinnati: South-Western College Pub.

Bickel, J. E. (2008). The relationship between perfect and imperfect information in a two-action risk-sensitive problem. *Decision Analysis*, 5, 116–128.

Bowman, E. H. and Moskowitz, G. T. (2001). Real options analysis and strategic decision making. *Organization Science*, 12, 772–777.

Chelst, K. and Bodily, S. (2000). Structured risk management: Filling a gap in decision analysis education. *Journal of the Operational Research Society*, 51, 1420–1432.

Clauss, F. J. (1997). The problem with optimal. *OR/MS Today*, 24(1), 32–35.

Covello, V. T. (1987). Decision analysis and risk management decision making: Issues and methods. *Risk Analysis*, 7(2), 131–138.

Coy, P. (1999). Exploiting uncertainty. *Business Week*, 7(3632), 118–123.

Engemann, K. J. and Miller, H. E. (1992). Operations risk management at a major bank. *Interfaces*, 22(6), 140–149.

Eppel, T. and von Winterfeldt, D. (2008). Value-of-information for nuclear waste storage tanks. *Decision Analysis*, 5, 157–167.

Eschenbach, T. G. (1992). Spiderplots versus tornado diagrams for sensitivity analysis. *Interfaces*, 22(6), 40–46.

Lavoie, B. and Sheldon, I. (2000). The comparative advantage of real options: An explanation for the U.S. specialization in biotechnology. *AgBioForum*, 3(1), 47–52.

Luehrman, T. A. (1994). Financial engineering at Merck. *Harvard Business Review*, 72(1), 94–97.

March, J. G. and Shapira, Z. (1987). Managerial perspectives on risk and risk taking. *Management Science*, 33, 1404–1418.

Myers, S. (1977). Determinants of corporate borrowing. *Journal of Financial Economics*, 5(2), 147–175.

Paddock, J., Siegel, D., and Smith, J. (1988). Option valuation of claims on physical assets: The case of offshore petroleum leases. *Quarterly Journal of Economics*, 103, 479–508.

Pickering, S. and Bickel, J. E. (2006). The value of seismic information. *Oil and Gas Financial Journal*, 3(5), 26–33.

Prange, M., Bailey, W. J., Couet, B., Djikpesse, H., Armstrong, M., Galli, A., and Wilkinson, D. (2008). Valuing future information under uncertainty using polynomial chaos. *Decision Analysis*, 5, 140–156.

Rothkopf, M. H. (1996). Editorial: Models as aids to thought. *Interfaces*, 26(6), 64–67, 22(6), 40–46.

Sender, G. L. (1994). Option analysis at Merck. *Harvard Business Review*, 72(1), 92.

Thackray, J. (1995). A Merck case study. *Planning Review*, 23(3), 47.

Trigeorgis, L. (1996). *Real Options: Managerial Flexibility and Strategy in Resource Allocation*. Cambridge, MA: MIT Press.

Chapter 12

Risk Attitude and Utility Theory

ENCO Ventures, an energy company, must select between two projects, both of which involve the development of a power plant abroad. For Project A, ENCO estimates that the investment cost will be $50 million and that the total revenue for the first 5 years of operation will be either $80, $90, or $110 million. There is, however, an estimated 20% chance that the local government will take over the operation of the plant once it is finished and repay ENCO only the original investment cost. Project B also requires a $50 million investment. Total revenue for the first 5 years of operation is projected at $66, $80, or $90 million. In this second country, there is no chance that the government will take over the project once it is completed.

While the expected value of Project A is $34.4 million and for Project B $28.8 million, there is a 20% chance that ENCO will not make a profit with Project A under the scenario in which the government takes over the plant. Should management choose Project A, which offers the larger expected revenue, or Project B, which offers a minimum of $16 million in net revenue after both investment and operating costs are included?

12.1 Goals and Overview

This chapter introduces the concept of translating an individual's attitude toward risk into a risk utility function. This function is then used in decision trees with the objective of maximizing the expected value of the utility function.

Individuals and organizations buy insurance to protect against the consequences of a catastrophe. Drivers are required to purchase insurance to cover the cost of being involved in a substantial personal injury accident. One of the most sought-after worker benefits an organization can provide is health insurance. Insurance companies gladly offer a wide range of policies to meet the demand for insurance. In all these instances, the expected value equation favors the insurance company. Why, then, is insurance so popular? The reason is that individuals and organizations are averse to taking major financial risks that could force a dramatic change in financial stability.

Most people would accept a gamble of winning $10 against losing $5 on a single toss of a coin. Increase the gamble to win $100 or lose $50 and fewer would accept the gamble. Increase the bet to winning $10,000 and losing $5,000 and few nonprofessional gamblers would take the chance. Yet in every case, the expected value of the gamble is positive. In essence, the subjective value to these individuals of winning $10,000 is not as great as the cost of losing $5,000.

Despite the tendency of people to be risk averse, the gambling industry is specifically designed to meet the needs of people who enjoy taking risks. Significant portions of state budgets for education are funded with government profits from lotteries. For people who gamble in the lottery, the value of $1 or $10 or even $100 they spend on lottery tickets with a high probability of losing is not as valuable as the infinitesimal probability of winning a fortune and changing their lives forever.

Managers of organizations also demonstrate risk aversion in their decisions. Although large companies may self-insure for small risks, they too buy insurance to deal with catastrophic risks. These risks may take the form of a physical catastrophe such as a massive explosion or earthquake that destroys a manufacturing facility or a massive oil spill that destroys the environment. Catastrophes could also arise as a result of a lawsuit related to product liability or patent infringement. Corporate risk aversion is also reflected in investment decisions involving risky projects. Surveys and interviews have documented common patterns of corporate risk aversion (Swalm 1966; Spetzler 1968; and Wehrung 1989). Others have analyzed actual decisions to infer the level of risk aversion (Walls and Dyer 1996; Walls 2005).

In previous chapters, we used expected monetary value to evaluate alternatives. Profit-based decisions maximize the expected profit, and cost-based decisions minimize the expected cost. We included the idea of a risk profile to at least provide a comparison of the respective range of risks. In this chapter, we introduce the concept of utility theory for decisions involving uncertainty. Utility theory involves developing a mathematical function that captures a decision maker's attitude toward risk. This function is used to transform every possible dollar outcome of a decision into a score between 0 and 1. We also include a logarithmic utility function outside the range of 0–1. The decision maker's utility function is applied to all outcomes of the decision. The decision tree methodology is then applied to maximize the expected utility (EU) score instead of the expected dollar value. By introducing a utility function, we can convert each decision's risk profile into a single numeric equivalent.

In this chapter, we develop the basic terminology and concepts of utility theory. We then apply it to several examples that are designed to illustrate why individuals and organizations purchase insurance or enter into risk-sharing partnerships. A case study involving Phillips Petroleum demonstrates their approach to developing consistent decision-making policies in the presence of significant uncertainty. We also explore how certain decision contexts produce excessive risk aversion that can distort decision making. The chapter concludes with a discussion of both the practical and theoretical challenges to the underlying axioms of utility theory.

12.2 Utility Theory: Concepts and Terminology

This section presents an overview of the basic terminology and concepts of utility theory.

12.2.1 Utility Function

The risk attitude of a decision maker can be assessed by eliciting a utility function through a structured interview. The utility function converts outcomes (e.g., cost, profit, and time) to utility units that range between 0 and 1. This function is a measure of the relative satisfaction with different outcomes. A utility function might be specified in terms of a graph, a table, or a mathematical expression.

12.2.2 Expected Utility (EU)

Once a utility function is specified, the decision tree analysis uses that function just as it would use dollar values. Instead of optimizing the expected dollar value, the decision tree optimizes the expected utility. Assume you are a contestant on the TV show *Deal or No Deal*, on which contestants are forced to choose between uncertain outcomes and a commensurate dollar offer to quit the game at hand. For instance, you have been offered $12,000 to stop playing the game. The gamble you face gives you a 50–50 chance of winning $5,000 or $25,000. The expected dollar value is $15,000, the average of the two amounts.

Assume for now that the utility function for this situation is of the form

$$U(x) = 1 - e^{-x/15000}$$

To determine the EU of this gamble, we insert \$5,000 and then \$25,000 into the equation.

$$U(x) = 1 - e^{-5000/15000} = 0.487$$

$$U(x) = 1 - e^{-25000/15000} = 0.736$$

The utility of \$5,000 is 0.487. The utility of \$25,000 is only 0.736 even though it is five times as much money.

The EU of this gamble is thus

$$0.5 * 0.487 + 0.5 * 0.736 = 0.611$$

We then compare this value to the utility associated with the offer of \$12,000.

$$U(x) = 1 - e^{-12000/15000} = 0.551$$

This last value (0.551) is less than the EU of the gamble (0.611), so you should continue playing the game. Now assume you are offered \$14,500.

$$U(x) = 1 - e^{-14500/15000} = 0.620$$

The utility of this larger amount is 0.620, which is higher than the gamble. Based on the EU, therefore, you should accept this offer and stop playing the game.

12.2.3   Certainty Equivalent

The certainty equivalent (CE) is the amount of money an individual would accept as equivalent to the risky decision. Any dollar amount offered above the CE is preferred to the risky decision. Offers of less money than the CE would lead the decision maker to stay with the risky decision. Let us refer back to the *Deal or No Deal* example. The hypothetical contestant would not accept \$12,000 to stop playing but would accept \$14,500. That means that his CE is a value between these two numbers. Because we have a mathematical function for his utility, we can invert this function to determine his CE for this gamble, which has a utility of 0.611:

$$CE = 15,000 * [-\ln(1 - 0.611)] = 14,182$$

The CE is thus \$14,182.

In a cost context, this is the maximum amount of money an entity would pay to avoid the risky outcomes. Imagine a company faces a 5% chance of a major loss of \$100,000 in a lawsuit each year. How much would it be willing to pay in insurance each year to cover that contingency? The expected value is \$5000. If a company were willing to pay up to \$7500 in insurance, then that is the CE for this risk.

12.2.4   Risk Premium

The Risk Premium (RP) is the difference between the expected value and the CE of a gamble. The RP is the amount of money an individual is willing to give up in order to avoid the risk. In the *Deal or No Deal* example, the individual was willing to accept $14,182 instead of the expected value of $15,000. The RP in this case is the difference, $812. In the lawsuit example, the difference is between an expected value of a $5000 loss and a definite insurance cost of $7500. That RP is $2500.

12.2.5   Risk Attitudes

How a decision maker deals with uncertainty depends ultimately on his attitude toward risk. A decision maker's risk attitude characterizes his willingness to engage in risky prospects. One of the fundamental axioms of utility theory is that rational decision-making requires individuals to be consistent in their risk attitude. Individuals and organizations are classified as risk neutral, risk averse, or risk prone. In practice, we find that individuals are not consistent, which has led to other ways to frame risk attitudes. These other models of risk attitude are explored in a later section.

Risk averse: A risk-averse individual or organization has a concave utility function, as illustrated in Figure 12.1. Risk-averse individuals or organizations are prepared to pay more than the expected value associated with an uncertainty to be sure costs do not become too great. Purchasing insurance is an example of risk-averse behavior. Risk aversion also applies to profits. In that case, a sure profit that is less than the expected value is preferred to the uncertainty associated with the alternative. Most individuals and organizations are risk averse when it comes to large potential losses. *Deal or No Deal* is built around the concept of risk aversion. At each stage, a contestant is offered a dollar amount to stop playing the game. That dollar amount is always less than the expected value of the values remaining on the board. Imagine a scenario with just two items left on the board, $1 and $500,000. Most people would accept an offer of $200,000 rather than proceeding with a gamble that has an expected value of $250,000.

Risk seeking: The typical gambler is risk-seeking or risk-prone. He buys lottery tickets even though the expected value of his winnings is half the price of the ticket. The Las Vegas gambler at a slot machine or blackjack table is playing against the house, which on average comes out ahead. A risk-seeking individual or organization has a convex utility function, as illustrated in Figure 12.1. Many entrepreneurs are risk prone. They repeatedly pursue ideas with a negative expected value with a small probability of major success.

Risk neutral: An individual is risk neutral if he is indifferent between the expected value of the uncertain consequences and the actual potential gamble. A linear utility function is used to reflect risk neutrality in Figure 12.1. For this type of individual, maximizing the expected value is the same as maximizing the expected utility.

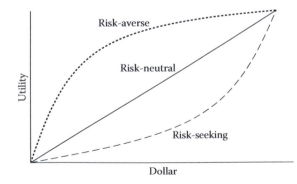

FIGURE 12.1:   Alternative utility curves.

1. **Activity:** Provide examples of risk-averse behavior.

   You pay to avoid loss (aside from buying standard insurance).

   _____

   You accept less profit but will be more certain of profit.

   _____

12.2.6 Risk Aversion Functions: Constant and Decreasing

Most people tend to be risk averse for all significant monetary levels. Managers in firms may be risk averse regarding company money. There are many kinds of utility functions that display risk aversion. In this section, however, we consider two specific functions that are widely applicable and very convenient to use. The first is an exponential utility function that exhibits a constant risk aversion, while the second, a logarithmic function, displays a decreasing risk aversion.

Constant risk aversion: Exponential function: An individual displays a constant risk aversion if he has the same positive RP for any two risky opportunities that have outcomes differing by only a constant amount. Intuitively, a constantly risk-averse individual would be anxious about taking a bet regardless of the amount of money he has in the bank. An exponential utility function represents constant risk aversion and takes the following form:

$$U(x) = 1 - e^{-x/R}$$

In this equation, e is the constant, 2.71828…, x is outcome of the event, and R is risk tolerance. The larger the R value, the flatter the curve, and the more the individual is able to tolerate risk. R, which divides x, essentially rescales the outcomes into multiples of R.

Table 12.1 shows a series of risky opportunities and their associated RPs. The R value used to create this table is $130. The first gamble is a 50–50 bet with two possible outcomes, $40 or $140. The expected value is $90. To determine the utility score in the next column, the values 40 and 140 are converted into numeric scores between 0 and 1 by using $U(x)$. Forty dollars has a utility of 0.265 and $140 has a utility of 0.659. The EU is just the average of these two values, 0.462. The CE of this EU is determined by inverting the equation for $U(x)$. The CE of $80.61 corresponds to an EU of 0.462. The difference between the expected value and CE is the RP of $9.39. Each successive row in Table 12.1 has outcomes that are $100 higher than the gamble listed just above it. In row after row, the expected values and CEs are $100 more. Since incremental values for expected value and CE are the same, there is a constant RP of $9.39.

Risk tolerance coefficient: The exponential utility function has one parameter, R, which is called the Risk Tolerance factor. There are many ways to determine the value of R (Delquié 2008). Here is a simple method that involves asking a series of related questions to determine the risk tolerance. Consider the following gamble:

Win Y dollars with a probability 0.5, or

Lose $Y/2$ dollars with a probability 0.5

TABLE 12.1: Example of constant risk aversion.

50–50 Gamble	Expected Value	50–50 Utility Scores	Expected Utility	CE	RP
40, 140	90	0.265 and 0.659	0.462	80.61	9.39
140, 240	190	0.659 and 0.842	0.751	180.61	9.39
240, 340	290	0.842 and 0.927	0.885	280.61	9.39
340, 440	390	0.927 and 0.966	0.946	380.61	9.39
440, 540	490	0.966 and 0.984	0.975	480.61	9.39

Would you be willing to take this gamble if Y were $100? Would you gamble $1000? Would you gamble $10,000? Increase the value of Y in the gamble until the decision maker is indifferent about gambling. The value of Y for which the decision maker is indifferent to taking or not taking the gamble is approximately equal to the decision maker's risk tolerance, R.

Some companies have a higher tolerance for risky ventures than others. For example, R may be $1 billion for a company with over $15 billion in sales; it may be $1 million for a small company with sales of only $20 million. Howard (1988) suggests certain guidelines for specifying a corporation's risk tolerance in terms of total sales, net income, or equity. Based on Howard's observations in the course of consulting with various companies, reasonable values for R are approximately 6.4% of total sales, or 1.24 times of net income, or 15.7% of equity. Walls et al. (1995) found the coefficient ranged between $20 and $33 million for an oil company involved in exploration with an annual exploration budget of $40 million. In their paper, they outline how to infer this parameter from a collection of corporate decisions.

Loss outcomes and constant risk aversion: The exponential utility function does not yield values between 0 and 1 when X is negative. Thus, if one of the outcomes involves a loss, there is a problem applying the exponential function. However, as pointed out earlier, the RP is unchanged when the same value is added to all possible outcomes. We can use this property of the exponential utility function to deal with losses or negative values. If one or more end points have negative values, use the following procedure:

- Determine the most negative end value "$-Y$"
- Add "$+Y$" to every end value
- Determine the optimal strategy and its CE
- Subtract "Y" from the CE

The constant risk aversion property of the exponential distribution means that this procedure will not change the decision maker's preferences.

2. Activity: Determine your personal risk aversion coefficient
   a. Would you take a 50–50 gamble: Win $10 or lose $5?
   b. Would you take a 50–50 gamble: Win $100 or lose $50
   c. Would you take a 50–50 gamble: Win $500 or lose $250?
   d. Determine R, the value at which you become ambivalent with regard to taking the 50–50 gamble: Win $R or lose $R/2?
3. Activity: Determine your organization's risk aversion coefficient. Determine R, the value at which your organization becomes ambivalent with regard to taking the 50–50 gamble: Win $R or lose $R/2?

Decreasing risk aversion: Many people and organizations tolerate greater risk as they become wealthier. For example, small companies often buy insurance against accidents while large companies with huge assets may self-insure against the same risks. Large trucking companies may self-insure against all traffic accidents not involving loss of life. They can tolerate these risks and resultant costs.

There are many utility curves that reflect decreasing risk-averse behavior. A simple and commonly used one is the logarithmic utility function. This function can be written as follows:

$$U(x) = \ln(x+A) \quad \text{for } x > -A$$

TABLE 12.2: Investment example for decreasing risk aversion for 60:40 gamble of $5,000.

A = Starting Wealth	Net Worth Outcomes	Expected Value	Utility Scores	Expected Utility	CE	RP
15,000	10,000 or 20,000	16,000	9.21 or 9.90	9.63	15,157	843
25,000	20,000 or 30,000	26,000	9.90 or 10.31	10.15	25,508	492
35,000	30,000 or 40,000	36,000	10.31 or 10.60	10.60	35,652	348
45,000	40,000 or 50,000	46,000	10.60 or 10.82	10.82	45,731	269
55,000	50,000 or 60,000	56,000	10.82 or 11.00	11.00	55,780	220

where

ln is the natural logarithm function

x is a monetary outcome of a risky opportunity

A is interpreted as total wealth at the time of the decision

Note that the logarithmic utility function is not scaled from 0 to 1. The larger the value of A, the lower the risk aversion. That is, the more cash reserves an individual has, the lower the RP he would be willing to pay to avoid or reduce risk.

We demonstrate decreasing utility function with an example that involves an investment opportunity that requires $5000. There is a 60% chance that the investment will earn an additional $5000. There is also a 40% chance that the investment will fail and the $5000 will be lost. The expected value of the net profit is $1000. We will now explore the logarithmic utility function by considering cash reserves ranging from $15,000 up to $55,000, as illustrated in Table 12.2.

Suppose the individual has $15,000 in a bank account. Then, if the investment fails, he is left with $10,000, but if it succeeds he has $20,000. The expected value of his net worth after investing is $16,000. The values in the first row of Table 12.2 are determined by using the logarithmic function and setting x equal to either ±$5,000 and A equal to $15,000. The net worth outcomes are just $(x+A)$. The natural logarithm of 10,000 equals 9.21 and of 20,000 equals 9.91. This 60–40 gamble has an EU of 9.63. We invert the natural logarithm function to find the CE, which is $15,157. This is $843 less than the expected value, and equals the decision maker's RP. In other words, when he has $15,000 in the bank, he is indifferent between a guaranteed profit of $157 and a 40–60 gamble with an expected net profit of $1,000.

Each subsequent row increases the person's cash reserves and each possible outcome by $10,000. The expected value increases by $10,000. However, the CE increases by more than $10,000. This causes the RP to decline as the person's cash reserves increase. For example, consider the case in which he has $55,000 in the bank. In this instance, the RP has decreased to $220. He is now indifferent between a guaranteed profit of $780 and a gamble with an expected net profit of $1000. In the first instance with $15,000 in the bank, if he were offered $500 instead of the investment opportunity, he would forgo the investment. In the second instance, if he were offered $500 to forgo the investment, he would gamble on the investment.

12.3 Utility Function Assessment

Different people have different risk attitudes; some are willing to take risk while others are sensitive to risk and avoid it. Hence, a utility function assessment is a matter of subjective preference. Earlier, we presented a series of questions used to estimate the risk tolerance parameter, R, in a utility function that follows the exponential distribution. Here, we present a methodology for assessing any utility function. This methodology uses CEs. In this approach, decision makers are usually asked to express preferences or indifferences between a series of 50–50 gambles and certain outcomes. The major steps of this method are outlined as follows (Samson 1988).

Step 1: Define the extreme values of the domain. This range should cover all the outcomes of the decision tree. It may, however, be wider than the current best and worst values to allow for the possibility that the data may change. Assign a utility score of 0 to the least preferred outcome and a 1 to the most preferred values.

Step 2: Obtain CE of utility values of 0.25, 0.50, and 0.75.

    a. Assess the CE that corresponds to a utility value of 0.50, CE50. The decision maker is presented with a 50–50 gamble that involves X_L, the minimum value, and X_H, the maximum value (Figure 12.2). The decision maker is asked to specify a certain amount that he would consider as equal to this gamble. In other words, a decision maker is asked for the amount for which he is indifferent between selling his opportunity to take this gamble and taking the money. People struggle with specifying a unique number and may only be able to specify an approximate value.

If we specify $U(X_L)=0$ and $U(X_H)=1$, then this gamble has a utility value of 0.50. In this instance, the decision maker specifies his CE50, such that

$$U(CE50)=0.5U(X_H)+0.5U(X_L)$$

$$= 0.5(1)+0.5(0)=0.5$$

    b. Assess the CE that corresponds to a utility score of 0.25, CE25. Specify a 50–50 gamble that involves the lowest value in the range with a utility score of 0 and the CE50, which by definition has 0.50 utility. Thus, the EU of this gamble is 0.25. The decision maker then specifies his CE for this gamble, CE25, which also has a utility 0.25 (Figure 12.3).

$$U(CE25)=0.5U(X_L)+0.5U(CE50)$$

$$= 0.5(0)+0.5(0.5)=0.25$$

    c. Assess the CE that corresponds to a utility score of 0.75, CE75. Specify a 50–50 gamble that involves the highest value in the range with a utility score of 1 and the CE50, which by definition has 0.50 utility. Thus, the EU of this gamble is 0.75. The decision maker then specifies his CE for this gamble, CE75, which also has a utility 0.75 (Figure 12.4).

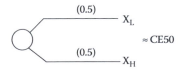

FIGURE 12.2:   CE for 0.5 utility.

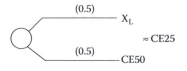

FIGURE 12.3:   CE for 0.25 utility.

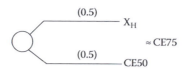

FIGURE 12.4: CE for 0.75 utility.

$$U(\text{CE75}) = 0.5U(X_H) + 0.5U(\text{CE50})$$

$$= 0.5(1) + 0.5(0.5) = 0.75$$

Step 3: Draw a curve through the five points on the utility graph with utility scores of 0, 0.25, 0.50, 0.75, and 1.

Step 4: Check the assessments and the graph for consistency. If the graph is not reasonably smooth, then check the assessments and make more assessments by designing some further gambles. We illustrate the assessment process with the data from the automation investment example introduced earlier. Boss Control (BC) is gearing up to manufacture an option to be offered on 1 million new cars worldwide. Initial estimates are that the take rate for the option could be as low as 30% or as high as 50%. The probability of a low take rate is 0.40. Relevant data for both alternatives are presented in Table 12.3.

The net profit ranges from a low of $0.8 million to a high of $10 million. As we assess the utility function, we allow for a slightly wider range of possible outcomes and set the lowest value to be 0 and the highest value to be $10 million.

Step 1: We assign the utility scores of 0 to the lowest value and 1 to the highest value $10 million. Thus, we define $U(0) = 0$ and $U(10) = 1$.

Step 2: This step specifies CE50, CE25, and CE75 that correspond to utility scores of 0.5, 0.25, and 0.75.

a. In this step, we construct a gamble in which there is a 50–50 chance to win either $0 or $10 million. We ask a decision maker the minimum amount for which he would be willing to sell his opportunity to play this gamble. There may be a need to pose various trial CEs and have the decision maker consider whether he prefers the suggested value or the gamble. For example, we can ask the decision maker whether he prefers this gamble or a guaranteed $6 million. If he prefers $5 million, we may try another (lower) CE in an attempt to find his point of indifference. Let us say that when offered $4 million, he still preferred the certainty. When offered $3.5 million, he preferred the gamble, but only slightly. This means that his CE for this gamble is somewhere between $3.5 and $4.0 million. Assume that the decision maker is indifferent between this gamble and $3.6 million. Thus, CE50 equals $3.6 million (Figure 12.5).

b. CE25 corresponds to a utility score of 0.25. Here, the decision maker has an opportunity to decide between the following gamble: there is a 50–50 chance that he will win either $0 or

TABLE 12.3: Data for BC automation investment.

Option	Investment Dollars ($Million)	Variable Cost	Take Rate	Net Sales Revenue ($Million)	Net Profit ($Million)	Expected Profit ($Million)
Low investment	8	$27 per option	30% 50%	9.9 16.5	1.9 8.5	5.86
High investment	13	$14 per option	30% 50%	13.8 23.0	0.8 10.0	6.32

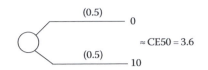

FIGURE 12.5:   BC automation investment CE for 0.5 utility.

$3.6 million. We again suggest possible values. When offered $2 million instead of the gamble, the decision maker preferred the certainty. When offered $1.5 million, he preferred the gamble. Eventually, the questioning resulted in an indifference value of $1.6 million. Thus, CE25 is $1.6 million (Figure 12.6).

c. CE75 corresponds to a utility score of 0.75. Here, the decision maker has an opportunity to decide between the following gamble: there is a 50–50 chance that he will win either $10 or $3.6 million. We again suggest possible values. When offered $6.0 million instead of the gamble, the decision maker preferred the gamble. When offered $7.0 million, he preferred the sure money to the gamble. Eventually, the questioning resulted in an indifference value of $6.2 million. Thus, CE75 is $6.2 million (Figure 12.7).

Step 3: We can draw a curve through the five points on the utility graph as depicted in Figure 12.8.

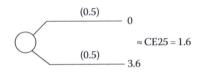

FIGURE 12.6:   BC automation investment CE for 0.25 utility.

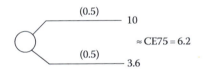

FIGURE 12.7:   BC automation investment CE for 0.75 utility.

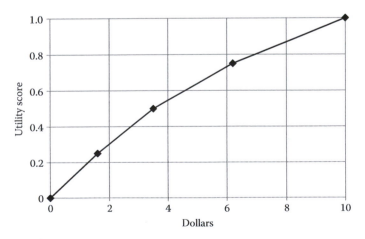

FIGURE 12.8:   Utility assessment for BC automation investment example.

TABLE 12.4: Utility scores for BC automation investment.

Value	**0**	0.8	**1.6**	1.9	**3.5**	**6.2**	8.5	**10**
Utility	**0**	0.13	**0.25**	0.29	**0.5**	**0.75**	0.90	**1**

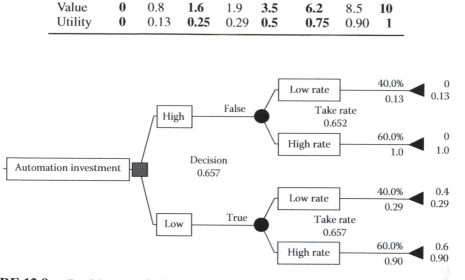

FIGURE 12.9: Decision tree for BC automation investment using utility function.

Step 4: Finally, we check the assessments and the graph (Figure 12.8) for consistency. Since the graph is reasonably smooth, we conclude that the assessment is consistent.

Using Figure 12.8, we can determine the utility score for each of the four outcomes presented in Table 12.4. The bolded values were determined through interviews. The nonbolded values were estimated by extrapolating from the graph.

Figure 12.9 shows a decision tree for a BC risk-averse decision maker. The EU scores for low and high investments are 0.657 and 0.652, respectively. The CE for the low investment is $5.19 million and for the high investment $5.14 million. Now, the low investment is preferred; however, the difference between the two CEs is only $0.05 million.

12.4 Change the Risk Equation: Insurance and Risk Sharing

Two common strategies for dealing with specific risks involve insurance and risk sharing. The impact and value of each of these strategies can be measured through the use of utility functions. We demonstrate these concepts with the ENCO Ventures case described at the beginning of this chapter. Figure 12.10 presents its decision tree.

The expected profit for each project is as follows:

$$E(\text{Project A}) = 0 * 0.2 + 30 * 43 \ (=30 * 0.30 + 40 * 0.40 + 60 * 0.30) * 0.8 = 34.4$$

$$E(\text{Project B}) = 16 * 0.30 + 30 * 0.40 + 40 * 0.30 = 28.8$$

The expected profit for Project A is $5.6 million more than for Project B. However, there is a 20% chance that the government will take over the plant; in that case, ENCO will earn nothing and just get its investment back. Project B has a minimum profit of $16 million that may prove attractive to a risk-averse manager. The respective risk profiles are presented in Figure 12.11.

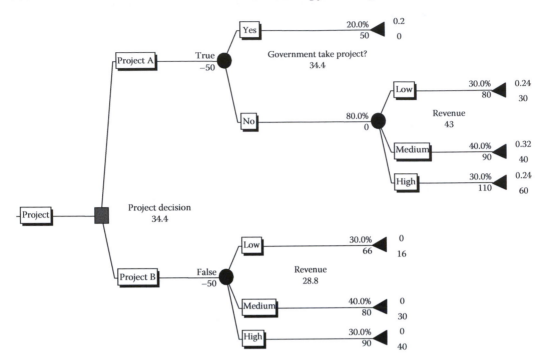

FIGURE 12.10: Decision tree for ENCO project selection.

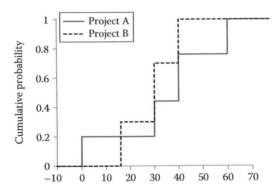

FIGURE 12.11: Cumulative risk profile for ENCO project selection. Redraw diagram without student version.

A senior manager was interviewed to determine his *R* value for use in an exponential utility function. R was found to be $30 million. In this decision tree, there are five distinct values for the outcomes. A utility score was calculated for each of the values.

$$U(0) = 1 - e^{-0/30} = 0$$

$$U(16) = 1 - e^{-16/30} = 0.413$$

$$U(30) = 1 - e^{-30/30} = 0.632$$

$$U(40) = 1 - e^{-40/30} = 0.736$$

$$U(60) = 1 - e^{-60/30} = 0.865$$

The EU values for Project A and Project B are as follows:

$$EU(\text{Project A}) = 0 * 20\% + (0.632 * 30\% + 0.736 * 40\% + 0.865 * 30\%) * 80\% = 0.595$$

$$EU(\text{Project B}) = 0.413 * 30\% + 0.632 * 40\% + 0.736 * 30\% = 0.598$$

In this analysis, Project B has a slightly higher EU, 0.598 as compared to Project A's score of 0.595. The CE of these scores is $27.322 million for Project B and $27.107 million for Project A. The CE for Project B was $1.278 million less than its expected value, which is its RP. In contrast, the RP for Project A was much larger, $7.293 million. This reflects the extreme possible outcome of 0 profit with a government takeover (Figure 12.12).

Impact of risk tolerance on certainty equivalent: Figure 12.13 shows the impact of risk tolerance on certainty equivalent (CE). Here, we use sensitivity analysis to study risk preferences. We vary the risk tolerance (R) in the exponential utility function to find the point where the decision changes. In the project selection example, as ENCO management becomes less risk-averse (as R increases), Project A becomes more attractive relative to Project B.

The current R value was set at $30 million. If the R value were higher than $31.2 million, Project A would become the best option. The optimum option is very sensitive to a small change in R value. Now, the management should be asked whether they would be willing to accept an investment in which they would have an equal chance of winning $31.2 million or losing $15.6 million. If they would not be willing to take this gamble, their risk tolerance must be smaller than $31.2 million; they should choose Project B.

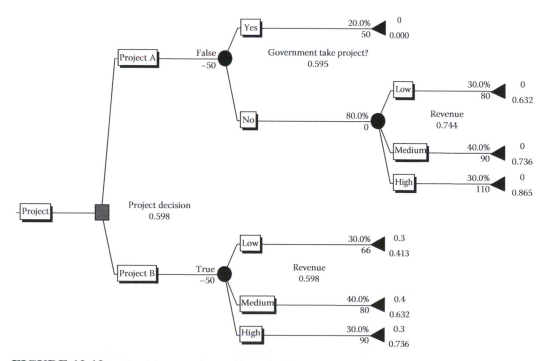

FIGURE 12.12: Decision tree for project selection example using EU.

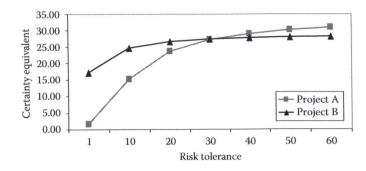

FIGURE 12.13:	Impact of risk tolerance on CE.

12.4.1   Buy Insurance

The overall attractiveness of Project A led ENCO management to explore buying insurance against a government takeover. Freud's of London was willing to offer a policy that guaranteed a $10 million payment if the government took over the project. The premium for the policy was $4 million. Figure 12.14 shows the decision tree, including the "buy insurance" option. The expected

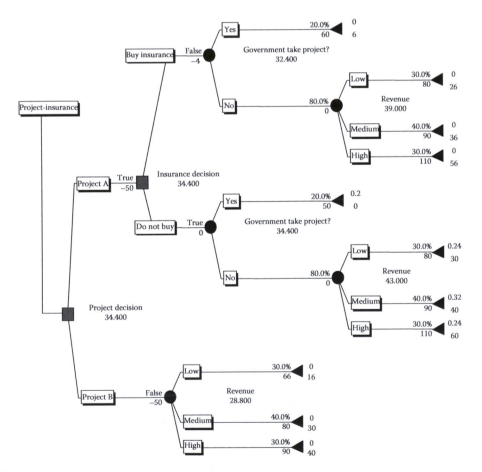

FIGURE 12.14:	ENCO decision tree with insurance option reporting expected value.

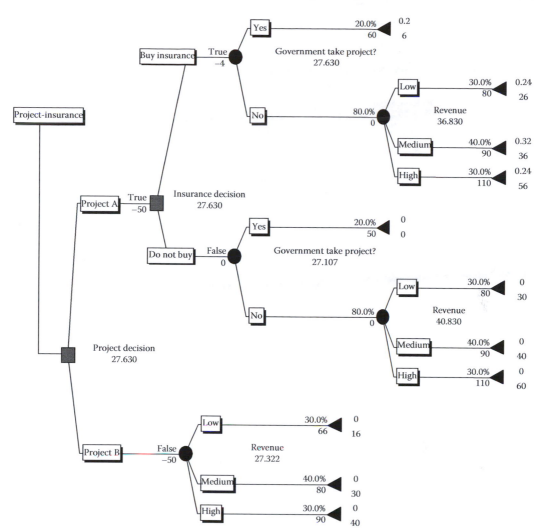

FIGURE 12.15: ENCO decision tree with insurance option reporting CE.

value of Project A with insurance is $32.4 million, which is $2 million less than no insurance. The advantage of this option is that, at least, ENCO will earn a profit of $6 million. ENCO management applied their utility function to this new tree as presented in Figure 12.15. They determined the CE for the insurance option to be $27.63 million. Without insurance, the CE for investment A was $27.107 million. Thus, alternative A with insurance is now the preferred alternative, as depicted in Figure 12.15.

12.4.2 Share Risk through Partnership

Risk sharing is a common strategy for large-scale projects involving significant uncertainty. The Alaskan pipeline was developed in the 1970s by a consortium of companies and is still owned and operated by a consortium. In 2002, Ford Motor Company and General Motors Corp. decided to invest $720 million to engineer and build a jointly developed six-speed automatic transmission. It is very common among insurance companies to reinsure and share risk of a large-scale catastrophic event. This section investigates risk sharing through partnership. In this partnership model, the company commits to only a percentage of the cost (liability) and accrues the equivalent percentage of revenue.

Consider the ENCO project selection decision. ENCO has an outside investor who is willing to invest in these two projects with the profit shared in direct proportion to the percentage of investment. Let us evaluate ENCO offering a 50% share to the outside investor. By sharing the investment, ENCO can now afford to participate in both Projects A and B. The bottom section of Figure 12.16 represents the new alternative. (To save space, parts of the tree have been compressed and are symbolized with a "+".) Its expected value is $31.6 million, which is simply the average of the revenue from each of the two projects, A and B.

When ENCO establishes a partnership with an outside investor, it improves the value of the worst case scenario from $0 to $8 million. The probability of the worst case scenario declines from 20% to 6%. This scenario occurs if the government takes away Project A (a 20% likelihood) and if Project B's revenue is low (a 30% likelihood). Even if the government takes over Project A, ENCO on average makes a profit of $14.4 million. The expected value of forming a partnership to invest in both options is $31.6 million, which is $2.8 less than investing in only Project A.

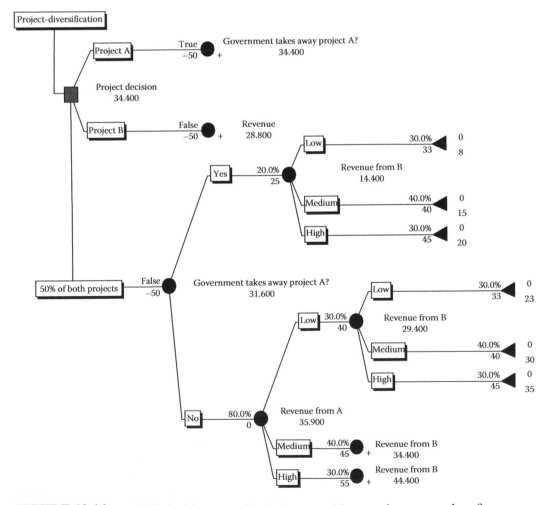

FIGURE 12.16: ENCO decision tree with 50% partnership reporting expected profit.

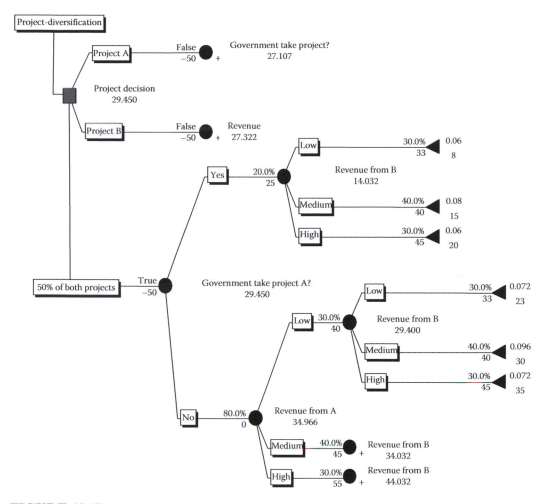

FIGURE 12.17: ENCO decision tree with partnership option reporting CE.

Figure 12.17 takes into account ENCO's risk attitude. The partnership model with a 50% share has the largest expected CE, $29.45 million. The CEs for Projects A and B are $27.107 and $27.322 million, respectively. The CE for the "buy insurance" option is $27.63 million.

12.4.3 Reduce Aggregate Risk through Diversification

The concept of diversification suggests that putting all your eggs in one basket is a risky decision. Diversification is widely used in finance as a risk management method to mix a wide variety of investments within a portfolio. Diversification minimizes the risk from any one investment as the fluctuations of a single security have less impact on a diverse portfolio. The concept of diversification is not limited to portfolios. Organizations diversify by competing in different markets, making unrelated products, and acquiring shares of other companies different from themselves. In addition, they diversify sources of supply, do business outside their home country, and outsource nonessential operations to other companies.

The ENCO decision to include a partner also offered an opportunity for diversification. With a partner, ENCO could afford to start projects in two different countries with distinct political environments. One country's political establishment was less stable and the economic payback was less certain but potentially higher. The second country was more stable but the potential rewards were lower.

12.5   Case Study: Phillips Petroleum and Onshore U.S. Oil Exploration

The management of Phillips Petroleum was an early adopter of a software system called Discovery (Walls et al. 1995). It was designed to enable them to use a consistent risk-taking policy to evaluate diverse exploration projects. Prior to the adoption of Discovery, "decision makers used informal procedures, rules of thumb, and intuition" that were not consistently applied across the corporation.

Finance theoreticians have developed the concept of a discount factor that is meant to capture the relative risk of different portfolios. Higher betas are assigned to projects with more risk and are supposed to be associated with higher average returns. However, Fama and French (2004) were unable to detect this positive association. Phillips chose not to use this approach to account for risk in their project decisions.

Discovery uses the concept of certainty equivalent to compare disparate projects. The projects to be evaluated range from those with high probability of success but low payoff to projects with low probability of success and high payoff. The management was especially concerned about projects that exposed the company to large financial losses. Discovery calculated a number of measures in addition to the CE of each project, and it estimated the expected net present value and cash flow as well as the after-tax cost if the project ended with a dry hole.

Discovery also was used to analyze different levels of investment in each project. The company could buy or sell partial interest in an exploration project as well as evaluate different royalty retention strategies. This enabled Phillips to determine the optimal level of investment in each particular project. Furthermore, Discovery evaluated the CE value of additional seismic information that might reduce risk. In some instances, they were able to determine that taking a smaller share of the project was a more cost-effective risk reduction strategy than gathering more seismic data.

The software evaluates individual projects rather than a portfolio of projects. This reflected the reality of Phillips's decision-making environment. It did not start each year with a set of projects to choose from. Rather, individual project opportunities arose throughout the year. Typically, they invested in four to six new drilling projects each year.

In modeling Phillips's corporate risk aversion, the analysts chose to use an exponential utility function. This function implies constant risk aversion. Therefore, the CE of a group of projects is just the sum of the CEs of the individual projects. The analysts were able to determine that the corporate risk aversion coefficient was between $20 and $33 million. Discovery's developers educated management about the role of risk aversion with charts that considered a range of risk aversion coefficients. In one analysis, they demonstrated that when risk aversion was low and R was more than $25 million, the optimal project investment was 100%. For R between $10 and $25 million, the optimal investment was 50%. For high-risk aversion with R less than $10 million, the optimal investment was 12.5%. (The lower the R value, the greater the risk aversion.)

Table 12.5 presents data on eight drilling opportunities for Phillips between the years 1990 and 1992. (The location names were changed to protect confidentiality.) The leftmost columns present an expected value analysis of 100% ownership of each project. The projects were ranked from highest to lowest expected value as measured in millions of dollars. The rightmost columns incorporate risk aversion and specify the optimal percentage ownership. The final column is the CE for that project assuming the recommended percentage ownership.

The South Louisiana project ranked first based on expected value but was ranked last when CE was used to rank projects. Discovery recommended that Phillips's share of the project be only 12.5%. A similar percentage was recommended for Norphlet and Frio, which had been ranked second and

TABLE 12.5:   Prospect ranking with R equal to $25 million.

Prospect	Expected Value Basis		CE Basis		
	Rank	EV 100% Share	Rank	Optimal Share (%)	CE
South Louisiana	1	18.6	8	12.5	0.6
Norphlet	2	16.5	6	12.5	0.8
Wilcox	3	11.8	5	25	0.8
Frio	4	10.8	7	12.5	0.7
Vicksburg	5	4.0	4	75	1.0
Yegua Deep	6	3.0	3	100	1.0
Smackover	7	2.5	1	100	1.8
Yegua Shallow	8	2.2	2	100	1.1

fourth based on expected value. The highest ranked project in terms of CE was Smackover. This project, along with Yegua Shallow and Yegua Deep, was recommended for 100% ownership.

12.6   Utility Theory: Practical and Theoretical Challenges

12.6.1   Practical Concerns

The application of utility theory to a real decision requires the determination of the decision maker's risk attitude function. This is typically accomplished through a structured interview. Unfortunately, decision makers are uncomfortable with the concept of uncovering their attitude toward risk. The concept of a utility function seems abstract to them. The elicitation process described earlier involves making decisions for hypothetical lotteries (Tocher 1977). Since these lotteries are not real decisions, the decision maker's judgments about the relative attractiveness of the lotteries may not reflect what he would really do.

One strategy for increasing the relevance of the interview is to embed the lotteries in examples that are directly relevant to the decision maker. In the Phillips study, the team used relevant examples in oil exploration to determine corporate risk attitude. They also reviewed prior Phillips decisions in order to infer a range of possible utility curves. In a different context, Wang (2008) developed case studies for automotive component decisions in his study of the risk attitude of American and Chinese automotive engineers.

Of greater concern is that the choice of utility assessment method can lead to different utility curves. Hershey et al. (1982) reported that the CE approach tends to result in more *risk-averse* responses than does the probability equivalent approach when the consequences are gains. On the other hand, the CE approach results in more *risk-seeking* behavior if the consequences are gains. Vrecko et al. (2009) carried out experiments that demonstrated the impact of the presentation of probability. Decision makers had different preferences when the uncertainty was presented as a probability density function as compared to when it was presented as a cumulative distribution function.

The interview process must also deal with a number of well-documented decision biases described later in this book. The biases most relevant to a utility function interview are the framing bias and the status quo bias. The framing bias points to people's greater willingness to take risk when the decision is framed in terms of gains; they are less willing to take the same risk when it is framed with reference to losses. The status quo bias can be seen in the context of buying or selling a lottery ticket. Research suggests that people tend to offer a lower price to buy the ticket than they would accept to sell it. There is a propensity to prefer the status quo; therefore, people are generally happier to retain a given risk than to take on the same risk (Thaler 1986).

Last, there is evidence a decision maker's attitude toward risk may not be a stable attribute. It could be influenced by outside factors irrelevant to the decision at hand. For example, a person's pessimistic or optimistic mood at the time of the decision could affect his risk aversion (Kliger and Levy 2003).

12.6.2 Theoretical Basis and Paradoxes

Early in the eighteenth century, Daniel Bernoulli responded to a paradox posed by his cousin Nicolas that first articulated the concept of risk-averse decision making. Bernoulli proposed the use of a nonlinear function to reflect the fact that in the presence of uncertainty, people make decisions that are not based simply on expected values. Two centuries later, von Neumann and Morgenstern (1947) developed the basic elements of utility theory. They used objective probability and lotteries to determine the relative utility of different payoffs. Savage (1954) introduced subjective probability into the equation as he developed the axioms of subjective utility theory.

These pioneering efforts attempted to model people's actual decision-making behavior while maintaining the concept of a *rational* decision maker. These theories include axioms of rationality. Most notably, decision makers should be consistent in their decision preferences. To be consistent, a decision maker's utility curve must be either risk averse, risk prone, or risk neutral. Unfortunately, for theoreticians, the reality of decision making cannot be modeled both accurately and rationally at the same time. There are a number of well-documented paradoxes, biases, and preferences that conflict with a rational decision maker who is assumed to maximize a consistent utility function. Ariely (2008) in his book *Predictably Irrational* highlights some of the more interesting irrational behaviors.

One of the earliest challenges is the Allais paradox (Allais 1953; Allais and Hagen 1979). To illustrate, suppose you were offered the choices of A (a guaranteed payoff of $1 million) and B (an 89% chance of receiving $1 million, a 10% chance of receiving $5 million, and a 1% chance of receiving nothing; see Figure 12.18). The studies show that as many as 82% of subjects prefer A over B. Apparently decision makers place a high value on the absolute certainty of option A.

Now consider the next decision (Figure 12.19), with options X and Y. This tree is modeled on the previous one in that it reduces the likelihood for each alternative of earning $1 million. The probability reduction in both cases is 0.89. For this new tree, however, 83% of the subjects prefer Y over X. The preference for Y is motivated by the significantly higher potential payout with only a small reduction in likelihood of earning a million dollars or more. But this preference for option A in decision 1 and option Y in decision 2 is contradictory, as demonstrated as follows.

Let $U(\$0) = 0$ and $U(\$5,000,000) = 1$; they are the worst and best outcomes. Then, the expected utilities for options A and B would be

$$E[U(A)] = U(\$1,000,000)$$

$$E[U(B)] = 0.10 * U(\$5,000,000) + 0.89 * U(\$1,000,000)$$

$$E[U(B)] = 0.10 + 0.89 * U(\$1,000,000)$$

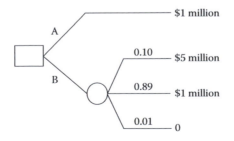

FIGURE 12.18: Certainty choice in the Allais paradox.

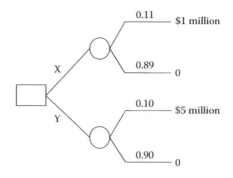

FIGURE 12.19: Uncertainty choice in the Allais paradox.

Since A is preferred to B,

$$U(\$1,000,000) > 0.10 + 0.89 * U(\$1,000,000)$$

$$0.11 * U(\$1,000,000) > 0.10$$

$$U(\$1,000,000) > 0.10/0.11 = 0.91$$

We can infer from the preference for option A that the decision maker assigns a utility greater than 0.91 to $1 million.

Now let us consider decision 2, with option Y preferred to option X.

$$E[U(X)] = 0.11 * U(\$1,000,000) + 0.89 * U(\$0) = 0.11 * U(\$1,000,000)$$

$$E[U(Y)] = 0.10 * U(\$5,000,000) + 0.90 * U(\$0) = 0.10$$

If Y is preferred over X, this implies that

$$0.10 > 0.11 * U(\$1,000,000)$$

$$0.91 > U(\$1,000,000)$$

This leads to the conclusion that the utility of $1 million is less than 0.91, which contradicts the earlier conclusion.

Ellsberg (1961) developed a different paradox that explored how people make choices when there is ambiguity with regard to probabilities. In this paradox, the contradiction revolves around the likelihood of an event occurring.

Kahneman and Tversky were pioneers in the exploration of a wide range of cognitive decision biases. Prospect theory (1979) was their initial attempt to better describe risk attitudes that take on different shapes for gains and losses. This work was later reformulated as cumulative prospect theory (1992), which better accommodated the observed tendency by decision makers to overweight a small likelihood of an extreme outcome. Starmer (2000) offers a comprehensive review of efforts to develop alternatives to EU theory that better describes decision makers' behavior.

In light of these findings, practitioners of decision and risk analysis face a fundamental dilemma. How far should they go in modeling a decision maker's actual risk attitude while recognizing that this could lead to inconsistent decisions?

12.7 Current Research in Utility Theory

Utility theory continues to be an active area of research in diverse disciplines: economics, psychology, finance, organization theory, operations research, and management science. A sample of journals in economics, psychology, and business that publish research on risk includes *The Journal of Risk and Uncertainty, Risk Analysis, The Journal of Economic Behavior and Organization, Experimental Economics, The Journal of Economic Psychology, The American Economic Review, Experimental Economics, The Journal of Mathematical Psychology, Psychological Science, Organizational Behavior and Human Decision Processes,* and *Theory and Decision.* The INFORMS journal *Decision Analysis,* launched in 2004, has become the main conduit for publishing research by operations researchers and management scientists.

One broad research topic involves the range of factors that can affect risk aversion in real-world decisions. This issue has been addressed through experiments and retrospective studies of actual decisions. One special focus involves demonstrating rank reversal of decision alternatives and identifying other forms of seemingly irrational decision choices. Another stream of research relates to the perception of risk. In classroom experiments, probabilities are provided and decision makers are asked to choose. However, real-world decisions rarely come with explicit statements of exact probability. Risky decisions can be influenced by risk perception surrounding the possible outcomes of the decision. Researchers struggle to separate the effects of risk attitude and risk perception. One area of active research involves identifying cultural differences in risk perception and risk attitude.

Baucells and Rata (2006) conducted a survey to investigate the factors that influence risk-taking behavior in real-world situations. The survey participants consisted of 77 undergraduate students, 131 MBA students, and 53 executives. The participants were asked to describe a recent decision they had made that included two alternatives: a sure alternative that did not involve uncertainty and a risky alternative that consisted of two uncertain outcomes.

The authors investigated a number of risk factors in the decision: reference dependence, domain, the default alternative, and the type of consequences. Reference dependence refers to the locus of the reference point relative to outcome. The authors then classified the decisions according to the perception of the sure outcome. Was the sure outcome perceived as a gain or loss or was it neutral? They grouped the decision into two domains, professional (human resources, job change, start MBA program), and private (safety, personal investments, buy or sell, and schedule activities).

The statistical analysis of the survey suggests that there is a strong relationship between reference dependence and risk-taking behavior as predicted by prospect theory. The logistic regression model showed that the risky option was chosen more often when the sure outcome was perceived as a loss. On the other hand, the subjects demonstrated more risk-averse behavior in gain gambles than in neutral gambles. In line with the predictions of prospect theory, the results supported higher rates of risk-taking behavior for losses as compared to either neutral or gains framing.

The model predicted that the subjects took more risks in professional decisions than in private decisions by suggesting that domain influences risk-taking behavior. In the professional domain, the decisions were made on behalf of a corporation. The observed risky behavior is consistent with decision theory, which ascribes a larger risk tolerance to corporations than to individuals.

Bickel (2006) developed three models to determine the effects on the degree of risk aversion. The models considered financial distress, external finance, and principal-agent relationship between shareholders and management. It compared these results to published risk-tolerance rules of thumb such as those noted earlier (Howard 1988).

One model explored the impact of financial distress on corporate risk tolerance. The analysis demonstrated that decreasing the cost of financial distress increased risk tolerance. Increasing portfolio variance tended to decrease risk tolerance. The results suggested that neither financial distress nor the threat of financial distress can explain the level of risk aversion reported in the decision analysis literature.

A principal-agent model was used to analyze the impact of the relationship between the shareholders and management on the corporation's risk tolerance. The results demonstrated that the corporate risk tolerance was quite sensitive to the CEO's investment level in the company and the CEO's relative risk aversion. For specific parameters, the principal-agent model could lend some support to Howard's market-value rule of thumb.

Gneezy et al. (2006) documented an unusual phenomenon in which some respondents placed a value on an uncertain lottery that was below the worst possible outcome. They labeled this the uncertainty effect and characterized these decisions as violating the internality axiom which states that decision makers should value a risky outcome as somewhere between the two extremes. Rydval et al. (2009) were concerned that misunderstandings may have contributed to this unusual preference. They therefore developed a physically more transparent protocol in which the lottery involved picking out of a closed bag one of two possible gift certificates for a local bookstore. The two certificates differed by a factor of two in their face value and had to be redeemed within 2 weeks. Similar experiments were repeated for hypothetical deferred payments. In their experiments, they did not find prevalent violations of the internality axiom.

Bruner (2009) carried out experiments to explore the different impact on a risk-averse decision maker of changing the probability of the reward or the value of the reward. The authors present their paper as "the most rigorous test, to date, of the expected utility prediction that risk averse people prefer changing the probability to changing the reward." Their experimental design confirmed this thesis. The authors also provide an overview of the potential ramifications of this research with law compliance and criminal behavior. These results would lean toward a greater deterrence effect from increasing the probability of apprehension than by increasing the level of punishment.

The manner in which individuals in power make decisions affects all of us. One issue that has been studied is whether individuals in power are more willing to take risks. Anderson and Galinsky (2006) documented that this tendency for risk taking among the powerful was linked to their optimism of achieving a positive result. Inesi (2010) designed a series of experiments to determine if activating a "power mindset" would reduce loss aversion in an independent, risky, follow-up decision. They attempted to induce a power mindset by asking the following:

> Please recall a particular incident in which you had power over another individual or individuals. By power, we mean a situation in which you controlled the ability of another person or persons to get something they wanted or were in a position to evaluate those individuals. Please describe the situation in which you had power: events, feelings, thoughts, etc.

The results of the experiment demonstrated that introduction of a power mindset did diminish loss aversion by reducing the negative value of anticipated losses. This mindset did not, however, increase the anticipated value of gains that would have been another factor in diminishing loss aversion.

Research into patterns of rank reversal is of special interest, as it continues to demonstrate a fundamental limitation of utility theory as a descriptor of decision behavior (Lichtenstein and Slovic 1971). Shaffer and Arkes (2009) carried out experiments involving cash and noncash rewards. They demonstrated that when considered together, participants preferred the cash reward alternative. However, when participants were asked to value each of the alternatives individually, they placed a higher value on the noncash alternative.

Loomes et al. (2010) were interested in assessing whether or not this type of "irrational behavior" decays with repeated exposure to markets. They developed a controlled experiment such that repeated experiences were not designed to extinguish this type of behavior. They were trying to remove the influence of a concept called "shaping." They found that rank reversal behavior did not significantly decay with repetitive involvement in markets. They also explored the decay of the gap between willingness to pay and willingness to accept. This gap is of specific concern to utility theory as it underscores a challenge as to how an individual's utility function is assessed. Unlike rank reversal, they were able to demonstrate significant decay in this gap through repeated exposure to markets.

The potential for regret and disappointment are two complementary concerns that can influence decision behavior and are discussed in greater detail in Chapter 14. Regret refers to looking back and noticing that if a different decision had been made, the outcome would have been better. Some decision makers adopt a strategy that is designed to minimize regret (Bell 1983). Disappointment is linked to expectations regarding the outcome of an uncertain event. Our satisfaction with a pay raise at the end of the year is linked to our expectations (Bell 1985; Loomes and Sugden 1986). Jia et al. (2001) expand on the work of Bell (1985) and develop a generalized risk-value model with just one parameter that is capable of modeling decisions involving lotteries with more than two outcomes. Delqui'e and Cillo (2006) are concerned with the core assumption of disappointment models in which an individual has one specific expectation against which his overall disappointment can be measured. They argue that in a lottery with multiple outcomes, each and every outcome could contribute to disappointment. Their model relaxes the assumption that a decision maker has a clearly defined prior expectation and instead assumes that the prior expectation is "fuzzy."

Exercises

Complete Chapter Activities

12.1 Provide examples of risk-averse behavior.

    a. You pay to avoid loss (aside from buying standard insurance).

    b. You accept less profit but will be more certain of profit.

12.2 Determine your personal risk aversion coefficient

    a. Would you take a 50–50 gamble: Win $10 or lose $5?

    b. Would you take a 50–50 gamble: Win $100 or lose $50

    c. Would you take a 50–50 gamble: Win $500 or lose $250?

    d. Determine R, the value at which you become ambivalent with regard to taking the 50–50 gamble: Win R or lose $R/2$?

12.3 Determine your organization's risk aversion coefficient. Determine R, the value at which your organization becomes ambivalent with regard to taking the 50–50 gamble: Win R or lose $R/2$?

Utility Theory Examples

12.4 Refer to Exercise 10.5, Down Home Restaurant. Assume the owner of the restaurant has an exponential utility function with a risk tolerance factor of 150,000:

    a. Calculate the utility of each decision when considering the competitor's action.

    b. Draw the associated decision tree.

    c. What decision should the company follow? Is it different from previous decision?

    d. What is the CE of the decision?

12.5 Refer to Exercise 10.6, the Red Hen Company. Assume the Red Hen company has a logarithmic utility function and its total wealth is 50,000,000.

    a. Draw the associated decision tree.

    b. What course of action should Red Hen follow in launching the new product if they want to maximize the utility value? Does risk aversion change the decision?

    c. Calculate RP and CE of the decision?

12.6 Refer to Exercise 10.7, T-shirt business. Assume the group has an exponential utility function with a risk tolerance factor of 40,000.

   a. Using utility values of three options, construct a decision tree to help them make their decision.

   b. Based on their utility function which option is the best for them? What is the EU if that decision is made? Does risk aversion change the decision?

   c. Calculate RP and CE of the decision?

12.7 Refer to Exercise 10.8, Buy and Wear. Assume Laila has an exponential utility function with a risk tolerance factor of 50.

   a. She is trying to decide about the $50 dress and she knows she can get the almost identical dress for $60 online. Based on the utility values, construct a decision tree to determine the week that she should buy the dress.

   b. Does risk aversion change the decision?

   c. Calculate RP and CE of the decision?

12.8 Refer to Exercise 10.9, Smart Quiz show. Assume that the contestant has an exponential utility function with a risk tolerance factor of 10,000.

   a. Draw a decision tree that can be used to determine how to maximize a contestant's EU. What is the best decision?

   b. Does risk aversion change the decision?

   c. Calculate RP and CE of the decision?

12.9 Refer to Exercise 10.10, U_Gov software. Assume U_Gov software has a logarithmic utility function and its total wealth is 20,000,000.

   a. Construct a decision tree to model this situation.

   b. Based on your decision tree, do you recommend them to submit a bid, and if so, what should they bid per installation?

   c. Under the optimal policy, what is the probability they will win the contract?

   d. What is the overall expected value if they bid on the contract?

   e. If they win the contract, what is their expected value of profit?

References

Allais, M. (1953). Le Comportement de l'Homme Rationnel Devant le Risqué: Critique des Postulats et Axioms de l'Ecole Americaine. *Econometrica*, 21, 503–546.

Allais, M. and Hagen, O., eds. (1979). *Expected Utility Hypotheses and the Allais Paradox.* Dordrecht, the Netherlands: Reidel.

Anderson, C. and Galinsky, A. D. (2006). Power, optimism, and risk-taking. *European Journal of Social Psychology*, 36, 511–536.

Ariely, D. (2008). *Predictably Irrational.* New York: Harper Collins.

Baucells, M. and Rata, C. (2006). A survey study of factors influencing risk-taking behavior in real-world decisions under uncertainty. *Decision Analysis*, 3, 163–176.

Bell, D. E. (1983). Risk premiums for decision regret. *Management Science*, 29, 1156–1166.

Bell, D. E. (1985). Disappointment in decision making under uncertainty. *Operations Research*, 33, 1–27.

Bickel, J. E. (2006). Some determinants of corporate risk aversion. *Decision Analysis*, 3, 233–251.

Bruner, D. M. (2009). Changing the probability versus changing the reward. *Experimental Economics*, 12, 367–385

Delqui'e, P. (2008). Interpretation of the risk tolerance coefficient in terms of maximum acceptable loss. *Decision Analysis*, 5, 5–9.

Delqui'e, P. and Cillo, A. (2006). Disappointment without prior expectation: A unifying perspective on decision under risk. *Journal of Risk and Uncertainty*, 33, 197–215.

Ellsberg, D. (1961). Risk, ambiguity, and the Savage axioms. *Quarterly Journal of Economics*, 75, 643–669.

Fama, E. F. and French, K. R. (2004). The capital asset pricing model: Theory and evidence. *The Journal of Economic Perspectives*, 18(3), 25–46.

Gneezy, U., List, J. A., and Wu, G. (2006). The uncertainty effect: When a risky prospect is valued less than its worst outcome. *Quarterly Journal of Economics*, 121, 1283–1309.

Hershey, J. C., Kunreuther, H. C., and Shoemaker, P. J. H. (1982). Sources of bias in assessment procedures for utility functions. *Management Science*, 28, 936–954.

Howard, R. A. (1988). Decision analysis: Practice and promise. *Management Science*, 34, 679–695.

Inesi, M. E. (2010). Power and loss aversion. *Organizational Behavior and Human Decision Processes*, 112, 58–69.

Jia, J., Dyer, J. S., and Butler, J. C. (2001). Generalized disappointment models. *Journal of Risk and Uncertainty*, 22(1), 59–78.

Kahneman, D. and Tversky, A. (1979). Prospect theory: An analysis of decision under risk. *Econometrica*, 47, 263–292.

Kahneman, D. and Tversky, A. (1992). Advances in prospect theory: Cumulative representation of uncertainty. *Journal of Risk and Uncertainty*, 5, 263–292.

Kliger, D. and Levy, O. (2003). Mood induced variations in risk preference. *Journal of Economic Behavior and Organization*, 52, 573–584.

Lichtenstein, S. and Slovic, P. (1971). Reversals of preference between bids and choices in gambling decisions. *Journal of Experimental Psychology*, 89, 46–55.

Loomes, G., Starmer, C., and Sugden, R. (2010). Preference reversals and disparities between willingness to pay and willingness to accept in repeated markets. *Journal of Economic Psychology*, 31, 374–387.

Loomes, G. and Sugden, R. (1986). Disappointment and dynamic consistency in choice under uncertainty. *Review of Economic Studies*, 53, 271–282.

Rydval, O., Ortmann, A., Prokosheva, S., and Hertwig, R. (2009). How certain is the uncertainty effect? *Experimental Economics*, 12, 473–487.

Samson, D. (1988). *Managerial Decision Analysis*. Homewood, IL: Irwin.

Savage, L. J. (1954). *The Foundations of Statistics*. New York: John Wiley & Sons.

Shaffer, V. A. and Arkes, H. R. (2009). Preference reversals in evaluations of cash versus non-cash incentives. *Journal of Economic Psychology*, 30, 859–872.

Spetzler, C. S. (1968). The development of a corporate risk policy for capital investment decisions. In: *The Principles and Applications of Decision Analysis*, Howard, R. A. and Matheson, J. E., eds. Menlo Park, CA: Strategic Decisions Group, pp. 665–688.

Starmer, C. (2000). Developments in non-expected utility theory: The hunt for a descriptive theory of choice under risk. *Journal of Economic Literature*, XXXVIII, 332–382.

Swalm, R. O. (1966). Utility theory-insights into risk taking. *Harvard Business Review*, 44, 123–136.

Thaler, R. H. (1986). *Illusions and Mirages in Public Policy in Judgment and Decision Making: An Interdisciplinary Reader*. Arkes, H. and K. Hammond., eds. Cambridge, U.K.: Cambridge University Press.

Tocher, K. D. (1977). Planning systems. *Philosophical Transactions of the Royal Society of London*, A287, 425–441.

Von Neumann, J. and Morgenstern, O. (1947). *Theory of Games and Economic Behavior*. 2nd edn. Princeton, NJ: Princeton University Press.

Vrecko, D., Klos, A., and Langer, T. (2009). Impact of presentation format and self-reported risk aversion on revealed skewness preferences. *Decision Analysis*, 6, 57–74.

Walls, M. R. (2005). Corporate risk-taking and performance: A 20 year look at the petroleum industry. *Journal of Petroleum Science and Engineering*, 48(3–4), 127–140.

Walls, M. R. and Dyer, J. S. (1996). Risk propensity and firm performance: A study of the petroleum exploration industry. *Management Science*, 42, 1004–1021.

Walls, M. R., Morahan, G. T., and Dyer, J. S. (1995). Decision analysis of exploration opportunities in the onshore U.S. at Phillips Petroleum Company. *Interfaces*, 25(6), 39–56.

Wang, G. (2008). Exploring cross-cultural differences in engineering decision making, Doctoral Dissertation, Wayne State University, Detroit, MI.

Wehrung, D. A. (1989). Risk taking over gains and losses: A study of oil executives. In: *Choice under Uncertainty, Annals of Operations Research*, Fishbum, P. C. and LaValle, I. H., eds. Vol. 19. Basel, Germany: J. C. Baltzer Scientific Publishing Company.

Part IV

Challenges to "Rational" Decisions

Chapter 13

Forecast Bias and Expert Interviews

Les Allais School District had been performing poorly on standardized math tests. Donald Quixote, the Superintendant of Les Allais, proposed to the school board a new mathematics curriculum for the six area high schools with 12,000 students. All 100 teachers were to be trained over the summer with the launch of the new curriculum in the fall. He predicted a significant impact within 18 months. After 2 years, there had been very modest gains. Dr. Quixote blamed the slow progress on the teachers he claimed had not fully embraced the new approach.

The VP of a German software company pushed to purchase a software support services company in Bangalore, India. The plan was to reduce the cost of global technical support for its major product lineup by eliminating its British-based English language support services. After 12 months, an analysis indicated that costs of support had gone down somewhat but that they had lost a number of major clients who had difficulty getting critical problems solved in a timely fashion.

13.1 Goals and Overview

This chapter addresses a core issue in decision trees, the accurate specification of subjective probabilities. The chapter explores a range of biases that undermine accuracy. It concludes with a description of an expert interview process designed to reduce the biases.

All decisions require data, estimates, and forecasts. Nevertheless, few complex decisions are identical. New product launches face different economic and competitive climates from year to year. New drugs address different medical concerns and entail varying risks. Each potential site for a new stadium or airport faces different traffic flow challenges and real estate development costs. Rarely are there enough data to resolve these concerns with certainty. As a result, estimates and projections rely significantly on expert opinions as well as subjective judgment based on years of experience. However, life and work experiences are contingent on what we call the "Law of Small Numbers." We all live but one life. Our individual and even our team's collective experiences are limited and likely to include a collection of rare but memorable events that can misguide the decision makers. These biases affect the way we pursue, receive, and interpret data. This chapter focuses on how these biases affect our forecasts and perceptions of the future. In the following chapter, we explore a complementary set of *decision making* biases that relate to how we make choices. The primary goal of these chapters is to teach the reader how to make estimates and decisions that are less biased and more robust by highlighting and clarifying the presence of common subconscious patterns that affect both individuals and groups.

When picking a kitchen contractor or a component supplier, there will be data on many key variables such as cost. However, estimates of quality, creativity, and timeliness of work will be a function of both objective information about past performance and subjective estimates of future behavior. Projections for demand for a new product will be based on existing data for similar products plus an assessment of the future competitive marketplace and economic factors. Similarly, estimates of how long it will take to deliver a nonstandard project or implement a new technology can at best be estimated using expert judgment infused with a wide range of past experiences.

Unfortunately, both the theoretical literature and business experience are replete with a wide range of biases that distort projections (Hammond et al. 1998). The theoretical literature is based primarily on structured experiments performed with students in a variety of settings. For example, the literature documents the widespread misunderstanding of basic concepts of probability theory, especially with regard to multiple random events and conditional probability. This is especially problematic as many decisions are made in the presence of multiple uncertainties. The literature also describes how a substantial majority of college students perceive themselves as above average relative to their peers. Sixty percent even rate their ability in the top 10%.

The business literature includes story after story of flawed decisions based on unrealistic projections. Who will ever forget the legend of the Big Dig in Boston? The original estimate was $2.6 billion, but ultimately the project cost more than five times as much, $14.8 billion. (Even when allowing for inflation, over the more than 20 years that elapsed from approval to completion, the final cost was more than triple the original estimate.) Lovallo and Kahneman (2003) report that 70% of new manufacturing plants in North America close within a decade. A RAND study of 44 chemical plants indicated that actual construction cost was double the original estimates and often the plants produced at less than 75% of their design capacity. Unfortunately, cost and time overruns on major projects—especially involving public infrastructure—are all too common. This is a global problem (Flyvbjerg et al. 2003).

Similarly, the literature on mergers and acquisitions documents the "delusional optimism" of executives who believe they will succeed where many others have failed under similar circumstances (Tetenbaum 1999; Cartwright and Schoenberg 2006). Every merger or acquisition comes with projections of savings, synergies, and short timelines that will be needed to achieve the benefits. All too often, however, these timelines prove to be unrealistic and the benefits are illusory (Sirower 1997). Three-quarters of mergers/acquisitions never achieve stated goals. More than half the time, the stock price drops immediately. Clearly, the collective judgment of the stock market disagrees with the executives' projections. In the extreme, as in the Daimler Chrysler merger, tens of billions of equity was destroyed before the merger was reversed and the two companies separated. The Chrysler saga became protracted when financiers with limited automotive experience believed that they could work magic where others had failed. These financiers bought Chrysler and a year later the company was headed toward bankruptcy.

In this chapter, we will review the wide range of factors that can distort and bias the forecasting process. We will then describe a structured interview designed to address most of these biases. Realistically, few readers are likely to use this structured process. We will, therefore, highlight the most critical and common concerns that should be taken into account even without a structured process. Hopefully, your organization can avoid the dysfunctional situation that arises when corporate executives do not believe the forecasts of technology or marketing experts, choosing instead to abide by some unannounced adjustment factor that has been projected by the latest market guru.

There is a long list of potential forecasting biases (Gilovich et al. 2002; Bazerman 2006). We have chosen to categorize them broadly as follows:

- Motivational and personal
- Point estimates and overly narrow ranges
- Errors in probabilistic thinking
- Availability and representative
- Confirmation

There is overlap among the categories mentioned earlier. For example, a misunderstanding of probability theory contributes to range estimates that are too narrow and is also a key factor in the representative bias.

13.2 Motivational and Personal Biases

Motivational bias is perhaps the most widespread and most difficult bias to address. This bias arises when the originally projected numbers are needed to approve a project or confirm an earlier decision. Few complex projects can move forward without a detailed forecast to justify the project's value on a financial basis or to meet some other organizational need. Many organizations, both public and private, have specific target values that must be matched or exceeded in order to qualify for project approval. Typically, a new product launch, especially one involving significant upfront capital investment, must exceed a specific corporate return on investment (ROI) for approval. The product champion and his dream team are motivated to overestimate sales in order to obtain approval. If cost and timing are also of concern, there is pressure to underestimate cost and time projections. Additionally, if the senior executives like the project for multiple reasons, they will simply not question the forecasts.

Large public sector projects are especially prone to motivational bias. Politicians seeking to justify a project for their constituency or who consider a project to be in the public interest fear that projecting accurate numbers will undermine public or peer support for their efforts. Mass transit projects, for example, routinely overestimate future ridership (Kain 1990; Love and Cox 1991). Their motivation bias is twofold. Higher projections of ridership mean that more people will benefit. Second, the number of prospective riders directly impacts projected revenues and reduces the need for long-term subsidies.

The other side of the equation is the motivation to underestimate the cost of a project, in terms of time and money. Massachusetts, for example, underestimated the time needed to complete the "Big Dig" and Jerusalem, likewise, seriously underestimated the time needed to complete their light rail project. This underestimation has two motivations. For one thing, people are more likely to support a project if they believe that the benefits will come sooner rather than later. In addition, many large public construction projects cause serious inconveniences during the construction phase, and it is in the interest of the backers of such projects to downplay this inconvenience. The light rail project of Jerusalem is disrupting downtown business traffic for 4 years instead of the projected two. The longer the disruption, the more likely a project will engender organized resistance from the people and organizations that are most inconvenienced. Interestingly, one approach to writing construction contracts is designed to address this. Many road repair projects come with clauses that provide bonuses for early completion and penalties for late delivery.

The same principles apply to R&D or product development projects in the private sector. The emphasis on short-term corporate goals places pressure on project initiators to unrealistically forecast early delivery of the finished product. In the process, they disregard all sorts of hurdles that will randomly occur. The Boeing Dreamliner is an extreme example of underestimating the challenges of working with both a new global supply chain and a totally new material for the frame of the plane. The original delivery forecast was for 2008 and first deliveries are now forecasted for 2011.

Motivational bias plays an interesting role in stock market vicissitudes. Short-term stock gains and losses are influenced by a company's performance relative to expectations. Thus, successful companies may tend to underestimate predicted performance in the next quarter, so that when they report actual performance they have exceeded expectations. Conversely, distressed companies may overstate projected losses. Then, when these losses turn out to be less significant than originally projected, their stocks experience a short-term boost. In a different vein, some executives, such as the corporate officers of Lehmann Brothers, have been known to cover up their companies' failings, biding their time while hoping desperately for a *deus ex machina* to put an end to the company's financial woes. Needless to say, sophisticated investors are not oblivious to these machinations and attempt to allow for them in their purchasing and selling patterns. In addition, federal rules officially call for a degree of integrity when making these forecasts.

1. *Activity*: Motivational bias

   Provide an example of a forecast within your organization or your community that illustrates the motivational bias. Describe the motivation that contributed to the bias.

Interestingly, one must also be aware of an unrealistic optimism bias that can develop and become compounded by success after success. This dynamic is characterized as the Law of Increasing Optimism. It was this dynamic that led NASA engineers to underestimate the risks associated with Shuttle launches. They had seen repeatedly that pieces of the outside foam insulation had broken off at launch. However, none of these events had ever caused catastrophic damage to the shuttle itself. That is, until this phenomenon damaged Space Shuttle Columbia leading to its disintegration upon reentry.

There is, of course, a positive value in optimism. Leaders believe that rallying the troops requires exuberance and optimism. Who wants to work on a project with only a 10% chance of success? How are you going to get people to work harder and longer hours on a project that is estimated to be completed 4 years down the road? What teachers will work on developing a new curriculum that will take 7 years to implement?

Another positive benefit of optimism involves the concept of a self-fulfilling prophecy, which, in the context of education and experimental design, has been labeled the Pygmalion Effect (Rosenthal and Jacobson 1992). (Pygmalion in Greek mythology created a statue of a beautiful woman that eventually came to life.) There is significant scientific data to indicate that teacher expectation of student performance and growth has a dramatic impact on the actual progress the students make over the course of the year. This is part of a much broader literature on the role of motivation in outstanding and even extraordinary achievement (McNatt 2000).

Entrepreneurs are known to have this optimistic bias (Fraser and Greene 2006). How many people would open a specialty restaurant if they believed the low probability of success. R&D researchers are a special breed as they work for the thrill of discovery even though the odds are against their work ever turning into a successful product. Interestingly, it has been noted that "unrealistic optimism" is a sign of personal good mental health. It also helps people persist in a difficult task (Taylor and Brown 1988, 1994; Taylor 1989). Consider, if you will the children's story, "the little engine that could."

Unfortunately, unrealistic optimism can be a counterproductive business process that may lead corporate executives to make truly bad decisions. The CEO of Federal Mogul, for example, bought a company that had a known asbestos problem and liability. He was sure he could control the magnitude of the product liability lawsuit and bought the company at what he considered a distressed price. In the end, however, the product liability costs overwhelmed the otherwise profitable Federal Mogul and drove it into bankruptcy.

Decisions often involve not only a forecast of external events and factors but also an internal assessment of a person's own ability and his team's capability to accomplish a particular task within a specific time period. In addition to the common optimistic bias noted earlier, there is an even more widespread bias when it comes to self-assessment (Brawley 1984; Brinthaup et al. 1991; Kramer et al. 1993). Many of us have a tendency to give ourselves more credit than we deserve, and the typical worker tends to assess his own performance as better than average (Diekmann 1997; Kruger and Dunning 1999; Bazerman 2006). If we believe we are above average in performance, then our forecast will reflect that assessment. We will be biased toward shortening our time completion estimates relative to the average and increasing our estimate of the probability of success. A corollary to the above is a tendency to denigrate our competitors' ability (Diekmann 1997). Thus, automotive suppliers must struggle with individual corporate estimates of car sales that, when added together, are likely to greatly exceed any realistic industry-wide sales volume.

2. *Activity*: Unrealistic optimism bias

Provides an example of an unrealistic optimism bias in your personal life, organization, or community.

13.2.1 Overcoming Motivational Bias

Motivational bias can be addressed at both the organizational and the project level. At the organizational level, this bias can be dealt with if there is a process in place to complete post reviews of all projects and compare them to the original forecasts. This review will quickly uncover systemic or individual patterns of motivational bias. Individuals or organizations that routinely provide unrealistic estimates will then be held accountable. In addition, a well-understood consistent process should be created to evaluate frequent categories of forecasts. Forecasts of new product demand might be assigned to a marketing group, project completion estimates to an engineering group, and financial estimates to a corporate wide finance group. Needless to say, these organizational forecasting assignments must include accountability as to the accuracy of the projections that are made. However, it should be easier to track and improve forecasts if one group is assigned responsibility. Later, in this chapter, we emphasize the need for forecasts to be expressed as ranges along with explanations rather than single point estimates coupled with justifications. Unfortunately, none of these strategies is a panacea in a corporation or organization that is rife with political maneuvering to get projects approved.

At the project level, an individual or group with no vested interest can be assigned to develop its own projections or at least critically review the basis for the current projection. The Congressional Business Office (CBO) is a classic example of an organization that was created for that specific purpose. In 2009, they played a critical role in vetting every proposal for a new health care initiative. In addition, at the national level, there are often institutes with competing political perspectives that provide alternative assessments—each with its own motivational bias. The hope is that a careful review of the conflicting views can provide better assessment of the value of the project in question. Unfortunately, few governments—at either the local or state level—can provide independent assessment of forecasts. In these contexts, it is often left to local news media to question the projections.

A change in leadership is a common means for overcoming many forms of forecast and decision bias. The new leader generally has a less vested interest in the original decision and the accompanying forecast. As long as this new leader does not arrive with his own preconceived view of the project or product, he can challenge the basis for the forecasts in a less biased manner.

13.3 Point Estimate and Narrow Ranges: Overconfidence

An organization has a forecasting problem if all estimates are expressed in single point numbers. Strangely enough, it requires less knowledge to create a single-point estimate than to produce a probabilistic range. A single-point estimate simply requires selecting specific assumptions and corresponding numbers that can be combined to justify the forecast. A probabilistic range or confidence interval, on the other hand, requires the forecasters to draw on a wide spectrum of experiences to define a range as well as to explore its associated probabilities. They also have to review

TABLE 13.1: Translate phrases into probabilities.

Phrase	Assign Probability	Phrase	Assign Probability
Likely	——	Small chance	——
Not likely	——	Reasonable chance	——
Poor chance	——	Most likely	——
Doubtful	——	Nearly certain	——
Perhaps	——	Possible	——

these experiences to determine underlying random fluctuations in order to determine how past experiences align with the current forecasting problem.

Corporate business systems may prove to be barriers to adopting a range approach to forecasting. For example, a particular company may use a complex spreadsheet to determine the capital investment that can be allocated to a project. Single-point estimates facilitate making this upfront allocation decision. It is more challenging to determine the investment in the presence of significant uncertainty as to the project's ROI. It requires a corporate culture and leadership that can tolerate and even embrace this ambiguity.

In some contexts, the forecast is not expressed as a quantitative value such as dollars or time, but rather in terms of the likelihood of meeting some goal or predetermined deadline. When discussing such issues, it is critical that team meeting participants refrain from using such terms as *likely*, *almost certain*, or *little chance*.

3. *Activity*: Translate words to probabilities

A project team has just reviewed plans for the launch of a new product. Each member of the team was asked to state in one word or phrase what they thought the chance of success was. Please interpret and assign a probability to each of the 10 words or phrases in Table 13.1.

In Appendix A, we present data on the range of values reported in a series of experiments (Beyth-Marom 1982). The term Likely had the widest range. The range for 80% of the respondents was from 0.42 to 0.81. However, because people routinely use verbal expressions of uncertainty, researchers have attempted to align these terms to specific probabilities (Wallsten et al. 1993; Timmermans 1994).

To illustrate the concept of a probabilistic range, complete Table 13.2. Each question asks you to create a 90% confidence interval around your knowledge or lack of knowledge of a specific value. The first question asks you to estimate Alexander Hamilton's age when he died. The 0.05 fractile value means that you believe there is only a 5% chance that Hamilton was younger than the number you specified. The 0.95 fractile means that you believe that there is 95% likelihood in your opinion that Hamilton was younger than the age you specified. The range between the two end fractiles is 90%. If you are accurate and consistent, when you look up the answers at the end of the chapter, you should, on average, find that 9 out of the 10 actual answers are within your range and one is outside it. If only five of the actual values are inside your range, then you were overly optimistic about your knowledge. If all 10 were within your range, then you were overly broad in your estimates. Obviously, if you specify 0 for 0.05 fractile and a trillion for the 0.95 fractile, all 10 values will lie within your range of estimates. Use this test to evaluate your assessment of your belief.

4. *Activity*: Specify your estimate of the low and high fractiles for each question (Table 13.2). Check your answers with Appendix B

The population of Turkey in 2010 is estimated to be 77 million. Provide a confidence interval on the population of Greece, its neighbor, with whom it is in long-term conflict over the division of Cyprus.

TABLE 13.2: Estimate confidence intervals.

	0.05 Fractile	0.95 Fractile
	(Low)	(High)
1. Alexander Hamilton's age at death	———	———
2. Volume of water in Lake Michigan (gal)	———	———
3. Population of United States in 1800 census	———	———
4. Number of chapters in Book of Psalms	———	———
5. Sun's volume as a multiple of Earth	———	———
6. Weight of empty Boeing 747 (lb)	———	———
7. Year of Da Vinci's birth	———	———
8. Average height of giraffe at birth (cm)	———	———
9. Air distance: Detroit to Buenos Aires (miles)	———	———
10. Depth of deepest point of Lake Superior (ft)	———	———

	0.05 Fractile	0.95 Fractile
	(Low)	(High)
2010 Population of Greece	———	———

There are more than 1 billion Muslims and more than 1 billion Christians in the world. How many Jews are there?

	0.05 Fractile	0.95 Fractile
	(Low)	(High)
2010 World's Jewish Population	———	———

The activities mentioned earlier are meant to illustrate several points. Researchers have found that approximately half of the actual answers lie within their estimated range and not 90% (Klayman et al. 1999). People tend to overestimate their ability to forecast and thus offer too narrow a range. The width of your range should reflect your state of knowledge. The less certain a person is, the wider the range that he should specify. The aforementioned example involves general knowledge. The overconfidence bias has been generally confirmed with data from experts such as doctors, lawyers, and businessmen (Koehler et al. 2002). However, there is evidence that people's confidence intervals are not as poorly calibrated when dealing with issues about which they are knowledgeable (Budescu and Du 2007). This distinction is critical since we are primarily concerned with interviewing experts in their areas of expertise.

The second example is designed to illustrate the concept of anchoring based on a somewhat irrelevant number. The population of Turkey should have no impact on an estimate of the population of Greece. Similarly, the estimates of the number of Muslims and Christians should not directly affect the estimate of the number of Jews in the world. Nevertheless, people tend to produce much higher estimates when these numbers are provided. In fact, there is much evidence to indicate that random numeric information can influence forecasts (Epley 2004).

Although the first example is often used to demonstrate the tendency of specifying too narrow a range, the questions themselves are not reflective of the process to be used in providing estimates. The problem with this example is that the typical participant has little or no knowledge to draw upon to make the range of estimates. In real-world decisions, we generally ask an "expert" to draw on a range of experiences to estimate, for example, the time it might take to complete a construction project or achieve a certain level of market penetration. Ideally, the expert would have extensive background that would include both bad and good experiences. This issue is elaborated on later as we describe the recommended interview process.

There are a number of specific biases that contribute to range estimates that are much narrower than experience would suggest. These biases include:

- Illusion of control
- Illusion of predictability and linear extrapolation
- Anchoring
- Misunderstood extremes
- How long completion of a project *should* take in contrast to how long it *will* take

13.3.1   Illusion of Control

Executives usually rise to their positions of authority as a result of a steady stream of successes. As a result, they tend to develop overconfidence in their ability to control events. The continuing saga of Boeing's delayed development and launch of the Dreamliner is a classic example of over-confidence. Top management recognized that there were challenging, almost unprecedented, elements in the Dreamliner's development and launch. The product development and manufacturing supply chain were significantly more global than anything the company had tried before. Most of the testing was to be carried out through computer-aided modeling rather than complete physical prototypes. Last, the designers were introducing widespread use of composite materials in the body of the airplane. Nevertheless, management believed that all of these challenges could be handled within an aggressive time frame. Their belief, however, was not realized. Instead, the aforementioned challenges have given lie to the company's overly optimistic forecast; the latest projection of delivery is already more than 2 years late. This unrealistic estimate has cost the company a fortune in order cancellations and penalties for late delivery.

Analogously, executives heading the merger of Daimler and Chrysler believed that they could control and efficiently manage the integration of two very different corporate cultures. The transformation was never successfully realized, and ultimately the merger was reversed. In 1998, Federal Mogul purchased the British firm, T&N plc, a manufacturer of brake materials. The buyers knew that T&N faced asbestos-related lawsuits but believed that Federal Mogul could manage the risk by setting aside funds to cover possible risk and liability. Instead, the mounting liability precipitated Federal Mogul's filing for bankruptcy in 2001. This illusion of control is quite widespread among those who undertake corporate mergers and acquisitions, and this is a key reason why most mergers do not deliver the projected synergies.

A common mistake made when developing a projected timeline is to focus on how long a project *should take* rather than how long it *will take*. Kahneman and Lovallo have labeled the *should take* forecast the *inside* view. The inside view focuses on all the required tasks and estimates how long each should take while identifying obstacles to overcome. It may even consider a few likely scenarios. Nonexperts looking at the same tasks are even less accurate in their estimates as they question why tasks should take so long.

The outside view steps away from the specifics of the situation at hand and reflects on how long similar projects have taken in the past. The expert draws on a wide array of factors that have delayed such projects. Although the expert cannot point to what specifically will delay the current project, he knows that there are a variety of reasons that a venture of this type tends to take 25% or 50% or even 100% longer than it should. Kahneman and Lovallo (1993) illustrate this point with a relevant anecdote.

> In 1976 one of us was involved in a project designed to develop a curriculum for the study of judgment and decision making under uncertainty for high schools in Israel. The project was conducted by a small team of academics and teachers. When the team had been in operation for about a year, with some significant achievements already to its credit, the discussion

turned to the question of how long the project would take. To make the debate more useful, I asked everyone to indicate on a slip of paper their best estimate of the number of months that would be needed to bring the project to a well-defined stage of completion of a complete draft ready for submission to the Ministry of Education. The estimates, including my own, ranged from 18 to 30 months. At this point I had the idea of turning to one of our members, a distinguished expert in curriculum development, asking him a question phrased about as follows: "We are surely not the only team to have tried to develop a curriculum where none existed before. Please try to recall as many such experiences as you can. Think of them as they were in a stage comparable to ours at present. How long did it take them, from that point to complete their projects?" After a long silence, something much like the following answer was given, with obvious signs of discomfort: "First, I should say that not all teams that I can think of in a comparable stage ever did complete the task. About 40% of them eventually gave up. Of the remaining, I cannot think of any that was completed in less than seven years, nor any that took more than ten." In response to a further question, he answered: "No, I cannot think of any relevant factor that distinguished us favorably from the teams I have been thinking about. Indeed, my impression is that we are slightly below average in terms of our resources and potential."

When it comes to personal situations, however, there is some positive news regarding inside and outside views. Many parents struggling with their teenager's acts of rebellion tend to look inwardly, at the specifics of their own child. They cannot imagine him or her ever becoming a productive member of society. Yet, an outside view indicates that the vast majority of teenagers do in fact grow up to become mature and responsible adults.

13.3.2 Illusion of Predictability

Another common misconception is the notion that random events are reasonably predictable. Thus, many people believe that linear extrapolation is an adequate forecasting tool for projects that include a wide range of contexts. U.S.-based auto executives, for example, moved major manufacturing facilities into Canada because of the continuing slide in the Canadian dollar that was worth $0.66 US in January 1999 and $0.63 US in January 2002. However, in October 2010, the Canadian and U.S. dollars were approximately equal in value. This resulted in nearly a 50% increase in the cost of manufacture and import from Canada into the United States. Similarly, fluctuations in the Euro confounded linear extrapolation. A single Euro was worth $1.17 US in January 1999, and its value slid to $0.89 US in January 2002. However, this downward trend reversed the following year, and the Euro was worth $1.40 US on the first of October 2010.

The illusion of control and predictability was a common bias in the George W. Bush administration's Middle East policy. They did little planning for peace in post war Iraq. The administration was sure it knew the will of the Iraqi people and predicted a smooth and rapid transition to democracy once the war was over. It was similarly confident of a favorable outcome when it pushed for popular elections in the West Bank and Gaza, only to find instead that extreme militants won a majority in Gaza as well as significant power in the West Bank.

One of the more commonly repeated examples of the illusion of predictability involves the high priced bidding for free agents and first-round draft picks in sports. One cannot help but be baffled by the highly competitive bidding for an individual who opts for free agency after completing one particularly outstanding year. It is not just the dollar amount offered but also the contract duration that seems to stretch belief. Athletes, after all, are prone to injuries that can dramatically curtail their performance and even end their careers prematurely in unpredictable ways. For example, the Detroit Tigers in 2009 had $32 million in high-priced contracts for individuals who ended up not playing due to injury or contributing only marginally to the team's success.

Even less understandable are the salaries offered to first-round draft picks. The transition from college to professional in almost every sport is dramatic and complex. As a result, outstanding college performance is at best a modest predictor of excellent professional performance. The seasons in professional baseball and basketball are several times as long, and the daily competition is incomparable to the college experience. The pounding taken in professional football is significantly worse on the body because professional athletes are, on average, much heavier, stronger, and faster than their college counterparts. A large proportion of first-round drafts fail to continue as starters through the end of their first contract. The Detroit Lions, for example, were notoriously poor in their selection of first-round drafts in the first decade of the twenty-first century.

During that same decade, the U.S. financial industry as well as many of its global partners lived and invested dangerously based on the belief of linear extrapolation. They extrapolated that the value of both the housing market and the stock market were headed for continuous growth. Although they recognized periodic bumps in the road with serious corrections, they assumed nevertheless that over any 10 year period growth would always justify any and all forms of risky investments in the future. In the deep recession of 2008–2009, the belief in linear extrapolation has been shredded as we all wait for the markets to return to levels of a decade ago. In the meantime, people nearing retirement are struggling to cope with the realization that their retirement savings are progressively diminishing in value.

5. *Activity:* Illusion of control

   Provide an example of an illusion of control bias in your personal life, organization, or community.

   _____

6. *Activity*: Illusion of predictability bias

   Provide an example of an illusion of predictability bias in your personal life, organization, or community.

   _____

13.3.3 Anchoring

One approach to developing a forecast calls for deciding upon a realistic estimate and proceeding to place ranges around that value. This approach, however, is likely to suffer from the bias of *anchoring*. Once a central value is established, there is a tendency not to move more than a modest percentage away from this target when developing a range. The forecaster becomes too anchored to the first estimate to develop a wide range that is reflective of actual dispersion. The preferred approach is to initially focus on estimating both good and bad extremes. This concept is developed later as we describe the seven steps involved in interviewing an expert to obtain a forecast range.

Salesmen are notoriously good at exploiting the bias of anchoring to frame negotiations. They set artificially high initial prices recognizing that the purchaser will be satisfied with negotiating the price down by a mere 5% or 10%, unaware that the original quote was more than 20% too high. In this way, a salesman distorts the buyer's estimate of the value of the product.

The principle of anchoring is best demonstrated by the story of how black pearls first established a market price. When new luxury items such as black pearls are introduced to the market, there is no cost basis for prices, nor can supply and demand help establish such a basis in the earliest stages. In this case, a savvy marketer started by displaying strings of black pearls in the window of a jewelry store right next to diamond necklaces. This strategic "contextual" placement led the consumer to believe that black pearls belong in the same price range as diamonds. The value of the black pearls was anchored by the presence of diamonds nearby (Ariely 2008).

13.4   Faulty Probability Reasoning

There is a conceptual reason for underestimating the range of extreme values. Few people have an intuitive understanding of the probability distribution of the minimum and maximum of a set of random variables. For example, let us assume that a component's lifetime is a random variable with a probability density function that is exponentially distributed with a mean of 2000 hours. Now, assume that 10 components are turned on simultaneously. Few individuals recognize that the mean time of the first failure is one-tenth the average and that this interval is a mere 200 hours. Worse yet, there is more than a 50–50 chance that the first failure will occur in less than 140 hours. Seeing this early failure, a typical observer would simply assume that one of the items was seriously defective and therefore unrepresentative. Using a dice analogy, were you to roll a set of four dice, there is a greater than 50–50 likelihood that the lowest value observed on any one die would be 1.

Unfortunately, there is a broad misunderstanding of basic concepts of probability theory. Sports fans are routinely bombarded by misstatements regarding the law of averages. If a baseball player is in a slump, the broadcaster will likely opine that the player is "due" for a hit. Conversely, basketball announcers routinely discuss streak shooters and coaches design strategies to get the ball to the player with the "hot hand." Yet, a detailed analysis of some of the more famous streak shooters indicates that their reputations do not stand up to careful data analysis (Gilovch et al. 1985).

Worse yet is the false impression that people develop while gambling, namely the sense that they are "hot" or "on a roll." It is heartbreaking to watch participants in a purely probabilistic television game show such as "Deal or No Deal." The suitcases with money values are assigned numbers at random. Yet, as their winnings accumulate, players will gamble hundreds of thousands of dollars that they admit would change their lives. They feel lucky and—for no good reason—are "sure" they have picked the winning number, only to lose the money that they might have taken home with them.

This activity will give you an honest assessment of your understanding of probability.

7. *Activity*: Each of 20 suppliers for complex components of a piece of equipment is highly reliable about meeting deadlines. Each has a track record of meeting the deadlines 95% of the time. What do you estimate is the probability that all of the components will be received on time? _____

One of the basic building blocks of probability is the multiplication rule. The probability of multiple *independent* events occurring is determined by multiplying the corresponding individual probabilities. In the activity listed earlier, the calculation involves raising 0.95 to the 20th power. This is 0.36. In our experience, few individuals intuitively recognize that this aforementioned probability is less than 50%. Instead, they seem to focus on the 95% reliability of all of their suppliers and think the probability of receiving all of the components on time is closer to 0.95, since all of the suppliers are so very reliable.

This point became clear to the writers of this text when the problem was presented to a class of automotive engineers. One group was getting ready to shut down an assembly plant over a weekend in order to build advanced prototype vehicles. They were optimistic that all of the parts would be there in time to build the vehicles. After seeing the aforementioned simple example and recognizing that they were dealing with far more than 20 suppliers, warning bells sounded. As a result, the engineers instituted a sophisticated tracking process for all of their suppliers scheduled to deliver prototype components. As the official build date approached, they focused more and more on the suppliers who were most at risk, calling them almost every day. For the final few days, they literally tracked the components to the point of their placement on a truck to assure that they would be delivered just in time for vehicle assembly to begin. Thanks to the steps that these engineers took, the story had a happy ending and the vehicle build proceeded on schedule.

One difficult class of personal decisions is health related. Testing is at the core of medical diagnostics. Some tests are performed as routine screening in the absence of any specific disease symptoms. These include mammograms to detect early stages of breast cancer and PSA as a predictor of prostate cancer. Other tests are performed to identify the causes of various symptoms. However, few if any test results are perfect predictors of disease. For example, mammograms identify unusual masses that require extensive follow-up tests to determine the nature of these masses. Similarly, TB tests reveal antigens in the bloodstream that indicate exposure to TB germs, but not necessarily the presence of active or contagious TB. There is active debate as to the value of mass screening given the high rate of false positive results and the accompanying anguish that is experienced.

Similarly, the next generation of parents will face the challenge of interpreting and acting on the results of genetic testing both pre- and post-marriage. Genetic screening is currently actively practiced prior to marriage in one segment of the Jewish community. Couples considering marriage are encouraged to see whether or not they are carriers of the Tay-Sachs recessive gene, which is commonly found among Jews of European ancestry. If both members of the couple are carriers, they have a one-in-four chance of giving birth to a Tay-Sachs child whose life expectancy does not exceed 6 years.

As researchers continue to unravel the mysteries of the human genome and related diseases, the issue of genetic testing will become a broader concern. Imagine a couple facing the following decision and dilemma posed by genetic test results from in-utero testing. Take this test to accurately assess your understanding. (You should do well on this activity if you already took the test as part of the Bayes Theorem discussion in an earlier chapter.)

8. *Activity:* A rare genetic disorder occurs once in every 50,000 individuals. The standard test is very reliable. If a person has this defect, there is 0.98 probability of the test detecting the disorder. That means that there is only a 2% rate of false negatives. In addition, if the genetic disorder is not present, there is 99% accuracy with only a 1% false positive result. The test comes back positive. What are the chances that the individual has this genetic disorder?

The overwhelming majority of people, even those who long ago studied probability in school, tend to focus only on test accuracy statistics. They fail to recognize the role played by the rarity of the disease. As a result, the common estimate is between 98% and 99%. The actual probability, however, is close to 0.002 or 1 in 500. The way to visualize this is to imagine 50,000 fetuses being tested and assume one has the genetic disorder. That individual is likely to be detected. However, there is a 1% false positive rate. One percent of 50,000 is 500. These will appear as false positives along with the one true positive. Thus, only 1 in 501 positive results has the disorder. The good news is that medical doctors in general and genetic counselors in particular are currently trained and required to understand this concept. They have their own vocabulary to describe this problem context, using terms like *disease prevalence* and *test specificity*. In standard probability texts, the aforementioned example illustrates Bayes Rule that was developed in the eighteenth century.

13.5 Availability and Representativeness

Availability and representativeness are two complementary cognitive biases originally explored by Kahneman and Tversky in their groundbreaking studies in the 1970s. Kahneman ultimately won a Nobel Prize in Economics in 2002 for his work; Tversky did not live to receive this award. Although several recent studies have called into question the nature of their experimental designs, their general thesis is widely accepted.

The availability bias relates to a tendency to overstate the likelihood of "high profile" events—such as earthquakes or airplane crashes—while downplaying the more common risks associated with everyday tasks—such as driving or crossing the street. For example, hundreds of people are killed each year in traffic accidents on the roads of Israel and far fewer die in terrorist attacks. Yet, people who live outside of Israel tend to consider the dangers of terrorism when thinking about the safety of travel to Israel and ignore the prevalence of traffic accidents.

Gambling establishments exploit the availability bias to encourage gambling. They highlight individual million dollar winners from slot machines, easily the worst gamble available, so as to lead people to overestimate their chances of winning similar amounts of money. The pervasive state lotteries do the same. Similarly, the purveyors of all sorts of safety devices and services, such as OnStar, trot out individual memorable stories so as to skew the prospective customer's perception of how frequently the device or service might save them in an emergency. The availability bias is alive and active in your organization if considerations of new ideas rapidly come to a halt because someone recalls a tragic, but unlikely, incident that took place long ago as proof that the idea in question will not work.

9. *Activity:* Availability or representative bias

   Provide an example of an availability or representative bias in your personal life, organization, or community.

Kahneman and Tversky also carried out controlled experiments to assess the representative bias which has multiple manifestations. One aspect of this bias is that people do not have an intuitive feel for the importance of sample size. Another key aspect is employing partial descriptive information in order to estimate the likelihood of a particular occurrence while not placing enough emphasis on other critical data. The controlled example Bazerman (2006) presents involves the following question

Mark is finishing his MBA at a prestigious university. He is very interested in the arts and at one time considered a career as musician. Where is he more likely to take a job?

a. In arts management

b. With a consulting firm

When responding, most do not consider the baseline data. More MBAs select consulting over arts management as a career. Although Mark may be representative of people who end up in arts management, few MBAs do so. In essence, people have difficulty distinguishing between the following two conditional probability statements.

1. P(someone like Mark | career in arts management)
2. P(career in arts management | someone like Mark)

With regard to statement 1, you may find that a significant proportion of people in arts management have an MBA and an interest in some aspect of the arts. However, the question posed earlier is equivalent to statement 2. Only a small proportion of people like Mark end up in arts management.

13.5.1 Limited Observation as a Predictor of Behavior and Long-Term Performance

Again, we are interested in some of the more obvious ways that people consider particular experiences and information as over-representative. Who has not projected his own limited life experience when judging how others might respond in a similar situation? One extreme example of this sort of projection involves a senior automotive executive who had recently bought an expensive sound system for his home. After test-driving a mid-priced vehicle, he pretentiously criticized the sound system, saying, "I just bought an expensive sound system and I know what a good system should sound like. This car's system is unacceptable." He then ordered the engineers to improve the system's design. Little did he realize that the average person in the market for mid-priced vehicle neither expects—nor is willing to pay for—a state of the art sound system. Similarly, when car companies use race car drivers to test out the ride and handling of a mass market vehicle, they too are employing the "limited observation" fallacy. The perception of a race car driver is totally unrepresentative of what a typical car buyer is likely to notice.

One critical element of most job applications is the interview. There is a tendency to place too much weight on the interview as compared to the individual's actual track record. The fact that an interviewee can give the right answer regarding how he might handle a difficult employee issue need not correspond to the way that he would actually behave when facing this issue. In truth, there is often no substitute for experience. It is absolutely stunning, for example, that the Detroit Lions hired Matt Millen to be President of the Detroit Lions based entirely on his interview and in total disregard of the fact that he had no front office experience. Not surprisingly, Millen proceeded to produce an unprecedented series of bad teams that, in 2008, won the Lions the dubious distinction of earning the first 0–16 record in the history of the NFL. In short, an interview is representative of how a person does on interviews and hypothetical situations; it is not necessarily representative of actual behavior or performance (Landy 1990). Similarly, an abundance of evidence indicates that a novice teacher's performance when giving a model lesson is a poor predictor of that teacher's ultimate performance and long-term growth (Wede 1996).

Parole boards face an especially tough decision making context in terms of predicting long-term behavior. When considering someone for parole, the board members have detailed information about the individual's prior criminal history. They may also have a psychological evaluation. They then have to decide whether or not the individual is ready to be released into society based on his behavior in the controlled environment of a prison and the way that he answers a series of questions. If all the weight were given to prior behavior, no prisoner would be released until he had served out his full sentence. Nonetheless, behavior in the controlled environment of prison is not necessarily predictive of future behavior outside of this environment. The answers to questions are not necessarily a true reflection of the parolee's attitude and belief. For this reason, halfway houses have been made available to help the parolee adjust to his new life, and parole officers are charged with the responsibility of supervising the prisoner on a frequent and regular basis. Limited observation is just that—*limited*—and the original job or parole interview must be supplemented by close monitoring of the candidate, even after he has succeeded in attaining the job he seeks.

10. *Activity:* Limited observation or interview bias

    Provide an example of a limited observation or interview bias in your personal life, organization, or community.

13.6 Confirmation and Interpretation Bias

One last bias involves the way we process or evaluate information. The confirmation bias denotes our tendency to give more credibility to information and data that support our preconceived ideas or the choices we are leaning toward and less credibility to contradictory evidence. Perhaps the most extreme example of this phenomenon involved the way that the George W. Bush administration misinterpreted the available data and the actions of Saddam Hussein to convince themselves and others that Iraq was concealing weapons of mass destruction in 2002. A less portentous example of this bias might be when sports fans watching a game are asked to judge the relative sportsmanship of the teams. Each team's fans tend to judge their own team as having played in a more sportsman-like manner.

An example of interpretation bias would be the *post hoc ergo propter hoc* fallacy, which relates to mistakenly assigning cause and effect to patterns that are merely sequential. Imagine a system that has changed over time and, in the interim, some specific event has occurred or an action has been taken. Too often, we will assume that the intervening event or action is responsible for the change in the system.

People tend to underestimate natural variability and the well-known tendency of regression toward the mean. Suppose, for example, that someone has the flu virus and takes an antibiotic; a few days later the flu is gone. There is no scientific reason to assume that the antibiotic helped to cure a virus. Similarly, let us imagine that over the course of a particular year, a person had to take an unusually large number of sick days. The following year, he takes mega doses of Vitamin C and has an average—or better than average year healthwise. Lacking additional evidence, one cannot prove that the Vitamin C prevented illness. Perhaps, the individual took other precautions, such as eating and sleeping properly or taking greater care with hygiene. Nevertheless, many would tend to assume that the Vitamin C was responsible for the individual's improved health. It is this very misinterpretation dynamic that requires drug companies to develop rigorous protocols to assess the relative effectiveness of a new drug or medical procedure.

The same dynamic applies to quality in a manufacturing plant. Let us assume that a highly visible new process or piece of equipment has just been installed in our plant. The next week, quality declines—or quality improves—and we quickly jump to the conclusion that this new process or piece of equipment either interfered with or improved performance. Similarly, if a new plant manager, sports team manager, or CEO has been appointed and, subsequently, the team or organization's performance improves, the new leaders are likely to be quick to claim credit for the change. Interestingly, when the reverse happens, these same leaders will often cite factors outside their control and assign cause and effect to those factors. They will point out that the economy is down, the currency exchange rate has changed, commodity prices are up, new competitors have arrived in town, or a key player was hurt. Unfortunately, few of us are equipped to question the cause and effect claims. Moreover, even with the best science, it can be extremely difficult to accurately assign cause. This issue also underlies such concerns as the ongoing argument over the impact of humans on climate change.

11. *Activity:* Confirmation bias

Provide an example of a confirmation bias in your personal life, organization, or community.

13.7 Expert Interview: How to Identify and Reduce Bias

In this section, we present a structured seven-step process that consultants in decision and risk analysis use to obtain ranges when interviewing experts in order to collect data regarding a particular project. In discussing this process, we point out some of the questions that should be asked and how these questions are intended to uncover and address biases. Shephard and Kirkwood (1994) provide an example of an actual interview along with a running commentary. The integrity of the interview process is a critical element in the decision analysis paradigm and is a continuing area of active research as discussed later in this chapter (O'Hagan et al. 2006; Abbas et al. 2008).

There are two types of interview goals. In one instance, we are exploring a random variable such as time or cost and attempting to develop a probabilistic range. We seek a 90% upper bound on how long a project might take or how much it might cost. We want the expert to specify an "X" value, so that according to his experience-based opinion, 90% of the time the task will be completed in less than X months. In other contexts, we may have a deadline in mind. For example, over the summer, a school building is undergoing extensive reconstruction. What is the probability that the building will be ready in time for classes to begin on September 7th?

This section works with two assumptions. First, we need to draw out critical information from the specific expert. Second, the preferred way to do so is through an extended, structured interview. Many decisions are assessed in group meetings that include individuals drawn from diverse organizations such as engineering, manufacturing, finance, and marketing. In general, it is a good idea to look at an issue from multiple perspectives, provided that these perspectives are offered only by those who are experts in their respective areas. However, problems and confusion arise when each individual is allowed to volunteer opinions on issues outside their areas of expertise. The egalitarian norms in American culture often mean that opinions once expressed must be given equal weight on each issue. This misdirected application of democracy can prove to be a problem when considering a new project.

For example, we are all consumers with likes and dislikes. In an open meeting, we may not hesitate to articulate what we believe the consumer wants in a product under discussion. We also can imagine how many more people will buy the product if only it had our favorite added feature. Similarly, we all have a sense of timing and task duration. We have trouble imagining outside of our area of expertise why it should take so long or cost so much to include this or that feature into the design. We have limited understanding of testing required to validate the performance of a product. Thus, we might not hesitate to suggest added features or expand the scope of a project while expecting the project team to hold the line on cost and delivery date.

A prerequisite step in establishing an expert interview plan involves understanding and dividing the overall project into distinct elements as well as identifying experts in each area to estimate relevant critical variables. Marketing would provide probabilistic ranges for demand. Engineering would estimate the range of time and cost required to deliver a fully validated product design. Manufacturing and finance would focus on the cost of manufacture and the time necessary to reach critical production or service levels. If the product or service involves installation at a customer site, customer service experts would estimate the time and training required.

If the product involves suppliers in a significant role, supplier technical experts would estimate the time and effort required to develop and sustain a reliable supply chain. This last issue can be a weak link, since senior executives may have little understanding of the complexity of these relationships. Boeing's experience is just one extreme example of top management underestimating the challenges of a global product development and manufacturing supply chain. Similarly, many automotive executives have misunderstood the challenges that come with outsourcing components to low cost, emerging markets. One senior executive announced that the following year, the company would outsource the manufacture of $1 billion dollars in parts to China. Almost nothing happened in the course of that year. It takes a long time to certify that a particular automotive supplier is capable of producing high volume parts for installation in a U.S. car that is subject to very expensive recalls if even 1 in 1000 bad parts ends up in vehicles.

The process for obtaining expert opinion involves an interview and not simply a survey form. A key aspect of the interview is for the interviewer to continuously probe and document the basis for the expert's estimates. He must also help the expert overcome many of the subconscious biases often exhibited when making projections. If a company develops a consistent process for obtaining these projections, it can continuously review the accuracy of these projected ranges and develop strategies to obtain more consistently accurate ranges.

The seven steps of the interview process and the goal of each step are listed as follows:

1. *Prepare to interview* → Learn enough beforehand to design intelligent interview
2. *Motivate interviewee* → Establish rapport and explore personal biases

3. *Structure interview* → Identify those critical variables that are most uncertain

4. *Condition interviewee* → Describe process for exploring ranges and potential cognitive biases

5. *Elicit and encode data* → Elicit probabilistic ranges for key variables and convert answers to probability distribution

6. *Verify answers* → Review answers and results to ensure that final distribution reflects expert's assessment of uncertainty

7. *Aggregate information* → Bring together all uncertainties

In discussing these steps, we will refer for illustrative purposes to a process we have used with regard to R&D as well as the detailed example provided in Shephard and Kirkwood (1994).

1. *Prepare to interview*

The starting point of any interview involves identifying credible and accepted experts within or perhaps outside the organization. Before meeting with the expert, the interviewer must bring himself up to speed on the critical issues in order to design an intelligent interview. We recommend sending a simple open-ended questionnaire to the expert in order to gain background information. Table 13.3 contains sample questions applicable for interviewing a researcher regarding an initiative within his expertise. Notice that the form asks for verbal descriptions that do not involve specifying any numeric values.

Questions one and two provide general background information. The third and fourth questions are designed to encourage the expert to think more broadly and provide a basis for the "outside" view of the range we discussed earlier. They are also designed to provide information for top R&D managers to enable them to explore alternatives to doing all the research in house.

Questions five and six are most directly linked to the formal interview that will follow. The interview questions will focus on the time and resources required and the likelihood of getting through each technical hurdle. One critical aspect of the interview is clearly defining the point at which some task is actually completed. The answers regarding future hurdles might also identify the need to interview other experts whose skills may be needed to surpass subsequent hurdles.

Questions seven and eight are designed to bridge the relationship between R&D, product development, and marketing. It is critical that the subsequent interviews with marketing regarding the range of potential demand be based on a common understanding of the product's features. Often R&D and marketing will have fundamentally different understandings of a product's key performance. For example, a particular automotive R&D group once worked on the concept of a heated windshield wiper rest. The R&D group was focusing on unfreezing the windshield wiper when a driver first sits in a car and gets ready to drive. Marketing, on the other hand, thought that the product was expected to solve the problem of ice building up around the wiper and windshield while

---

TABLE 13.3.   Open-ended questions in R&D prior to expert interview.

R&D Pre-Interview Questionnaire

1. Describe in two paragraphs the overall nature of the research project
2. Discuss current specific focus (or foci) of research
3. What is the current capability of other companies, or other organizations with regard to this project and potential products?
4. What opportunities exist for technical collaboration?
5. What is the next technical hurdle(s) to overcome before moving on to a subsequent phase?
6. What other major technical hurdles lie ahead?
7. What end product would use this technology?
8. What are some key performance characteristics of the proposed product or process improvement?

driving in a snowstorm. Only when R&D demonstrated, in a cold room, exactly what the system was capable of delivering, did marketing suddenly realize their mistake and understand that, for this reason, market potential was limited.

Question seven, regarding technology integration, is also designed to uncover other key elements and barriers along the path leading to initial sales. Surely one cannot compare the integration challenges posed by a technology that draws electric current from a battery or interfaces with the powertrain to those posed by the need to provide different shades and colors of tinted glass when manufacturing an automobile.

In preparing for the interview, it is important for the interviewer to gain background knowledge about the expert. It is important to know if the expert has certain predispositions to various aspects of the decision under consideration. For example, the expert you choose to interview may have been the key sponsor for technology that is at the core of the decision. It would be useful to ascertain whether this involvement is likely to influence his forecasts. We do not want the interviewee to discard his experience, but, at the same time, we do not want this experience to carry a disproportionate amount of weight in the forecasting process.

2. *Motivate interviewee*

Before moving to the specifics of the interview, the interviewer needs to establish rapport with the expert. It is important to explain how the results of this interview will fit into the bigger picture of decision making and that the goal of the interview is to explore uncertainty. It would probably be a bad idea, however, to suggest to the interviewee that you are exploring the expert's uncertainty rather than what the expert knows for sure. At this stage, it is critical that the expert take the process seriously and agree to be thoughtful and explanatory when responding to your questions.

Having established a rapport with the interviewee, the interviewer should next explore the expert's background and recent experiences in an attempt to identify potential biases. The biggest challenge to the person conducting the interview involves uncovering motivational and availability biases. Shephard and Kirkwood (1994) made note of official forecasts of demand that likely affected the expert's ability to honestly assess an objective range of demand.

In R&D contexts, we often interview technology experts who may be proponents of specific technologies. Their respective proclivities are likely to bias their estimates of both the speed with which a project can be implemented as well as the likeliness of its success. Take, for example, the interviewers doing a study for the U.S. postal service who sought probabilistic ranges on the rate of adoption of the extra four digits to be added to the zipcode system (Ulvila 1987). At the end of the process, interviewers felt unable to overcome the employees' overly optimistic bias of the rate of diffusion. As a result, they explored the decision's sensitivity to these estimates. In the end, all that could be accomplished was raising the interviewee's awareness of his own biases, and the interviewer could do no more than note where specific biases were likely to affect the forecast.

12. *Activity:* Non-meaningful survey

   Describe a survey you recently completed wherein you felt that your answers were not meaningful.

The interviewee should also explore the expert's understanding of probabilistic concepts. The output of the interview is a probabilistic range, and the expert needs to understand the meaning of a confidence interval. In addition, if, for example, the timeline involves multiple events occurring, it may be important to state the multiplication rule of independent events.

In summary, all that can be accomplished at this point is to raise the interviewee's awareness of his own biases. The interviewer will need to identify what specific biases are likely to affect the estimates. In those areas, the interviewer may have to dig more deeply as the interview proceeds.

He must also take care to avoid becoming overly aggressive, so that he does not risk turning off and disengaging the interviewee.

3. *Structure interview—Define variables or events*

The core of the interview revolves around specific measurable random variables or random events. It is critical that the random variables are clearly defined and meet what decision analysts call the clarity test. For example, a statement that there is only a 10% chance that a particular project will be completed in 10 months or less must include a clear definition of "project completed." To meet the clarity standard, years from now the powers that be would have to be able to look back and point to the exact time when the project was completed. If that completion date may be fuzzy even after the fact, then it is not possible to interpret the expert's estimate of the probability of completion within 10 months.

Similarly, if a marketing expert predicts the probability that the launch of a new product will be a success, the expert would need to define specific measures of success. These could be total sales, total revenue, or market share. In the expert interview, we are trying to avoid the dynamic that too often occurs when management after the fact declares something a success irrespective of having met the original measurable goals.

Earlier, we suggested the example of a school undergoing extensive summer reconstruction and spoke of its being ready for classes to start on September seventh. The term "ready for classes" would not meet the standard of clarity. A building could be in various stages of disarray with ongoing construction and still be used for teaching. One clearly defined standard of readiness would be securing a building occupancy permit. In this case, a clear definition of readiness is not only needed for an expert assessment; it is also helpful in clarifying the contractor's agreement to have the school "ready" for classes to start.

13. *Activity:* Clarity standard

   Describe a context in which two individuals or groups had radically different understandings of task completion.

In defining specific time variables, it is often useful to divide the total time it will take to complete a project into clearly defined stages. For example, the publication of this book has a long timeline. The first piece is the time needed to deliver a completed manuscript, including homework problems, to the publisher. It then must go through an editing process that may involve one or more cycles of interaction between editors and the authors. Eventually, it is prepared in camera-ready or print-ready format. Finally, an initial publication run is printed, and the authors receive printed copies in their hands. The estimated date of receipt of actual copies must pass the clarity test. Rather than estimating the probability distribution of the total timeline, he should focus on the randomness associated with each of the aforementioned phases: (1) delivering a finished manuscript; (2) editing and rewriting the manuscript; (3) preparing final layout, final review, and signoff for printing; and (4) printing and delivering copies to authors. We have defined manuscript completion as the date on which the book is sent to the editor rather than when we think we will have finished *writing* the draft. The latter point is often fuzzy since a "finished" manuscript is continually rewritten. However, it is easy to look back and say when the manuscript was *sent*.

Analogously, when attempting to estimate the probability of an event occurring at all, it may be necessary to decompose the event into a series of smaller, consecutive events. Clearly, it is difficult to offer valid projections regarding extremely rare events that occur, for example, once in a 100 years or affect 1 in a million people. Few of us have experiences with extreme rare event randomness. In addition, when testing products under development, we rarely can run our tests long enough to uncover the 1 in 10,000 risk of failure. Fault tree analysis (Fischhoff et al. 1978) is a tool that experts use to deconstruct a rare event, such as a nuclear power plant failure that leads to a radiation leak.

In some instances, we can determine the event indirectly. For instance, it may be difficult to determine directly the likelihood of a dam failure or levee failure on a specific date, although history shows that these failures are likely to occur once in 50 years. However, if the most likely triggering event is extreme weather such as a stage 5 hurricane, we can review 100 of years of weather records to estimate the frequency and predictability of this particular weather event. It is just such thinking that underlies the Netherlands planning of dikes to protect their country from a once in several thousand year weather event. In other instances, the rare event is a complex product of other events that, likewise, can be calculated using probability theory. For example, the probability that all suppliers will meet a deadline can be determined from estimates of each of the individual supplier's likelihood of meeting the deadline.

The primary responsibility for articulating measurable variables lies with the expert and not the interviewer. It is the expert who is best able to articulate the various stages and provide clear definitions and measures that meet the clarity standard. The expert must be totally comfortable with the definition of the random variable or event. The interviewer is there only to facilitate clear thinking.

The pharmaceutical industry, a major user of decision and risk analysis tools, works within a framework of clearly defined steps that facilitate expert assessments. The U.S. Federal Drug Administration (FDA) and its partners around the world have defined the various stages of development and testing that a drug must go through before sale to the public. Each stage has well-defined starting and ending points as well as specific measures of success that must be met before progressing to the next stage.

14. *Activity:* Describe a project or task timeline with concrete start and end dates. Then, identify critical stages within the timeline that also have clearly defined end dates that would pass the clarity test.

_____

_____

During this stage, the interviewer could explore some of the simplified cognitive rules experts use that can lead to biased estimates. In particular, it is important to understand if the variable might require the addition or multiplication of probabilities to determine a probability value. One issue of special concern would be a variable that involves conditional probability. For example, few of us have accurate intuition regarding how long an on-going phenomenon will last. This applies both to winning and losing streaks as well as to the number of years of life remaining for someone who is 85 years old.

4. *Condition interviewee: Explore ranges*

At this stage, the expert is encouraged to conceptualize the outside view that involves recalling a wide range of relevant experiences. It is explained that in the next step, he will be asked to recall a similar project that was completed over an unusually long period of time. Conversely, he will also be asked to recall projects wherein almost everything went as planned and that were completed much sooner than anticipated. As part of this discussion, the interviewer might explain the natural tendency to place narrow bounds around an estimate.

During this phase, the interviewer should be looking for evidence of availability and representative biases. If the expert continually refers to one extreme fiasco, this is an indication of these biases. A related bias is the imaginability bias. If the expert imagines all sorts of extremes occurrence that could occur, he will probably have a tendency to overestimate their likelihood. One last bias to look out for is the illusion of control. There is a tendency for experts and executives to believe that they can control uncertain events. If the expert suggests that he can handle every contingency, this would reflect the illusion of control.

At this time, the interviewer should also employ any facilitating tools that he plans to use in the formal questioning. For example, to illustrate a 90th percentile, the interviewer might offer one Canadian penny alongside nine U.S. pennies. In the subsequent questions, the expert will be asked to compare the likelihood of the project taking longer than X months to the chances of randomly selecting the Canadian out of the pile of 10 coins. If the ultimate question involves specifying the probability of an event occurring, the interviewer may choose to introduce a probability wheel with a spinner. The expert would be asked to adjust the range on the wheel to represent the likelihood of the event.

As a prelude to the formal questioning, it is appropriate to explore the expert's current mood. Moods of optimism and pessimism can affect forecasts and create their own biases as discussed in the next chapter. The interviewer would want to know if anything extraordinary has recently occurred in the expert's job or life that might color his forecast of random events—especially relatively rare ones.

5. *Elicit and encode data: Probabilistic ranges*

In working with the interviewee to assign probabilities 0.1 and 0.9, it is worthwhile concretizing the concepts. For example, a 0.1 probability can be represented by the likelihood of selecting one Canadian penny out of a batch of nine U.S. pennies and one Canadian penny. The interviewer might begin by exploring extreme case—but not necessarily worst case—scenarios. He must recall that the target value he is seeking aligns with the 90th percentile and not the 99th percentile. If, however, the expert strongly desires to start with a most likely scenario, this would be evidence of an anchoring bias that will be hard to overcome.

The questioning might proceed as follows:

What are some of things that have gone wrong on similar projects that have delayed completion significantly? Please describe and discuss these factors as they affected analogous situations. Now imagine it is 5 years hence and you are looking back at this project. You tell yourself that this project took much longer than you ever imagined it would take. How long did this extremely long and delayed project take?

Suppose the expert's answer is 18 months. This is NOT YET the number we are seeking. At this stage, we do not know where 18 months fits on the distribution of randomness. Is this an extreme 99th percentile value or a more moderate 80th percentile? At this point, the 18 months are to be compared to the one coin in ten example. Ask the expert which of the following is more likely:

a. Picking the Canadian coin out of the pile of 10

b. The project taking 18 months or longer to complete

If option "a" is more likely, then taking longer than 18 months has a lower than 10% chance of occurrence. In that case, the expert would be encouraged to select a value *less* than 18 months, such that the chances of taking longer would approach 10%. If, however, option "b" is more likely, then the project taking longer than 18 months has a greater than 10% of occurrence. In this case, the expert would be encouraged to select a value more than 18 months such that the chance of taking longer than that has only a 10% chance of occurrence.

The process is then repeated with an analysis of a highly optimistic perspective with a discussion as follows:

What are some of things that have gone very well on similar projects that have speeded completion significantly? Please describe and discuss these factors as they influenced analogous situations. Now imagine it is 5 years hence and you are looking back on this project. You tell yourself that this project went far smoother than you ever imagined. How long did this extremely successful project take?

Suppose the expert's answer is 9 months. Once again, this is NOT YET the number we are seeking. To align this number to the 10th percentile, the expert would be led through the comparison of which of the following is more likely?

a. Picking the Canadian coin out of the pile of 10

b. The project taking 9 months or less to complete

It is only after exploring the extreme possibilities that the expert is asked to provide a middle estimate. The reason for this is to avoid the anchoring bias. Anchoring is the tendency to not move too far away from a middle value once it has been set. The questioning can take one of two forms. The expert can be asked about a "most likely value" or the median value, the 50th percentile. In this instance, the expert set the median value at 12 months. Notice that responses are not symmetrical. The median is much closer to the 10th percentile, which is 9 months than to the 90th percentile, which is 18 months.

The aforementioned interview dynamic relates to a continuous random variable and specifying a value X that corresponds to a fixed percentile. At other times, you may be seeking a probability value. For example, your company wants to estimate its chances of winning a major contract in a particular context. When interviewing an expert, you would start by discussing factors that in the past have affected the company's experience in winning or not winning competitive contracts. The interviewer would then ask the expert to characterize the current situation: is it quite typical or are there factors that would increase or decrease the chances of winning? He might also ask the expert one or more of the following questions: Are there more competitors than usual? Are they more desperate and therefore willing to cut their price to win the contract? Is the company going up against an incumbent with a good track record?

6. *Verify* → Check expert's belief in results

After going through this intensive interview, it is important to review the answers with the expert. The expert would be shown the resultant cumulative distribution function and asked to confirm that it reflects his expect judgment (Figure 13.1 from Shephard and Kirkwood 1994). The expert should be encouraged to clarify his comfort level with the results and be allowed to revise his forecasts. At every step, the expert should be asked to explain his answers without feeling that he is being challenged as to their correctness. After all, there are no objectively correct answers, just probabilistic estimates ideally drawn from a wide range of similar experiences.

The interview focused on a limited estimation of the probability distribution, two extreme points and one in the middle. In more complex studies that involve major projects, such as a nuclear

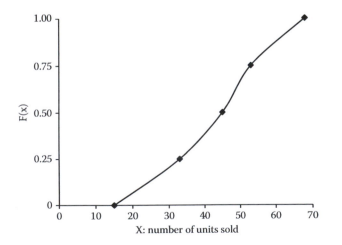

FIGURE 13.1:   Confirm cumulative distribution curve.

installation or an entire weapons system, decision analysts often invest more time, so that they can develop a broader description of the probability distribution. They will likely include multiple experts whose judgments will need to be combined.

7. *Aggregate* → Bring together uncertainties

In the final stage, it is important to bring together all of the relevant uncertainties. In complex projects, multiple experts might be interviewed with regard to the same uncertainty. There is no one theoretically best approach to aggregating multiple opinions (Arrow 1951). However, simply averaging the independent forecasts has been shown to improve forecast accuracy (Ariely et al. 2000; Hora 2004). The preferred strategy is to work toward achieving expert opinion consensus by discussing the differences. Alternatively, management may choose to accept a single expert's perspective. The analyst would then use the sensitivity analysis phase of the project to determine whether or not the decision is sensitive to the differences in expert assessment.

Figure 13.2 illustrates the results of expert interviews for a specific R&D project. The top section briefly defines the project's objectives and anticipated benefits. We also noted the technical champion for the project as well as the business champion. In this instance, the main uncertainty was when the product would be on vehicles sold to the public, Job1. The expert interviewed was an experienced R&D researcher. Time was measured until the product was on a vehicle to be sold to a customer. This passes the clarity test as it will be clear when such an event happens. There was a 10% chance this could happen in less than 3 years, but there was also a 10% chance it could take longer than 7 years. The second section summarized the business uncertainties: investment cost, projected sales, incremental revenue, and incremental cost. Finance staff in manufacturing were interviewed to forecast investment costs, and marketing experts were interviewed to estimate sales. This information as well as similar information on all major R&D projects was used by top management of an automotive supplier to support R&D project prioritization. MAUT was used to rank the various R&D projects.

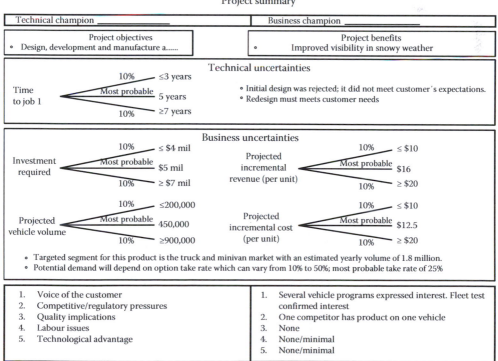

FIGURE 13.2: R&D project summary and uncertainties.

The last section summarizes the various elements that went into the multi-objective analysis: voice of the customer, competitive pressures, quality implications, labor issues, and technological advantage.

13.7.1 Final Task: Convert for Decision Tree Input

Advanced decision tree software is able to take as input a cumulative distribution function. However, it is useful to understand in a simple problem context the conversion from the cumulative to a discrete approximation. In a standard tree, each uncertain branch has a probability associated with the likeliness of that branch occurring. The expected value is calculated by taking a weighted sum of the individual values. This cannot be applied directly to the cumulative distribution. Keefer and Bodily (1983) suggest the following conversion of the cumulative to the discrete as summarized in Table 13.4. They recommend different norms when the interviewer asks for the median or the mode.

Many people initially confuse this conversion because they do not understand why the 10th percentile would have a probability of 0.30 instead of 0.10. The illustration in Figure 13.3 from the Normal distribution may help. The 10th percentile value, −1.28 in the unit normal, must not only reflect the area on its left tail but also some of the area to the right of it. Similarly, the 90th percentile value, $z = 1.28$, is used to represent the area to its right plus some of the area closer toward the middle. Keefer and Bodily suggest that each will represent the 0.1 in the tail plus another 0.20 on the other side. The median will be left with 40% of the total area.

To see how this might appear in a tree, we refer back to our original project question. The tree on the left in Figure 13.4 contains the information as gathered in the interview. The tree on the right illustrates how probabilities are assigned to the branches of the tree with values equal to 9, 12, and 18.

Our discussion placed the expert interview in the context of providing values for a decision tree. In our experience, the interview is also valuable in a much broader range of contexts. Every R&D group should be required to regularly undergo an expert interview process to ensure that valid less-biased estimates are obtained for both the technical and marketable aspects of the projects. It is surprising how little effort generally goes into asking experts their opinion; all too often experts are not asked to explore the basis for their estimates or the potential for bias. Unfortunately, in 2008, the whole world witnessed and came to learn the truth about the biases of financial forecasters first hand.

TABLE 13.4: Convert cumulative to discrete approximation.

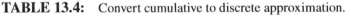

	Specify X_i	Convert Cumulative to Discrete		
Question	$P(X \leq X_i)$	0.10	0.50 (median)	0.90
Approximation	$P(X = X_i)$	0.30	0.40	0.30
	Specify X_i	Convert Cumulative to Discrete		
Question	$P(X \leq X_i)$	0.10	Mode (most likely)	0.90
Approximation	$P(X = X_i)$	0.25	0.50	0.25

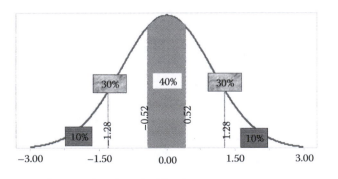

FIGURE 13.3: Normal approximation clarification.

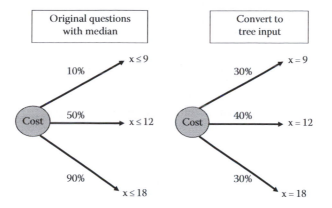

FIGURE 13.4: Conversion of cumulative probabilities to three-point discrete approximations.

13.8 Research into Probabilistic Forecasts

Each of the biases discussed earlier continues to be studied. Researchers explore overconfidence to clarify its various dimensions with regard to different types of tasks (Klayman et al. 1999; Soll and Klayman 2004; Teigen and Jørgensen 2005; Speirs-Bridge et al. 2010). In medicine, doctors demonstrate a mix of overconfidence and underconfidence depending upon the underlying base rate of the illness at hand and its treatment success (Koehler et al. 2002). Overconfidence with respect to the time needed to complete a task is so pervasive that it has earned its own label, planning fallacy (Buehler et al. 2002). Students demonstrate this overconfidence on their projects, as do business executives; witness Boeing having repeatedly delayed the launch of the Dreamliner. A number of researchers have focused on improving the estimates of time, effort, and cost required to complete software development projects (Jørgensen et al. 2004; Jørgensen 2004; Jørgensen and Shepperd 2007). Among their recommendations is asking estimators to justify and criticize their numbers and integrating independent estimates from different experts based on various approaches.

Poor probabilistic reasoning is one source of bias, especially with regard to compound events. Yechiam and Budescu (2006) explored how students adjust their estimates of the likelihood of an occurrence as the time horizon expands. For example, when asked the likelihood of receiving a speeding ticket within the next week and within the next 8 weeks, students did not adjust their probabilities appropriately to reflect the different time periods. Tentori et al. (2004) clarified the prevalence of the conjunctive fallacy, noting that it could not be explained away as a simple misunderstanding of ambiguous terms.

Merely educating people about a variety of forecasting biases may have only a modest impact (Wilson et al. 2002; Larrick 2004). However, specific training that helps individuals understand probabilistic reasoning, such as the relationship between relative frequency and probability, has been shown to have a more consistent debiasing effect (Sedlmeier 1999). Soll and Klayman (2004) suggested specifying the 10th and 90th percentiles in separate steps in order to reduce the overconfidence bias. Speirs-Bridge et al. (2010) suggested a four-step approach for creating confidence intervals. Participants in their study were asked for a realistic lower and upper limit as well as a best guess. Only afterward, in the fourth step, were they asked to specify the size of the confidence interval. This contrasts with the standard approach when the width of the confidence interval has been specified.

The expert interview process presented earlier specified fixed probabilities (FP); the expert was asked to determine the value of the random variable that matched that cumulative probability. An alternative approach involves specifying a fixed value (FV) and asking the expert to specify the cumulative probability. The two approaches are equally valid, but the FV method is preferred since people are commonly asked about the likelihood of an event occurring (Abbas et al. 2008). (See O'Hagan et al. 2006 for a comprehensive review of all aspects of the interview process.)

In many contexts, forecasts are obtained from diverse experts and sources. Often, there is a combination of hard data and expert opinion. The challenge is to combine these into a coherent estimate (Genest and Zidek 1986). Clemen and Winkler (1990, 1993) presented a Bayesian approach for developing point estimates from disparate sources of information. In other research on probability distributions, Clemen and Winkler (1999) compared mathematical and behavioral aggregation and found the different approaches performed comparably.

Recent research has focused on combining experts from non-commensurate sources (Wallsten et al. 1997). Predd et al. (2008) developed a methodology for when experts offer forecasts that are not probabilistically consistent. They also modeled the integration of experts who abstain from forecasting one or more parts of a related set of random events. A strategy for eliciting a forecast involving a complex random variable is to ask the expert to partition the variable into logical components or partitions. However, if the experts represent different disciplines or organizations, their partitions will not necessarily align. Bordley (2009) has presented a general process for combining these distinct partitions.

Appendix 13. A: Phrases: Bad Alternative to Actual Quantification

In the experiment, Beyth-Marom (1982) used 30 different words. Table 13.5 contains the summary of the responses f or the 10 words we used in the activity. There are two sets of columns. One column summarizes the range of 25%–75% and 50% of respondents fell within this range. The next set of columns summarizes the wider range of 10%–90% and covers 80% of the respondents. The words and phrases are sorted by the lower limit. The term *Not Likely* had a value of 5 for its low value and 15 for its high value when considering the 50% range. The range was 2–18 when covering 80% of the respondents. The term *Likely* had the widest range when considering 80% of the respondents. The responses ranged from 0.42 to 0.81, from less than half to more than four-fifths. *Reasonable Chance* displayed a similar pattern. The narrowest ranges were for the terms *Not Likely*, which was always below 0.2, and *Possible*, which was centered around 0.50.

TABLE 13.5: Verbal expressions and probabilities.

Sorted by Lower Limit	Range of 25%–75% 50% of Respondents Fell within These Limits C25–C75		Range of 10%–90% 80% of Respondents Fell within These Limits C10–C90	
Verbal Expression	**Limits**	**Range**	**Limits**	**Range**
Not likely	5–15	10	2–18	16
Poor chance	11–25	14	4–33	29
Doubtful	16–33	17	11–39	28
Small chance	22–36	14	17–42	25
Perhaps	37–59	22	28–58	30
Possible	51–63	12	42–61	19
Likely	53–69	16	42–81	39
Reasonable chance	54–69	15	49–81	32
Most likely	78–92	14	72–97	25
Nearly certain	83–96	13	76–99	23

TABLE 13.6: Answers to confidence interval questions.

Answers	
1. Alexander Hamilton's age at death	47
2. Volume of water in Lake Michigan (gal)	1.3 quadrillion gallons (or 173.7 trillion cubic feet)
3. Population of United States in 1800 census	5.3 million
4. Number of chapters in Book of Psalms	150
5. Sun's volume as a multiple of Earth	109
6. Weight of empty Boeing 747 (lb)	128,730 lb
7. Year of Da Vinci's birth	1452
8. Average height of giraffe at birth (cm)	1800 cm (or 1.8 m)
9. Air distance: Detroit to Buenos Aires (miles)	5500 miles
10. Depth of deepest point of Lake Superior (ft)	1332

Appendix 13. B

Compare answers in Table 13.6 and with your confidence interval ranges specified in Table 13.2.
Population of Greece: 11.3 million
World Jewish Population: There is an estimated 13–14 million Jews in the world with approximately half living in Israel.

Exercises

Complete Chapter Activities

13.1 Provide an example of a forecast within your organization or your community that illustrates the motivational bias. Describe the motivation that contributed to the bias.

13.2 Provide an example of an unrealistic optimism bias in your personal life, organization, or community.

13.3 A project team has just reviewed plans for the launch of a new product. Each member of the team was asked to state in one word or phrase what they thought the chance of success was. Please interpret and assign a probability to each of the following ten words or phrases. Table 13.1.

13.4 Specify your estimate of the Low and High fractiles for each question in Table 13.2.

13.5 Provide an example of an illusion of control bias in your personal life, organization, or community.

13.6 Provide an example of an illusion of predictability bias in your personal life, organization, or community.

13.7 Each of 20 suppliers for complex components of a piece of equipment is highly reliable about meeting deadlines. Each has a track record of meeting the deadlines 95% of the time. What do you estimate is the probability that all of the components will be received on time?

13.8 A rare genetic disorder occurs once in every 50,000 individuals. The standard test is very reliable. If a person has this defect, there is 0.98 probability of the test detecting the disorder. That means there is only a 2% rate of false negatives. In addition, if the genetic disorder is not present, there is 99% accuracy with only a 1% false positive result. The test comes back positive. What are the chances that the individual has this genetic disorder?

13.9 Provide an example of an availability or representative bias in your personal life, organization, or community.

13.10 Provide an example of a limited observation or interview bias in your personal life, organization, or community.

13.11 Provide an example of a confirmation bias in your personal life, organization, or community.

13.12 Describe a survey you recently completed wherein you felt that your answers were not meaningful.

13.13 Describe a context in which two individuals or groups had radically different understandings of task completion.

13.14 Describe a project or task timeline with concrete start and end dates. Then identify critical stages within the timeline that also have clearly defined end dates that would pass the clarity test.

References

Abbas, A. E., Budescu, D. V., Yu, H., and Haggerty, R. (2008). A comparison of two probability encoding methods: Fixed probability vs. fixed variable values. *Decision Analysis*, 5, 190–202.

Ariely, D. (2008). *Predictably Irrational, Revised and Expanded Edition: The Hidden Forces That Shape Our Decisions*. New York: HarperCollins.

Ariely, D., Au, W. T., Bender, R. H., and Budescu, D. V. (2000). The effects of averaging subjective probability estimates between and within judges. *Journal of Experimental Psychology: Applied*, 6, 130–147.

Arrow, K. J. (1951). *Social Choice and Individual Values*. New York: John Wiley.

Bazerman, M. H. (2006). *Judgment in Managerial Decision Making*, 6th edn., Hoboken, NJ: John Wiley & Sons.

Beyth-Marom, R. (1982). How probable is probable? A numerical translation of verbal probability expressions. *Journal of Forecasting*, 1(3), 257–269.

Bordley, R. F. (2009). Combining the opinions of experts who partition events differently. *Decision Analysis*, 6, 38–46.

Brawley, L. R. (1984). Unintentional egocentric biases in attributions. *Journal of Sport Psychology*, 6, 264–278.

Brinthaupt, T. M., Moreland, R. L., and Levine, J. M. (1991). Sources of optimism among prospective group members. *Personality and Social Psychology Bulletin*, 17, 36–43.

Budescu, D. V. and Du, N. (2007). Coherence and consistency of investors' probability judgments. *Management Science*, 53, 1731–1744.

Buehler, R., Griffin, D., and Ross, M. (2002). Inside the planning fallacy: The causes and consequences of optimistic time predictions. In: *Heuristics and Biases: The Psychology of Intuitive Judgment*, Gilovich, T., Griffin, D., and Kahneman, D., eds. Cambridge, U.K.: Cambridge University Press.

Cartwright, S. and Schoenberg, R. (2006). Thirty years of mergers and acquisitions research: Recent advances and future opportunities. *British Journal of Management*, 17, S1–S5.

Clemen, R. T. and Winkler, R. L. (1990). Unanimity and compromise among probability forecasters. *Management Science*, 36, 767.

Clemen, R. T. and Winkler, R. L. (1993). Aggregating point estimates: A flexible modeling approach. *Management Science*, 39, 501.

Clemen, R. T. and Winkler, R. (1999). Combining probability distributions from experts in risk analysis. *Risk Analysis*, 19, 187–203.

Diekmann, K. A. (1997). Implicit justifications and self-serving group allocations. *Journal of Organizational Behavior*, 18(1), 3–16.

Epley, N. (2004). A tale of tuned decks? Anchoring as accessibility and anchoring as adjustment. In: *Blackwell Handbook of Judgment and Decision Making*, Koehler, D. J. and Harvey, N., eds. Oxford, U.K.: Blackwell.

Fischhoff, B., Slovic, P., and Lichtenstein, S. (1978). Fault trees: Sensitivity of estimated failure probabilities to problem representation. *Journal of Experimental Psychology: Human Perception and Performance*, 4, 330–344.

Flyvbjerg, B., Holm, M. K. S., and Buhl, S. L. (2003). How common and how large are cost overruns in transport infrastructure projects? *Transport Reviews*, 23, 71–88.

Fraser, S. and Greene, F. J. (2006). The effects of experience on entrepreneurial optimism and uncertainty. *Economica*, 73, 169–192.

Genest, C. and Zidek, J. (1986). Combining probability distributions: A critique and annotated bibliography. *Statistical Science*, 1, 114–135.

Gilovich, T., Griffin, D., and Kahneman, D., eds. (2002). *Heuristics and Biases: The Psychology of Intuitive Judgment*. Cambridge, U.K.: Cambridge University Press.

Gilovich, T., Vallone, R., and Amos Tversky, A. (1985). The hot hand in basketball: On the misperception of random sequences. *Cognitive Psychology*, 17, 295–314.

Hammond, J. S., Keeney, R. L., and Raiffa, H. (1998). The hidden traps in decision making. *Harvard Business Review*, September–October, 76, 47–58.

Hora, S. C. (2004). Probability judgments for continuous quantities: Linear combinations and calibration. *Management Science*, 50, 597–604.

Jørgensen, M. (2004). A review of studies on expert estimation of software development effort. *Journal of Systems and Software*, 70(1–2), 37–60.

Jørgensen, M. and Shepperd, M. (2007). A systematic review of software development cost estimation studies. *IEEE Transactions on Software Engineering*, 33, 33–53.

Jørgensen, M., Teigen, K. H., and Moløkken, K. (2004). Better sure than safe? Overconfidence in judgment based software development effort prediction intervals. *Journal of Systems and Software*, 70, 79–93.

Kahneman, D. and Lovallo, D. (1993). Timid choices and bold forecasts: A cognitive perspective on risk taking. *Management Science*, 39, 17–31.

Kain, J. F. (1990). Deception in Dallas: Strategic misrepresentation in rail transit promotion and evaluation. *Journal of the American Planning Association*, 56, 184–196.

Keefer, D. L. and Bodily, S. E. (1983). Three point approximations for continuous random variables. *Management Science*, 29, 595.

Klayman, J., Sll, J. B., Gonzalez-Vallejo, C., and Barlas, S. (1999). Overconfidence: It depends on how, what, and whom you ask. *Organizational Behavior and Human Decision Processes*, 79, 216–247.

Koehler, D. J., Brenner, L., and Griffin, D. (2002). The calibration of expert judgment: Heuristics and biases beyond the laboratory. In: *Heuristics and Biases: The Psychology of Intuitive Judgment*, Gilovich, T., Griffin, D., and Kahneman, D., eds. Cambridge, U.K.: Cambridge University Press.

Kramer, R. M., Newton, E., and Pommerenke, P. L. (1993). Self-enhancement biases and negotiator judgment: Effects of self-esteem and mood. *Organizational Behavior and Human Decision Processes*, 56, 113–133.

Kruger, J. and Dunning, D. (1999). Unskilled and unaware of it: How difficulties in recognizing one's own incompetence lead to inflated self-assessments. *Journal of Personality and Social Psychology*, 77, 1121–1134.

Landy, F. (1990). *Psychology of Work Behavior*. New York: Dorsey.

Larrick, R. P. (2004). Debiasing. In: *Blackwell Handbook of Judgment and Decision Making*, Koehler, D. J. and Harvey, N., eds. Malden, MA: Blackwell Publishing.

Lovallo, D. and Kahneman, D. (2003). Delusions of success. How optimism undermines executives decisions. *Harvard Business Review*, 81(7), 56–63.

Love, J. and Cox, W. (1991). False dreams and broken promises: The wasteful federal investment in urban mass transit. *Cato Institute Policy Analysis*, No. 162.

McNatt, D. B. (2000). Ancient Pygmalion joins contemporary management: A meta-analysis of the result. *Applied Psychology*, 85, 314–322.

O'Hagan, A., Buck, C. E., Daneshkhah, A., and Eiser, J. R. (2006). *Uncertain Judgments: Eliciting Experts Probabilities*. West Sussex, England, U.K.: John Wiley & Sons.

Predd, J. B., Osherson, D. N., Kulkarni, S. R., and Poor, H. V. (2008). Aggregating probabilistic forecasts from incoherent and abstaining experts. *Decision Analysis*, 5, 177–189.

Rosenthal, R. and Jacobson, L. (1992). *Pygmalion in the Classroom*. New York: Irvington.

Sedlmeier, P. (1999). *Improving Statistical Reasoning: Theoretical Models and Practical Implications*. Mahwah, NJ: Erlbaum.

Shephard, G. G. and Kirkwood, C. W. (1994). Managing the judgmental probability elicitation process: A case study of analyst/manager interaction. *IEEE Transactions on Engineering Management*, 41(4), 414–425.

Sirower, M. L. (1997). *The Synergy Trap: How Companies Lose the Acquisition Game*. New York: Free Press

Soll, J. B. and Klayman, J. (2004). Overconfidence in interval estimates. *Journal of Experimental Psychology: Learning, Memory, Cognition*, 30, 299–314.

Speirs-Bridge, A., Fidler, F., McBride, M., Flander, L., Cumming, G., and Burgman, M. (2010). Reducing overconfidence in the interval judgments of experts. *Risk Analysis*, 30, 512–523.

Taylor, S. E. (1989). *Positive Illusions: Creative Self-deception and the Healthy Mind*. New York: Basic Books.

Taylor, S. E. and Brown, J. (1988). Illusions and well-being: A social psychological perspective on mental health. *Psychological Bulletin*, 103, 93–110.

Taylor, S. E. and Brown, J. (1994). Positive illusions and well-being revisited: Separating fact from fiction. *American Psychologist*, 49, 972–973.

Teigen, K. H. and Jørgensen, M. (2005). When 90% confidence intervals are 50% certain: On the credibility of credible intervals. *Applied Cognitive Psychology*, 19, 455–475.

Tentori, K., Bonini, N., and Osherson, D. (2004). The conjunction fallacy: A misunderstanding about conjunction. *Cognitive Science*, 28, 467–477.

Tetenbaum, T. J. (1999). Beating the odds of merger & acquisition failure: Seven key practices that improve the chance for expected integration and synergies. *Organizational Dynamics*, 28(2), 22–35.

Timmermans, D. (1994). The roles of experience and domain of expertise in using numerical and verbal probability terms in medical decisions. *Medical Decision Making*, 14, 146–156.

Ulvila, J. W. (1987). Postal automation (zip+4) technology: A decision analysis. *Interfaces*, 17(2), 1–12.

Wallsten, T. S., Budescu, D. V., Erev, L., and Diederich, A. (1997). Evaluating and combining subjective probability estimates. *Journal of Behavioral Decision Making*, 10, 243–268.

Wallsten, T. S., Budescu, D. V., and Zwick, R. (1993). Comparing the calibration and coherence of numerical and verbal probability judgments. *Management Science*, 39, 176–190.

Wede, R. J. (1996). *Teacher Selection: Use of Demonstration Lessons*. Des Moines, IA: Diss. Drake University.

Wilson, T. D., Centerbar, D. C., and Brekke, N. (2002). Mental contamination and the debiasing problem. In: *Heuristics and Biases: The Psychology of Intuitive Judgment*, Gilovich, T., Griffin, D., and Kahneman, D., eds. Cambridge, U.K.: Cambridge University Press.

Yechiam, E. and Budescu, D. V. (2006). The sensitivity of probability assessments to time units and performer characteristics. *Decision Analysis*, 3, 177–193.

Chapter 14

Decision Bias

> You and your spouse attend a charity auction and purchase $100 tickets for the premiere of a movie. After the first 15 minutes, you realize that you are not enjoying the movie. Would you ask your spouse if he feels likewise and consider walking out? Would your decision be affected if many of your friends were at the same show? Would your decision be affected had you not bid for the tickets but rather received them as a gift for donating $200 to an organization?

> Janie inherited a $2.5 million portfolio of stocks and bonds from her recently deceased mother. She quickly reviewed the portfolio, which was mainly composed of blue chip stocks. She was surprised and impressed at how large a portfolio her mother had developed over the last 20 years, notwithstanding that 25% of the stocks were performing poorly. The portfolio was also not aligned with the successful strategy Janie herself had used the last few years. Should she let the inherited portfolio stay as is? Or, if she wants to make changes, how long should she wait to do so?

14.1   Goal and Overview

The goal of this chapter is to develop an understanding of how to overcome cognitive decision biases that are antithetical to good decision making.

In the previous chapter we discussed a number of cognitive biases that affect the accuracy of forecasts and predictions. These inaccuracies obviously undermine good decision-making processes. There are other well-documented cognitive biases that are closely linked not just to the information input but to decisions themselves, such as tendencies to select non-optimal alternatives and reject good ideas. In this chapter, we explore these decision biases and discuss ways of overcoming each of them:

- Sunk cost, escalation, and de-escalation of commitment
- Framing
- Status quo and omissions
- Regret
- Fairness
- Mood
- Groupthink, optimism, and miscellaneous biases

Several of the biases overlap or are closely linked to one another. To isolate the effect of a specific bias, researchers develop creative scenarios for their experimental subjects.

In exploring these biases, we discuss some of their underlying psychological explanations. However, it is impossible to present all the nuances within the extensive literature on each bias. We have chosen to avoid many of these subtleties so as to simplify the presentation of the core concepts. Racial, ethnic, or sexual biases as well as conflict of interest are discussed in a later chapter on ethical decision making.

One may not completely control the decision environment in any given situation, but it is nevertheless critical to increase awareness of one's personal tendencies that can affect decisions via bias. Such awareness can also help one develop a better sense of how bias affects others in their decisions. Toward these ends, we identify actions that can be taken to overcome specific biases.

14.2 Sunk Cost and Escalation of Commitment

According to basic economic principles, future investment should be judged by estimates of future returns without regard to how much has already been invested. The sunk cost bias undermines this principle, because people have a tendency to factor into future investment decisions how much has already been spent—even if it seems that this would be like throwing good money after bad. A psychological explanation for this cognitive bias is rooted in an unwillingness to admit that an investment has been wasted. If the decision at hand is within an organizational or political context, then to discount prior investments means publicly admitting the error of prior decisions.

Politicians are notorious for continuing to invest public funds for projects that have so far failed to live up to expectations and are unlikely to achieve their goals (Arkes and Blumer 1985). This bias becomes even more tragic when lives are at stake. Sunk cost bias is at work when officials say a war must continue, because to stop now would mean the lives already lost will have been wasted. The primary issue should be whether the realistically achievable goals of the war are worth the investment of additional lives. It is tragic for lives to have been wasted in retrospect, but it is even more tragic to invest additional lives in a cause that is likely to be lost.

On a dramatically smaller scale, the sunk cost bias is on display when we continue to watch a movie that we're not enjoying. At the minimum, we have already sunk time into viewing the movie. If we rented a video to watch at home or bought a ticket at a movie theater, we have sunk money into it as well. In the latter situation, especially in the presence of friends, the issue of appearing wasteful has a more public dimension. The same bias is also alive and well when we order dessert, take one bite, and continue to eat in spite of not liking it. The situation may be compounded if a person is dining with someone he does not know well.

Several researchers have documented the sunk cost bias with data from specific decision contexts. McCarthy et al. (1993) determined the effect of sunk cost bias in the investment behavior of entrepreneurs who started their own firms. In a study of professional basketball teams, Staw and Hoang (1995) demonstrated the effect of sunk cost bias regarding the amount of playing time given to high draft choices.

The sunk cost bias has been demonstrated numerous times by researchers in simple questionnaire surveys. One of the first experiments involved asking students to imagine the following scenario:

> You have bought tickets for a weekend skiing trip in Michigan for $100. You later buy $50 tickets for a similar vacation in Wisconsin. Based on past experience, you anticipate that you will enjoy the weekend in Wisconsin even more than the weekend in Michigan. But then you realize that the tickets are for the same dates. Your only option is to choose one weekend over the other.

In a survey of 61 students, 54% chose Michigan over Wisconsin as the preferred ski trip (Arkes and Blumer 1985). From a future perspective, an individual should choose the weekend he will like more regardless of how much he paid for each. Yet more than half of these students demonstrated a sunk cost bias. Making a decision that appeared to waste only $50 by throwing away the Wisconsin ticket was preferable to wasting $100 by throwing away the Michigan ticket.

These same researchers demonstrated sunk cost bias with an experiment involving personal money. Individuals were randomly charged different amounts for season tickets to a university theater season of 10 shows. At the end of the study, they found that individuals who paid the largest amount had a higher attendance rate in the first five shows of the season than those who had been given a discount. However, the effect did not seem to continue into the second half of the theater season: all individuals cut their rate of attendance by half.

The sunk cost effect becomes harder to isolate with more financially complex scenarios. For example:

- Participants were asked whether, as president of an airplane manufacturing company, they would invest the last $1 million of research funds in an R&D project that was considered unlikely to outperform a recently released competitive product. If told that $10 million had already been invested, more than 80% of the respondents chose to spend the final $1 million. With no prior investment specified, 84% chose not to invest.

- Participants were asked to decide, as president of a printing company, whether to invest $10,000 to significantly improve productivity by buying equipment at a steep discount from a company going bankrupt. The decision is to be made in the context of two different previous improvements. In one, the company already invested $200,000 in productivity improvements unrelated to printing. In the second, the company recently purchased new printing equipment that is not as good as that now being offered. Under the first scenario, 76% chose to buy the new printing equipment; in the second case, 47% chose to buy.

The primary explanations for the sunk cost bias are social and psychological, such as saving face and not wanting to appear wasteful (Arkes and Blumer 1985) or a desire for self-justification (Staw 1976; Brockner 1992). There may also be structural barriers in the form of organizational pressure, inertia, or pride that prevent pulling the plug on a failed hire, new product, or delinquent loan. The specifics of the project may also be a factor: the decision maker may perceive high benefits for completing the project and high cost for withdrawal.

With regard to investment decisions, however, the sunk cost bias may result from a misunderstanding as to how to carry out financial analyses. Decision makers may not realize that decisions should be based on marginal analysis of future investment and future returns.

1. Activity: Provide an example in your *organization* of a decision that represents a sunk cost bias. _____

2. Activity: Provide an example in your *personal life* of a decision that represents a sunk cost bias. _____

A failure to understand the concept of sunk cost can also lead to a reverse effect, the de-escalation bias (Heath 1995; Fennema and Perkins 2008). This could occur if the sunk cost dwarfs future payback. Imagine having invested $9 million in a project with no return at all on the investment so far, the investment has been a total loss. Now assume that if you were to invest another $1 million, the projected return would be $2 million. A marginal analysis would warrant proceeding since there is a projected 100% return on the new investment. However, many people use a *mental budget* model that looks at the total of $10 million and compares it against the $2 million payback. They would choose not to make the additional investment.

The sunk variable can also be defined in time rather than money. Navarro and Fantino (2009) created scenarios involving a copper mining project, varying the time sunk into the project from

0 to 60 days until the group discovered a small copper vein. They also manipulated the difficulty of the work and projected returns. Participants' decisions to persist in the project ranged from a low of 34% to a high of 68%, depending on the level of sunk time.

3. Activity: Provide an example in your *organization* or *personal life* of a decision that represents a sunk time bias unrelated to cost. _____

The sunk cost bias as applied to the ski trip did not involve any future investments. If the sunk cost leads to a continuing series of investments, however, it touches on a related phenomenon. When the investments in money or lives grow, the dynamic can be characterized as an escalation of commitment to a failing course of action (Garland 1990; Bazerman et al. 1984). Studies have analyzed this effect with regard to the continued employment of incompetent employees and managers (Drummond 1994), failed loans (Staw et al. 1997), and failed new products (Schmidt and Calantone 2002). Gamblers who increase their bets after a series of losses demonstrate a similar tendency.

4. Activity: Provide an example of escalation of commitment in your *organization* or *personal life* after experiencing losses that should have led to stopping or reducing the commitment.

_____

The escalation of commitment has a parallel in the foot-in-the door technique used by sales people and others (Freedman and Fraser 1966). The foot in the door describes the situation in which an individual agrees to a small request and is later asked for more (Burger 1999). Phone call solicitations from charitable organizations often follow this pattern. Once they reach an individual who is willing to make a small donation, they turn up the charm and immediately ask if the individual could do more. Teenagers are skilled practitioners of this technique with their parents. They may ask first for a small sum of money to buy something modestly priced, then, upon approval, follow with a more substantial request. Neighbors do the same thing when they ask first to borrow a hand tool, then later an expensive power tool, and eventually a car.

14.2.1 Overcoming Sunk Cost Bias

The magnitude of the sunk cost bias is believed to be linked to the level of responsibility the current decision maker feels for the original decision (Schulz-Hardt et al. 2009). If it is the same individual making both decisions, then changing course means admitting an error. If the current decision maker is a peer within the same organization of the original decider, there may still be a shared sense of responsibility that would discourage pulling the plug. In contrast, a new decision maker recruited from outside the organization would more readily assess the future without carrying the baggage of the past. This phenomenon regularly plays out both in the public and corporate domains.

Richard Nixon could more readily accept an inglorious end to the Vietnam War than Lyndon Johnson or even Hubert Humphrey. Ford CEO Allan Mullaly, recruited from Boeing, found it easier to sell off Astin Martin, Jaguar, Land Rover, and Volvo than had the Ford executives who had been with the company over the decade in which the brands had been accumulated. The leadership of General Motors took years to eliminate the Oldsmobile brand even though it was widely viewed as redundant. As GM went through bankruptcy, President Obama and his advisors felt that career GM executives were unable to shed even more redundant brands without significant outside pressure; it fell to a selected board of directors dominated by non-GM types to eliminate Pontiac, Hummer, and Saturn, when no suitable purchasers could be found.

The issue of sunk cost bias and the urge to escalate commitment arises naturally when making loans to businesses. Imagine the dilemma of a loan officer who made a substantial loan to a company that is now struggling to survive. The company executive comes in for another loan to stave off bankruptcy. Should the loan officer refuse and anticipate writing off the original investment, or should he gamble by offering more funds? The hope, of course, is that not only can the new loan be paid back but also the earlier one. In a macro analysis of banking practices, Staw et al. (1997) found that high turnover rates among bank executives led to increased write-offs of delinquent loans and increased set-asides to cover loan losses. This suggests that bank executives would be wise to have a policy that when loan repayment seems to be at risk, all requests for additional loans should be processed by a loan officer who had no responsibility for the earlier loan. Another policy that has been found effective in reducing these biases involves increased monitoring of performance (Kirby and Davis 1998). However, the threat of removal from a loan portfolio and increased supervision may actually have an unintended side effect. To avoid removal or increased oversight, loan officers may simply understate risks of failure as a company gets into deeper financial straits (McNamara et al. 2002).

The responsibility effect, although widely recognized in industry, is hard to replicate in student surveys. Presenting a student with a hypothetical scenario and asking him or her to imagine being responsible cannot possibly replicate a true sense of responsibility for a decision. There would also be no embarrassment associated with going back on a prior decision.

5. Activity: Provide an example in your *organization* or *personal life* in which the sunk cost bias was overcome. What do you believe facilitated the willingness to stop investing in a product, service, or relationship? _____

14.2.2 Awareness, Education, and Justification

We believe that sunk cost bias can be reduced by raising awareness of the concept. Similarly, economics and business education should also have a positive effect on reducing this bias. One professor of accounting is reported to have drilled home the irrelevance of sunk cost by writing on the board at the beginning of each class, "Sunk cost is sunk cost is sunk cost." And yet a number of studies have reported sunk cost bias among MBA students (Conlon and Parks 1987) and accounting undergraduates (Shanteau and Harrison 1991). One study (Tan and Yates 1995) could find no difference in investment behavior between accounting and non-accounting students with regard to sunk costs.

Fennema and Perkins (2008) focused their study of education and sunk cost on the mental budgeting model. In this context, instead of using marginal economic analysis, decision makers reject future investments that are profitable because the investment cannot earn enough to cover the already sunk cost. In their study, education was indeed found to mitigate the sunk cost effect regarding future investments. Whereas psychology majors were given credit for correctly answering only one half of one question (out of three) on a test on the subject, MBA students scored 1.5 on the same test; even so, a score of 50% is hardly overwhelming. In a parallel condition, researchers asked subjects to justify their decisions with both calculations and a narrative explanation. With this requirement, MBA students' performance improved to 2.1 correct answers.

14.3 Framing Bias

Framing bias was one of the earliest biases identified as affecting decision makers (Tversky and Kahneman 1981; Kahneman and Tversky 1984; Gilovich et al. 2002), though it is far from obvious. One form of framing bias is linked to an individual's attitude toward risk. Kahneman and Tversky demonstrated that there is a greater propensity for risk taking when a decision is framed as a choice regarding losses. When the decision is framed in terms of gains, there is greater risk aversion. Their experiment presented two different scenarios in which 600 people were expected to die in an epidemic.

Scenario I: Choose the program you would prefer to implement.

A: Two hundred people will be saved.

B: There is a one-third probability that 600 will be saved and a two-third probability that no one will be saved.

Scenario II: Choose the program you would prefer to implement.

A: Four hundred people will die.

B: There is a one-third probability that no one will die and two-third probability that 600 people will die.

Individuals offered scenario I with a choice between A and B selected A 72% of the time. Yet 78% of the individuals presented with scenario II preferred B, even though the alternatives in each scenario are mathematically identical. The only difference is that scenario I is framed in terms of gains and scenario II in terms of losses. This led the researchers to develop an alternative to utility theory that they called Prospect Theory.

The New Coke decision fiasco in 1985 can be explained in part by this bias (Whyte 1991). Executives at Coca-Cola focused on their losses in market share and were prepared to take risks to recover them. They should have noted instead that they were still the market leader. As the market leader, it makes little sense to tamper with the core taste of your product.

Medical practitioners are especially concerned about framing bias when discussing risks with patients. Describing a procedure as having a 90% survival rate is more likely to lead to its acceptance than when the procedure is characterized as having a 10% fatality rate (Malenka et al. 1993; Gordon-Lubitz 2003).

The framing bias can also appear in conjunction with overvaluing perceived certainty. An insurance policy that provides 100% coverage in one set of circumstances is perceived as more valuable than one that provides 99% coverage in slightly more circumstances. Similarly, a vaccine that works 100% of the time on half the viruses is preferred to one that is 50% effective against twice as many viruses. In both cases the chance of infection is the same.

Johnson et al. (1993) explored the practical relevance of the framing bias in the design of automotive insurance policies and disability policies. They found that potential customers preferred insurance policies with rebates over insurance policies with high deductibles even though the former were significantly more expensive.

The core concept of a framing bias involves uncertainty. However, an analogous concern arises when there is a 4% difference between cash and credit card payments for the same purchase. Credit card companies prefer that the base price be the credit card price and that customers can get a discount by paying with cash. The alternative is that the base price is cash and the purchaser must pay a premium to use a credit card. The companies believe that customers are more willing to use a credit card if the credit price is considered the baseline rather than if it were viewed as a premium option.

The bias of framing can appear in other forms. Economic theory suggests that we should be willing to invest the same amount of effort to save $100 whether the savings relates to the purchase of a $500 item or a $10,000 item. Yet many of us frame our decision in terms of a percentage of savings, in this case 20% versus 1%, with the latter not worthy of effort.

Political leaders and corporate executives often use the concept of framing when they strive to downplay expectations. Candidates running in presidential primaries often set low goals for their vote share in early contests. Regardless of the absolute numbers, they strive to define success as exceeding expectations and not necessarily as winning the primary. Similarly, executives worried about stock prices are concerned that not meeting expectations will lead to a decline in share price irrespective of the absolute value of their profit performance.

6. Activity: Provide an example in your *organization* or *personal life* in which you experienced some form of framing bias. _____

There is no obvious way to overcome the framing bias other than to be aware of it and to seek out alternative frames of reference (Frisch 1993; Park and Rothrock 2007). Several studies indicate that approaches that encourage participants to think more clearly about a strategic decision can mitigate this bias (Hodgkinson et al. 1999; Wright and Goodwin 2002). In general, when decision makers are presented with options to reduce risks in a narrow range of circumstances, they should explore the broader array of risks. In addition, whenever savings are presented as percentages, the decision maker should also look at the absolute values to see whether one's perspective on the issue would change (Hammond et al. 1998).

14.4 Status Quo and Omission Bias

The status quo bias is explored in detail in a classic article by Samuelson and Zeckhauser (1988). They document this bias through a wide range of controlled experiments in a number of decision contexts, some personal, some business related, and some involving public policy decisions. In each context, participants are presented alternatives in three different positions: status quo, no status quo, and alternate status quo.

For example, a classic decision context involves selecting a preferred strategy for investing in a portfolio of alternatives. In one scenario, the individual is simply provided with a sum of money to invest and there is no status quo. In the second scenario, the individual inherits a portfolio of equivalent worth and is asked to select a future investment strategy. The status quo of the inheritance significantly changes his preferences for future investments.

The status quo bias is also clearly demonstrated in choosing the color of a car in short supply. Initially, subjects playing the role of car buyers are told they have to accept whatever color is available. This is the status quo. At the last minute, however, they are told that three other colors are available as well from which they may choose. A similar scenario is also played out with a control group in which four cars of different colors arrive at once. With no color as status quo, 22% preferred red when given the choice as the cars arrive. But with red as the status quo, 50% stuck with the original color even when offered the last-minute opportunity to switch colors.

In a public policy example, survey participants were asked to allocate water between townspeople and farmers in a time of drought. The subjects were given three status quo alternatives: 100,000, 200,000, and 300,000 ac-ft. In this case, the impact of the status quo can be quantified in absolute terms. The corresponding future allocations aligned with the increasing status quo. When the original allocation considered the status quo was 100,000 ac-ft, the average future allocation was 153,000 ac-ft. For those told the original allocation was 200,000 ac-ft, they allocated 183,000 ac-ft. With 300,000 as the status quo, the recommended allocation was 200,000 ac-ft.

The same authors also studied health plan choices and investment allocation strategies for TIAA-CREF retirement fund participants. They compared the choices of continuing participants in these programs with the choices that newcomers picked. With the health insurance example, fewer than 4% changed policies in any given year in their dataset (covering the first half of the 1980s). A variety of insurance plans were selected four times more often by new enrollees than those already enrolled in a plan. Although it is clear that people stay with their first choices, the authors had to adjust for a number of factors to identify the effects of status quo bias. With regard to investments in TIAA-CREF, only 28% changed their ratio of investment allocations even though there was absolutely no cost for changing. This is even more surprising given that the majority of participants stated that their initial allocation was not based on any substantial research.

Samuelson and Zeckhauser present a range of contextual and psychological factors that may contribute to the status quo bias. In their experimental design, they were careful to ensure there was no transition cost with selecting the non-status quo alternative. In addition, identical information was given for each alternative. In contrast, in real-world decisions, there is often a transition cost in switching. This may be an actual startup cost as in the case of using a new supplier. At the minimum, there is a time cost in identifying the other alternatives and gathering information about them. This may be followed by the time cost of analyzing the alternatives. Even without transition costs, the status quo bias increases with the number of alternatives to the status quo. This effect is thus compounded when each of the alternatives require extensive analysis before making a decision.

One explanation for the status quo bias is that people are likely to experience more regret if something goes wrong after taking action than if something goes wrong after failing to act. (We discuss the regret bias in the following section.) Another driver for status quo bias is the inherent uncertainty with non-status quo alternatives; therefore, risk aversion associated with an unknown alternative becomes a factor. This element is present when it comes to product brand loyalty (Jeuland 1979; Chernev 2004). It requires a significant level of dissatisfaction before many buyers will leave their current brand. Unfortunately for the U.S. car industry, many purchasers switched to Japanese cars because of their dissatisfaction with the quality of American cars. U.S. car companies face a daunting status quo bias as they try to woo back American consumers.

This loss aversion and regret can play an interesting role in maintaining the status quo in multi-objective decisions. Imagine two alternatives that are compared on two objectives with each strongest on one of the two objectives. Replacing the status quo means taking a loss on one measure in exchange for a possible gain on another. Our greater tendency to avoid losses than to seek gains would lead the decision maker to stay with the status quo even when there may be more to gain than lose (Kahneman and Tversky 1984).

In many decision contexts, the status quo bias is supplemented by a sunk cost bias if the status quo involves an initial investment of money or time. In an organizational setting, changing the status quo may involve reversing an original decision that could undermine the reputation of the originator of the status quo.

In the majority of instances, maintaining the status quo requires taking no action. The tendency not to act has been labeled the omission bias (Ritov and Baron 1992; Schweitzer 1994). Medical researchers documented this tendency in a study of pulmonary physicians (Aberegg et al. 2005). They were presented with alternative case management strategies in a number of critical care patients. The survey results documented the use of suboptimal patient management as a result of the status quo and omission bias.

When the status quo involves something that is owned, there can also be an endowment effect (Kahneman et al. 1991; Huck et al. 2005). One aspect of this effect is the well-documented difference between selling and purchasing prices. Individuals who are asked the price for which they would sell an item they have just been given quote a price higher than they would be willing to pay themselves to purchase the same item (Morrison 1998). Similarly, when two groups of individuals are randomly given gifts of approximately equal value, a large percentage would be unwilling to exchange their gift for the alternative (Knetsch 1989).

7. Activity: Describe an instance in which you switched from a brand or store that you had been using. Was the motivating factor a negative experience with the status quo or a positive attraction for the new brand or store? _____

The concept of a no-cost trial offer for a product is designed to leverage the status quo bias once the new customer has tried the product. The bias is even stronger when individuals are induced to make a purchase with a money-back guarantee. Similarly, magazine publishers offer a subscription with the first two issues free that can be cancelled at any time.

Public policies are especially prone to the status quo (Pokrivcak et al. 2006). Entitlements once entrenched in law are almost impossible to change. In 2010, state governments were wrestling with long-term deficits, with public sector pension plans a significant contributing factor. The status quo in some states included an automatic cost-of-living adjustment, which seems to be almost untouchable.

There are many forces arrayed against any substantial changes to the status quo of public policies (Pokrivcak et al. 2006). In 2010, Congress failed to pass a law with regard to inheritance until after the election; during a lame duck session, it passed laws that would apply for a few years. The status quo based on a law passed years earlier was that in 2010 no estate of any size would be subject to an inheritance tax. Everyone expected some action on the part of Congress to plug this legal gap, but instead the status quo held for all people who died in 2010. Because Congress recognizes its own inability to change something once it has been put in motion, they sometimes include in legislation a sunset clause that forces changes unless action is taken. The Bush tax cuts of 2001 came with such a clause.

14.4.1 Overcoming the Status Quo

The primary mechanism for addressing this bias is to encourage the decision maker to recognize that the status quo should be treated equally with the other options. It is important not to overestimate the transition cost when considering alternatives. Also, it is important not to justify inertia by complaining of too many alternatives to consider (Hammond et al. 1998).

To encourage a broader review, some companies adopt policies to force serious consideration of change. They seek, for instance, to avoid automatically renewing contracts that may no longer be competitive or quietly retaining employees who no longer perform at an acceptable level. Companies may thus require the purchasing department to request bids from several potential vendors before renewing an existing contract. In personnel decisions, General Electric upsets the status quo by requiring managers to give a grade of C to 10% of their employees. Employees with successive C grades are forced to leave the company.

The concept of zero-based budgeting was once a popular strategy for forcing management to honestly review the status quo. Normally, organizational budgets start with the current year and use it as a baseline for adjustments, with only the adjustments requiring justifications. Zero-based budgeting in theory requires management to justify all expenditures each year and not just changes to the baseline.

There is nevertheless a positive social benefit from the status quo bias. This tendency contributes to individuals maintaining social relationships in times of crisis or conflict rather than reassessing whether they should stay in the relationship. In the past, many states created significant transition costs to discourage rapid divorces. However, the recent widespread adoption of no-fault divorce laws has dramatically reduced this cost and is a contributing factor to rising divorce rates. In addition, the social cost associated with divorce has also been minimized as divorce has become more acceptable in society.

14.5 Regret

Regret is an emotional state ranked second only to love as the most commonly mentioned emotion (Shimanoff 1984). If only we had chosen differently, we often think, the results would have been better. In some cases regret can be a positive influence. You might choose to work late or travel on business and as a result miss an important activity of your children. Regret might lead you to choose differently the next time. We learn from our mistakes, which can inform and improve future decisions. However, when a strong retrospective emotion of regret distorts accurate assessment, the emotion becomes a bias. This important link between regret arising from past experiences and its impact on future decisions has not been adequately studied (Zeelenberg et al. 2002).

Regret aversion can play a role in purchase decisions (Simonson 1992; Zeelenberg and Pieters 1999; Inman and Zeelenberg 2002). It is well established that extended warranties on appliances are not cost effective. However, imagine that within a week of the end of the warranty period on a refrigerator, a major component fails. The owner pays hundreds of dollars for repairs that could have been covered by an extended warranty. If, as a result, the owner purchases an extended warranty on all future appliances, regret has distorted his decision making.

A similar logic can apply to high deductibles for auto collision insurance. Most buyers of a new car can afford to absorb damages up to $1000 and should accept this risk. However, imagine a car owner regretting the decision the first time he is in a collision and has to pay the first $1000 before the insurance covers the rest. This regret bias can be compounded by the representativeness and availability biases discussed in the previous chapter. This one bad experience becomes viewed as overly representative of the risk the owner faces, and he may recall it the next time he makes a similar decision.

Regret is not just a retrospective phenomenon; it can be anticipated. Imagine trying to decide whether to carry around an umbrella all day because there is a 25% chance of thunderstorms. If you carry it around and it does not rain, you will likely regret your decision. If you choose not to take it and are caught in a thunderstorm, you will regret it even more. Savage (1954) framed this decision as minimizing the maximum loss or minimax. Bell (1982) and Loomes and Sugden (1982) developed alternative formulations to utility theory in order to capture the way anticipatory regret can lead decision makers away from maximizing a standard utility function.

Regret aversion should not be confused with risk aversion. Experiments in which there was more anticipatory regret associated with the less risky decision than with the riskier decision have demonstrated the presence of regret aversion in these cases (Zeelenberg 1999). The effect, however, seemed more pronounced when regret aversion and risk aversion were both associated with the same alternative.

There are several factors that contribute to the magnitude of the regret emotion. One of the most important is responsibility for the decision (Ordonez and Connolly 2000). The more responsibility one bears for the final decision, the greater the regret if things go wrong. The issue of responsibility is significant when one is faced with a choice between the dominant market product and an alternative that outperforms it but has not yet been widely accepted. For example, IBM dominated the mainframe computer business through the 1970s. IT professionals knew that there would be little accountability and correspondingly less regret if they purchased an IBM, even if problems arose later.

One area of continued research and debate is the role of action and inaction in both retrospective and anticipatory regret. Following a bad outcome, do people experience more regret from taking action or from not taking action? One suggestion is that action and inaction play out differently in short-term and long-term regret. Short-term (or hot) regret arises in the immediate aftermath of a decision or non-decision when the results turn out badly. Long-term (or wistful) regret arises when we reflect back over our lives and ponder what might have been. In surveys that explored long-term retrospective regret, significantly more respondents regretted actions not taken than actions taken. These often included not making the most of educational opportunities or failing to seize an opportunity. In contrast, studies of short-term regret indicate greater regret with actions gone bad than with inaction gone bad. Several studies (Shimanoff 1984; Gilovich and Medvec 1994; Zeelenberg et al. 2002) suggest the real differentiator is not action or inaction but rather the normative decision in the specific context. Following the normative decision, whether it is action or inaction, generates less regret. For example, faced with a patient at immediate risk of dying, the normative decision is to take some action. In this situation there would be more regret about not trying to save the person's life than doing something that made the situation worse.

8. Activity: Describe something that you regret doing or not doing in the past month. Will this in any way affect any future decisions? _____

9. Activity: Describe something that you regret doing or not doing more than 5 years ago. Did this in any way affect your decisions since then? _____

Regret aversion is significantly affected by the potential for feedback (Zeelenberg 1999). There is greater anticipatory regret for the alternative with feedback than for one without feedback. In the umbrella decision mentioned earlier, the individual knows the results at the end of the day. However, in the case of the medical emergency, although eventually the patient's outcome is known, there may be no way of linking the decision to the final outcome. For many decisions, we only know the results for the path taken and not for the choice not made. We do not know if we would have had problems or have been satisfied with the cell phone plan, car, or college we did not choose. One interesting real-world study involved an unusual Dutch post code lottery in which a person knows whether or not he would have won had he purchased a ticket. In that lottery, the winner is not an individual but a post code. All people in that chosen post code who bought a lottery ticket win. Thus, non-buyers in this type of lottery experience more regret than those in standard lotteries (Zeelenberg and Pieters 2004).

The phenomenon of anticipatory regret can be manipulated to influence behavior. Car rental companies attempt to trigger regret avoidance by encouraging the renter to purchase expensive collision insurance. Who would want to have to pay a few hundred dollars when bringing back a car with a fresh dent? Simonson (1992) carried out an experiment in which some of the participants were encouraged to consider regret before making a purchase decision. One decision focused on purchase timing, buying something on sale now or waiting for an even better sale a month later, even though the item might not be on sale at all next month. A second decision explored the difference between a name brand and a less expensive, lesser known brand. When buyers were encouraged to think about regret, the likelihood of purchase increased 15%–20%.

Richard et al. (1996) explored the potential of raising regret awareness to increase safe-sex practices. Respondents who were induced to focus on their anticipated, post-behavioral regret showed an increased expectation to reduce their risk in future interactions. A study of actual safe-sex behavior 5 months after the study also showed a modest impact on male behavior.

Lastly, regret avoidance can impact the readiness to even make a decision at all (Anderson 2003). Beattie et al. (1994) documented the role of regret in avoiding decisions in a set of medical contexts. Some of the decisions focused on health insurance and others explored treatment options.

14.6 Fairness

The first thing that comes to mind when juxtaposing bias and fairness is social justice. Too often there are biases in our system of justice and economics that lead to unfair outcomes. Our system of justice is supposedly blind, but it is obvious that wealthy individuals can afford legal counsel that increases their chances of winning in court. They can also afford higher quality health care, something that many others consider a basic right of all individuals. The elite American universities, both private and public, have a disproportionate percentage of individuals from families with annual incomes of more than $100,000. However, this bias against fairness or equity is not our focus here. Our discussion is about a bias toward fairness, also labeled inequity aversion, that contradicts fundamental assumptions of game theory. This fairness bias violates the assumption that the decision maker seeks to optimize his or her financial utility function. The concern for fairness suggests creating a social utility function that goes beyond economic self-interest.

The fairness bias can be demonstrated with the ultimatum game described by Guth et al. (1982) (See also Camerer and Thaler 1995). An individual is given $10 and must allocate some part of it to an individual who is a stranger and is aware of the situation. If the proposed offer is accepted, the allocator and recipient keep their respective amounts. If the recipient rejects the offer, they both receive nothing. Game theory argues that the recipient should be willing to accept any amount because he has nothing to lose. Knowing this, the allocator should offer as little as possible. Hoffman et al. (1996) repeated the experiment with $100. In this case, three of four offers of $10 were rejected and two of five offers of $30 were rejected. In numerous other iterations of the game, offers under $20 are often rejected and offers of $40 or more are almost always accepted. The most common offer is to split the amount equally. Variations on the basic ultimatum game continue to be a research topic.

Thaler (1988) placed the ultimatum game in an interesting context. Imagine a vacationer lying on a hot beach when a friend offers to bring back a beer from a nearby location, but neither knows what it will cost. How much is the vacationer willing to pay for a cold beer? This question was placed within two distinct contexts: a fancy resort hotel and a small run-down grocery store. On average, the imagined vacationers were willing to pay 75% more for the same purchase from the fancy hotel. Paying that same price at the grocery store, however, was considered a rip-off.

10. Activity: Describe a context in which you felt you were ripped off on a purchase. Did you buy the item anyway? _____

One of the most common areas of concern is wage fairness (Kahneman et al. 1986; Akerlof and Yellen 1990; Fehr et al. 1993). Who is not upset seeing a peer in the same organization receive a higher pay raise for the same level of performance? How many of us would be happy working alongside someone being paid 20% more to do the same job? During the U.S. Civil War, several African–American regiments refused to accept any pay until the federal government set their pay scale equal to that of white soldiers. Yet one way companies and governments deal with budget and revenue shortfalls is by establishing lower pay and benefits packages for recent hires and incoming employees. There is evidence that this type of wage structure leads to higher rates of employee turnover. In another study, workers preferred a lower wage to work alone rather than a somewhat higher salary to work together with somebody doing the same job who is paid even more.

Labor-management negotiations have at times reached an impasse over labor's unwillingness to accept an unfair wage offer. Companies have insisted on wage concessions with the threat that without them the company will move operations outside the country. This is an extreme example of the ultimatum game and leaves labor with the unsavory choice between unfair wages and no jobs. Whatever the final decision, this type of situation can lead to long-term protracted negotiations or an extended strike before labor ultimately accepts the inevitable or loses out completely.

Exponential growth in CEO salaries has attracted growing concern over fairness (Cremer and Dijk 2005). In 1965, CEO salaries averaged 24 times the average salary of a worker. By 1978 this number had grown to 35 and by 1989 to 71. By 2005, CEO salaries were a staggering 262 times the average worker's salary. In one especially egregious case, an automotive supplier CEO was rewarded with a multi-million dollar bonus after negotiating a wage reduction for his unionized workforce under threat that he would move the jobs out of the country unless they accepted cutbacks.

11. Activity: Describe a context in which you observed wage unfairness. Did you or a colleague consider a job change as a result? _____

The primary response to unfairness considered so far involves refusing to accept the situation: rejecting the ultimatum, not paying for the overpriced item, or seeking another job. There is also an entirely different stream of research involving reciprocal behavior. One specific response is to work less hard if wages are deemed unfair. Automotive suppliers have stated that, in response to their perceived unfair treatment at the hands of U.S. automotive companies, they would present their ideas for innovation to Japanese companies first.

There are other fairness stories beyond those involving unfair wage offers that maximize a company's short-term benefit. There is widespread evidence that companies do not consider it fair to their workers to reduce wages in the midst of a recession with high unemployment (Kahneman et al. 1986). This was true even in the recession of 2008, which was the worst since the 1930s. However, there is some indication that companies lower their salary offers to new employees. As a result, there is growing concern that these same companies will pay for this policy with higher turnover as the job market picks up.

It is human nature to perceive certain actions or situations as unfair. Kahneman et al. (1986) asked survey participants to categorize 18 different actions as acceptable or unfair. These included multiple situations involving price increases, rent increases, or wage reductions. The first scenario involved a price increase in snow shovels immediately after a snowstorm; 82% of the respondents declared this unfair. The most egregious action according to respondents involved a rent increase for someone whom the landlord knew had special reasons for not moving. Ten out of eleven respondents declared this to be unfair. An interesting scenario involved a toy store that decided to auction off its last, highly-sought-after doll to the highest bidder. Seventy-four percent found this auction unfair. However, when the proceeds of the auction were designated for charity, those deeming the auction unfair dropped to 21%.

The fairness bias is distinct from the others in this chapter. There is generally nothing implicitly wrong with the bias that warrants discussion of strategies to overcome this bias. However, there are many contexts in which it is necessary for individuals to accept the reality of an unfair situation. One factor that can reduce the sense of inequity is whether the unfairness is intended or unintended (Falk et al. 2008). Politicians use this excuse to sell the public on compromises that they claim they had no choice but to accept. When there is no alternative to the unfair situation, politicians cannot be accused of intentional bias. In late 2010, President Obama had to use this excuse to appeal to members of his party to support his tax cut compromise even though many Democrats felt it was unfair to continue tax breaks for wealthier individuals. It was clear to all Democrats, however, that the incoming Republican-controlled House of Representatives would not pass a tax bill they would consider fairer. Nevertheless, there was significant sentiment to not accept this ultimatum and simply forgo a tax cut deemed inequitable. In the end, the compromise passed.

14.7 Mood

The concept that mood influences behavior and decision making is as ancient as the Hebrew Bible and a topic of discussion among Greek philosophers. Of particular interest to philosophers and ethicists is the struggle between desire and reason. They wonder why intense desire can cause people to disregard risk or improprieties that under cooler conditions would lead to different choices (Ditto et al. 2006).

One emotion of particular interest is anger, which can be viewed both positively and negatively. It is a term used to describe God when He takes aggressive action in response to major ethical lapses of the Israelites. Moses was angry when he broke the tablets with the Ten Commandments. Yet it is also noted that anger can lead to rash action soon to be regretted. Laurence Peter wrote, "Speak when you are angry—and you'll make the best speech you'll ever regret." Aristotle (2007) also framed the issue: "Anyone can become angry—that is easy, but to be angry with the right person at the right time, and for the right purpose and in the right way—that is not within everyone's power and that is not easy." In a modern context, we are all best advised not to send an e-mail written while angry. There is nothing wrong with writing an e-mail while angry; it is clicking the send button that the writer will usually regret.

Davidson et al. (2003) offer a comprehensive overview of research into mood, emotions, and behavior in the *Handbook of Affective Science*. Our focus is limited to decision making and a few examples of the most prominently researched relationships. In particular, we are interested in the role that *incidental* mood or emotion plays in decision making, one that is unrelated to the specific decision at hand. Thus, we explore how a negative or positive mood might affect an investment decision involving risk, but not, for example, how a negative mood brought on by a catastrophic illness might affect decisions related to that illness. In the latter instance, the mood is *integral* to the decision. The regret bias discussed earlier is almost always integral to the decision.

In developing experimental design, researchers often explore how mood both directly and indirectly affects the ultimate decision. Mood may affect what data is sought, how carefully that data is processed and interpreted, and finally the preferences for different outcomes (Peters et al. 2006).

Researchers exploring the relationship between mood and making a decision use a number of strategies to ensure that a specific affect is present at the time of decision. One strategy involves asking participants to recall and then dwell upon an incident that is likely to create the mood being studied. Alternatively, they may provide written scenarios or videos to generate the emotion of interest. In some instances researchers have randomly selected individuals and asked them to assess their current mood before engaging in decision making (Ito et al. 1998).

Early research was dichotomous, simply classifying people as being in either a negative or positive mood. Later, researchers strove for more detailed descriptions of mood to understand the effect on decision making. The most general observation is that people in a happy mood are more prone to approach a decision with a non-systematic heuristic strategy. They will tend to rely on their current knowledge and not focus on details. People who are sad will tend to be more systematic and look for critical data to be reviewed carefully (Schwarz 2000).

One stream of research focuses on decisions involving uncertainty. When studying the impact of mood, researchers design experiments that differentiate between risk perception and risk attitude, both of which can impact risk aversion. One study reported that participants experiencing anxiety preferred low risk/low reward gambles, while sad participants tended toward high risk/high reward choices (Raghunathan and Pham 1999). One decision involved a choice between two jobs: a high salary with low job security versus average salary with high job security. Seventy-eight percent of the sad participants chose the high salary with low security, while only 32% of the anxious participants chose the same job. Participants whose mood was neutral fell in the middle of these two groups; they chose the high salary job 56% of the time.

Anger is the most widely studied emotion and typically leads to aggressive behavior. Angry decision makers tend to consider few alternatives. Although generally grouped with negative emotions, it has many aspects that align it with positive emotions. It generates an eagerness to act as well as an optimistic view of the chance for success when taking action. In decisions involving uncertainty, angry people will be risk prone (Lerner and Keltner 2000; Lerner and Tiedens 2006). The aforementioned observations apply also to situations in which the anger is unrelated to the decision; it is simply a carryover effect of being in an angry mood.

Van Winden (2007) explored the role anger played in an individual's assessment of the unfairness of a proposed allocation of resources. The angrier the respondent, the more likely the individual was to destroy resources rather than allow someone else to take an unfair share.

Good mood is associated with risk aversion (Kliger and Levy 2003). The primary psychological explanation is that people in a good mood want to maintain the mood and are leery of taking risks that could produce a bad outcome and thus a mood change. This aversion to a mood-changing loss even outweighs the natural optimism of a good mood that leads to lower risk perception (Nygren et al. 1996). Mano (1994) included arousal as another categorization of mood as he studied attitudes toward taking risks and buying insurance. Although anger and sadness are both considered negative, anger is categorized as high arousal and sadness as low arousal leading to distinct decision behaviors with regard to risk taking.

A number of the articles have explored how moods impact other biases discussed earlier. Lin et al. (2006) showed that the endowment effect was present when people were induced to be happy but not when induced to feel sad. Yen and Chuang (2008) found that a positive mood increased the strength of the status quo bias, with 75% preferring the status quo; a negative mood had the reverse effect, with only 47% preferring the status quo. Although not a scientific study, the congressional elections of 2010 occurred while the U.S. electorate was in a generally bad mood, which led to heavy losses for incumbents.

12. Activity: Describe a recent context in which you made a decision in an aroused emotional state. Do you think you would have made the same decision if you had been calmer?

There is evidence that the impact of incidental affect can be reduced by raising the decision maker's awareness of his accountability for the decision. In one study, an emphasis on accountability reduced the tendency for a happy person to use high-level heuristics such as stereotypes in making decisions (Bodenhausen et al. 1994). In a study of anger, accountability reduced the carryover tendency of angry people to be punitive to other unrelated decisions (Lerner et al. 1998).

14.8 Groupthink, Optimism, and Miscellaneous Biases

Dysfunctional processes in group decision contexts encompass an entire separate stream of bias research. The most prominent theory is the concept of groupthink, which arose out of an effort by Janis (1971, 1972, 1982) to explain how groups of bright people could make terrible decisions leading to major fiascos (Janis and Mann 1977). These include the Kennedy administration's decision to approve a planned invasion of Cuba at the Bay of Pigs, Johnson's escalation of the Vietnam War, and Nixon's Watergate. More recently, it has been used to explain the Bush administration's belief that a military victory in Iraq would be followed by U.S. troops being greeted with cheers as heroes and with a productive democratic government installed quickly thereafter. This linkage with widely known fiascos has given the concept of groupthink mass credibility that is not generally justified by actual research into group decision making.

There are two components to the groupthink hypothesis. The first focuses on the conditions or antecedents that lead to a groupthink decision fiasco. These include a highly cohesive group that is insulated from outside experts. The group operates under strong directed leadership to address a high-stress decision with no clearly good alternatives to the choice under consideration. The second describes the symptoms of a flawed decision process. These symptoms include an illusion of invulnerability, stereotypes of outsiders, and belief in the inherent morality of the group. The associated behaviors that contribute to a bad decision process are poor information search, selective information processing, self-censorship, and failure to really understand the risks.

The core key research questions regarding groupthink are "How many of these antecedents are critical to producing groupthink, and are the resultant symptoms a necessary part of the groupthink phenomenon?" In a special 1998 issue of *Organizational Behavior and Human Decision Processes*, a number of researchers presented their critical reviews of groupthink and related research (Esser 1998; Fuller and Aldag 1998; Paulus 1998; Turner and Pratkanis 1998; Whyte 1998). The discussions suggest the overall theory is questionable at best, with only parts of it supported by solid research. Researchers are especially interested in documenting which factors contribute to bad decisions. For example, there is serious doubt about the role of group cohesion as a negative factor (Bernthal and Insko 1993). In many other contexts, group cohesion can lead to effective, high-quality decisions that are efficiently implemented.

Baron (2005) suggests that the notion of groupthink retains its power because decision groups often experience symptoms of the theory that contribute to bad decisions, even though the antecedents are not present. In essence, important elements of groupthink are ubiquitous, which is what people remember even if the causal theory is flawed. Baron proceeds to review the research on group decision processes in each of the following areas: conformity and suppression of dissent, self-censorship, illusion of consensus, failure to consider risks, and out-group vilification. This last issue, was highly visible within each of the major political parties during the U.S. elections in 2010.

Janis (1982) also suggested vigilant decision making for overcoming groupthink. Vigilance involves utilizing many of the strategies described here: defining a range of objectives, extensively searching for relevant information, and developing contingency plans. The primary goal is to reduce the pressures of the group to conform to a quickly achieved consensus. Peterson et al. (1998) reviewed the decisions in a number of successful and unsuccessful organizations in order to identify which of the elements of vigilance were most commonly found in successful decision groups. One area of disagreement is that Peterson et al. found that a strong leader articulating his preferences was commonly present in these successful groups. In contrast, Janis suggested that the leader should set a tone of impartiality (Hart 1998).

One of the earliest and most widely documented biases is that of optimism, which leads people to underestimate the risks they face. Branstrom et al. (2006) documented the tendency for people to believe they are less susceptible to specific diseases than other people are. This optimism bias is a contributing factor to why individuals often engage in risky behavior, whether related to exposure to the sun, unprotected sex, or drinking (Weinstein 1980, 1987; Harris et al. 2008). It also explains why people routinely underestimate task completion times (Newby-Clark et al. 2000).

Other biases include the *not invented here* syndrome which is a barrier to organizational transfer of knowledge. Katz and Allen (1982) studied this phenomenon in the context of R&D projects. This syndrome is also a barrier for manufacturing and service providers to adopt ideas and processes that have worked well elsewhere. Conversely, many proposals are shot down because they are deemed similar to something that supposedly failed earlier. This criticism is hard to combat without all the facts regarding how serious the previous effort was and if in fact the situations in question are really parallel. This is reflective of the representative bias discussed in the previous chapter.

The use of stereotypes to judge people and behaviors is a broad-based bias (Devine 1989). Included are biases related to race, gender, ethnicity, religion, handicap, and many others. This bias can lead to biases in hiring and promotion, loan approval, punishment, or investigation, and is a leading factor in witness mis-identification. Some of these behavioral biases have been made illegal in the United States, but it is impossible to legislate against cognitive processes that may unknowingly influence decisions.

In summary, there is a wide array of hidden and not-so hidden traps that we as individuals and groups can stumble over as we make routine and complex decisions. This chapter, coupled with the preceding chapter about forecasting bias, has been designed to raise the reader's awareness in order to help reduce dysfunctional cognitive tendencies.

Exercises

Complete Chapter Activities

14.1 Provide an example in your *organization* of a decision that represents a sunk cost bias.

14.2 Provide an example in your *personal life* of a decision that represents a sunk cost bias.

14.3 Provide an example in your *organization* or *personal life* of a decision that represents a sunk time bias unrelated to cost.

14.4 Provide an example of escalation of commitment in your *organization* or *personal life* after experiencing losses that should have led to stopping or reducing the commitment.

14.5 Provide an example in your *organization* or *personal life* in which the sunk cost bias was overcome. What do you believe facilitated the willingness to stop investing in a product, service, or relationship?

14.6 Provide an example in your *organization* or *personal life* in which you experienced some form of framing bias.

14.7   Describe an instance in which you switched from a brand or store that you had been using. Was the motivating factor a negative experience with the status quo or a positive attraction for the new brand or store?

14.8   Describe something that you regret doing or not doing in the past month. Will this in any way affect any future decisions?

14.9   Describe something that you regret doing or not doing more than 5 years ago. Did this in any way affect your decisions since then?

14.10 Describe a context in which you felt you were ripped off on a purchase. Did you buy the item anyway?

14.11 Describe a context in which you felt you felt or observed wage unfairness. Did you or a colleague consider a job change as a result?

14.12 Describe a recent context in which you made a decision in an aroused emotional state. Do you think you would have made the same decision if you had been calmer?

References

Aberegg, S. K., Haponik, E. F., and Terry, P. B. (2005). Omission bias and decision making in pulmonary and critical care medicine. *Chest*, 128, 1497–1505.

Akerlof, G. A. and Yellen, J. L. (1990). The fair wage-effort hypothesis and unemployment. *Quarterly Journal of Economics*, 105, 255–283.

Anderson, C. J. (2003). The psychology of doing nothing: Forms of decision avoidance result from reason and emotion. *Psychology Bulletin*, 129(1), 139–167.

Aristotle. (2007). Nicomachean Ethics. (W. D. Ross Trans.) Sioux Falls, SD. NuVision Publications. (Original work published 350 B.C.E.).

Arkes, H. R. and Blumer, C. (1985). The psychology of sunk cost. *Organizational Behavior and Human Decision Processes*, 35, 124–140.

Baron, R. S. (2005). So right it's wrong: Groupthink and the ubiquitous nature of polarized group decision making. *Advances in Experimental Social Psychology*, 37, 219–253.

Bazerman, M., Giuliano, T., and Appleman, A. (1984). Escalation of commitment in individual and group decision making. *Organizational Behavior and Human Performance*, 33, 141–152.

Beattie, J., Baron, J., Hershey, J. C., and Spranca, M. D. (1994). Psychological determinants of decision attitude. *Journal of Behavioral Decision Making*, 7, 129–144.

Bell, D. E. (1982). Regret in decision making under uncertainty. *Operations Research*, 30, 961–981.

Bernthal, P. R. and Insko, C. A. (1993). Cohesiveness without groupthink: The interactive effects of social and task cohesion. *Group and Organizational Management*, 18, 66–87.

Bodenhausen, G. V., Kramer, G. P., and Susser, K. (1994). Happiness and stereotypic thinking in social judgment. *Journal of Personality and Social Psychology*, 66, 621–632.

Branstrom, R., Kristjansson, S., and Ullen, H. (2006). Risk perception, optimistic bias, and readiness to change sun related behavior. *European Journal of Public Health*, 16, 492–497.

Brockner, J. (1992). The escalation of commitment to a failing course of action: Toward theoretical progress. *Academy of Management Review*, 17, 39–61.

Burger, J. M. (1999). The foot-in-the-door compliance procedure: A multiple-process analysis and review. *Personality and Social Psychology Review*, 3, 303–325.

Camerer, C. and Thaler, R. H. (1995). Anomalies: Ultimatums, dictators, and manners. *Journal of Economic Perspectives*, 9, 209–219.

Chernev, A. (2004). Goal orientation and consumer preference for the status quo. *Journal of Consumer Research*, 31, 557–565.

Conlon, E. J. and Parks, J. M. (1987). Information requests in the context of escalation. *Journal of Applied Psychology*, 72, 344–350.

Cremer, D. and Dijk, E. (2005). When and why leaders put themselves first: Leader behaviour in resource allocations as a function of feeling entitled. *European Journal of Social Psychology*, 35, 553–563.

Davidson, R. J., Scherer, K. R., and Goldsmith, H. H., eds. (2003). *Handbook of Affective Science*. New York: Oxford University Press.

Devine, T. (1989). Stereotypes and prejudice: Their automatic and controlled components. *Journal of Personality and Social Psychology*, 56, 5–18.

Ditto, P. H., Pizarro, D. A., Epstein, E. B., Jacobson, J. A., and MacDonald, T. K. (2006). Visceral influences on risk-taking behavior. *Journal of Behavioral Decision Making*, 19, 99–113.

Drummond, H. (1994). Escalation in organizational decision making: A case of recruiting an incompetent employee. *Journal of Behavioral Decision Making*, 7, 43–55.

Esser, J. K. (1998). Alive and well after 25 years: A review of groupthink research. *Organizational Behavior and Human Decision Processes*, 73, 116–141.

Falk, A., Fehr, E., and Fischbacher, U. (2008). Testing theories of fairness—intentions matter. *Games and Economic Behavior*, 62, 287–303.

Fehr, E., Kirchsteiger, G., and Reidl, A. (1993). Does fairness prevent market clearing? An experimental investigation. *Quarterly Journal of Economics*, 108, 437–460.

Fennema, M. G. and Perkins, J. D. (2008). Mental budgeting versus marginal decision making: Training, experience and justification effects on decisions involving sunk costs. *Journal of Behavioral Decision Making*, 21, 225–239.

Freedman, J. L. and Fraser, S. C. (1966). Compliance without pressure: The foot-in-the-door technique. *Journal of Personality and Social Psychology*, 4, 195–202.

Frisch, D. (1993). Reasons for framing effects. *Organizational Behavior and Human Decision Processes*, 54, 399–429.

Fuller, S. R. and Aldag, R. J. (1998). Organizational Tonypandy: Lessons from a quarter century of the groupthink phenomenon. *Organizational Behavior and Human Decision Processes*, 73, 163–184.

Garland, H. (1990). Throwing good money after bad: The effect of sunk costs on the decision to escalate commitment to an ongoing project. *Journal of Applied Psychology*, 75, 728–731.

Gilovich, T., Griffin, D. W., and Kahneman, D. (2002). *Heuristics and Biases: The Psychology of Intuitive Judgment*. Cambridge, U.K.: Cambridge University Press.

Gilovich, T. and Medvec, V. H. (1994). The temporal pattern to the experience of regret. *Journal of Personality and Social Psychology*, 67, 357–365.

Gordon-Lubitz, R. J. (2003). Risk communication: Problems of presentation and understanding. *JAMA*, 289, 295.

Guth, W., Schmittberger, R., and Schwarze, B. (1982). An experimental analysis of ultimatum bargaining. *Journal of Economic Behavior and Organization*, 3, 367–388.

Hammond, J. S., Keeney, R. L., and Raiffa, H. (1998). The hidden traps in decision making. *Harvard Business Review*, 76(5), 47–52.

Harris, P. R., Griffin, D. W., and Murray, S. (2008). Testing the limits of optimistic bias: Event and person moderators in a multilevel framework. *Journal of Personality and Social Psychology*, 95, 1225–1237.

Hart, P. T. (1998). Preventing groupthink revisited: Evaluating and reforming the groups and government. *Organizational Behavior and Human Decision Processes*, 73, 306–326.

Heath, C. (1995). Escalation and de-escalation of commitment in response to sunk costs: The role of budgeting in mental accounting. *Organizational Behavior and Human Decision Processes*, 62, 38–54.

Hodgkinson, G. P., Bown, N. J., Maule, A. J., Glaister, K. W., and Pearman, A. D. (1999). Breaking the frame: An analysis of strategic cognition and decision making under uncertainty. *Strategic Management Journal*, 20, 977–985.

Hoffman, E., McCabe, K. A., and Smith, V. L. (1996). On expectations and the monetary stakes in ultimatum games. *International Journal of Game Theory*, 25, 289–301.

Huck, S., Kirchsteiger, G., and Oechssler, J. (2005). Learning to like what you have—explaining the endowment effect. *The Economic Journal*, 115, 689–702.

Inman, J. J. and Zeelenberg, M. (2002). Regret in repeat purchase versus switching decisions: The attenuating role of decision justifiability. *Journal of Consumer Research*, 29, 116–128.

Ito, T., Cacioppo, J. T., and Lang, P. J. (1998). Eliciting affect using the international affective picture system: Trajectories through evaluative space. *Personality and Social Psychology Bulletin*, 24, 855–879.

Janis, I. L. (1971). Groupthink, *Psychology Today*, 5, November, 43–46 and 74–76.

Janis, I. L. (1972). *Victims of Groupthink*. Boston, MA: Houghton Mifflin.

Janis, I. L. (1982). *Groupthink*, 2nd edn. Boston, MA: Houghton Mifflin.

Janis, I. L. and Mann, L. (1977). *Decision Making: A Psychological Analysis of Conflict, Choice, and Commitment*. New York: Free Press.

Jeuland, A. P. (1979). Brand choice inertia as one aspect of the notion of brand loyalty. *Management Science*, 25, 671–682.

Johnson, E. J., Hershey, J., Meszaros, S., and Kunreuther, H. (1993). Framing, probability distortions, and insurance decisions. *Journal of Risk and Uncertainty*, 7, 15–36.

Kahneman, D., Knetsch, J., and Thaler, R. H. (1986). Fairness as a constraint on profit seeking: Entitlements in the market. *American Economic Review*, 76, 728–741.

Kahneman, D., Knetsch, J. L., and Thaler, R. H. (1991). Anomalies—The endowment effect, loss aversion, and the status quo bias. *Journal of Economic Perspectives*, 5, 193–206.

Kahneman, D. and Tversky, A. (1982). The psychology of preference. *Scientific American*, 246, 160–173.

Kahneman, D. and Tversky, A. (1984). Choices, values and frames. *American Psychologist*, 39, 341–350.

Katz, R. and Allen, T. J. (1982). Investigating the not invented here (NIH) syndrome: A look at the performance, tenure, and communication patterns of 50 R & D project groups. *R&D Management*, 12(1), 7–20.

Kirby, S. L. and Davis, M. A. (1998). A study of escalating commitment in principal-agent relationships: Effects of monitoring and personal responsibility. *Journal of Applied Psychology*, 83, 206–217.

Kliger, D. and Levy, O. (2003). Mood-induced variation in risk preferences. *Journal of Economic Behavior and Organization*, 52, 573–584.

Knetsch, J. L. (1989). The endowment effect and evidence of nonreversible indifference curves. *American Economic Review*, 79, 1277–1284.

Lerner, J. S., Goldberg, J. H., and Tetlock, P. E. (1998). Sober second thought: The effects of accountability, anger, and authoritarianism on attributions of responsibility. *Personality and Social Psychology Bulletin*, 24, 563–574.

Lerner, J. S. and Keltner, D. (2000). Beyond valence: Toward a model of emotion-specific influences on judgment and choice. *Cognition and Emotion*, 14, 473–493.

Lerner, J. S. and Tiedens, L. Z. (2006). Portrait of the angry decision maker: How appraisal tendencies shape anger's influence on cognition. *Journal of Behavioral Decision Making*, 19, 115–137.

Lin, C.-H., Chuang, S. C., Kao, D. T., and Kung, C. Y. (2006). The role of emotions in the endowment effect. *Journal of Economic Psychology*, 27, 589–597.

Loomes, G. and Sugden, R. (1982). Regret theory: An alternative theory of rational choice under uncertainty. *Economic Journal*, 92, 805–824,

Malenka, D. J., Baron, J. A., Johansen, S., Wahrenberger, J., and Ross, J. M. (1993). The framing effect of relative and absolute risk. *Journal of General Internal Medicine*, 8, 543–548.

Mano, H. (1994). Risk-taking framing effects, and affect. *Organizational Behavior and Human Decision Processes*, 57, 38–58.

McCarthy, A. M., Schoorman, F. D., and Cooper, A. C. (1993). Reinvestment decisions by entrepreneurs: Rational decision-making or escalation of commitment? *Journal of Business Venturing*, 8(1), 9–24.

McNamara, G., Moon, H., and Bromiley, P. (2002). Banking on commitment: Intended and unintended consequences of an organization's attempt to attenuate escalation of commitment. *Academy of Management Journal*, 45, 443–452.

Morrison, G. C. (1998). Understanding the disparity between WTP and WTA: Endowment effect, substitutability, or imprecise preferences? *Economics Letters*, 59(2), 189–194.

Navarro, A. D. and Fantino, E. (2009). The sunk-time effect: An exploration. *Journal of Behavioral Decision Making*, 22, 252–270.

Newby-Clark, I. R., Ross, M., Buehler, R., Koehler, D. J., and Griffin, D. (2000). People focus on optimistic scenarios and disregard pessimistic scenarios while predicting task completion times. *Journal of Experimental Psychology: Applied*, 6(3), 171–182.

Nygren, T. E., Isen, A. M., Taylor, P. J., and Dulin, J. (1996). The influence of positive affect on the decision rule in risk situations: Focus on outcome (and especially avoidance of loss) rather than probability. *Organizational Behavior and Human Decision Processes*, 66, 59–72.

Ordonez, L. D. and Connolly, T. (2000). Regret and responsibility: A reply to Zeelenberg et al. (1998). *Organizational Behavior and Human Decision Processes*, 81, 132–142.

Park, S. and Rothrock, L. (2007). Systematic analysis of framing bias in missile defense: Implications toward visualization design. *European Journal of Operational Research*, 182, 1383–1398.

Paulus, P. B. (1998). Developing consensus about groupthink after all these years. *Organizational Behavior and Human Decision Processes*, 73, 362–374.

Peters, E., Vastfjall, D., Garling, T., and Slovic, P. (2006). Affect and decision making: A 'hot' topic. *Journal of Behavioral Decision Making*, 19, 79–85.

Peterson, R. S., Owens, P., Tetlock, P. E., Fan, E. T., and Martorana, P. (1998). Group dynamics in top management teams: Groupthink, vigilance, and alternative models of organizational failure and success. *Organizational Behavior and Human Decision Processes*, 73, 272–305.

Pokrivcak, J., Crombez, C., and Swinnen, J. F. M. (2006). The status quo bias and reform of the common agricultural policy: Impact of voting rules, the European commission and external changes. *European Review of Agricultural Economics*, 33, 562–590.

Raghunathan, R. and Pham, M. T. (1999). All negative moods are not equal: Motivational influences of anxiety and sadness on decision making. *Organizational Behavior and Human Decision Processes*, 79, 56–77.

Richard, R., van der Pligt, J., and de Vries, N. (1996). Anticipated regret and time perspective: Changing sexual risk-taking behavior. *Journal of Behavioral Decision Making*, 9, 185–199.

Ritov, I. and Baron, J. (1992). Status-quo and omission bias. *Journal of Risk and Uncertainty*, 5, 49–61.

Samuelson, W. and Zeckhauser, R. (1988). Status quo bias in decision making. *Journal of Risk and Uncertainty*, 1, 7–59.

Savage, L. J. (1954). *The Foundations of Statistics*. New York: John Wiley and Sons.

Schmidt, J. B. and Calantone, R. J. (2002). Escalation of commitment during new product development. *Journal of the Academy of Marketing Science*, 30, 103–118.

Schulz-Hardt, S., Thurow-Kröning, B., and Frey, D. (2009). Preference-based escalation: A new interpretation for the responsibility effect in escalating commitment and entrapment. *Organizational Behavior and Human Decision Processes*, 108, 175–186.

Schwarz, N. (2000). Emotion, cognition, and decision making. *Cognition and Emotion*, 14, 433–440.

Schweitzer, M. (1994). Disentangling status quo and omission effects: An experimental analysis. *Organizational Behavior and Human Decision Processes*, 58, 457–476.

Shanteau, J. and Harrison, P. (1991). The perceived strength of an implied contract: Can it withstand financial temptation? *Organizational Behavior and Human Decision Processes*, 49, 1–21.

Shimanoff, S. B. (1984). Commonly named emotions in everyday conversations. *Perceptual and Motor Skills*, 58, 514.

Simonson, I. (1992). The influence of anticipating regret and responsibility on purchase decisions. *Journal of Consumer Research*, 19, 105–117.

Staw, B. M. (1976). Knee-deep in the big muddy: A study of escalating commitment to a chosen course of action. *Organizational Behavior and Human Performance*, 16, 27–44.

Staw, B. M., Barsade, S. G., and Koput, K. W. (1997). Escalation at the credit window: A longitudinal study of bank executives' recognition and write-off of problem loans. *Journal of Applied Psychology*, 82, 130–142.

Staw, B. M. and Hoang, H. (1995). Sunk costs in the NBA: Why draft order affects playing time and survival in professional basketball. *Administrative Science Quarterly*, 40, 474–494

Tan, H. and Yates J. F. (1995). Sunk cost effects: The influences of instruction and future return Estimates. *Organizational Behavior and Human Decision Processes*, 63, 311–319.

Thaler, R. H. (1988). Anomalies: The ultimatum game. *Journal of Economic Perspectives*, 2, 195–206.

Turner, M. E. and Pratkanis, A. R. (1998). Twenty-five years of groupthink theory and research: Lessons from the evaluation of a theory. *Organizational Behavior and Human Decision Processes*, 73, 105–115.

Tversky, A. and Kahneman, D. (1981). The framing of decisions and the psychology of choice. *Science*, 30, 453–458.

Weinstein, N. D. (1980). Unrealistic optimism about future life events. *Journal of Personality and Social Psychology*, 39, 806–820.

Weinstein, N. D. (1987). Unrealistic optimism about susceptibility to health problems: Conclusions from a community-wide sample. *Journal of Behavioral Medicine*, 10, 481–500.

Whyte, G. (1991). Decision failures: Why they occur and how to prevent them. *Academy of Management Executive*, 5(3), 23–31.

Whyte, G. (1998). Recasting Janis's groupthink model: the key role of collective efficacy in decision fiascos. *Organizational Behavior and Human Decision Processes*, 73, 185–209.

van Winden, F. (2007). Affect and fairness in economics. *Social Justice Research*, 20, 35–52.

Wright, G. and Goodwin, P. (2002). Eliminating a framing bias by using simple instructions to 'think harder' and respondents with managerial experience: Comment on 'breaking the frame.' *Strategic Management Journal*, 23, 1059–1067.

Yen, H. R. and Chuang, S. C. (2008). The effect of incidental affect on preference for the status quo. *Journal of the Academy of Marketing Science*, 36, 522–537.

Zeelenberg, M. (1999). Anticipated regret, expected feedback and behavioral decision making. *Journal of Behavioral Decision Making*, 12, 93–106.

Zeelenberg, M. and Pieters, R. (1999). Comparing service delivery to what might have been: Behavioral responses to regret and disappointment. *Journal of Service Research*, 2, 86–97.

Zeelenberg, M. and Pieters, R. (2004). Consequences of regret aversion in real life: The case of the Dutch postcode lottery. *Organizational Behavior and Human Decision Processes*, 93, 155–168.

Zeelenberg, M., van den Bos, K., van Dijk, E., and Pieters, R. (2002). The inaction effect in the psychology of regret. *Journal of Personality and Social Psychology*, 82, 314–327.

Part V

Decisions with Multiple Perspectives

Chapter 15

Value-Added Negotiations

Hal Stack

A college student with a limited budget needs to purchase a car in order to get to school and work. A small business person wants to sell his car in order to get a larger one that will enable him to carry the equipment needed for his business. How should each person prepare for negotiating the sale and purchase of the car? What factors are likely to influence the outcome?

A major league catcher, nearing the end of his career, is hoping to negotiate a contract with a major league team in his hometown. The same team finds itself in need of an experienced catcher. How should the parties prepare for these negotiations? What issues are likely to arise?

A hospital finds itself in a dispute with an IT supplier over the implementation of a new electronic record system. Failure to resolve this dispute will cost the hospital several million dollars, and the hospital is already on the verge of bankruptcy. Without appropriate resolution, the IT company will damage its credibility in a rapidly growing market and lose much of the money it has invested in the project. How should the parties proceed to resolve their dispute?

15.1 Goal and Overview

The primary goal of this chapter is to develop the knowledge and analytic skills for making negotiated decisions that culminate in value-added agreements. More specifically, this chapter will first increase your ability to recognize and avoid the biases and psychological traps that limit the effectiveness of negotiator decision making. It will then enhance your ability to apply the principles and practices of highly effective negotiators. Ultimately, the chapter will increase your capacity to apply a systematic framework for improving negotiation outcomes.

The primary focus of the book until this point has been on decisions for which an individual or team has total responsibility. Although the decisions might involve multiple interests groups within a corporation, the assumption has been that each group has a general concern for the overall success of the entire organization. Their personal corporate responsibilities and life experiences will color their views and priorities, but they should all be considered motivated to work toward the organizational good. In a decision that involves the markets or the public, the decision makers must factor in those interests, but in the end it is their decision to make and implement. The same applies to public sector managers. They, too, have primary responsibility for the decisions while needing to be sensitive to the needs of the people they serve. Bad decisions may lose them the support of their constituencies and cost them their jobs, but the same is true of business decisions.

This chapter, in contrast, addresses decision contexts where one side or the other cannot make a unilateral decision other than to walk away. When buying a car, the decision maker can specify a price he is willing to pay, but he cannot conclude a purchase without the other side agreeing. In a conflict over fulfillment of a contract, one side can demand and threaten, but that still does not produce a decision as to how to resolve the dispute and implement a solution. Similarly, a country negotiating a treaty cannot simply define the articles in the treaty. In each case, multiple perspectives are at the core of the decision. Thus, it is critical that an effective decision maker fully understand and appreciate the other side's decision-making perspective.

Another critical difference is that negotiated decisions are more dynamic than those discussed in previous chapters. Negotiations involve a process of give and take that must be constantly updated as the decision makers receive information and insights from the other side in the negotiations. Thus, it is critical that the decision-making process involve a degree of flexibility that is not necessary in unilateral decisions. It is for these reasons that the chapter is titled "Value-Added Negotiations" because it is possible in many contexts to improve upon the outcomes for all sides by better understanding their respective positions.

Following a discussion of the nature and structure of negotiations, the chapter examines the psychological traps and obstacles that limit our effectiveness as negotiators. This is followed by a discussion of a comprehensive framework for managing the negotiation process, which enables negotiators to avoid these obstacles and produce lasting agreements with superior outcomes. The chapter also discusses the difference in approaches to negotiating a deal and negotiating the resolution of a dispute. Also discussed is the complexity of negotiating through agents and managing multiparty negotiations. The chapter concludes with an examination of the challenges of negotiating across borders and negotiating ethically.

15.2   Understanding Negotiations

Everyone negotiates. Negotiation is used every day to resolve differences and achieve our goals (resources, information, cooperation, support, etc.). It occurs between colleagues working on a project, home owners and contractors, unions and employers, corporations and their suppliers, and even nations. It also occurs with spouses, children, and strangers. Negotiation skills are essential to anyone who works with and through people to achieve objectives. For both individuals and organizations, negotiation is a core competency.

Negotiation is an interactive, interpersonal decision-making process by which two or more interdependent parties attempt to agree on a mutually acceptable outcome in a situation where their interests are or appear in opposition. This may be a conflict of preferences, a conflict of priorities, or a conflict over resources. It involves not only tangible issues like price, but also intangible issues like fairness, precedent, or principle.

1. *Activity*—Your experience with negotiations: The negotiation examples you provide here also provide the basis for activities later in the chapter.

   a. List several negotiation experiences you have had over the last few weeks. These may be workplace related, such as a difference with a customer, a contract with a supplier, or a difference with a member of your project team. They may have involved reaching agreement on an issue or resolving a dispute or conflict. Consider also your personal life. These may include the purchase of a home, reaching agreement on curfew with a teenager, or resolving a dispute over payment for work done by a contractor.

Workplace experiences   **Personal experiences**

_____   _____

_____   _____

_____   _____

b. Now select a situation or situations you feel were successfully addressed. What made it successful? Consider both the outcomes and the process.

_____

_____

c. Now select a situation or situations you feel were unsuccessfully addressed. What factors contributed to these failures? Again, consider both outcomes and the process.

_____

_____

Some negotiations involve a one-time interaction between individuals who will likely never again be in contact, such as a used car salesman and buyer. More often, however, negotiations involve continuing relationships, as when negotiating a job offer or a collective bargaining agreement. In either case, the negotiation process involves a series of individual and joint decisions that cumulatively determine the value of the agreement or whether there is an agreement at all. Ultimately, we are seeking an agreement that addresses our interests; can be implemented effectively; and lays the foundation for an ongoing, productive relationship. The structured decision process discussed in earlier chapters of this book can help but are not sufficient by themselves.

2. *Activity*—Negotiations with and without an ongoing relationship.

a. Describe a recent negotiation experience that involves an ongoing relationship.

_____

b. Describe a recent negotiation experience that does *not* involve an ongoing relationship.

_____

Negotiations often involve both tangibles (e.g., the price or the terms of an agreement) and intangibles (e.g., fairness, reputation, an important principle, maintaining a precedent). Intangibles are the underlying psychological motivations that may directly or indirectly influence the parties. They affect our judgment about what is fair or right or appropriate in reaching an agreement.

In selling a long-cherished home, the owner may choose to sell to someone offering a somewhat lower purchase price because that party valued the home in its existing state, while another potential buyer indicated that he planned to knock out walls and turn the home into a bachelor pad. Clearly, the psychological attachment for one's home is an intangible that can influence the final agreement. Similarly, a union's commitment to the principle of seniority will influence its response to management proposals on promotion and scheduling.

3. *Activity*—Intangible issues in negotiation. Review the list of negotiation experiences created in Activity 1. Identify and explain two intangible issues that arose in one or more of these negotiations.

a. Intangible 1. _____

b. Intangible 2. _____

Negotiation is an interactive process of give and take during which the parties attempt to move from initial and often exaggerated positions to common ground and agreement. Central to this give and take is the communication process—the ability to communicate our interests and positions and to understand those of the other party. While we can always "take it or leave it," we negotiate to improve our outcome through a process of dialogue and discussion.

The challenge in negotiation is to craft a deal that not only addresses your interests but also meets the needs of the other party better than a no-deal option. Unlike the decisions discussed in the preceding chapters, no final decision regarding a negotiation can be implemented without reaching an agreement with the other party. Consequently, it is in your interest to understand the other party's problem and to help them solve it. This is contrary to the common view of negotiation as a contest over who will claim the most of a fixed pie. While such purely distributive negotiations exist, most negotiations involve more than one issue; furthermore, there are usually opportunities to make the pie bigger before you divide it. In integrative negotiations, the parties cooperate to create value before they decide how that value will be divided. Most negotiations, then, include two subprocesses: creating value and claiming value. The inherent tension between the two must be managed carefully.

In addition to these two subprocesses, we can distinguish two forms of negotiation: deal making and resolving disputes. In deal making, the parties focus on reaching an agreement that will define their future relationship. A contract between a supplier and Wal-Mart Corporation, for example, opens up a huge new market for the supplier but also imposes new requirements in cost control and technology utilization. In resolving disputes, the parties focus on a dispute arising under an existing agreement, as in the IT example at the beginning of the chapter. How this dispute is resolved will determine if the current relationship will continue. In these negotiations, the interests of the parties are linked and the parties often come to the table angry. A homeowner dissatisfied with the quality of work promised in the remodeling of a kitchen may withhold payment, threaten to sue, or place a lien on the contractor's business. At the same time, there is a risk that the contractor might declare bankruptcy and be unable to address the homeowner's concerns. These negotiations pose special challenges.

15.3 Challenges to Effective Negotiation

Many factors converge to limit our effectiveness as negotiators. Some are rooted in our mental model of negotiations; others are found in the psychological traps and biases that distort our judgment in decision making. Many of these are the same biases discussed in earlier chapters but are more critical now. Negotiation is a real-time process during which there may not be time to reflect on and overcome these cognitive biases.

How we think about negotiation affects our choice of strategy and tactics. Many negotiators assume that their interests directly conflict with those of the other party. They see all negotiations as a distributive, zero-sum game in which their gain is the other party's loss. For them, negotiation is a test of wills that, ideally for both sides, can culminate in splitting the difference.

Distributive negotiations usually involve a single issue in which one person gains at the expense of the other. Most negotiations, however, involve more than one issue and the parties value the issues differently. Think of a company and its suppliers. Price is important but so are quality, support services and delivery time. As a result, an agreement may be found that is better for both parties than what they could have achieved through a win-lose, distributive approach. Assuming that every negotiation is a fixed-pie, win-lose negotiation may result in no agreement or missed opportunities for trade-offs that could benefit both sides. This is reflected in former congressman Floyd Spence's view of the Strategic Arms Limitation Treaty negotiations between the United States and the Soviet Union. "I have had a philosophy for some time in regard to SALT, and it goes like this: the Russians will not accept a SALT treaty that is not in their best interest, and it seems to me that if it is in their best interest, it can't be in our best interest" (Bazerman and Neale 1992).

Similarly, our effectiveness is limited by our tendency to see negotiations as an argument over positions. A position is one party's solution to a problem. It does not address the concerns of the other party and is often a suboptimal solution, even from the point of view of the party proposing the position. The logic of positional bargaining is that negotiators spend their time arguing the merits of their position while discrediting the position of the other party. Even acknowledging the other party's concerns is seen as weakening your position. This was clearly in play in the negotiations between Republicans and Democrats on health care reform during 2009. Focusing only on whose position should prevail means that other and potentially superior solutions are never considered. The solution may be one extreme or the other or, more typically, splitting the difference between the two final positions rather than accepting a solution to address the real interests of the parties.

Fisher and Ury (1991) illustrate this in their book *Getting to Yes*. In negotiating the return of the Sinai to Egypt, Israel and Egypt could not agree on where to draw the boundary separating the two countries. It seemed to be a classic zero-sum negotiation in which every square mile lost to one party was the other party's gain. When the negotiations became value added and the parties focused on their real interests, however, the dispute was resolved. For Egypt, the critical interest was sovereignty over the Sinai, while Israel's top concern was its security. The solution involved creating a demilitarized zone under the Egyptian flag. Getting beneath the positions to the underlying interests of the parties was critical to reaching a breakthrough. This approach, when combined with inventive solutions, led to an agreement superior to the position of either party.

Interests are the needs, concerns, or fears underlying your position. As negotiators, we often start the process with a definite position on a specific issue. A potential employee, for example, may initially request $56,000 (her position) for her salary. However, the individual may not have clearly defined the underlying interests that are much broader. These might include financial security, status, start date, paying off a college loan, location, and improved career prospects. By failing to consider these interests and focusing only on salary, she may reject a job offer that does not meet her position of $56,000 but which would better address her broader interests. The potential employer's interests include maintaining a consistent salary schedule for new hires and thus may be unwilling to offer a salary higher than $50,000. The employer can, however, offer generous moving expenses and a signing bonus and accommodate her preference to relocate to Chicago. Understanding her interests and those of the potential employer can lead to a deal that will meet both parties' needs. By focusing on interests, we can develop a better understanding of mutual concerns and invent solutions acceptable to both sides.

One way to get at underlying interests is to ask why. Why is this important to you? Why do you want this solution? In the Israeli and Egyptian negotiations over the status of the Sinai, it was the recognition that Israel's real interest was security and not land that led to an agreement. In the course of these negotiations, a skilled negotiator asked the Israeli representatives why they insisted on keeping some of the Sinai. It was this simple question that produced the eventual solution.

4. *Activity*—Interests versus Positions: A position is one party's solution to a problem. An interest is the underlying need or reason for taking a position. For the following issue, indicate the parties' respective interests.

Position of Employer: Employees must pay more of the cost of health insurance.
Interest of Employer: _____
Position of Union: The employees will not pay more.
Interest of Union and Employees: _____
Given the respective interests of the parties, what might be a solution that would address both their interests? _____

5. *Activity*—Think about the negotiations in which you have been involved. For one of these negotiations complete the following questions:

Position—What did they say? _____
What were their underlying needs and interests? _____

Closely related to the use of positional bargaining is a tendency to ignore the other side's problem. The other side's interests and concerns are seen as their problem, not ours. But if the goal is agreement, negotiators must understand that any agreement must necessarily satisfy some of the other party's interests and priorities. It is to our benefit, then, to understand the other party's interests and priorities and to help them solve their problem.

Appreciating the other party's interests is complicated by the problem of partisan perceptions, a tendency to "see" what is in our self-interest to see. Partisan perceptions lead each party to see his or her demands as fair and reasonable, and the other person's as one-sided and unreasonable. This partisan perception leads us to reactively devalue any proposal put forward by the other side. This bias is illustrated by two groups who were asked to assess an arms reduction proposal. One group was told Gorbachev was the author and the other was told that it was Reagan who was the author. Table 15.1 presents each groups' assessment of which country benefited most from the proposal. From Group I, believing the proposal came from Gorbachev, 56% assessed the proposal as favoring the USSR. Only half as many made this assessment when they believed the proposal was authored by Reagan (Bazerman and Neale 1992). Similarly, it is not unusual for a proposal by labor to be dismissed by management even when the proposal is objectively good for management, and vice versa.

Another obstacle is the challenge of managing the tension between creating value (making the pie bigger) and claiming value (getting a bigger slice of the pie). Information drives this tension. Discovering options that might create value and potentially make both sides better off requires sharing information about one's preferences, interests, and priorities. But if this openness is not reciprocated, the disclosing party risks being taken advantage of. As a result, efforts to claim value tend to drive out moves to create it. This tension between efforts to create value and competitive efforts to gain individual advantages is central to the negotiation process (Lax and Sebenius 1986).

Negotiating effectiveness is further limited by decision-making biases that blind us to opportunities for better outcomes. We escalate our commitment to our initial position even when it is no longer the best solution. We stay committed to an initial position or course of action even when the data will no longer support it. We throw good money after bad rather than admit making a mistake. Commitment biases our perceptions and judgments. We seek out data that supports our decisions while rarely searching for data that challenges them. This confirming-evidence bias leads us to give too much weight to supporting evidence and too little to conflicting information. (This would be the kindest interpretation of recent Wall Street decision making.) This irrationality, combined with our unwillingness to admit failure, to appear inconsistent, or to recognize that time and resources already invested are "sunk costs," often leads to irrational escalation. We see this play out in corporate mergers and acquisitions, price wars, and, tragically, in military actions.

One well-documented bias is that we anchor our judgments on an initial starting point and then adjust upward or downward; this occurs even when the initial anchor is obviously arbitrary. The mind gives disproportionate weight to the first information it receives. In negotiation,

TABLE 15.1: Partisan perceptions of arms reduction proposal.

	Believed Author of Arms Reduction Proposal	
	---	---
Group Evaluation	Group I Gorbachev	Group II President Reagan
Favored Russian	56%	27%
Favored US	16%	27%
Favored both sides	28%	45%

opening offers will act as anchors and will often have a powerful impact on our judgments. In the purchase of a new car, for example, the salesperson will attempt to anchor the negotiations around the manufacturer's sticker price while the purchaser will attempt to anchor around the invoice price.

We allow our judgment to be affected by how information is presented or framed. A proposal framed as a potential gain for the other party is more likely to be accepted than one seen as a loss. In collective bargaining, for example, the union representative can view any management counter offer as either a gain relative to the existing contract or a loss relative to the union's initial proposal. Whether the offer is viewed as a loss or a gain will significantly impact the party's willingness to accept it. Similarly, framing with different reference points will impact the response of the other party. Many people will decline when offered a 50–50 chance of either losing $300 or winning $500. Yet, they are more likely to agree to the equivalent gamble when asked if they would prefer to keep their checking account at $2000 or accept a 50–50 chance of having either $1700 or $2500 in their account. Both offers are the same but have different reference points. The second frame emphasizes the real financial impact of the decision (Hammond et al. 2004).

Often we enter into negotiations overconfident that our position will prevail; and we fail to consider better alternatives. Because of this overconfidence, we underestimate the strength and validity of the other side's position and are less willing to compromise. On other occasions, we are under-confident and settle for less than we should (Malhotra and Bazerman 2007).

15.4 Managing the Negotiation Process

Negotiation success, then, depends upon our ability to avoid common traps and to optimize outcomes. The case in the following involves the purchase of a used car, a common two-party negotiation in which a seller and a potential buyer agree upon a price. In this negotiation, the seller is a private individual who plans to replace his current vehicle with one more suited to his needs. The sale illustrates a distributive bargaining situation in which the gain of one party comes at the expense of the other party. Consider the information in Table 15.2.

In preparing for the negotiation, both parties have researched the market value of the car in an effort to establish a "fair" price. Each will attempt to determine how the other person values the car. As part of the preparation, each will establish a target and reservation price for the car. (A reservation price is the lowest price a seller is willing to accept and the highest price a purchaser is willing to pay.) For the seller, the target price—the price he hopes to get—is $3750. The seller's reservation price is $3350, the trade-in price offered by the car dealer. This reservation price is also the seller's

TABLE 15.2: Negotiating the purchase of a used car.

Seller's Information	Buyer's Information
The seller needs a larger vehicle that will enable him to expand his business, in which he delivers and sets up computers for a local computer store	The buyer, a commuter student at the local college, has $3450 to spend. Borrowing funds from family and friends is not possible
The Blue Book retail value of the car is $4000 and a dealer has offered $3350 as a trade-in on a new car	The buyer wants a small, fuel efficient car for getting to school and to his part-time job
A local mechanic has certified that the car is in good condition except for the wear on the tires	The buyer hopes to purchase the vehicle for $3100 and have sufficient money remaining for insurance.
	The seller has advertised the car for $3750

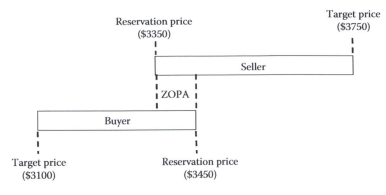

FIGURE 15.1: Zone of possible agreement (ZOPA).

Best Alternative to a Negotiated Agreement (BATNA). If the buyer is unwilling to match or beat the car dealer's offer, the seller's alternative is to sell the car to the dealer. The buyer's reservation price is $3450, as defined by his bank account. The buyer's target price is $3300. As illustrated in Figure 15.1, determining the parties' reservation and target prices indicates whether there is a Zone of Possible Agreement (ZOPA). This zone is the region between the two reservation prices. There is no ZOPA if the buyer's reservation purchase price is less than the seller's reservation selling price.

While this negotiation has a positive zone of agreement between $3350 and $3450, there still remains the question of how the $100 will be divided. Typically, this is decided through a process of offer and counteroffer (often called the "negotiation dance") until the parties settle on a price within this range. Often the final price will be the result of a compromise where the parties "split the difference." In this scenario, a likely agreement would be $3400.

Alternatively, the parties might find a way to make the deal bigger and create value by adding issues to negotiation. The seller, for example, might allow the buyer to purchase the car with a down payment of $3100 and 4 monthly payments of $100 to enable him to purchase car insurance. Alternatively, the parties might discover an opportunity for the student to assist the seller in the computer set-up business, with his pay applied to the car purchase. Ultimately, a deal will occur because each party has prepared carefully and gained relative to their BATNA or reservation price. How much they gain depends on how they agree to allocate the value represented by the $100 zone of agreement. Additionally, the parties, by adding issues, have the potential to create additional value.

Successful negotiation requires careful management of the key elements of the negotiation process, as illustrated in Table 15.3. Value creation is an important part of this process. We often view negotiation as a test of wills, a contest over the division of the pie. While negotiation does involve claiming value, it also includes the possibility of creating value. As noted earlier, the key to successful negotiation is the ability to manage the tension between value creation and value claiming. This requires careful preparation.

15.4.1 Preparation

Preparation for negotiations is perhaps the most important element of the process. It involves a careful assessment of your goals and strategic situation, a preliminary assessment of the other party's goals and strategic situation, and an analysis of the context in which the negotiation is occurring. As reflected in Table 15.2, preparation requires thoughtful assessment of self, the other party, and the situation or context in which the negotiation is occurring.

In preparing for negotiation, it is important to identify the goals you hope to achieve in the upcoming negotiation, considering both outcome and future relationship goals. In addition, one should identify and prioritize the issues that must be addressed in achieving these goals. For each issue, clarify the underlying interests and brainstorm potential solutions. If you hope your initial

TABLE 15.3: Process of managing negotiations.

Preparation	Value Creation	Value Claiming	Closure
Assessment of self	**Strategies**	**Strategies**	**Strategies**
Goals	Problem solving	Determine BATNAs	Take it or leave it
Issues	Define the issue	Establish reservation and	Appeal to "fair" standard
Interests and priorities	Identify interests	target price	Deal with objections
Target and reservation price	Brainstorm options	Probe valuation and	Summarize agreements
Supporting arguments	Consider criteria	information differences	and concessions
Information needs	Select best solution(s)	Estimate ZOPA	Package issues
Your BATNA	Beneficial trades	Manage negotiation dance	Make equivalent proposals
	Add issues	When to begin?	Split the difference
Assessment of other party	Fractionate issues	Who goes first?	Mediation
Other party	Leverage differences	How much to offer?	Fact finding
Goals	Contingent contracts	How to react to offers?	Arbitration
Issues	Cost-cutting	How to make	Use power strategically
Interests and priorities		concessions?	Post-settlement settlement
Target and reservation price	**Tactics**		
Their BATNA	Take problem-solving approach	**Tactics**	
	Ask diagnostic questions	Open aggressively	
Assessment of situation	Build trust	Reciprocal concessions	
Importance of relationship	Share information	Concession size and timing	
Linkage effects	Solve the other party's problem	Make first offer	
Time pressures	Provide reasons for disagreement	Avoid ranges	
Difference in power		Provide rationale	
Need for change		Label your concessions	
		Save face	

offer will serve as a strong anchor, prepare supporting arguments with relevant facts and data. Also identify the information or data you might need to address the issues intelligently. Where appropriate, establish your target and reservation price. Finally, identify and prioritize your BATNA, your alternative courses of action to reaching agreement with the other party.

If I am negotiating for a job and have been out of work for 6 months, I have a far weaker BATNA than if I am currently employed. The better my BATNA, the more power I have at the bargaining table. I should therefore seek ways to improve my BATNA. This means exploring all possible courses of action. In the job scenario, this might involve continuing the job search, taking a part-time job, or reducing my cost of living. Skilled negotiators always keep their alternatives open and attempt to improve upon them. Revealing your BATNA, particularly if it is very weak, dramatically reduces your bargaining power. And even where you have a very strong BATNA, revealing it means that you will not receive any offer better than that.

Assessing the other party's situation requires a similar analysis, beginning with a basic understanding of who you face at the bargaining table (Lax and Sebenius 2006). What is their reputation? What authority do they have? Much of your assessment of the other party's interests, priorities, and BATNA will be little more than your best guesses. However, careful research and continued probing during the negotiations will enable you to be a better prepared negotiator. Many negotiators find it helpful to use a planning document, such as one developed by Jeanne Brett of Northwestern University (Table 15.4). In column 1, the table lists the issues to be negotiated, followed by your priority ranking, position, and interests for each issue in column 2. In column 3, you list the other party's priority ranking, position, and interest. The respective priorities are not likely to be the same. For example, a job prospect may want a salary of $86,000. His interests are security and paying off loans. For him, salary is his first priority. The employer's salary position, however, is $72,000 and his primary interest is fairness. Salary is ranked only third on his list of issues.

Because most negotiations involve multiple issues and differing priorities, effective negotiators are constantly considering different ways to package these issues. A union, for example, may be willing to agree to changes in work rules in return for a wage increase. These work rule changes may be of low cost to the union but of high value to the employer. Typically, it is the ability to create a mutually acceptable package that will lead to a deal.

A related issue is sequencing, or the relative order in which the issues should be discussed. Many negotiators prefer starting with easy issues and then moving to difficult ones as the parties build positive momentum. Other negotiators prefer tackling the hard issues first and then wrapping up the easy ones. Still others like to take all the issues together. The challenge is not to be too linear. In

TABLE 15.4: Planning document template to fill in.

Issue	Your Perspective		Other Perspective	
	Priority	Position	Priority	Position
Description		Interests		Interests
Salary	1	$86,000	3	$72,000
		security, pay off loans		fairness
Issue X	...	...	...	...
		...		...
Issue Y	...	...	...	...
		...		...
BATNA	Accept other job offer		Offer position to another person	
Reservation price	$78,000		$77,000	

most cases, your ability to bring negotiations to a successful conclusion will depend on your ability to identify potential trades and create a package of issues that effectively address the priority issues of each side. This may require looping back to earlier issues and making adjustments to put together a winning package. Unfortunately, once an issue has been resolved, it is unlikely to be revisited and incorporated into a value-creating package.

6. *Activity*—Managing issues in negotiation: Use the planning document in Table 15.4 to construct a chart for one of the negotiations identified in Activity 1.

In assessing the situation or context of the negotiations, it is helpful to consider the following:

Relationships: What is the current relationship with the other party and what do you want that relationship to be in the future? Is the other party someone you will want to do business with again? For example, does the supplier you are negotiating with have a technology critical to your product?

Linkages: Are there linkage effects between these negotiations and other negotiations? Will these negotiations establish precedents likely to influence a negotiation with another party? For example, will the wage increase you negotiate with union A set a precedent for your upcoming negotiations with union B?

Time pressure: Are there any time pressures that may impact you or the other party? Do these negotiations need to be completed before a certain date such as a new product launch? Does the teachers' union contract end on the first day of school?

Power: What is the relative power of the two parties? Does the other party have an attractive alternative to doing business with you?

Pressure to change: Also, is there a need for change? Are your costs out of line with your competition? Are new technologies disrupting your markets?

Ground rules: What ground rules do you want to establish to guide the negotiations? Where will the negotiations occur? What will be the bargaining calendar? At what point can new issues not be added without the consent of the other party? Are the parties willing to extend the current contract if necessary? How will communication with constituents or the press be handled?

Opening statement: Opening statements are your opportunity to set the tone for the negotiations while signaling your basic themes. What do you want to accomplish at the start of negotiations? Should you address significant cost issues or the other party's responsibility for product engineering?

Even when your message contains unwelcome issues for the other side, be sure your opening statement is delivered without hostility or threats. Do not frame your opening statement as a set of demands but as a way of starting discussion. The opening statement can relieve the tension that is often present at the beginning of negotiations. Frame the negotiation as a joint effort that will benefit both parties and emphasize your openness to the other party's issues and concerns.

7. *Activity*—Opening statement: Drawing upon one of the negotiations identified in Activity 1, identify the key points you would have made in your opening statement.

The opening bargaining session is crucial in getting the negotiations off on the right foot and in laying the foundation for a positive outcome. As in any meeting, it is helpful to have a clear, agreed-upon agenda. Similarly, the meeting should end with agreement on the next steps and the agenda for the next meeting. Often a flip chart can be helpful in keeping the bargaining teams focused and on track. Although some negotiations may begin with an exchange of specific proposals, these can quickly transform into fixed positions and rigid demands. A better approach is to focus on understanding and clarifying the issues that will need to be addressed. Potentially controversial data might best be collected jointly by a subcommittee consisting of members from both teams. This will save time and reduce conflict over "the facts."

15.4.2   Value Creation

Effective negotiators recognize that most negotiations involve opportunities to create value, to make the pie larger. Value-creating strategies involve creative problem solving, the use of beneficial trades, leveraging differences, the use of contingent contracts, and ways to cut the cost of agreement.

Creative problem solving challenges the parties to seek solutions for mutual gain rather than engage in prolonged arguments over position. While different problem-solving models exist, everyone agrees on the importance of defining each issue carefully, considering a wide array of possible solutions, and evaluating those possible solutions against objective standards or criteria. In the context of negotiation, this problem-solving approach is combined with the principles popularized by Fisher and Ury in *Getting to Yes*. The first is to "focus on interests, not positions." In problem-solving negotiations, the definition of the issue or problem requires an understanding of the interests of the parties. Further, in focusing on interests rather than positions, the parties avoid prematurely limiting the range of solutions to their initial positions. In identifying possible solutions, Fisher and Ury recommend "inventing options for mutual gain." This means not only brainstorming as many solutions as possible but also seeking solutions that address the interests of both parties. In selecting the optimum solution, they recommend using "objective criteria" or standards such as market value. In negotiating over the price of a used car, the parties might rely on market value information available, such as on Edmunds.com, to determine a fair price. Further, in engaging in joint problem solving, Fisher and Ury advise being "hard on the problem and soft on the people." As in any problem-solving process, optimal results require rigorous analysis and a group process where the parties listen carefully, ask probing questions, and respect differing viewpoints. Particularly where the relationship is important, it is foolish to damage a relationship that will be important to the successful implementation of the deal. In any case, you are much less likely to get a favorable deal if you have attacked and angered the other party.

Combining these insights leads to the problem-solving model illustrated in Table 15.5. Use of this "interest-based problem-solving" model will increase the likelihood of identifying a solution superior to those originally conceived by the parties. In other cases, the increased clarity of the issue and the parties' interests may enable the parties to identify future beneficial trades.

8. *Activity*—Interest-based problem solving: Apply the interest-based problem-solving model in Table 15.5 to one of the issues involved in one of the negotiations identified in Activity 1.

TABLE 15.5:   Interest-based problem solving.

Define the issue: What is the issue and why is it important to us?
    What is the current situation?
    Who is affected and how are they affected?
    What has contributed to making this an issue?
    What will happen if nothing is done?
Identify stakeholder interests involved: What is at stake for you and other key stakeholders on this issue?
Generate options: Using brainstorming, generate as many ideas as you can to address the issue. Wild and crazy ideas are welcome.
Establish objective criteria: Are there criteria to consider in evaluating solutions? Cost of implementation? Market value? Accepted standards? Fairness?
Evaluate options: How well do the options address the issue, respond to the interests of the parties, and meet your criteria?
Select best solution(s): What solution or combination of solutions will work?
Reduce the solution to writing.

Beneficial trades between the issues on the table are yet another way to add value. Most negotiations involve more than one issue, and most negotiators have different preferences across issues. Thus, there is potential for trade-offs to create mutual gain. In some cases, it may be possible to add issues and increase the potential for beneficial trades. When the negotiation seems to be about a single issue and therefore with no opportunity for a beneficial trade, it may be possible to "fractionate" the issue. Fractionating (also called "unbundling") an issue involves dividing the issue into component issues that may then create the basis for a mutually beneficial trade. Car dealers, for example, fractionate the issue of purchasing a car to include not just price but also financing options, service maintenance, warranties, and accessories.

Negotiation is often viewed as a search for common ground, but often agreement is possible because of differences among the parties. These may be differences in interests, priorities, attitudes toward risk, or forecasts about future performance. Effective negotiators actively search for differences among the parties as a source of value creation in negotiation. Different beliefs about future sales, inflation, or energy costs can be the basis for mutually beneficial contingent contracts. Contingent contracts are if-then agreements in which one thing happens only if another thing happens before it. In collective bargaining, for example, wage increases are linked to the cost of living index, which protects workers from the impact of inflation.

Differences in attitudes toward risk provide another opportunity to create value. In some situations, value can be created by allocating more of the risk to the less risk-adverse party and compensating that party with more of the potential returns. Differences in tax status may create opportunities for joint gain in a divorce negotiation where the government policy treats "family support" differently from "alimony." The individual with the highest income and higher tax rate can provide more money in the form of family support at a lower cost than alimony.

Finally, negotiators can create value by reducing the cost to reaching agreement. This may involve reducing the transaction costs (time and money) or making a concession that will make it easier for the other side to agree. A union, for example, may agree to changes in certain work rules that will make it easier for the employer to more easily agree to a wage increase.

In creating value, be sure to approach the negotiation as a joint problem-solving exercise. Ask diagnostic questions to better understand the other party's interests and priorities. Share information to help the other side better understand your interests and priorities. It is important to note that initiating the sharing of information and the reciprocity that often follows contributes to building trust. Keep in mind that it is in your interest to help the other side solve their problem. Finally, be prepared to provide reasons for your proposals and reasons for your disagreement with the other party's proposals.

15.4.3 Value Claiming

Once the parties have succeeded in enlarging the pie, they must still divide it. The challenge is to claim your share of the pie while not undermining the cooperation required to make the pie bigger. Negotiators, then, must effectively incorporate both cooperative and competitive strategies as they necessarily practice "mixed motive" bargaining.

In claiming value, the parties begin by reviewing their respective BATNAs and reservation and target prices. The parties probe each other's valuation of the issues and explore information differences that the parties may have about the issues and their values. Based on the information available, the parties determine if they will make the first offer and how much they will offer. If they are well prepared, they will make the first offer and try to anchor the negotiations around their first offer.

Concessions will be made in small increments, slowly, to avoid overshooting the other party's reservation price. While it is good to open with an ambitious goal, the offer must be discussable. In making concessions, one should make reciprocal concessions, not unilateral concessions. Offer to agree with the other party on issue 1 if they would be willing to agree to issue 2. Note that small concessions and long times between them signal to the other side that you are unlikely to make additional concessions. In making offers, avoid stating ranges, such as a pay increase of 2%–3%. The other party will focus on the end of the range that favors their interests. Be prepared to provide the rationale for your offer and clearly label concessions you have already made. Finally, enable the other party to save face. Through this process of give and take, the parties narrow the gap between them to the point where they recognize that they are close to closure.

15.4.4   Closure

At some point the parties must bring the negotiations to closure. When there is no zone of agreement, the parties will end the negotiations. When the parties feel they are close to an agreement, there are a variety of strategies for achieving closure. A take-it-or-leave-it approach may be seen as an ultimatum and escalate into a power struggle resulting in a suboptimal outcome and increased transaction costs. An appeal to "fairness" may increase the other party's understanding of your interests and priorities, but typically the parties will have different views of what is fair. Sometimes dealing with the other party's objections and summarizing both the agreements and the concessions the parties have made will convince the other party that it is time to end the "negotiation dance" and agree to a settlement. Packaging the issues with their beneficial trades will often bring the negotiations to closure. Some negotiators find it helpful to present equivalent packages and let the other party select the one it prefers. A common approach, of course, is to "split the difference" (Lewicki and Hiam 2006).

When the parties cannot reach an agreement, there may be interventions available such as mediation or arbitration. Mediators can help the parties put together an agreement, but the parties decide whether to accept it. Arbitration differs from mediation in that the arbitrator determines the final outcome. Both options, however, have their limits. Mediators may focus on getting an agreement, any agreement, to the detriment of one or both parties' best interests. Arbitrators are often accused of simply splitting the difference between the parties' final offers.

While the exercise of power may be used to bring the negotiations to closure, it is not without cost. By choosing to use power, you have increased the probability that the other side will use power as well. The result may well be a damaged relationship, increased legal fees, delays, loss of market share, or worse. Power should be exercised strategically: you should avoid idle bluffs and target high-priority interests. At the same time, your purpose in exercising power is to bring the other party back to the table and negotiate an agreement. Thus, you must leave the other party a way back to the table that enables him or her to save face. Your best source of power, of course, is a strong BATNA, which will enable you to walk away from the negotiations as needed.

Even after the parties reach agreement, there may be opportunities for increasing the value of the agreement through the use of a "post-settlement settlement." The concept of the post-settlement was first developed by Howard Raiffa (2003) as a way to optimize negotiation outcomes. After an initial agreement has been reached, the parties agree to reopen the negotiations in search of an agreement that is better for both sides. If a better agreement is found, then both sides share in the gain. If a better agreement acceptable to both sides is not found, then the parties stay with the original agreement. Under the terms of the post-settlement settlement, the parties can only improve their respective outcomes by improving the other party's outcomes. Seventy-five percent of the time when negotiators use post-settlement settlement, they achieve a better agreement (Thompson 2008).

15.5   Negotiating a Deal

The dynamics of this negotiation process can be seen more concretely in the case that follows. This case involves negotiations between a catcher and a baseball team. Consider the following information in Tables 15.6 and 15.7 (adapted from Barrett, 1989).

TABLE 15.6: Catcher's and team's interests and goals.

Catcher	Baseball Team
The catcher, near retirement, is a free agent who has recovered from a knee injury	The team's starting catcher is out for the year due to injury
It is 2 years since he has played after a failed negotiation with another team last year	The team is interested in this catcher but is concerned about his knees
The catcher is interested in playing in his home town as a first-stringer for a contender	The team needs an experienced catcher who knows the batters and is a steady hitter
The catcher's lifetime batting average is .285	The team sees itself as a contender and hopes to fill the stadium and increase revenue
The catcher is interested in a career as a broadcaster and possibly in the team's front office after his retirement	The average pay for a first-string catcher is $1.9 million; the team's current catcher receives $1.7 million
Pay is not as important to him as a broadcasting career when he retires from playing. His pay 2 years ago was $2.3 million	Salary negotiations with other players on the team begin next spring
Goal. A 3 year contract with an opportunity to be the starting catcher and to move into broadcasting after retirement.	*Goal.* Secure an experienced catcher capable of helping the team reach the playoffs

TABLE 15.7: Summary of interests and BATNA.

Catcher	Baseball Team
Salary	Salary
Length of contract (3 years)	Length of contract (1 year)
Broadcasting opportunity	Fitness to play
Starting as catcher	Performance level
Playing for a contender	Increased revenue
Playing in hometown	Precedent
	Catcher who knows the batters
BATNA. Pursue catching or a broadcasting opportunity elsewhere.	**BATNA.** Sign a younger and less experienced catcher.
Reservation/target price: $1.9/$2.5 million	Reservation/target price: $2.3/$1.7 million

These negotiations involve multiple issues that the parties value differently. Consequently, a range of solutions may create value and leave both parties better off than striking no deal at all. In anticipating the team's likely concern about the fitness of the catcher's knees, the catcher's agent would stress the catcher's recovery and conditioning.

In anticipating the return of their incumbent catcher, now injured, the team may be interested in signing the veteran catcher to a 1 year contract, but it may also recognize the value of having an experienced hand behind the plate in the event the injured player does not fully recover. Listening carefully to his other interest in developing a broadcasting career in his hometown, the team might consider a creative contract that could ease him into doing color commentary on television. Working with the catcher to sponsor baseball camps and set up speaking engagements in the community will enhance the catcher's visibility and could help to build attendance at games.

Uncertainty about the catcher's continued fitness to play might be addressed through a contingency that links the catcher's pay with variables such as batting average and games played. Discussion over salary could lead to a contingent contract that ties total salary to team performance measures, such as reaching the playoffs and attendance. A contingent contract with a base salary $1.7 million and contingencies that reach $2.6 million might be acceptable to the catcher while at the same time allowing the team to avoid a precedent that would influence future negotiations with other players. Similarly, the team might pay the catcher a separate salary for his work as a broadcaster while keeping his salary as a player competitive with the rest of the team. The salary may not be as important to the catcher as the opportunity to get back to his hometown and launch his career in broadcasting. Nevertheless, there remains the intangible issue of self-image and concern that the salary be appropriate to his status as an experienced, nationally known player.

15.6 Negotiating a Dispute

Many negotiations result from disputes or rejected claims. The dispute might be with a contractor over the quality of work performed, with a car dealer over the warranty on a new car, or with a supplier over the terms of a contract. Disputes may be approached in different ways. A common approach is to resort to a rights-based approach and sue the other party or take the issue to arbitration. Alternatively, the parties might attempt to use a power-based approach such as a hostile takeover, a strike, or switching suppliers (Mnookin 2010). Both of these approaches have high costs associated with them, resulting in winners and losers and producing suboptimal outcomes. A better approach is to take an interest-based approach and negotiate an agreement (Ury 1991).

> 9. *Activity*—Interests, rights, and power in resolving disputes: Consider a negotiation identified in Activity 1 that involved a dispute. Select one of those negotiations where a rights- or power-based strategy was used and describe the outcome. How might an interest-based approach have been used?

Now consider the case developed by Goldberg and Brett (2004). A hospital experiencing annual losses of $3 million has been advised by its consultant to invest in a point-of-care clinical information management system. This system would utilize hand-held computers that the consultant projects would generate net savings of $7.5 million per year. Accordingly, the hospital has awarded a $6.8 million contract to install the system over a 1 year period. The company receiving the contract had pioneered the development of hand-held wireless devices such as used in the shipping industry and has been anxious to move into the healthcare field.

Implementation proceeded on schedule, with the hardware fully installed in 3 months and the medical decision support system in 4.5 months. It was at this point that the hospital notified the IT company that it was expected to write data entry software for the clinical information interface system. The goal was for the new system's electronic forms to look like the hospital's paper forms and thereby reduce the cost of training staff. The IT company, however, insisted that it was not contractually required to develop this new software and that to do so would take 9 months and an additional $1 million, whereas it would be faster and less costly to provide a generic version of the software. The hospital indicated that it was only interested in software that modeled its forms and thus stopped payment and sued the company for the $2.6 million already paid, as well as $30 million in damages. After reviewing the contract with the hospital, the IT company counter-sued for breach of contract for the $4.2 million it had not yet received. The company's attorneys were confident that they would prevail in court but noted that the legal fees would be $275,000. The attorneys also noted that winning the suit might push the hospital into bankruptcy and therefore recommended exploring an out-of-court settlement. Thus, although the IT company was in a strong legal position, its bargaining power was considerably weakened due to the hospital's financial position.

In negotiating a settlement, the parties identified their issues and interests, summarized in Table 15.8. The key to such negotiations is finding a solution that furthers both parties' interests

TABLE 15.8: Interests of the hospital and the IT company.

Hospital	Computer Company
Working system	Profitability
Increased net revenue	Future sales to health systems
System in place quickly	Tested generic data entry software
Reputation	Reputation
Staff acceptance and use of system	Staff acceptance and use of system

more than the legal alternative. This might include the hospital agreeing to use the IT company's generic software program and the IT company using the hospital as a beta site for its wireless clinical information system. This arrangement would enable the IT company to establish a leadership position in a key market and would enable the hospital to quickly gain financial solvency. In addition, the partnership would enable both parties to gain a marketing advantage from promoting the new system.

Distributive issues remain before the parties can close the deal. The IT company has already invested $250,000 in the development of the generic date entry software and estimates that it would take 3 months and another $250,000 to complete the development. How much, if any, of this cost should the hospital pay? Using the generic software would increase training costs for the hospital by $250,000–$300,000. Would the IT company help defray this cost? The parties might also agree to a profit-sharing arrangement on the generic data entry software, or agree on linking the hospital payments to the savings that will accrue with the implementation of the system. Whatever the final form of the agreement, both parties have significant reasons to ensure its success.

15.7 Agents and Multiparty Negotiations

The negotiations examined thus far have been two-party negotiations carried out directly. Many negotiations, however, are conducted by agents or third parties. Agents can bring specialized knowledge and expertise to the table. Think of the agent negotiating on behalf of an athlete or the real estate agent representing the seller of a house. They understand their industry, its rules and regulations, and market values, and they may possess far more information about the negotiation than one or both of the parties involved. Agents, however, must be compensated. This means that the bargaining zone between a buyer and a seller is reduced when an agent is involved. More important, an individual employing an agent should be aware of the agent's goals and interests relative to their own interests. While the seller is interested in selling the home at the highest price, the real estate agent's interest is for the buyer and the seller to reach an agreement. Since the seller's agent typically gets half of the 6% commission, each additional $1000 in the sale price is worth only $30 to the agent. Understanding the agent's incentive structure helps to inform the seller's bargaining strategy (Mnookin et al. 2000).

Negotiations can also involve more than two parties, a phenomenon that is becoming more common within and among organizations, involving budget negotiations, product teams, regulatory decisions, or treaty negotiations. Consider, for example, the different stakeholders involved in the congressional tobacco negotiations. They included state attorneys general, the tobacco companies, tobacco growers, the American Lung Association, the American Cancer Association, the FDA, the Clinton administration, and various members of Congress.

Multiparty negotiations typically make it more difficult to reach an agreement. They complicate social interactions, increase information processing demands, and can lead to formation of coalitions. Coalitions enable the coalition partners to have a greater influence over outcomes, often to the disadvantage of other groups and the overall organization (Thompson 2009).

To effectively manage these negotiations, the parties should use the same interest-based problem-solving process as in two-party negotiations, but they should be aware of the traps often used to simplify the process of reaching agreement. A common approach is majority vote. While majority rule is easy and efficient, it provides little opportunity or reason to learn of others' interests and priorities. Without this information, it is harder to trade off issues and find integrative agreements. A better approach is to strive for unanimous agreement. While time consuming, unanimous agreement forces the parties to find trade-offs that will satisfy the interests of all parties. While majority rule may make sense because it helps to forestall an impasse, it should be avoided whenever possible.

Another consideration is how to organize the discussion. There is a natural tendency to address issues serially; issues are considered individually and not visited again once the group has moved on to a new topic. This strict issue-by-issue agenda limits the parties' abilities to discuss issues simultaneously, which enables the identification of beneficial trade-offs. To minimize this, keep the agenda itself open to discussion and be willing to reconfigure it if doing so will facilitate beneficial trade-offs.

A key challenge in multiparty negotiations is the formation of coalitions (Watkins 2002). Coalitions occur when parties seek to add the resources or the support of others to increase the likelihood of achieving their individual outcomes. While coalitions are one way otherwise weak group members can marshal a greater share of resources, they are inherently unstable. They often lead to no agreement or agreements that are not in the best interest of the organization. The 2003 World Trade Organization negotiations in Cancun, Mexico, collapsed when the United States and its usual partners from the developed nations were not prepared to respond seriously to the concerns of developing nations that they reduce farm subsidies.

Instead of working to reach an integrative agreement that serves all groups and the best interests of the overall organization, coalitions try to get what they want using majority rule. This reinforces the importance of requiring unanimity or consensus rather than majority rule in any multiparty negotiations. Effective multiparty negotiations require coordination, establishment of decision rules, and anticipation of the possible formation of coalitions.

15.8   Negotiating across Border

Negotiating across borders is increasingly important and inherently challenging. A number of factors account for this, including different political and legal systems, currency fluctuations, the different roles of government regulation and bureaucracy, varying degrees of political and social stability, differing ideologies, and language and cultural differences. Here our focus is on language and cultural differences, which can influence negotiation in significant and unexpected ways. Sometimes this is the result of ignorance of etiquette and deportment rooted in fundamental differences in national cultures. [See Table 15.9 which is adapted from Sebenius (2002), "The hidden challenge of cross-border negotiations."]

Culture consists of the values, norms, and ideologies shared by members of a group and the social, economic, political, and religious institutions that regulate social interaction. Cultural values define what is important. Cultural norms define what behaviors are appropriate. Cultural ideologies provide shared standards for interpreting situations. Finally, cultural institutions preserve and promote values, norms, and ideologies.

While cultural differences can affect negotiations even within the United States, they play a critical role in cross-border negotiations. People bring their culture to the negotiation table, and this means that the cross-cultural negotiator cannot take common knowledge and practices for granted. First, culture affects the interests and priorities that underlie the negotiators' positions on the issues. Culture, then, helps explain why negotiators take the positions they do, or why one issue is more

TABLE 15.9: Cross-cultural etiquette.

Issue	Concerns
Greetings	How do people greet and address each other? What role do business cards play?
Degree of formality	Will I be expected to dress and interact formally or informally?
Touching	What are the attitudes toward body contact?
Eye contact	Is direct eye contact polite?
Emotions	Is it inappropriate to display emotions?
Silence	Is silence awkward? Respectful?
Body language	Are certain gestures or forms of body language rude?
Punctuality	Is punctuality expected? Are agendas adhered to?
Communication	Is communication direct or indirect?

important than another. Negotiators from cultures that value tradition over change, for example, may resist economic development proposals that threaten traditional ways of life (Brett 2001).

Second, culture may also affect negotiation strategies. Culture affects whether we confront directly or indirectly our motivations and also the way we use information and influence others. These strategies, in turn, create patterns of interaction in negotiation that may facilitate agreement or lead to suboptimal outcomes. An experienced observer once characterized cross-cultural negotiation as a dance in which one person does a waltz and the other does the tango.

While there are many dimensions to culture (Table 15.10), here we focus on four dimensions that directly affect negotiation:

- Language and nonverbal communication
- Individualism versus collectivism
- Egalitarianism versus hierarchy
- Direct versus indirect communication

TABLE 15.10: Negotiation and culture.

Aspect	Range of Cultural Perspectives
Focus of negotiation	Substance ←→ relationship
Negotiation process	Sequential ←→ simultaneous
Negotiation process	Specifics first ←→ general principles first
Goal of negotiation	Maximize individual gain ←→ maximize collective welfare
Motivation	Economic gain ←→ social capital gain
Information collection	Questioning ←→ inferences from proposals and counterproposals
Information sharing	Concise, direct ←→ extensive, detailed
Interaction style	Assertive ←→ polite
Persuasion	Facts and reason ←→ appeal to social good
Influence	Information and BATNA ←→ deference to superiors
Reach agreement	agree on specifics first ←→ agree on general principles first
Form of agreement	Detailed contract ←→ broad agreement on general principles
Implement agreement	Letter of the contract ←→ contract as starting point
Communication	Low context (direct, explicit) ←→ high context (indirect, implicit)
Conflict management	Direct confrontation ←→ indirect confrontation
Long-term orientation	Low ←→ high
Protocol	Informal ←→ formal
Risk propensity	High ←→ low
Emotionalism	High ←→ low

TABLE 15.11:   Errors in translation of advertising slogans.

Original	Translation (Country)
Finger-lickin good	Eat your fingers off (China)
Schweppes tonic water	Schweppes toilet water (Italy)
It won't leak in your pocket and embarrass you	It won't leak in your pocket and make you pregnant in a Parker Pen advertisement (Mexico)
Salem—feeling free	When smoking Salem, you feel so refreshed that your mind seems to be free and empty (Japan)
Come alive with the Pepsi generation	Pepsi will bring your ancestors back from the dead (Taiwan)

Language problems can be substantial in cross-cultural negotiations. Problems in translation are common. Several examples of errors in translation are presented in Table 15.11 (Lewicki et al. 2010). Even when the language used is English, it may be the second language of many of the negotiators at the table. Further, even native speakers from Great Britain, India, and the United States often have trouble understanding one another.

In Japan the word for yes, "hai," can have different meanings depending on the context. A Japanese yes in its primary context simply means the other person has heard the speaker and is contemplating a response. This is because it would be considered rude to keep someone waiting for an answer without immediate acknowledgment. In another context, "hai" may mean "I understand your wish and would like to please you but unfortunately ..." where "unfortunately" is implied but not said. Communication is further complicated by differences in nonverbal communication (tone of voice, loudness, eye contact, facial expressions, periods of silence, and gestures). Direct eye contact is expected in Europe but should be avoided in Southeast Asia until the relationship is firmly established. How things are said is often more important than what is said.

Individualism and collectivism generate cultural differences in motivation. Individualistic cultures such as the United States, Great Britain, and the Netherlands emphasize personal goals, even at the expense of those of work groups or society. People regard themselves as free agents whose accomplishments are to be rewarded and whose individual rights are to be protected by the legal system. Negotiators from individualistic cultures tend to use competitive bargaining tactics and bluffing to increase their bargaining power.

Collectivist cultures emphasize the welfare of the group, and individuals regard themselves as members of groups. Where individualistic cultures focus on influence and control, collectivist cultures (e.g., Colombia, Pakistan, Japan, and South Korea) emphasize harmony, interdependence, and social obligations. Legal institutions in these cultures place the greater good above the rights of the individual. In negotiations, collectivists tend to be more cooperative and more concerned with preserving relationships. While members of individualistic cultures are more likely to handle conflicts directly through competition and problem solving to resolve the issue, collectivist negotiators handle conflict indirectly in an attempt to preserve the relationship. This is illustrated in Table 15.12 in the context of negotiations with a Korean company (Tinsley et al. 1999).

Egalitarianism and hierarchy reflect the means by which people influence others and the basis of power in relationships. Egalitarian cultures such as Denmark, Israel, Austria, and the United States expect people to participate in decision making and to be treated equally. And while everyone is not of equal status or power, social boundaries are permeable. Individuals in egalitarian cultures are empowered to resolve conflicts themselves. In negotiations, one's BATNA and information are key sources of power, not status or rank.

In hierarchical cultures (e.g., India, China, France, and Venezuela), deference is paid to status, subordinates are expected to defer to superiors, and superiors are expected to look out for subordinates. High-status members are not to be challenged, and conflicts between subordinates are

TABLE 15.12: Negotiation issues in South Korea.

Issue	Korean Perspective
Financial practices	Korean firms favor the accounting standards of the International Accounting Standards Committee (IASC), as opposed to those of the Generally Accepted Accounting Practices (GAAP) used in the United States. Because IASC standards are less detailed and seen as too loose, they make it difficult to evaluate "true" financial performance and are a liability for publicly traded firms. This issue must be negotiated
Government role in industry	In Korea the government is an implicit player in negotiations. As a result, a Korean firm will try to ensure that any deal addresses government interests as well. These may include access to product technology, equity ownership, and export growth. At the same time, the Korean government can assist the Korean firms in their negotiations with foreign companies through tax incentives and tariffs
Intellectual property rights	In Korea the firm is seen as a vehicle to serve the national interest rather than exclusively the interests of consumers or shareholders. The government's activist role encourages the growth of export industries and technology sharing among firms. Believing that industry cooperation is the best way to promote development, the Korean government does not believe that technological innovations should be protected at the expense of industry development. Thus, the Korean patent system does not provide the same protections to the individual inventor as the American system. In negotiations, Korean firms will make it a priority to gain access to the American firm's product technology and will be unsympathetic to appeals from U.S. firms to protect their technology or inventions

handled by deference to a superior rather than by direct conflict. In hierarchical cultures, power is associated with position and rank, and it is an insult to send a lower-rank individual to meet with or negotiate with one of higher rank. Similarly, negotiations often require several levels of approval, and negotiators attempt to secure a deal that is clearly in their favor so that it will be easier to convince higher authorities that their side won the negotiations.

Direct and indirect communication refers to different norms about information sharing. In a direct communication culture, such as the United States, Germany, and Scandinavia, messages are transmitted explicitly and directly. Individuals ask direct questions, and the meaning of the communication will be the same regardless of the context. In other cultures, such as Japan, China, and Korea, people communicate in an indirect, discrete fashion. The meaning of the communication is inferred rather than directly stated. In indirect communication cultures, negotiators will not ask questions but rather make multiple proposals from which inferences may be drawn as to priorities and points of concession. For example, Japanese negotiators prefer to share information indirectly, often through stories. Further, in direct cultures, the process of deal making comes first; in indirect cultures, the relationship comes first and provides a context for making deals.

Negotiating across cultures poses many challenges, particularly when each culture expects the other to adapt its style of negotiating. When negotiating across cultures, Brett (2001), Salacuse (2003) and others recommend the following:

- Understand your own culture's negotiation practices and organize this information into a profile as in Table 15.13 (Weiss 1994).

- Anticipate differences in strategy and tactics that may cause misunderstandings. This also prepares the negotiator to better create and claim value. It aids the negotiator in avoiding negative attributions about the other party that are due to different cultural styles.

TABLE 15.13: Negotiator profile.

Element	Dimensions	Continuum of Orientation
General model	Basic concept	Distributive bargaining ←→ integrative bargaining
	Focus	Substantive ←→ relationship
	Process	Specifics first ←→ general principles first
	Process	Sequential ←→ simultaneous
Role of the individual	Selection of negotiators'	Knowledge/expertise ←→ personal attributes/ status
	Individuals goals	Individual gain ←→ welfare of collective
	Decision making in groups	Consensual ←→ authoritative
Interactions: dispositions	Orientation toward time	Monochronic ←→ polychronic
	Risk-taking propensity	High ←→ low
	Bases of trust	External sanctions ←→ reputation
Interactions: process	Concern with protocol	Informal ←→ formal
	Communication	Low context (direct, explicit) ←→ high context (indirect, implicit)
	Nature of persuasion	Logic (facts and reason) ←→ emotion (appeal to social responsibility)
Outcome	Formal contract	Detailed (specific) ←→ implicit contract (general principles)

- Analyze cultural differences to identify differences in values that may expand the pie. Understanding differences in beliefs, values, risks, and expectations can help the negotiator identify opportunities for joint gain, such as value-added trade-offs and contingency contracts.

- Recognize that the other party may not share your view of what constitutes power.

- Find out how to show respect in the other culture, and do not assume that their way of showing respect is the same as your culture's. Failure to appropriately show respect for the other party is rarely forgiven; showing respect in a culturally appropriate manner is sure to be appreciated.

- Choose your representative carefully. In hierarchical cultures, power is associated with one's position and rank, and it is insulting to be asked to negotiate with an employee of lower rank.

- Understand the network of relationships. In hierarchical cultures, negotiations often require several levels of approval. Thus, hierarchical negotiators will attempt to achieve a deal that is clearly in their favor so that it will be easier to get approval from higher authorities.

- Consider using an agent or advisor. When there is very low familiarity with the other party's culture, consider hiring an agent or advisor who is familiar with the cultures of both parties.

- Find out how time is perceived in the other culture. Cultures largely determine what time means and how it affects negotiations. In monochronic cultures such as the United States and Western Europe, time is linear and once used is never replaced. These cultures focus on one thing at a time; schedules and deadlines are important. Monochronic negotiators are more likely to process issues sequentially and negotiate in a highly organized fashion. Polychronic cultures, such as those in Asia, Africa, South America, and the Middle East, see time as circular in nature. People do many things simultaneously, and to the extent these activities interfere with completing a task, schedules and deadlines are unimportant. In negotiations, polychronic negotiators prefer to discuss issues all at once and then discuss them again without reaching a decision on any of the issues. They prefer to think about the whole package before committing to any part of it.

- Commit the time to building and maintaining relationships.

While it is important to appreciate the importance of cultural differences in negotiations, it is also important to avoid stereotyping by being too quick to lump together people from the same culture. There is a great deal of diversity within a culture. Further, other factors such as international experience, organizational culture, regional background, and gender can all be important factors as well. Finally, cultures are dynamic. They change and grow. Effective international negotiators must get to know the people they are working with, not just their culture and country.

15.9   Negotiating Ethically

All negotiators, whatever their culture, face a core contradiction when they think about bargaining ethics. How does one reconcile the use of deception in negotiation with the need to maintain personal integrity in dealings with others at the bargaining table? When a supplier says that he cannot accept anything below $76 a unit and you indicate that you cannot go a penny over $53 a unit, both sides are lying, and both sides know that they are lying. This is euphemistically referred to as "puffing" and is accepted in negotiations.

Although some misrepresentations are considered acceptable "puffing," others are clearly inappropriate, if not illegal. It is not always easy to draw the line between acceptable and inappropriate statements, but a useful test is to ask yourself how you would feel if your opponent were to make the misrepresentation you are contemplating. If you would consider your opponent dishonest, then you should not engage in this conduct.

At a minimum, negotiators must obey the law. While bargaining laws differ between countries and cultures, all share basic principles of fairness and prudence in bargaining conduct. While U.S. law does not require "good faith" in negotiating commercial agreements, it does presume that no one has committed fraud. In negotiations, a bargaining move is fraudulent when a party makes a knowing misrepresentation of a material fact on which the other party reasonably relies and that results in damages. A car dealer is obviously committing fraud when he resets a car's odometer and sells it as a new car.

Effective negotiators not only obey the law but negotiate consistently using an ethical code of conduct. Shell (1999) presented three frameworks for thinking about ethical issues in negotiation: the Poker School, the Idealist School, and the Pragmatist School. Consider these frameworks as you develop your own code of conduct. This will help you increase your confidence and comfort at the bargaining table.

The Poker School of ethics sees negotiations as a "game" with "rules." The rules are defined by the law. But while poker has rules about not hiding cards or reneging on bets, you are in fact expected to deceive others about your hand. In negotiations, then, you must not commit outright fraud, but anything short of fraud is permissible. A car salesperson using this ethic would not turn back the odometer, but he might not tell a potential buyer that the car had its odometer replaced at 27,000 miles.

The Idealist School sees bargaining as an aspect of social life, not a special activity with its own set of rules. The same ethical behaviors that apply at home should apply in the negotiation process. While idealists prefer to be candid and honest at the bargaining table, they do not entirely rule out deception in special situations, such as bluffing, not volunteering information when not asked directly, distracting the other party to avoid answering a question, or declining to answer a question. For idealists, negotiation is not a game but a serious communication process used to resolve differences in society. An idealist car salesperson would tell the buyer that the odometer reading was off by 30,000 miles. A limitation of the Idealist School is the potential to be taken advantage of by other negotiators who do not share their values. Idealists need to maintain a healthy skepticism about the way other people negotiate.

The third school of bargaining ethics, the Pragmatist School, combines elements of the other two schools with a pragmatic concern for the impact of questionable tactics on one's credibility and on present and future relationships. Pragmatists agree with the Poker School that deception is a necessary part of negotiation, but like the idealists they prefer not to mislead or lie to the other party if there is a practical alternative. Pragmatists recognize the importance of credibility both in preserving working relationships and protecting one's reputation in the community. Unlike idealists, pragmatists tend to be more flexible with the truth. A pragmatist car salesman will not mislead the buyer about the car's odometer, but will be willing to mislead a customer as to his reservation price or the rationale for the car's price.

To test your understanding of your ethical thinking, answer the following questions.

10. *Activity*—Ethical thinking in negotiation: Assume that you are selling your home and the other party asks you if you have another offer. In fact, you do not have an offer. Which answer in the following comes closest to your answer? Which school of ethics does each represent?

    a. I have no offer at this time but I am hopeful that I will receive an offer soon.

    b. Yes. A party presented an offer for $450,000 this morning and I have 48 h to respond to it.

    c. What other houses are you considering?

15.9.1 Coping with Questionable Tactics

Whatever your ethical standards, you must be able to protect yourself from questionable or unethical tactics; some of the most frequent are listed in Table 15.14 (Shell 1999). These are particularly common in situations where the stakes matter and the relationship doesn't, and when there are significant power differences. To avoid being the victim of these tactics, consider the following:

- Research your bargaining partner. What is the person's reputation? Who else has dealt with this person? How important is the continuing relationship with your partner? Are future deals likely?

TABLE 15.14: Common questionable bargaining tactics.

Lowballing. "Too good to be true" offers usually are. The other side is getting you to commit to the deal before revealing the true cost to you

Phony issues. One side adds phony or "red herring" issues and then pushes hard on all other issues before relenting on the phony ones in exchange for major concessions on the issues that really matter

Authority ploys. A negotiator claims to have authority she does not have. Or a negotiator denies having authority when in fact she does. Avoid dealing with agents and whenever possible make your offers directly to those who have the power to say "yes" or "no"

Overcommitment. One party drags out the negotiation process and then raises or lowers the price at the last minute. The assumption is that you have too much invested to lose and will say yes

Good cop/bad cop. The bad cop introduces outrageous demands while his teammate, the good cop, becomes your advocate. You bond with the good cop and end up agreeing to less outrageous, but still unfavorable, demands

Consistency traps. The other negotiator gets you to agree to an innocent-sounding standard and then shows you that her proposal is the logical consequence of this standard

The nibble. At the close of the negotiations, the other party asks for a small concession; not wanting to upset the negotiations, you agree without trading for it. The other party has achieved a gain at no cost

- Probe thoroughly when you suspect deception. Clarify the other party's offer with probing questions and investigate the other party's claims when they raise suspicion.

- Maintain your standards even when the other party uses unethical tactics. Avoid the temptation to respond in kind. It is important to maintain your reputation and not sink to their level. Once you do, you lose any moral or legal advantage you might have.

15.10   Conclusion

Negotiation is not only a core competency for individuals but is a critical capability for organizations (Movius and Suskind 2009). The failure to negotiate effectively results in lost opportunities, increased transaction costs, and damaged relationships and reputations. This chapter has provided a research-based model of the negotiation process that can be used to decide what to do and what not to do across the negotiation process. This negotiation model is often referred to as a mutual gains or interest-based approach. It stresses the following:

- Effective preparation that distinguishes positions from interests
- Value creation through creative problem solving that uses trade-offs across issues and contingent agreements
- Value claiming based on objective criteria and clearly defined reservation and target prices
- Closure through the effective packaging of issues and splitting of differences

Finally, there is considerable evidence that this approach, with its emphasis on working to understand various interests and on building relationships, is quite consistent for bargaining effectively across cultures. The list of references also include additional resources not cited in the book.

Exercises

Complete Chapter Activities

15.1 Your experience with negotiations: The negotiation examples you provide in the following will also provide the basis for future activities.

a. List several negotiation experiences you have had over the last few weeks. These may be workplace related, such as a difference with a customer, a contract with a supplier, or a difference with a member of your project team. They may involve reaching agreement on an issue or resolving a dispute or conflict. Consider also your personal life. These may include the purchase of a home, reaching agreement on curfew with a teenager, or resolving a dispute over payment for work done by a contractor.

b. Now select a situation or situations you feel were successfully addressed. What made it successful? Consider both the outcomes and the process.

c. Now select a situation or situations you feel were unsuccessfully addressed. What factors contributed to these failures? Again, consider both outcomes and the process.

15.2 Negotiations with and without an ongoing relationship.

a. Describe a recent negotiation experience that involves an ongoing relationship.

b. Describe a recent negotiation experience that does not involve an ongoing relationship.

15.3 Intangible issues in negotiation: Review the list of negotiation experiences created in Activity 1. Identify and explain two intangible issues that arose in one or more of these negotiations.

15.4 Interests versus positions: A position is one party's solution to a problem. An interest is the underlying need or reason for taking a position. For the following issues indicate the parties' respective interests.

Position of Employer: Employees must pay more of the cost of health insurance.
Interest of Employer: _____
Position of Union: The employees will not pay more.
Interest of Union and Employees: _____
Given the respective interests of the parties, what might be a solution that would address both their interests?

15.5 Think about the negotiations you have been involved in. For one of these negotiations, complete the following questions:

a. Position—What they said?

b. What were their underlying needs and interests?

15.6 Managing issues in negotiation: Use the planning document in Table 15.4 to construct a chart for one of the negotiations identified in Activity 1.

15.7 Opening statement: Drawing upon one of the negotiations identified in Activity 1, identify the key points you would have made in your opening statement.

15.8 Interest-based problem solving: Apply the interest-based problem-solving model in Table 15.5 to one of the issues involved in one of the negotiations identified in Activity 1.

15.9 Interests, rights, and power in resolving disputes: Consider a negotiation identified in Activity 1 that involved a dispute. Select one of those negotiations where a rights- or power-based strategy was used and describe the outcome. How might an interest-based approach have been used?

15.10 Ethical thinking in negotiation: Assume that you are selling your home and the other party asks you if you have another offer. In fact, you do not have an offer. Which answer in the following comes closest to your answer? Which school of ethics does each represent?

a. I have no offer at this time but I am hopeful that I will receive an offer soon.

b. Yes. A party presented an offer for $450,000 this morning and I have 48 h to respond to it.

c. What other houses are you considering?

Additional Exercises

15.11 Interests versus positions exercise

Definition

- Position: A statement of one party's solution to an issue.
- Interest: A statement of one party's concern about an issue.

Directions
Under each issue in the following are two statements. One is a position statement on the issue. The other is an interest statement about the issue. Mark the position statement with a "P" and the interest statement with an "I."
a. Issue: Subcontracting

Statements

- "There will never be any language allowing management the right to subcontract in this contract." _____
- "The job security of our members should not be adversely affected as a result of subcontracting." _____
  b. Issue: Wages

Statements

a. "It is critical that our labor costs become consistent with industry practice if we are to retain market share." _____
b. "There can be no wage increase for the next 3 years." _____

15.12 Apply the interest-based problem-solving model in Table 15.5 to the following cases and recommend a solution.

1. Church construction project. A board of the local church was polarized for nearly a year over who should build its new church. Several board members favored hiring a contractor to do the job; others preferred that the church do the job itself by relying on the skills of the parishioners.

2. Conflict over customers. A large clothing store was faced with constant conflict among the sales force. The sales personnel, who were paid on a commission basis, fought over customers and were reluctant to do the necessary stock work in the back of the store. (The minimal amount of stock work didn't require a regular stock clerk. The clerks resisted the stocking chore because it kept them off the sales floor where their commissions were made.) The manager, having tried everything to secure peace and efficiency, finally decided to let the sales staff meet as a group to resolve their problems. If you were in the sales staff meeting, what would you propose to achieve a resolution of this longstanding conflict?

3. Conflict over a dam. A utility company is anxious to build a dam in order to meet the state's requirement that the utility produce a certain percentage of its energy from green sources. This is being vigorously opposed by both farmers concerned about reduced water flow below the dam and environmentalists concerned about the destruction of habitat for an endangered bird species.

15.13 A colleague posts the following situation on the company negotiation website asking for advice. What advice would you give your colleague?

We decided to outsource our copy centers to a third party to lower costs and free up on-site space. In its request for proposal, the company provided average monthly volumes for the preceding 24 months and requested a price based on the volume forecast. The outsource company submitted a competitive bid based on those volumes but conditioned their pricing on our company agreeing to minimum volume commitments. Our company responded by saying, "We don't know what is going to happen either, but that is their risk of doing business." The supplier said, "We are simply asking you to stand by your numbers. If you can't, and we need to absorb this added risk, we will need that built into the up-front price."

15.14 Consider the case of the Disney Company negotiating to launch the EuroDisney theme park outside Paris. Disney was surprised to find that the small villages in rural France where the theme park was to be built were vigorously resisting having the park in their area through demonstrations and blocking roads to the site. From Disney's perspective, the villagers stood to benefit from increased property values and creation of jobs for their children. The villagers saw things differently and were demanding "voluntary payments" to each of the villages before they would agree to accept EuroDisney into their area. What did Disney miss? How do you explain the resistance of these French farmers to EuroDisney? Why might they resist increasing property values and seeing their children go to work at the theme park?

15.15 Consider the cases in the following and indicate whether you believe the action is legal and why or why not. If you believe it is legal, indicate what negotiation school is reflected in the negotiator's behavior and the basis for your decision.

   a. You are preparing to sell your car and a friend mentions to you that he would give you $2000 for it if he were in the market for a car. Later, when a potential buyer asks you how much you are asking, you tell the potential buyer that you already have an offer for $2000.

   b. In selling you laptop computer, you decide not to tell prospective buyers that the computer occasionally crashes without warning and that the hard drive seems likely to fail soon.

   c. In negotiating with a supplier you state that your company must have a 10% reduction in cost or you will use another supplier. In truth, you would be delighted to get an agreement with a 5% reduction in price.

   d. In negotiating a new collective bargaining agreement, you tell the union that the company will close the plant unless it gets significant concessions on wages and work rules.

   e. In fact, the company is planning to place a new product in the plant.

References

Acuff, F. (2008). *How to Negotiate Anything with Anyone Anywhere around the World*, 3rd edn. New York: AMACOM.

Ariely, D. (2008). *Predictably Irrational: The Hidden Forces that Shape Our Decisions*. New York: HarperCollins.

Arvuch, K. (1998). *Culture and Conflict*. Washington, DC: United States Institute of Peace.

Barrett, J. T. (1989). "Unbenching Bobby Bench." Mimeo.

Barrett, J. T. and O'Doud, J. (2005). *Interest-Based Bargaining: A Users Guide*. Victoria, British Columbia, Canada: Trafford Publishing.

Bazerman, M. H. and Neale, M. A. (1992). *Negotiating Rationally*. New York: Free Press.

Bazerman, M. H. and Watkins, M. (2004). *Predictable Surprises*. Cambridge, MA: Harvard Business School Press.

Brett, J. M. (2001). *Negotiating Globally*. San Francisco, CA: Jossey-Bass.

Brown, S. (2003). *How to Negotiate with Kids*. New York: Viking Press.

Cialdini, R. B. (2001). *Influence: Science and Practice*. New York: Allyn & Bacon.

Ertel, D. and Gordon, M. (2007). *The Point of the Deal*. Boston, MA: Harvard Business School Press.

Fisher, R. and Brown, S. (1998). *Getting Together: Building Relationships as We Negotiate*. New York: Penguin Books.

Fisher, R. and Ury, W. (1991). *Getting to Yes*, 2nd edn. New York: Penguin Books.

Goldberg, S. B. and Brett, J. M. (2004). *Brookside Hospital vs. Black Computer Systems, Inc.: Negotiation Version*. Evanston, IL: Dispute Resolution Research Center, Northwestern University.

Hammond, J., Keeney, R. L., and Raiffa, H. (2004). *Smart Choices: A Practical Guide to Making Better Decisions*. Boston, MA: Harvard Business School Press.

Lax, D. and Sebenius, J. (1986). *The Manager as Negotiator*. New York: The Free Press.

Lax, D. and Sebenius, J. (2006). *3D Negotiation*. Boston, MA: Harvard Business School Press.

Lewicki, R. and Hiam, A. (2006). *Mastering Business Negotiation*. New York: John Wiley & Sons.

Lewicki, R., Sanders, D. M., and Minton, J. W. (2010). *Negotiation*, 6th edn. New York: McGraw-Hill/Irwin.

Malhotra, D. and Bazerman, M. H. (2007). *Negotiation Genius*. New York: Bantam Books.

Mnookin, R. (2010). *Bargaining with the Devil: When to Negotiate, When to Fight*. New York: Simon & Schuster.

Mnookin, R., Peppet, S. R., and Tulumello, A. S. (2000). *Beyond Winning: Negotiating to Create Value in Deals and Disputes*. Cambridge, MA: The Belknap Press.

Movius, H. and Susskind, L. (2009). *Built to Win: Creating a World-Class Negotiating Organization*. Boston, MA: Harvard Business School Press.

Raiffa, H. (2003). *Negotiation Analysis: The Science and Art of Collaborative Decision Making*. Cambridge, MA: Belknap Press.

Requejo, W. H. and Graham, J. (2008). *Global Negotiation: The New Rules*. New York: Palgrave Macmillan.

Salacuse, J. (2003). *The Global Negotiator*. New York: Palgrave Macmillan.

Shell, R. (1999). *Bargaining for Advantage*. New York: Viking Press.

Thompson, L. (2008). *The Truth about Negotiations*. Upper Saddle River, NJ: FT Press.

Thompson, L. (2009). *The Mind and Heart of the Negotiator*, 4th edn. Upper Saddle River, NJ: Prentice-Hall.

Tinsley, C. H., Curhan, J. J., and Kwak, R. S. (1999). Adopting a dual lens to examine the dilemma of differences in international negotiation. *International Negotiation: A Journal of Theory and Practi*ce, 4, 5–22.

Ury, W. (1991). *Getting Past No: Negotiating with Difficult People*. New York: Bantam.

Watkins, M. (2002). *Breakthrough Business Negotiation*. San Francisco, CA: Jossey-Bass.

Watkins, M. and Rosegrant, S. (2001). *Breakthrough International Negotiations*. San Francisco, CA: Jossey-Bass.

Weiss, S. E. (1994). Negotiating with "Romans"—Part 2. *Sloan Management Review*, 35, Spring, 85–99.

Chapter 16

Ethical Decisions

Dean W. Pichette

> Tim has worked for his present employer for 10 years. During this time, he has advanced from trainee to senior engineer. Until approximately 5 months ago, Tim's performance had been exceptionally strong but then changed dramatically. Assignment deadlines were missed and those that were turned in were either incomplete or of poor quality. Team members were becoming frustrated because of Tim's lack of participation. His boss is considering a variety of forms of progressive discipline.

> As a grandmother aged, she began exhibiting signs of dementia. It was not a significant issue for quite some time. She could still function and take her daily walks in a town she knew very well. Neighbors were helping out in the meantime. Her children began thinking about what they would do as her situation deteriorated and the burden on neighbors became more substantial. Should she move in with the oldest daughter or should she be placed in a nursing home?

16.1   Goal and Overview

The primary goal of this chapter is to raise the awareness level as to a wide range of ethical issues that routinely arise when making decisions. It is also to understand that often these ethical issues conflict.

- You have been assigned to a process improvement project to reduce your company's costs. Would you help develop and implement a solution that could eliminate your job? What about a solution that would eliminate tens of jobs of your coworkers?

- Your company is facing significant cost pressures. Should you propose a plan to take work away from long-time suppliers and move the work to low-cost emerging markets?

- Your company is in an industry in economic crisis. Should it renegotiate existing contracts to squeeze out suppliers and reduce their prices?

- A close friend has a job opening in his company. Should you press him to hire your son, who is looking for a job?

- A social worker is trying to decide whether to recommend the removal of a child from a dysfunctional but loving home. How should she balance all the factors involved?

477

- A coworker is being unfairly treated by your boss. Under what circumstances would you go to the boss and express your concerns? When might you go over the boss's head or have a discussion with human resources?

- Your manager has told the project head that your team's piece of the project is on schedule. Yet everyone on the project, including the manager, knows that the project is significantly behind schedule. What should you do?

All these decision contexts represent common dilemmas involving multiple ethical issues in conflict with one another. The primary focus of this chapter is the type of dilemma that arises routinely out of interactions with coworkers, bosses, senior management, clients, suppliers, and family.

In Chapter 1, we explored the difficulty of decision making, taking into account the involvement of competing goals, multiple objectives, and uncertain outcomes. In this chapter, we explore yet another factor that can complicate our ability to make a good decision: ethics. We define a good ethical decision as one that recognizes and evaluates the ethical issues involved and, if necessary, strikes an appropriate balance between conflicting ethical values. Poor ethical decisions often result from a failure to recognize the existence of ethical issues rather than from a lack of ethical values. A second factor that contributes to unethical decisions is the pressures often surrounding critical decisions. These may be time pressure, peer pressure, or organizational pressure. Last, there is an array of cognitive biases that limit our ability to perceive the ethical dilemmas we face.

This chapter begins by discussing a number of common ethical values that arise in a wide range of decisions. The goal is to raise the awareness level of the reader in making decisions. Next, we describe a number of cognitive biases that distort our ethical reasoning. Then, we explore some of the pressures that can compromise a person's commitment to ethical decision making.

A major theme of this chapter is that decisions often involve multiple ethical values in conflict. We approach this issue from two perspectives. First, we discuss some of the most common conflicting values. Second, we use diverse case studies to illustrate the ethical issues in common decision contexts and point out the need to deal with conflicts.

Before proceeding, it is useful to ground the discussion with several definitions of ethics. The Collaborative International Dictionary of English (v.0.48) defines ethics as "the science of human duty; the body of rules of duty drawn from this science; a particular system of principles and rules concerning duty, whether true or false; rules of practice in respect to a single class of human actions; as, political or social ethics; medical ethics."

Ethics investigates and creates theories about the nature of right and wrong. Ethical theories include such concepts as "doing the right thing," "doing no harm," "telling the truth," "not interfering with the rights of others," and "observing the golden rule." But doing the right thing from one perspective may result in doing the wrong thing from another perspective. Would you tell the truth to a sick friend who is asking you how she looks? Would you have told the truth about hiding a slave in the antebellum period or hiding a Jew during the Holocaust? Not interfering with the rights of others may cause serious harm to someone else. For example, a social worker's or judge's decision to take a child from a dysfunctional home engages a range of ethical issues from multiple perspectives.

The implicit thesis of this chapter is that people strive to make ethical decisions. We also presume that it is generally in the best interest of businesses to make ethical decisions as people tend to hold ethical enterprises in high esteem. Conversely, customers and business associates would be wary of engaging in activities and relationships with organizations with poor ethical records. (If, however, you believe that there is no such thing as business ethics, perhaps you should skip this chapter!)

Even if you strive to make ethical decisions, it is easier said than done. This chapter attempts to raise awareness of ethical issues by defining particular ethical values as well as potential sources of conflict between these values. We also explore some of the common barriers to making ethical decisions, such as biases and pressures. The chapter concludes with examples of common ethical decisions to better understand why people who strive to be ethical face many difficulties.

While corporate policy statements, codes of ethics, and laws forbidding corrupt practices attempt to guide ethical decision making, they cannot prevent breaches (Guy 1990). These can only supplement the values that are within the individual, and how they are applied to each decision. People attempting to make moral decisions may ask such questions as: "Am I addressing the right problem?" "Am I aware of all of the issues?" "Who will be harmed by this decision?" "What is the right thing to do?" "Will I regret this decision later?" "Would I be embarrassed if my decision became public knowledge?" "Do the long-term benefits outweigh the short-term losses?" These questions bring an individual's core values into a decision. However, choices are rarely made directly between values. More often, choices are made between alternatives that differ in the extent to which values are reflected and weighted. One way that moral standards vary among people is in the different weights assigned to a particular ethical value in a complex decision.

16.2   Ethical Decision-Making Framework

All decisions are made within some context and situation. In Figure 16.1, the decision maker is depicted in the center, surrounded by the various factors that impact upon the decision. These include but are not limited to available information, pressures from interested parties and competing interests, and the decision maker's own knowledge, values, intentions, and goals.

Ethical decision making integrates a decision maker's preferences, utilities, costs, benefits, goals, and objectives (Guy 1990). When applying ethical analysis, the challenge is to operationalize and quantify one's moral values. Cost and benefit analysis needs to be supplemented by consideration of many ethical values simultaneously. Ethical decision making is the process of identifying a problem, generating alternatives, and choosing among them, so that the alternative selected maximizes the most important ethical values while achieving the intended goal. Not all values can be maximized simultaneously, which means that some values must be compromised. Compromising among competing values means pursuing a decision path that will permit satisfaction at some specified level of ethical need.

Dialectical inquiry is an approach that examines a decision completely and logically from two different and opposing views. It is a good method for ethical decision making because it identifies hidden assumptions that are examined from both sides of the issue. The conflict that this approach engenders serves to provide a deeper analysis of assumptions, interpretations, and the range of options.

One strategy to assist in making ethical decisions is to involve a diverse group in the decision-making process. When possible, the group should include females and minorities, young and old,

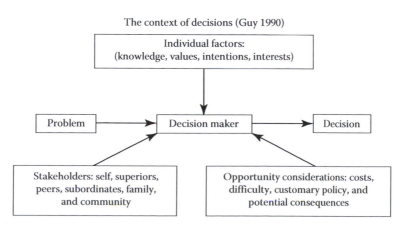

FIGURE 16.1:   Decision contexts.

and members of different functions within the organization. Including diverse team members will both increase the likelihood that all ethical issues will be identified and also assist in evaluating how the decision will be perceived by a broader representation of those affected.

Another strategy for integrating ethics into decision making involves asking four questions (Badaracco 1992):

1. Which course of action will do the most good and the least harm?
2. Which alternative best serves others' rights?
3. What plan can I live with that is consistent with my basic values and commitments?
4. Which course of action is feasible in the world as it is?

The first question focuses on consequences. It asks decision makers to examine the full range of consequences that will result from different ways of resolving the issue. The basic question can be broken into subquestions. Which groups and individuals will benefit from different ways of resolving the issue and how greatly? Who will be put at risk or suffer? How severe will the suffering be? Can the risk and harm be alleviated? One caution is that there are no universal definitions of good or harm. Much will depend upon particular circumstances, institutions, and legal and social arrangements.

The second question focuses on rights. The Declaration of Independence states that human beings have inalienable rights to life, liberty, and the pursuit of happiness. Badaracco suggests that we accept similar ideas in everyday life. All individuals should have the right to be treated with respect, to have promises kept, to be told the truth, and to be spared unnecessary injury.

The third question focuses on values, which are discussed in detail in the next section. On a personal level, the question is as follows: What best serves my commitments and aspirations in life? At the organizational level, a business or government agency should articulate its mission and values to clearly define the type of community it is or that it aspires to become. These can be used as ethical guides as challenges arise.

The fourth question is purely pragmatic. Of the possible alternatives identified, which ones are feasible? In a business context, feasibility may relate to a decision maker's actual power in the organization. It may also reflect a company's competitive, financial, and political strength. It should include the likely costs and risks of various plans of action, and the time available for action.

16.3 Values

The list of ethical values in the following bears the influence of Western culture. We believe that the biggest differences among cultures, however, is less about specific values and more about how much weight is given to values when they conflict. For example, many cultures accept the importance of both the individual and community. The difference may be in whether a society expects the individual to sacrifice his needs for the benefit of the community or places greater emphasis on individual rights. Similarly, many Eastern cultures place a high value on avoiding embarrassment to oneself and to others. The Talmud praises those who take personal risks to avoid placing a neighbor in an embarrassing situation.

Table 16.1 lists many such ethical values and also identifies which values reinforce others and the potential sources of conflict between different values within the decision-making process (Guy 1990).

Caring for people in general means treating people as ends in themselves and not as a means to an end. It entails displaying compassion, courtesy, and kindness, and treating people with dignity, tolerance, and mercy while avoiding harm to others. In other words, it bespeaks adherence

TABLE 16.1: Values, their reinforcement, and potential conflict.

Values	Reinforce	Po
Caring for people in general	Promise keeping, prioritizing the needs of others in a social network, harmony and avoidance of conflict, respect for others, fidelity, and self-discipline	Pursuit o... sustain, and u... self-discipline
Prioritizing the needs of others in a social network	Caring, respect for others, fidelity, responsibility, and responsible citizenship	Responsibility, pursuit of excellence, and survive, sustain, and thrive
Respect for others	Caring, responsible citizenship, fidelity, responsibility, and self-discipline	Pursuit of excellence, and survive, sustain, and thrive
Harmony and avoidance of conflict	Caring, and respect for others	Honesty, integrity, and pursuit of excellence
Honesty	Integrity and fairness	Caring and survive, sustain, and thrive
Fairness	Honesty and integrity	Caring and survive, sustain, and thrive
Integrity	Honesty	Caring, survive, sustain, and thrive, and respect for others
Fidelity	Caring, pursuit of excellence, respect for others, survive, sustain, and thrive, and responsibility	Promise keeping and responsibility
Self-discipline	Pursuit of excellence, harmony, respect for others, and survive, sustain, and thrive	Caring and pursuit of excellence
Promise keeping	Caring, respect for others, responsible citizenship, and fidelity	Pursuit of excellence, survive, sustain, and thrive, and fidelity
Responsibility	Pursuit of excellence and respect for others	Survive, sustain, and thrive
Pursuit of excellence	Responsible citizenship, survive, sustain, and thrive, fidelity, responsibility, and self-discipline	Caring, promise keeping, respect for others, and responsible citizenship
Survive, sustain, and thrive	Pursuit of excellence, fidelity, and self-discipline	Caring, honesty, promise keeping, integrity, respect for others, responsible citizenship, and fidelity
Responsible citizenship	Promise keeping, respect for others, and responsibility	Pursuit of excellence, and survive, sustain, and thrive

to the Golden Rule. In some cultures, however, caring for others may be limited to those in one's own group.

Potential sources of conflict with this value can be found in an employment contract that enforces a strict approach to job performance issues. There also may be a conflict between what is considered best for the employee and what is best for the organization. For example, a supervisor may terminate an employee either for poor performance or in order for the business to survive a downturn. It is difficult to demonstrate caring when firing someone, and it may appear insincere and hypocritical.

The value of caring is unusual in that it can result in internal ethical conflicts when applied to multiple individuals. Caring for one person, for instance, could negatively impact others. An ethical dilemma arises when a respected and reliable colleague begins experiencing personal problems that

terfering with his ability to effectively interact or complete assignments with his team mem-
. The supervisor may have to take action that involves deciding whether it is more important to
re for the individual or the work group.

Caring for others can even conflict with your own personal responsibilities. Imagine a situation in which you are facing time pressures to complete an important assignment. What do you do
when a colleague on a tight deadline requests information that only you can provide but that will
require several hours to gather, tabulate, and analyze? Should you stop working on your assignment to provide your colleague with the requested information? Or should you complete your
assignment first?

Prioritizing the needs of others in a social network involves family, friends, colleagues, and
members of one's community. Such priorities vary according to culture. Scandinavians, for example, prioritize family and the Japanese prioritize colleagues at work. In contrast, Americans attempt
to balance various social networks, which leads to more conflict. And yet, there are consequences
among cultures with clear priorities, as in Japan, where work is all-consuming for the husband and
family issues fall to the wife. This has produced a society in which many women are unable to
develop long-term careers. A secondary consequence is that Japanese women tend to delay marriage and have children much later than women in Western societies. As a result, the Japanese now
have a declining population with dire demographics looming.

Some U.S. companies attempt to balance the needs of their employees for work and home by
providing day care facilities onsite or near the workplace. There are also companies that attempt to
balance work life and community needs. Ford Motor Company, for example, has instituted community service days in which employees are encouraged to perform some form of community service
at company expense, such as packing and delivering food to the needy, cleaning up a state park, or
working at a local food bank. Programs such as these may still benefit the company in various ways,
but their impetus derives in part from a perceived need to serve and balance the interests of workers
and the community at large.

Such accommodations cannot resolve every situation, of course. A day care facility may have no
provision for a child who is ill, and work schedules many conflict with a planned community service
day. An organization trying to balance priorities will inevitably come up against opposing forces
that cannot be balanced.

Respect for others means recognizing each person's right to be treated with dignity. It includes
being courteous, prompt, decent, and providing others with information they need to make decisions. Potential sources of conflict include a process reduced by time pressures and conflicting
priorities. One example would be a situation where a subordinate has failed to perform a critical
task that the supervisor promised a peer would be completed by a certain date. When the supervisor
finds out, does he criticize the subordinate immediately or does he wait for a time when the criticism
can be done privately? Another example would be a situation where a supervisor needs to lay people
off to meet a budget objective. Is it better to give the people advance notice, so that they can plan
their next steps and exit with dignity or to escort them out of the building immediately? Many businesses escort terminated employees out of the building immediately because of perceived security
concerns. It has also been touted as a humane practice when terminating employees because those
leaving and those staying will not have time to dwell on an uncomfortable situation. But many say
that such attempts at humaneness are artificial and that most would prefer to have an opportunity to
say goodbye to their coworkers with dignity.

Harmony and Avoidance of Conflict is similar to *Caring* and *Respect for Others*, but we have
chosen to list them separately because they are significant elements in both personal life and geo-
politics. We all might prefer to live in a more harmonious family, community, or world. Yet conflict
is built into many American institutions. The U.S. legal system functions on the belief that truth
emerges from conflict. Similar logic is often used in constructing decision-making teams from
people who have conflicting views. Team-based decision making with multiple viewpoints may thus
lead to more ethical awareness if the team does not defer to a strong personality to avoid conflicts

that may arise from debate. This deference can be due to a team member's belief that harmony is paramount. In some cases, this deference can be due to organizational or national culture where team members do not challenge those of higher stature or authority. Also, some people are simply timid: they value peace and harmony over potential conflict even if the group decision is not aligned with their values. They are likely the ones to say, "Don't rock the boat." This places the avoidance of conflict in opposition to issues of integrity and striving for excellence.

Honesty is defined as being truthful. It also includes being willing to admit error. Caring for another individual, however, can often conflict with this value, as when an honest response may hurt an individual's feelings. Married couples, for example, are familiar with the perils of being brutally honest with a spouse. And how truthful should a physician be with a patient who is diagnosed with a fatal disease? While American medical practice requires a level of honesty in such cases that is not what is expected in Japan. A similar quandary arises concerning the appropriateness of lying to save an individual. The Quakers, who were key members of the Underground Railroad that helped runaway slaves in the nineteenth century, faced this truth dilemma whenever they were confronted with the question, "Are you hiding a slave?" The same situation occurred with those who hid Jews during the Holocaust era.

Negotiations often challenge our commitment to honesty as both sides engage in puffery, posturing, and anchoring. Similarly, a purchaser may understate his willingness to pay for an item when attempting to get the best price from a seller. In addition, U.S. rules about information sharing for publicly traded companies may force senior executives to not discuss their plans with their employees unless they are ready to tell the whole world. Thus, while secret negotiations over a merger are being pursued privately, they may have to publicly deny such activity for fear of undermining their corporate relationships and stock value.

Fairness is defined as conformity with rules or consistent standards, the ability to make judgments free from discrimination, and treating people equally. In a workplace setting, fairness means applying consistent standards with regard to salary, promotions, and work distribution. Fairness can be challenged when, for example, a supervisor develops social relationships outside of work with certain subordinates and then may favor those employees with a lighter workload or a bigger percentage of the funds allocated for pay increases. Almost any pay raise strategy can be perceived as unfair.

Fairness is one of the most difficult values to follow consistently. It is often perceived differently by a giver and receiver. How often have we all heard the complaint, "But this is not fair!" If everyone is paid equally, the better performers will say it is unfair. If the higher performers are given bonuses, the nonrecipients will declare that it is not fair. Children learn at a very early age to use the fairness argument against their parents to complain about their treatment relative to that of a sibling.

Integrity is defined as exercising good and consistent judgment. It suggests adherence to ethical principles and using independent judgment, avoiding conflicts of interest, and acting on one's convictions. Integrity is closely aligned with honesty but is primarily about a consistent commitment to both personal and professional values. Maintaining professional integrity may require an individual to speak up in opposition to others in his group if in his judgment a decision may be compromising a core value such as safety or excellence.

On a personal level, integrity may be challenged as a result of potential conflicts of interest. One example is a supplier who offers an expensive holiday gift at a time when gift giving is traditional. Would you accept the gift, telling yourself that it will not have an impact on awarding future business? Would you return the gift at the risk of hurting the supplier's feelings? What if this was in a culture in which gift giving is even more important than in the United States?

Fidelity is defined as faithfulness to clients, allegiance to the public trust, loyalty to one's employer, firm, or agency, and loyalty to the profession. Potential sources of conflict include public relations activities that involve the evaluation of values, principled compromise, and conflicts of interest. One example is an executive who takes advantage of his employees' fidelity and does not adequately reward them for their work. Another is a situation where an employer pressures a professional to slant the truth in order to win a contract.

Self-discipline is defined as acting with reasonable restraint and not indulging in excessive behavior. Conflict may arise when a supervisor deals with an employee who is failing to perform and is either unwilling to change his behavior or unresponsive to coaching and counseling. Can the supervisor handle the situation without losing his temper? Lack of self-discipline was also manifest in a different form during the recent collapse in the financial industry, as executives took huge bonuses that they said were promised or owed to them even while their companies faced disastrous losses. They also displayed a total lack of concern for fairness.

Promise keeping is keeping one's word and fulfilling one's commitments. The most frequent challenge to promise keeping is when circumstances change and it becomes difficult to keep a promise without compromising other values. A changing environment and not being in complete control of situations can make it hard to fulfill promises. One example is promising employees job security during good economic times and then laying them off when the economy takes a sudden turn for the worse, as with the start of the financial industry collapse in 2007. Another is a supervisor promising someone career advancement and then being rotated to a new position where the promise cannot be fulfilled.

In addition, if you are not the top boss, a commitment you made in good faith may be beyond your authority to keep. You may have built a long-term relationship with a supplier only to find a finance official insisting that you must go with a lower-cost supplier. Alternatively, you might be forced to go back to a supplier and insist on a price change after signing an agreement because of pressures from top management. Unless you are prepared to quit every time you are forced go back on your word, there may be little choice but to go back on your word.

Another conflict regarding commitments may arise when a commitment is conditional upon performance. Oftentimes, the two parties involved perceive the performance in question differently and therefore disagree as to whether the commitment must be honored.

Responsibility means reliability and dependability. It suggests accountability, including accepting the consequences of one's actions, accepting responsibility for one's decisions, setting an example for others, and displaying trustworthiness. Potential sources of conflict include time pressures that can truncate the decision-making process by limiting options or only focusing on one objective and not having total control of the decisions that one needs to make. One example is a situation where an individual caves into team peer pressure knowing full well that the decision is being made using inaccurate information. Another is the need to make a complex decision within a timeframe that does not allow the proper time to evaluate an adequate number of alternatives.

Pursuit of excellence is defined as striving to be as good as one can be. It includes being diligent, industrious, committed, well informed, and well prepared. In some cultures, especially where teams or group behaviors are heavily weighted, it may not be a value in industry. Potential sources of conflict include time pressure that requires a decision by a specified deadline, leading to a compromise in quality. There may also be pressure to meet other objectives such as cost in which the best choice is too expensive to meet the organization's budgetary constraints. Too often, U.S. automotive companies sacrificed excellence in quality because they believed it was too expensive to achieve.

Unfortunately, pursuit of organizational excellence often seems to conflict with caring for the individual. Some companies have a reputation for aggressively eliminating employees who do not achieve the highest standards. Law firms may force junior lawyers to compete against one another as each pursues the goal of achieving partner. Some universities have a similar reputation regarding the treatment of their students.

Survive, sustain, and thrive refers to the fundamental right to survival. It includes the ability to retain a certain lifestyle or remain in business, and the ability for a business or individual to grow and prosper. Potential sources of conflict arise in difficult economic situations, as when one must decide whether to move manufacturing to a lower-cost region, which would result in terminating local employees. Another is reducing staff during an economic downturn even if your company is not facing potential bankruptcy.

Celestial Seasonings includes a statement regarding ethical trade on their tea boxes: "We're passionate about the people and places that produce our ingredients. We support fair wages and sustainable harvests in more than 35 countries." This statement indicates that, beyond their concern for corporate profits, Celestial Seasonings is also concerned about their employees and the environment. Although the print on the box is small, presumably, consumers who also hold these values would consider choosing these products over others even if the price is slightly higher.

Responsible citizenship means acting in accord with societal values. It suggests obeying just laws, protesting unjust laws through accepted means, voting, and expressing informed views. Ideally, it means abiding by the spirit of the law and not just the letter of the law. Potential sources of conflict include pressure to make money and issues related to the risk of survival.

This issue of responsible citizenship is a major challenge for global companies who work in nondemocratic countries. What does it mean for Google to be a responsible corporate citizen in China, a country that spies on its citizens and strives to suppress dissent? Environmental concerns are a growing element of responsible citizenship. What does it mean to care for the environment when U.S. standards are much higher than third-world standards? Similarly, should a multinational have the same standards for worker safety around the globe? Risks that are unacceptable in the United States may be the norm elsewhere.

Sometimes, responsible citizenship can include an implicit dilemma with regard to local versus national needs. All congressional officials are elected locally and therefore have a duty to represent their interests. Is it therefore unethical for an elected official to steer an appropriation toward his or her constituency when the appropriation would have been better spent in another region or not spent at all?

When assessing ethical issues, it is often useful to think of values in pairs. For example, a decision related to moving an operation to a lower-cost region should result in the comparison and evaluation of values such as *Responsible Citizenship* and *Survive, Sustain, and Thrive*. A decision related to establishing a bid price should result in a comparison and evaluation of values such as *Honesty* and *Integrity*. A decision regarding whether or not to remain employed with a company whose corporate goals are in conflict with individual values should result in the comparison and evaluation of values such as *Fidelity* and *Survive, Sustain, and Thrive*.

Ethical concerns are usually complex. They may include dilemmas, paradoxes, inconsistencies, and differing expectations. For example, corporate executives want to build companies that are independent, strong, growing, and vital. They want to attract and retain the best employees and provide them with a rewarding place to work. Hard moral choices are therefore at times inescapable (see, e.g., Badaracco 1992). How do you fire someone that you have worked with for many years? When is it right to violate someone's right to privacy, for example, if he or she has a substance abuse problem and needs help? What about the repercussions to a local workforce and community when an executive decides to move a business operation to a low-cost, overseas site? Can you have a clear conscience when you commission a team to work on a project that will result in a leaner, more efficient organization knowing that the team will be working themselves out of a job? In some situations, there are no win–win solutions. Sometimes, the "best" way to resolve a dilemma may severely test an executive's sense of integrity. Responsible, thoughtful, practical-minded people will often disagree on the right path to take when faced with this type of decision.

1. *Activity*: Describe a situation (other than the ones identified earlier) in which dilemmas, paradoxes, inconsistencies, or differing expectations may result in a decision with ethical ramifications.

16.4   Biases, Myopia, and Don't Want to Know

One of the primary reasons that ethical decision making can be difficult is that each of us brings biases into the process. These may have developed based on our upbringing, environment, work experience, or social circles. Other biases arise because our brains routinely make associations and see patterns that may or may not exist. Biases are likely so engrained into our being that we may be blind to the concept that there are ethical issues associated with the decisions we make and actions we take.

Decision making may include biases that help mask ethical evaluation (Banaji et al. 2003). Three related biases that create ethical challenges are (1) implicit prejudice, (2) in-group favoritism, and (3) conflict of interests. Implicit prejudice emerges from one's unconscious beliefs. A compounding factor is that we may simply be oblivious to the ethical issues surrounding a decision. Alternatively, we may incorrectly assess the impact of our decision on others, thereby mitigating the potential ethical concerns.

Implicit prejudice is widespread and persistent because both in our personal and professional lives we learn to associate things that commonly go together and expect them to inevitably coexist. This skill to perceive and learn from associations often serves us well, especially in emergencies when time is critical; but it is rarely applicable to every situation. We associate thunder and rain but they do not necessarily always go together. This bias toward associations creates an ethical challenge when people judge others according to their own personal linkage of values.

Managers may associate commitment to the job and overall performance by the way an individual sets his or her priorities. Leaving to participate in a child's celebration or attend to a child's needs might be interpreted as a lack of commitment even if the worker puts in long hours on other days to get the job done or is more efficient at getting the job done in the normal workday. At least with regard to medical issues, federal law under the Family Medical Leave Act (FMLA) prohibits penalizing a worker for placing his family first. However, the law can do little to affect colleagues' perceptions and the impact on a person's career. Rankings of companies as good places to work include this employee accommodation as a major element of corporate culture.

Another element of corporate culture is the perception that affirmative action has led to a watering down of academic requirements and on-the-job performance evaluations for minorities. In some quarters, every minority who succeeds is assumed to be the beneficiary of affirmative action and thus not as qualified as other employees in the workplace.

> 2. *Activity*: Describe a situation in which you made an association regarding things that you believed went together but later recognized did not.

A second common bias is in-group favoritism that favors your group. This form of bias arises when one helps a friend, a relative, or a colleague get a useful introduction, admission to a school, or a job. However, this bias presents an implicit dilemma, since caring about your social network is also an appropriate value. It stems from placing high value on the needs of family, friends, community, and colleagues. However, this value can lead to ethical conflicts. When is it appropriate to do more favors for those we know and those who tend to be like us? Although it seems innocent to help people we know, the result is that this behavior effectively discriminates against those who are different from us. In-group favoritism gives "extra credit" for group membership. Few would question this preference by the owner of a family business; most would argue against this preference by a government official. What about all the other situations of hiring or contracting? This type of bias was a critical element in keeping good middle-class jobs closed to minorities who were not part of the in-group controlling entrance to these careers. This kept minorities out of many construction unions as well public sector jobs such as firefighting.

However, from a practical standpoint, there are benefits to developing lasting relationships that involve doing reciprocal favors and giving preferences. The Japanese have the term *keiretsu*, an interlinking of corporations to form horizontally integrated alliances across many industries. They view this as a critical element of an effective way to conduct business. Where *keiretsu* companies supply one another, the alliances are vertically integrated as well. These valuable relationships are especially critical in times of crises. For example, on February 1, 1997, a fire destroyed an Aisin factory, one of Toyota's biggest suppliers. Aisin was the sole source for a p-valve, which is an essential brake part used on all Toyota vehicles worldwide. Toyota was using 32,500 p-valves per day. Because of Toyota's Just in Time (JIT) system, only 2 days' inventory of parts were available in the entire supply chain. After 2 days, every Toyota assembly plant worldwide would shut down. To respond to the crisis, 200 suppliers self-organized to help Toyota. Sixty-three different firms took responsibility for making the parts by piecing together engineering design information, using some of their own equipment, and rigging together temporary assembly lines. These relationships almost seamlessly kept Toyota's assembly lines up and running (Liker 2004).

Culture is a significant factor in defining how a society positively and negatively views in-group favoritism. "Culture is a shared pattern of categorizations, attitudes, beliefs, definitions, norms, values, and other elements of subjective culture" (Triandis et al. 2001). In individualist cultures, these elements are centered on the individual; in collectivist cultures, they are centered on the in-group. In collectivist cultures, individuals are defined by the groups to which they belong. In these cultures, the group's goals take priority over individual goals. People who live in collectivist cultures therefore tend to prioritize the in-group's goals over their personal goals. Thus, global companies face a special ethical challenge when they operate in diverse cultures that have divergent perspectives about in-group favoritism.

3. *Activity*: Describe a situation in your experience where in-group favoritism was ethically justifiable.

_____

4. *Activity*: Describe a situation in your experience where in-group favoritism was not ethically justifiable.

_____

A third pervasive bias involves a conflict of interests that leads decision makers to favor those who can benefit them. Lawyers who earn fees based on their clients' awards and settlements can create a conflict of interests. This arrangement facilitates people being given the opportunity to have their day in court that they otherwise would not be able to afford. However, do they know that their attorney's decision to settle out of court as opposed to pursuing a jury trial was in their best interest? Similarly, there is potential conflict of interest as CEOs negotiate mergers. They pursue a merger because they believe a merger will generate significant synergies that will result in more market value than if the companies remained as individual entities. However, an important portion of the negotiations may focus on how the CEO and his peers will benefit from the merger. What will his role be in the new company? Will he get a large settlement if he is asked to leave?

Yet another example is an outside employee auditing a company that could be a source of future revenue or employment. It is a common concern that arises when congressional staffers help craft legislation for industries where they have worked in the past and may work in the future. Politicians create obvious conflicts of interests when they curry favors from lobbyists and receive large donations from a small group of constituents. In many states, legislators work part time as state senators or representatives while retaining their previous jobs in a range of industries. Conflict arises as they design legislation and vote in ways that advance one group's needs at the expense of the best interests of the broader population.

The massive oil spill in the Gulf of Mexico triggered a comprehensive review of conflicts of interest in the regulation of the oil drilling industry. A May 2010 report from the Interior Department's inspector general found that from 2000 to 2008, employees from the former Minerals Management Service accepted lunches, football tickets, hunting trips, and other gifts from the oil and gas companies they were charged to regulate (Dlouhy 2010). Several actions were taken to remediate this issue. Interior Secretary Ken Salazar dismantled the Mineral Management Service and created a new bureau with three separate divisions in its place, the Bureau of Ocean Energy Management, Regulation, and Enforcement. This resulted in the separation of regulatory functions from the roles of developing offshore energy resources and collecting royalties from oil and gas produced on federal property.

On August 30, 2010, the Obama administration imposed a conflicts of interest policy to put greater distance between inspectors and the offshore platforms and rigs they police. Bureau employees must now tell supervisors about any potential conflict of interests and submit formal requests not to be assigned inspections or other official duties when those conflicts arise. Employees must also ask to step down when their inspections or official duties involve a company employing a family member or close personal friend. This new policy is directed toward the most clear-cut potential conflicts of interest. It acknowledges that drilling regulators along the Gulf Coast may live next door to rig workers and supervisors they see in the field. The guidelines do not require disclosure as long as the neighbors have limited personal knowledge of each other and only share general conversations.

The health care industry is rife with opportunities for conflict of interests. Surgeons, who are paid a fee for services, must decide on a daily basis whether or not to recommend surgery. To limit this potential conflict, insurers will routinely pay for a second opinion from a physician who would not do the surgery. At the national level, there is a recognized concern about the interaction between doctors and pharmaceutical firms. A 2007 study reported that 94% of physicians have "a relationship" with the pharmaceutical, medical device, or related industries (PEW Charitable Trusts 2008; Caputo 2009). The study found that the pharmaceutical industry spends an estimated $28–$46 billion each year marketing its products or $35,000 for each physician. More than 100,000 pharmaceutical sales representatives visit U.S. physicians on a regular basis, providing free lunches, gifts, medication samples, and medical literature to promote their products. The Pew-initiated Prescription Project is concerned that aggressive marketing to physicians creates real and perceived conflicts of interest for doctors and raises questions about the treatments chosen. They reference a June 2008 report issued by the American Medical Student Association, which found that only 21 of 150 medical schools surveyed have strong policies to address conflicts of interest caused by such marketing.

A consumer survey conducted by the Pew Prescription Project in 2008 found that 68% of respondents supported legislation that would require public disclosure of financial relationships between physicians and industry. Seventy-eight percent believed that accepting gifts from the pharmaceutical industry influences their doctor's prescription choices, but only 34% said they would ask their doctors about financial ties. The belief is that patients feel it would be difficult or awkward to do so and may lead to an antagonistic relationship with their doctor.

Medical research universities are struggling with the ethical dilemma of potential conflicts of interest. Harvard and Yale medical schools, for example, have recently established new guidelines for their faculty. However, the pharmaceutical companies are moving more of their applied research to individual private practitioners who are outside the scope of these guidelines. Conflicts of interest are also an issue that research journals must address. They have expanded requirements to include disclosure statements about potential conflicts of interest within the reported research.

In March 2010, as part of the overall health care bill, the Physician Payments Sunshine Act was signed into law by President Obama. It requires companies to begin recording any physician payments that are worth more than $10 in 2012 and to report them on March 31, 2013. That includes stock options, research grants, knick-knacks, consulting fees, and travel to medical conferences. The details will be posted in a searchable database starting September 30, 2013.

> 5. *Activity*: Describe a situation in which you have observed a conflict of interest in your surroundings at work or in local government.

We may not be aware that the decisions we are making involve ethical issues. Sometimes, this is the result of self-deception, defined as being unaware of the processes that lead us to form our opinions and judgments (Tenbrunsel and Messick 2004). Sometimes, this is the result of using euphemisms, such as "right-sizing" a business, "collateral damage" in a military campaign, and "accommodation" regarding civil rights issues.

Self-deception leads to framing decisions in such a way that either eliminates negative ethical characterizations or converts them into positive ones. We may then view the decision as ethically colorless. In other words, we do not frame the decision as an ethical one; rather, we categorize it in other terms such as a business, economic, personal, or a legal decision. For example, in our quest to "right-size" a business, we may frame the decisions in terms of keeping the company viable but may not recognize that one result is that people will lose their jobs.

> 6. *Activity*: Describe a situation where a euphemism was used to describe a situation or objective that may lead to an ethically colorless decision.

Many corporate codes of conduct prohibit employees receiving large gifts from suppliers for obvious reasons but allow small gifts with the belief that they are nonproblematic and help build needed business relationships. Research, however, indicates that even small gifts subtly affect the way the receiver evaluates claims made by the gift giver (Dana and Lowenstein 2003). The influence can be so subtle that the gift receiver is not aware of the bias and will therefore not attempt to correct it or avoid it in the first place. Their research (Moore et al. 2005) indicates that a policy of limiting gift size is unlikely to eliminate bias. Their conclusion is as follows: "The sheer ubiquity of trinkets given by pharmaceutical companies is evidence of their effectiveness; why else would profit-minded companies continue to provide them? Thus, policies against gifts should not be limited to large gifts." Another source states: "Physicians I have spoken to are quick to dismiss the chance that a 10¢ pen, or even a $30 basketball ticket, could ever influence their prescribing patterns" (Moore et al. 2005). They ignore the likelihood that when they accept even these small gifts from sales representatives, they will feel even some small obligation to maintain a relationship. They might subconsciously think more favorably of that company than they otherwise would have.

> 7. *Activity*: Describe an organizational situation in which you or a colleague received a small gift from someone. What do you think the giver was hoping to accomplish?

Another potential source of bias arises when we inaccurately estimate the consequences of a decision. One classic study looked at whether it was appropriate to tell a lie in a romantic relationship (Gordon and Kaplar 2004). A total of 122 undergraduate students documented lies in former romances. Each individual reported on one lie told and another received. The results were consistent: when on the delivering end, people viewed the circumstances of the lie far more favorably than when on the receiving end. Of those reporting having told such a lie, 32% said they lied to avoid upsetting the receiver. When on the receiving end of a lie, however, only 4% believed that the lie was told to avoid upsetting them. When on the telling end, 62% said that the lie was justified, but when on the receiving end, only 8%. On the telling end, 8% felt that the receiver's anger was justified, but on the receiving end, 57%.

One approach to ethics is based on utilitarian theory, in which one balances the benefits and harms of a decision's consequences. Utilitarians base their reasoning on the claim that actions are morally acceptable when the resulting consequences maximize benefit or minimize harm. The challenge in applying utilitarian ethics is that the decision maker must estimate the overall consequences to diverse groups of one's actions before making a decision (Mazur).

Another ethical challenge to the utilitarian philosophy is the question "Does one have the right to hurt one group to maximize the majority?" Unfortunately, public officials often justify a utilitarian approach when sacrificing the minority with the least political power. For example, the poor have a disproportionate share of toxic waste sites in their neighborhoods. A 1983 U.S. General Accounting Office study found that three of four toxic waste landfills in the southeastern United States were located in communities where the population of racial minorities exceeded that of whites (Withgott and Brennan 2009). This occurred even though minorities accounted for only 20% of the region's population. The authors also cite a 1987 United Church of Christ Commission for Racial Justice study, which found that the percentage of minorities in areas with toxic waste sites was twice that of areas without toxic waste sites.

8. *Activity*: Describe a decision in which you poorly estimated the consequences to others. What was the impact on the recipient? What was the recipient's reaction to the decision?

---

9. *Activity*: Describe a situation in which you know a lie was told to you or information was withheld but you wish you had been correctly informed.

---

10. *Activity*: Describe a situation in which you felt it was ethically correct to withhold information or lie. How do you think the other person would have felt if he had known the truth?

Ethical dilemmas often arise when ethical boundaries and legal enforcement based on societal norms are fluid. Few people are concerned about driving 59 mph in a 55 mph zone. What about 75 mph? At what point would you draw the line? Similarly, how far can you go with helping your child with his or her school work? Would you consider it acceptable to assist your sixth grader in writing a paper for English? What about help writing his or her college application essay? If you answered "yes," how much assistance would you consider acceptable? Your answer probably falls in a range between doing nothing to editing the finished paper—although some parents have gone so far as to write the essay themselves.

When it comes to white lies, we are all challenged in a variety of situations. The Hebrew Bible records the first instance of a white lie—and it was told by God. When God informed Sarah she would have a child, she expressed incredulity because her husband was so old. When God retold Abraham about Sarah's incredulity, God said that Sarah questioned the prophecy because she said she was too old. How would most of us respond when our significant others ask if their outfit looks good on them? Would your response depend on where they were going? Would your response be different if the destination were a social event or a job interview? What would your response be to the question "How do I look?" if you were asked by a convalescing patient you are visiting in the hospital?

The American Medical Association (AMA) Principles of Medical Ethics (2001) and an AMA Opinion (8.082) based on the medical ethics principles deals with the topic of physicians withholding information from patients: "Withholding medical information from patients without their knowledge or consent is ethically unacceptable." The opinion does, however, recognize that patients may have different preferences regarding how they would like the information shared. The opinion further states: "Physicians should honor patient requests not to be informed of certain medical information or to convey the information to a designated proxy, provided these requests appear to genuinely represent the patient's own wishes."

The opinion also informs a physician that all information does not need to be communicated to the patient immediately or all at once. It recommends that physicians assess the amount of information a patient is capable of receiving at a given time and, if necessary, delaying the remainder to a later, more suitable time. Finally, it recommends that consultation with the patients' families, colleagues, or an ethics committee may help in assessing the balance of benefits and harms associated with delayed disclosure. It is interesting to note that this norm of truth telling is broadly supported in Western culture but is not universally accepted in others.

11. *Activity*: Describe a situation where you allowed some leeway or told a white lie that had ethical implications. How did you justify this decision to yourself?

16.4.1 Don't Want to Know and Won't Bother to Investigate

There are several reasons why a person with supervisory responsibility may choose not to take action if he/she believes an activity going on in the workplace is worth investigating. One primary reason may be the difficulty of correcting the problem were the supervisor to become aware of the situation. For example, a department chair in a university may choose not to learn more about the poor teaching performance of a tenured faculty member since there are usually few readily available options for corrective action. Unfortunately, the same can apply with more devastating consequences with tenured teachers in K-12 schools. This situation also arises with a supervisor of union-represented employees: there may be strict time-consuming procedures required to document poor performance that an arbitrator hearing an appeal of the disciplinary action is likely to scrutinize.

A related concern is that it may be difficult to prove a suspicion. Imagine that there have been rumors that a lower-level supervisor is engaging in sexual harassment or abusing his or her subordinates. If hard evidence does not exist, it would now fall to the supervisor to spend a significant amount of time to develop potential evidence. Yet another reason is that becoming involved in an investigation will likely lead to uncomfortable discussions and confrontations. The supervisor may simply decide that it is easier to give an employee the benefit of the doubt rather than seek a confrontation. Recall that one of the values cited earlier was avoiding conflict.

There can also be a much broader organizational concern for not seeking to know the truth early on. This significant issue, were it to become public, may undermine the reputation of the organization. The hope may be that the issue will go away or resolve itself, so that the organization does not need to face public embarrassment. If the significant issue later becomes public, however, it can be devastating to the organization that was trying to hide it in the first place. For example, the Catholic Church's decision not to investigate and deal with sexual abuse by priests for decades resulted in hundreds of millions of dollars in fines once the abuse was uncovered and made public. A similar dynamic arose with the abuse of Iraqi prisoners in Abu Ghraib prison.

The failure to investigate can also involve complex public health issues and conflicts of interest. There have been studies in the developing world that quantify the impact of higher concentrations of fluoride on the IQ of children (Xiang et al. 2010). Instead of attempting to carry out scientific studies of these effects in the United States, fluoridation promoters here and in other countries have invested their energy in attacking these findings. They also argue the studies are totally unrelated to the impact of extra doses of fluoride for children in the developed world and thus there is no need to even study the issue.

12. *Activity*: Describe a situation where you became aware of a potentially suspect activity. What action, if any, did you take? What made the decision difficult?

16.4.2   Strategies to Counter Biases and Blindness

To overcome the effects of classic racial and other stereotypical biases, decision makers have three general strategies: collect data about yourself and your organization, shape the environment, and broaden the decision-making process (Banaji et al. 2003).

The first is to collect data about yourself and others in your organization to reveal the presence of bias. One method is to take an Implicit Association Test (IAT). IAT is an experimental method within social psychology designed to measure the strength of automatic association between mental representations of objects (concepts) in memory. The IAT requires the rapid categorization of various stimulus objects, such that easier pairings (and faster responses) are interpreted as being more strongly associated in memory than more difficult pairings (slower responses). The main biases that the IAT is looking for include race, gender, and age. (These tests are available at: https://implicit. harvard.edu/implicit/demo.) The authors warn that the IAT is an educational and research tool, so one should consider the results to be private information. The tool provides information related to the magnitude and pervasiveness of biases that can help someone direct attention to areas of decision making that are in need of careful examination and reconsideration.

The next is to shape the environment by exposing oneself to images and social environments that challenge stereotypes. Curtis Harden and colleagues at UCLA used the IAT to study whether race bias would be affected if the test were administered by a black investigator. One group of students took the test with a white administrator and another with a black administrator. The study concluded that the mere presence of a black administrator reduced the subjects' implicit antiblack bias. This strategy would suggest selecting a decision-making team with diverse representation (e.g., male, female; black, white), perspectives, and backgrounds.

The third is to broaden one's decision-making process. This approach is based on philosopher John Rawls's concept of the "veil of ignorance," which posits that only a person ignorant of his own identity is capable of a truly ethical decision. For example, imagine you are evaluating a policy that would lower the mandatory retirement age, eliminating some older workers but creating advancement opportunities for younger ones. Now imagine that, as you make your decision, you do not know which group you belong to. You will eventually find out, but not until after the decision has been made (Banaji et al. 2003). Would you be willing to risk being in the group disadvantaged by your own decision? How would your decision differ if you could make it wearing various identities not your own?

Many U.S. companies encourage their rising executives to take extended overseas assignments so as broaden their cultural sensitivity. In general, Europeans are naturally exposed to cultural diversity because of the diverse languages and work ethics that abound on the European continent. However, Europeans are struggling with the extension of cultural sensitivity to Islamic societies.

Diversifying a decision-making group has practical benefits as well. For example, Ford Motor Company formed a Women's Marketing Committee to meet and review each new vehicle in the product development phase. They have provided useful feedback that is incorporated into the design prior to the initiation of product engineering. One example of an issue that was identified and corrected early in the product development phase involved console switches that were well styled but presented functional challenges to drivers with long fingernails.

16.5   Pressures Undermine Ethical Balance

A primary reason that ethical decision making can be difficult is that many of our decisions are made under pressure. Research has indicated that a manager's ethical behavior is negatively influenced by external pressures of time, scarce resources, competition, or personal costs (Trevino 1986). Pressures tend to force a person to think only about immediate needs and therefore to close out all other considerations. The pressure to meet performance objectives, to save money, or to make a quick decision in all likelihood leads to overemphasis on one goal without sufficient regard to the impact on other goals. Often, critical decisions are made at a time of perceived crisis, forcing a quick decision. We may also experience pressures related to emotional well-being, friendships, or the need to support and protect our family. Competing pressures make it difficult to consider the ethical implications of a decision.

16.5.1 Time Pressure

One of the most common and broadest pressures we face daily is time pressure. For example, people under great time pressure and engrossed in their assigned task are less likely to notice the needs of others (Trevino 1986). Time pressure coupled with a drive to meet stated goals can cause people to short-cut processes and ignore or consciously disregard information that could be critical to decision making.

On January 28, 1986, the space shuttle Challenger exploded in midair. Millions of television viewers worldwide watched the explosion and fiery death of six astronauts and a school teacher aboard the shuttle. President Ronald Reagan appointed a special commission, headed by former secretary of state William Rogers, to investigate the cause of the accident. The commission's report cited the technical cause of the disaster as the failure of an "O-ring" seal in the solid-fuel rocket coupled with unusually cold weather (Greene). More important, the commission found serious flaws in NASA's decision-making process that led to the decision to launch under poor weather conditions. The most important factor that undermined safety concerns was NASA's goal to become an economically self-sufficient cargo hauler. This put NASA in the business of launching communication satellites for a wide variety of customers.

Pressures developed because of the need to meet commitments. This translated into a requirement to launch a certain number of flights per year and to launch them on time. The report stated: "It is evident, then, that NASA was subjected to external pressures to accept very ambitious goals." Due to this pressure, at one point, NASA proposed that they could launch 714 flights between 1978 and 1990. These external pressures were internalized as organizational goals, which resulted in pressure on individual decision-makers.

Time pressure can also lead to long work hours and fatigue, such that workers are more prone to errors. Medical interns, for example, are on duty 24 h straight and truck drivers may operate their vehicles beyond their endurance for driving safely. Recent international research regarding driver fatigue has suggested that it is underrepresented in accident statistics, and some estimates show that it could be a contributing factor in 24% of fatal crashes (SmartMotorist.com 2010). A study conducted by the Adelaide Centre for Sleep Research indicated that drivers who have been awake for 24 h exhibit driving performance equivalent to that of a person with a blood alcohol content of 0.1 g/100 mL (the legal limit as of 2011 in the United States is 0.08 g/100 mL) and is seven times more likely to have an accident.

The federal government has recognized that the economic pressure to work long hours undermines safety. As a result, there are regulations that govern hours of driving for truck drivers, shift schedules for medical interns and residents, and hours of flight time for airline pilots and their crews. In addition, the United States long ago established laws for paid overtime that at least reduce the economic incentive for more hours for hourly employees. Unfortunately, there have been several high-profile lawsuits of companies forcing workers to work unreported and uncompensated hours.

In December 2008, Wal-Mart, the largest U.S. employer, agreed to pay at least $352 million and possibly as much as $640 million to settle dozens of class-action lawsuits across the country in which Wal-Mart was accused of illegally forcing employees to work unpaid hours off the clock, erasing hours from their time cards, and preventing workers from taking lunch and other breaks guaranteed by state laws (Greenhouse and Rosenbloom 2008).

Pressure to work long hours can have direct impact on the ethical decisions associated with achieving a balance between family and work. Japanese culture values employees who work long hours. This has put the burden of child-rearing and other family-related issues almost exclusively on the wife. It is not clear if Japanese males consider this a significant issue, but the cultural norm results in the Japanese male sacrificing his family life for his job.

13. *Activity*: Describe a situation where you experienced time pressure to complete an assignment that led to cutting corners that undermined other ethical values.

16.5.2   Cost Pressure

Cost and profit pressure can put business relationships at risk. A large company may be in a position to leverage their power to squeeze suppliers for cost concessions. With the recent economic downturn, for example, some companies have opened existing contracts to further reduce their costs. A case study later in this chapter explores this particular issue; here, we focus on examples of the negative effects of cost and profit pressure.

Stanford University conducted a study related to the impact of "production pressure" on anesthesia patient safety (Healzer et al. 1998). The survey was conducted to assess the impact of intense emphasis on cost cutting and efficiency in medical care delivery. The study focused on the functioning of operating suites based on what has been perceived as the maniacal drive to "do more with less" and to "go faster, no matter what the risk." Among the key findings among respondents: 63% reported a heavy workload when on call, 75% felt fatigued at work, 80% had witnessed a surgeon do something that appeared to be unsafe, and 59% had witnessed an attending physician do something that appeared to be unsafe.

The pressure of running an automotive assembly plant while maintaining quality is handled completely differently in the United States than it is in Japan. In *The Toyota Way* (Liker 2004), the author relates a story about Russ Scaffede, who had worked for decades at General Motors (GM) before becoming the vice president of powertrain for Toyota. At GM, the golden rule of automotive engine production was very simple: Do not shut down the assembly plant! Managers there were judged strictly by their ability to deliver the numbers. The culture was to get the job done no matter what. Building too many engines was fine; building too few sent you to the unemployment line. When Fujio Cho, Toyota's president, noticed that Scaffede had not shut down the assembly plant once in a whole month, Scaffede's response was: "Yes sir, we had a great month, sir. I think you will be pleased to see more months like this." Cho's response was: "Russ-san, you do not understand. If you are not shutting down the assembly plant, it means that you have no problems. All manufacturing plants have problems. So you must be hiding your problems. You will shut down the assembly plant, but you will also continue to solve your problems and make even better-quality engines more efficiently."

Another example of cost pressure relates to cutting corners to save money. Profit margins on eggs that retail for 8¢ apiece are slim. To make money, producers must think in terms of massive volume and be ruthless about controlling costs. This has led to some disturbing findings in some cases. The Food and Drug Administration (FDA) recently completed a postevent investigation of the Wright County and Hillandale egg operations that resulted in 550 million salmonella-tainted eggs, which sickened at least 1500 people (Phillpott 2010). This investigation found many incidences of cutting corners at the expense of meeting health standards to save money. One significant finding was feed mill contamination related to inadequate grain storage. Another was cross-contamination between the laying houses due to employees either not wearing or not changing protective clothing when moving from house to house.

Cost-cutting pressure is also seen in the pharmaceutical industry. Recalls of prescription and over-the-counter drugs are surging, raising questions about the quality of drug manufacturing in the United States (Kavilanz 2010). The FDA reported more than 1742 recalls in 2009, a significant increase from 426 in 2008, according to the Gold Sheet, a trade publication on drug quality that analyzes FDA data. The high rate of drug recalls continued in 2010, with 296 reported in the first half of the year.

One of the reasons cited for the recalls is cost cutting that goes too close to the bone. Drug makers, facing intense price competition, are trimming manufacturing investment or outsourcing production. Prabir Basu, executive director of the National Institute for Pharmaceutical Technology and Education, states: "It is very expensive to make drugs. It also costs a lot of money to maintain adequate quality controls." And since generic and over-the-counter drugs are not as lucrative for drug makers as prescription drugs, companies may not be investing enough resources to make high-quality, safe products. The Gold Sheet report also stated that 165 recalls last year, up 58% from 2008, were of products manufactured or believed to have been manufactured abroad.

14. *Activity*: Describe a situation where you were faced with cost pressure that had an impact on some other measure in your organization. How did you deal with the trade-offs associated with this pressure?

16.5.3 Peer Pressure

From the time children start school, they begin experiencing peer pressure. Peers play a large part in a young person's life and typically replace family as the center of a teen's social and leisure activities. Teenagers have various peer relationships and interact with many peer groups. Some kids give in to peer pressure because they want to be liked, to fit in, or because they worry that other kids may make fun of them if they do not go along with the group. Others may go along because they are curious to try something new that others are doing. The idea that "everyone's doing it" may influence some kids to leave their better judgment behind. Peer influences have been found to be among the strongest predictors of drug use during adolescence. One piece of good news is that peer pressure related to drug use can be moderated by the influence of family. The study found that some children are more likely than others to fall for peer pressure just because of how and where they were raised.

It happens to adults, as well. The main reason we give in to peer pressure is because of our need for companionship and to be accepted. This is human nature, of course, but it becomes an issue in the present context when we become willing to sacrifice who we are. For example, 118 business school students at a major southeastern university, ranging in age from 20 to 38, participated in a study in which they were told the investigator was examining whether groups or individuals perform better in a situation that involves pay for performance. Both individuals and groups of three or four were asked to play a game; as an incentive, the participants had the opportunity to earn money at the same time. Both the individuals and groups were asked to keep track of the number of times they played the game in order to earn their money; they were told that no other record was being kept regarding how many times they played. Unbeknownst to the participants, however, the experimenters did keep track of the number of times each individual and group played the game.

The experimenter told the participants that they had 15 min to play and then left the room. When the 15 min elapsed, the experimenter returned and asked the participants the number of times they had played the game. Of the 25 individuals working alone, none inflated the number of attempts. Of the 23 groups, however, 5 lied to make more money. Given the design of this experiment, individuals had to make an active, conscious decision to lie to the experimenter in order to increase their reward. In the group condition, however, individuals had to make another decision as well—whether to object to others in the group who proposed that they should all lie. Individuals had to consider whether other members of the group would ridicule them for being a "straight arrow" or "not cool" if they objected. They also had to consider what the consequences would be if they decided to go against the group and report the lie to the experimenter. Such considerations and psychological costs were absent from the individual condition. The real or even imagined group pressure and the resulting conformity explain the results of this study. The authors go on to state that individuals in groups may lose their individuality and thus the ability to monitor and self-regulate their own behaviors. In such situations, the group takes over and makes moral decisions for the individual. Accordingly, individuals in a group may have not made a conscious decision to lie, but just allowed the group to make the moral judgment for them.

In the corporate world, scandals involving profit are seemingly commonplace, and yet even companies involved in such high-profile scandals as Enron, WorldCom, Arthur Andersen, and Lucent did not start out as deceitful organizations (Clark et al. 2003). Temptations such as a high-risk financial stake or pressures such as a project deadline or the expectation for ever-greater earnings can all increase the propensity for deception (Fleming and Zyglidopoulos 2008). An executive may also sanction, condone, or order employees to lie, which can make participation seem legitimate and perhaps even desired. One author states: "Honest employees can be converted into wrongdoers in a number of ways, but the process often begins with peer pressure or a supervisor's direct request" (Cialdini et al. 2004). This process can propagate to the point where lying becomes the norm (Erez et al. 2005).

In the case of Enron, WorldCom, and Lucent, peer pressure drove these organizations to a tipping point after which most of the organization was involved in the deception (Fleming and Zyglidopoulos 2008). Once lying is institutionalized, it affects much of day-to-day procedure, both directly and indirectly. Deception then forms a background that positively reinforces more lying. Organizational level deception encourages newcomers and previously honest members to begin lying. Socialization techniques like peer pressure will encourage deception at the individual level. Failing to lie will then risk detection, censure from superiors, and negative peer appraisal. It is also easier to engage in wrongdoing if employees feel they have little choice in the matter, because "everybody is doing it." Peer pressure incorporates more people in the organization's culture and thus increases the pervasiveness of lying. This rationalization was so pervasive at Enron that many of the employees portrayed themselves as "heroes" committed to saving the company by going along with the lies. The deception was also powerful enough that Enron was able to enlist the auditing firm of Arthur Andersen to participate.

Another famous case involved Frank Serpico, a New York City police officer who testified against police corruption in 1971 (Wikipedia 2010, Oct 15). He was made famous in a book by Peter Maas and a movie starring Al Pacino. After 12 years in the department, he was assigned to work as a plainclothes officer, where he encountered widespread corruption within the system. His career as a plainclothes officer was short lived precisely because he avoided taking part in corruption. Risking his own safety, Serpico exposed those who did. In 1967, he reported "credible evidence of widespread, systematic corruption." The police bureaucracy slowed down his efforts, however. Feeling that he had no place else to go, Serpico became a source for a *New York Times* front-page story in 1970 on widespread corruption in the New York police department. This forced Mayor John V. Lindsay to take action by appointing a five-member panel, headed by Whitman Knapp, to investigate police corruption.

In Serpico's 1971 testimony before the Knapp Commission, he was quoted as saying: "Through my appearance here today... I hope that police officers in the future will not experience the same frustration and anxiety that I was subjected to for the past five years at the hands of my superiors because of my attempt to report corruption.... We create an atmosphere in which the honest officer fears the dishonest officer, and not the other way around.... The problem is that the atmosphere does not yet exist in which honest police officers can act without fear of ridicule or reprisal from fellow officers."

15. *Activity*: Describe a situation where you faced peer pressure. How did you deal with the pressure? Why did you choose to deal with the situation in this way?

16.5.4 Family Pressure and Work-Life Balance

Another form of pressure may come from family over the issue of balancing one's home and work obligations. For example, this pressure can result in conflicts between the values of *promise keeping* to family members and *pursuit of excellence* related to a work obligation. It can bring into question the *fairness* with regard to sharing the workload of your department Family pressure can also result in a unique dilemma of *prioritizing the needs of others in a social network* where both a family and a work social network are involved.

Such complaints from spouses, parents, and children stem indirectly from increased job pressure in the United States, with Americans having one of the highest workloads among those in advanced societies (Wikipedia 2010, Oct 9). According to a study by the National Sleep Foundation, the average employed American works a 46 hours week, with 38% of respondents working more than 50 hours per week (Tenenbaum 2001). Sometimes, these pressures lead to absurd situations like the following true story. Imagine a "Quality of Work Life" meeting held at 9:00 p.m. on a Thursday, because it was the only time that this company could get executive management together to discuss the issue and recommend ways to balance work and family. As Arnold Zack, a Massachusetts lawyer said to his friend Paul Tsongas who was suffering from lymphoma, "No one on his deathbed ever said 'I wish I had spent more time on my business.'" (Keyes 2006.)

Women face a special challenge because they tend to be the primary caregivers and thus face added pressure to balance work and family. The United States has attempted to address this issue (not only for women) with the FMLA. It allows any person who has worked at least 1250 hours in the last year to take off up to 12 work weeks in any 12 month period following the birth or adoption of a child, or to care for a family member, or if that person has a serious medical condition. The law by itself does not mean that someone exercising its provisions will be free of work-life balance pressure. People may still feel that their career is passing them by or that they are disadvantaged in terms of rewards and career advancement. Employers may also view these people as less than fully committed to their work.

Twenty-five years ago, *Working Mother* began listing the 100 best companies to work for in terms of being family-friendly: those companies that provide benefits and programs to help working moms balance career and family (Owens 2010). Each one of the companies in the 2010 list offers a menu of benefits, including flex time, telecommuting, and temporary part-time work options. Two companies, IBM and Johnson & Johnson, have been on the list since its inception. IBM offers programs such as special care for children who have developmental issues and wellness action plans that include financial incentives. Johnson & Johnson offers programs such as free college coaching to help employees' children pick the right school and compressed work weeks or telecommuting for primary caregivers.

The U.S. workplace is not as family-oriented as that in many other wealthy countries and even in many middle- and low-income countries. For example, American workers average approximately 10 paid holidays per year; in contrast, British workers average 25 and German employees 30.

16. *Activity*: Describe a situation where you had to make a decision between work and family commitments. How did you deal with the pressure?

16.5.5 Competitive Pressure

In December 2008, Siemens, the German engineering giant, agreed to pay a record total of $1.6 billion to U.S. and European authorities to settle charges that it routinely used bribes and slush funds to secure contracts around the world (Dougherty and Lichtblau 2008). The company also pleaded guilty in federal court to charges that it violated a 1977 U.S. law banning the use of corrupt practices in foreign business dealings. To put this settlement in perspective, the previous high for a foreign corruption case set in 2007 was $44 million in a case involving Baker Hughes, an oil conglomerate. Joseph Persichini, Jr., head of the Washington office of the FBI at the time, stated: "Their actions were not an anomaly. They were standard operating procedures for corporate executives who viewed bribery as a business strategy." Mathew Friedrich, the acting head of the Justice Department's criminal division, noted that the case was part of a noticeable spike in the department's foreign corrupt practices investigations, with 44 cases having been brought in the previous 4 years, compared with 17 in the 4 years before that.

Companies that produce commodity products with low profit margin and that require large capital investment can choose to compete or cooperate with their competitors. When they choose to cooperate, the result is price fixing. On March 10, 2009, Hitachi agreed to plead guilty to fixing prices of liquid-crystal display panels sold to Dell and pay a fine of $31 million (Gullo 2009). Court documents stated that Hitachi and unnamed co-conspirators worked together from 2001 to 2004 to set prices charged to Dell for the panels, which are used in desktop monitors and notebook computers. Hitachi was the fourth company to plead guilty in a global U.S. display panel price-fixing investigation. LG Display, Chunghwa Picture Tubes, and Sharp all agreed to plead guilty in November 2008 and pay $585 million in criminal fines.

Intellectual property and copyright infringement activities also stem from economic pressures (Editor 2010). The International Intellectual Property Alliance maintains a Priority Watch List that identifies countries, which pose a grave concern regarding all forms of piracy, which acts as a trade barrier that inhibits corporate growth. China is on the list because it consistently fails to deal with widespread piracy. The incentive to do so is usually low, and pirates themselves face meager fines in the unlikely event that they are caught and brought to trial. Surprisingly, perhaps, Canada is also on the list because it has not modernized its copyright laws to address theft via the Internet. According to a recent study, intellectual property theft in Canada doubled in 2009 over the previous year, which cost the average Canadian company an estimated $834,139 in 2009 (Edited 2010). Another study surveyed 800 U.S. companies on intellectual property theft, estimating that a combined $4.6 billion in losses due to thefts of intellectual property in 2008 alone.

Often, a company will pursue a strategy in the name of good business practice that violates anticompetition laws. One such course of action is to attempt to drive competitors out of business via predatory pricing (as opposed to hurting the competition by offering a better product or better value). Government officials in both Wisconsin and Germany, for example, accused Wal-Mart of pricing goods below cost with the intent to drive competitors out of the market (New Rules Staff 2000). The complaint in Wisconsin claimed a total of 352 violations of predatory pricing laws. Many such laws prohibit below-cost pricing because small businesses would be driven out of business and consumers would be left with fewer options and, ultimately, higher prices. Crest Foods in Oklahoma also filed a predatory pricing lawsuit against Wal-Mart, charging that Wal-Mart employees and executives regularly visited the Crest store to monitor prices and targeted below-cost price cuts in order to undermine Crest Foods.

Microsoft released its Internet Explorer web browser for free by bundling it into the Windows operating system package. This quickly led to Internet Explorer becoming the leading web browser in the world. It also forced Microsoft's primary competitor in the browser market, Netscape, to make its Navigator browser available for free as well, and hastened Netscape's demise (United States v. Microsoft).

In 2008, the French government ordered Amazon.com to stop offering free shipping to its customers, because the policy was in violation of French predatory pricing laws (Shannon 2008). After Amazon refused to obey the order, the government proceeded to fine them €1000 per day. Rather than ending its policy of free shipping, which had led to a huge boost in sales, Amazon chose to continue paying the fines.

17. *Activity*: Describe a situation where you faced competitive pressure. How did you deal with the pressure? Why did you choose to deal with the situation in this way?

16.6 Short Cases

The seven cases discussed in the following deal with various ethical issues and value conflicts, using a format adapted from Mary E. Guy (Guy 1990). Table 16.2 identifies which ethical issues arise in each of the cases.

CASE 16.1: The Employee Performing Poorly of Late (Personal Crisis)

Tim has worked for his present employer for 10 years. During this time, he has advanced from trainee to senior engineer. Until approximately 5 months ago, Tim's performance had been exceptionally strong. All his assignments were of high quality and were always turned in on time or ahead of schedule. Tim had also volunteered for extra assignments and had been a strong contributor to the team.

About 5 months ago, Tim's performance changed dramatically. Assignment deadlines were missed and those that were turned in were either incomplete or of poor quality. Tim had also stopped volunteering for extra assignments, and team members were becoming frustrated because of Tim's lack of participation.

Several months after Tim's performance began to slide, Tim's immediate supervisor, Bob, met with him to discuss his performance. Tim told Bob that he was going through a rocky period in his marriage, but thought that he and his wife could work things out if he cut back to a 40 h week to spend more time at home. Initially, Bob thought that this was a reasonable approach and delegated extra assignments to other team members. Bob continued to meet with Tim regarding his lack of performance, but his performance did not improve. Nevertheless, Tim continued to assure Bob that things would get better.

Bob was losing patience with Tim's performance. In one meeting, he lost his temper trying to explain that the company was becoming increasingly unwilling to tolerate Tim's unacceptable performance. Each week Bob was growing more concerned about how this situation was affecting the rest of the team, many of whom relayed to him that they felt unfairly burdened carrying Tim's load. Further, many decisions that Tim needed to make were being left unresolved. The team was concerned that Tim's lack of performance was now affecting their performance as well.

Ethical issues

Tim violated *pursuit of excellence* because his lack of performance was now affecting the team's performance. Bob's *caring* for Tim was violating *prioritizing the needs of others in a social network*. If the situation continues, Bob's organization would be letting the company down and may result in lost work or potential loss of jobs for those employees diligently trying to fill the gap. Another concern is the issue of *fairness*. Is it fair for other workers to have to work harder because a member of the group is not completing his share of the work?

Bob was also trying to balance *respect for others* and *caring*. Tim had a long history of strong performance and was going through a rough time. He felt that Tim had a right to privacy as he was working through his personal problems. Seeing Tim's drawn face and fatigued appearance also brought out feelings of compassion and mercy. On the other hand, Tim was letting the team down. Bob knew that he was violating *self-discipline* when he lost his temper but was exasperated at the lack of improvement. Bob felt he needed to look past Tim's problems and do the right thing for the team.

Alternatives

Several alternatives were available to Bob. He could have simply bypassed Tim's roles and responsibilities and appointed another senior engineer to do Tim's job until the situation at home

TABLE 16.2 Values incorporated in seven short cases.

Values	Short Cases						
	An Employee Performing Poorly of Late	Placing a Relative in a Managed Care Facility	Squeezing Suppliers and Renegotiating Contracts Crisis	Pressure to Achieve and Ignoring Future Problems—Stretch Objectives	Corporate Culture versus Personal Ethics	Selecting a Supplier	Becoming a Lean Organization
Caring for people in general	x	x		x		x	x
Prioritizing the needs of others in a social network	x						
Respect for others	x	x				x	
Harmony and avoidance of conflict				x			
Honesty			x	x	x	x	x
Fairness	x		x			x	x
Integrity				x	x		
Fidelity					x		x
Self-discipline	x						
Promise keeping		x	x				
Responsibility		x	x		x		
Pursuit of excellence	x			x		x	x
Survive, sustain, and thrive			x		x		x
Responsible citizenship						x	

was resolved. He could also have put Tim on special assignment and let him work things out from the sidelines. Finally, he could have started the process of formally documenting Tim's performance, either forcing improvement or leading to termination if Tim's performance did not improve.

Resolution

Bob decided to discuss Tim's performance issue with Shelly in the human resources department. Shelly was concerned that Bob had waited so long to speak with her because Tim's performance issue could affect the morale and performance of Bob's group. Bob told her that this was in fact the reason he felt he could wait no longer to deal with Tim.

Shelly reviewed each of Bob's alternatives and offered another choice. The company offered a program free of charge that would provide confidential professional counseling for many personal problems. Shelly gave Bob the information package on the program and encouraged him to meet with Tim. Bob should let Tim know that his personal problems were resulting in performance issues that could lead to demotion or termination. If Tim decided to start the program, he could take a personal leave during which he would continue to be employed. Any discussions with the counselor would be held in strict confidence. Tim decided to enroll in the program, and several months later returned to work. He also returned to his exceptionally strong performance level.

CASE 16.2: Placing a Relative in a Managed Care Facility

Austin and Irma, an elderly couple lived in a small town where everyone knew and cared about their neighbors. Their daughter Ruth and her husband Walt lived in a large suburban neighborhood hundreds of miles away. They visited each summer and brought along the grandchildren. During these periodic visits, Ruth spent most of her time cleaning their home and ensuring that an adequate supply of staples would continue to be on hand until the next visit. Irma was beginning to exhibit signs of dementia. It was not a significant issue because she could still function and take her daily walks in a town she knew very well. Austin and their neighbors kept an eye on her and helped her back home when she would occasionally become disoriented. But when Austin became terminally ill, Ruth began thinking about what she would do when her father could no longer care for her mother. Both she and her husband were medical professionals and told the grandmother that, when the time came, she would move in with them. Irma had often said she never wanted to live in a nursing home and found it comforting to know that Ruth would take care of her.

When Austin died, Irma moved to her daughter's home. Without the benefit of her familiar surroundings and her husband's assistance, however, it became obvious that Irma's dementia was worse than previously thought. Most of the responsibility of caring for her fell to her daughter. Since Irma was still physically fit, she was up and around much of the day and sometimes at night. Often she tried to go outside for walks at night. The first few months were very stressful, with Ruth no longer able to get a full night's sleep. As a result, the children were not benefiting from their grandmother being in their home as they watched her struggle with daily life. Ruth and Walt felt forced to reconsider their decision regarding Irma's care.

Ethical issues

Ruth and Walt would be violating *caring* and *promise keeping* by moving Irma into a nursing home. Many of Ruth's relatives still lived in the small hometown, and aging relatives were often cared for there by the resident extended family. Typically, when grandparents were no longer able to take care of themselves, they moved in with their children. For Ruth, this would have been a break in her family's tradition and a violation of the promise that she had made to her mother to continue the tradition. They also violated *respect for others* by moving her into a nursing home, given that they had heard from friends and neighbors that a nursing home might not respect the grandmother's human dignity. Finally, they would be violating *responsibility* if they continued to spend the majority of their time with Irma rather than raising their children. Ruth and Walt were living in a situation commonly referred to as the "sandwich generation."

Alternatives

Several alternatives were available to Ruth and Walt. They could have brought in a caregiver to assist them during the day. This choice would not have helped at night, however. They could have hired a live-in nurse to provide care for more hours of the day. This choice would have required significant rearrangement or expansion of an already crowded house. Finally, they could have moved her into a nursing home. In investigating this possibility, they considered both the perceived quality of care and the nearness to their home.

Resolution

Ruth and Walt chose the nursing home option. They spent a significant amount of time researching and visiting nursing homes within a reasonable distance from their house. They selected the one they thought would provide the best overall care and respect its residents' dignity. They recognized that this solution provided Ruth the relief she needed to raise her family, but often she was unsure about the quality of care her mother was receiving. To address this latter concern, they hired private aids to work a few hours in the morning and at the end of the day. This ensured that Irma started the day off right, was dressed appropriately, and had a quality breakfast. A second aid helped out at dinner and stayed until bedtime. The costs for the aids were significant but less than would have been required to support her if she lived with her daughter. Ruth visited several times a week and Irma spent most Sundays and holidays with the family.

CASE 16.3:   Squeezing Suppliers and Renegotiating Contracts in Crisis

In 2009, as the economy continued its downward spiral, John heard the CEO publicly announce his company's third consecutive quarterly loss. Within the hour, Tony, a purchasing manager and John's boss, notified his staff of a mandatory 7:00 a.m. meeting the next day. The subject was supplier cost reductions. John and his peers were now requested to provide information by the end of the day regarding how much the company could save if the cost of each of its commodities were reduced by 10% or 15%. This information was to be discussed further at the morning meeting.

John scrambled to gather the latest volume data, so that he could build the spreadsheet for each of his commodities with the two alternative cost reduction scenarios. As he began building the spreadsheet, it dawned on him that the company was analyzing potential cost reductions for commodities that had existing contracts. He was not a lawyer, but he knew that the company had used at least some of these contracts in the past to deny supplier-requested cost increases. He was pretty sure that these contracts worked both ways. His first priority was to prepare for the next day's mandatory meeting, but he was certain that this topic would be discussed then.

The purchasing team gathered in the large conference room. Tony started the meeting by telling the team that each purchasing manager had received a cost reduction target with a short implementation deadline. He told the team that the vice president of purchasing had already sent letters to the suppliers stating that the company had no alternative but to reduce supplier payments. He also told them that they were to clear their calendars, so that they could focus on negotiating the cost reductions. The strategy was to negotiate a 10% cost reduction from suppliers where alternative suppliers were not available and a 15% cost reduction from the remainder.

John stated that, for his commodities, he would be renegotiating existing contracts. He asked if this was allowed within present terms and conditions. Tony responded that the whole team was in the same boat and that "this was a matter of survival. The company cannot continue to bleed cash" and the "suppliers profited in the good years; they now needed to share the burden in the lean years." He acknowledged that this was not a pleasant task, but they had no choice.

Ethical issues

John recognized that his company had a *right to survive* but renegotiating contracts violated *promise keeping*. After all, his company held tight to the contract when suppliers requested well-documented cost increases. Was it *fair* to demand that they accept cost reductions, because his company was losing money? Was it *fair* that they were holding suppliers responsible for their poor product decisions and overcapacity?

John knew that it would be difficult for him to meet face to face with his suppliers during these negotiations. This approach violated *responsibility*. He was also concerned about how strong his position could be if the suppliers refused to cooperate. Even though, for most of his commodities, there were alternative suppliers, he wondered if resourcing was a viable option for engineering. Also, was it *fair* to treat suppliers differently because alternatives existed? This threat of lost future business violated *honesty*, whether or not it was being used as a bluff.

Alternatives

John discussed a couple of alternatives with Tony. He suggested that they approach suppliers with the target objective and let them figure out how to meet it. Tony's response was that the approach they were taking amounted to the same thing. Suppliers were free to decide how they could recover the price reduction within their own shops. John also suggested engaging engineering to sort out how viable it was, and what work would be required, to resource commodities if the suppliers rejected the price reduction mandate. Tony said that this would slow down the process: "Our company needs the cost reductions now!" As the head of a family, John felt trapped. The objective was clear and, other than quitting his job in a poor economy, he could not identify any other viable alternatives.

Resolution

John started the process of meeting with each of his suppliers to begin implementing the price reductions. Most suppliers complied with the mandate (likely because they felt they had no choice). Some required escalation meetings. In the end, all his suppliers complied with the mandate.

18. *Activity*: Would you do anything differently if your company was making a profit but not hitting its profit target 2 years in a row?

CASE 16.4:   Pressure to Achieve and Ignoring Future Problems

The design of a powertrain control module includes both hardware and software dimensions. The automotive company decided to develop a common hardware architecture that could be used on all corporate powertrains and would be the means for tailoring the performance to the specific powertrain families. The goal of this powertrain control module strategy was to achieve significant hardware cost reduction via economies of scale. The target cost for the module was $200 and Sam, a newly appointed engineering supervisor, had a stretch objective of $180.

Dan, a highly experienced senior electrical engineer, was just appointed manager of the powertrain controls department. Sam, who has been with the company only 5 years and has been on a fast management track, was given the challenge to lead the development of the hardware for the new powertrain control strategy. Sam's co-lead for software was Jim, a senior software engineer who had worked closely with Dan on numerous projects. Based on the data, the team settled on a hardware architecture that they felt could accommodate all corporate powertrains. Both Sam and Jim felt that software development would be able to work with this physical architecture to tailor the module to specific needs. Initial cost estimates left Sam with a comfortable margin to his stretch objective of $180.

During design validation testing, several issues required hardware design modifications. The redesigned module subsequently passed all design verification tests, and Sam now felt confident that he could deliver the stretch objective. He communicated this information to Dan, who was more than pleased to hear the news.

Now it was up to Jim and his team to develop the software. They began their tests on existing powertrains and encountered a number of significant problems. Jim suggested to Sam that hardware changes would make the software design task easier. Jim was also concerned that, even if the software could be developed to address the current powertrain problems, the probability was high that it would not work with the new powertrains that were 4 or 5 years out in the planning cycle. Sam's response was that he had already communicated the hardware cost of the module to Dan, who had relayed the good news to the chief engineer. Sam was concerned that if they changed the design now, it would open up the module to retesting and possibly even require a second round of testing to resolve all issues. This would push back module introduction and delay the projected cost savings of millions of dollars. He also felt if he brought this information to his manager, Dan would lose confidence in his team. Sam further argued that the software problems currently experienced were not representative of the likely challenges that would arise with the newer powertrains that were just on the drawing board.

Ethical issues

Sam violated *pursuit of excellence*. This project had long-term objectives that were being ignored in lieu of short-term targets. Sam also violated *honesty*. There were significant concerns that the present design would work well with existing powertrains and a good likelihood that the design would not work with future powertrains. Hardware changes could be incorporated at a cost that would keep the module below the $200 objective but not the stretch objective. Yet Sam chose not to disclose the issues to his manager. Sam took advantage of the fact that Jim placed a high value on *harmony and avoidance of conflict*. Sam also violated *integrity*. He had progressed in the company based on a solid engineering track record. It now seemed that he was willing to sacrifice good engineering judgment for the chance at rapid career advancement. Jim also perceived Sam's behavior as the antithesis of *caring*. He was taking advantage of Jim's reputation and using him as a stepping stone to advance his own career.

Alternatives

Jim believed that the module required hardware changes to be robust with the most challenging existing powertrains. The module, as designed, would not be capable of working with the new powertrains planned over the next several years. Jim's team could struggle with the present hardware design and possibly achieve a software solution that would work in the short term. Years later, when new powertrains were delivered, the hardware would need to be changed, software would need to be rewritten, and the new design revalidated on all corporate powertrains. Jim projected this cost to be significantly more than the short-term cost savings of several million dollars that would be achieved with the present hardware design.

Jim could discuss his concerns with Dan, but he was convinced he would not take action because upper management had put Sam on a fast track. Jim saw they had both already declared victory in meeting a stretch objective on the most significant project in the department. Jim could go over Dan's head, but he was uncomfortable with this idea since both Dan and Sam were highly respected in the organization. Also, Jim foresaw that the way the project was progressing, it was likely that Sam would be promoted again and become his boss.

Resolution

Jim worked through the issues with the existing powertrains and the controller was successfully launched. He did not disclose to Dan the discussion he had with Sam regarding having hardware drivers in the initial design. Sam was promoted to a position outside of Powertrain Controls after successfully delivering the next generation control module to the stretch cost target. Two years later, the module required hardware redesign, so that it would be capable of working with the new powertrains that were delivered according to the plan. Redesign costs were approximately twice the cost savings achieved with the initial design.

CASE 16.5: Corporate Culture versus Personal Ethics

Parks Corporation decided to start bidding on R&D contracts because companies winning the R&D and qualification phases had the edge on being awarded the highly profitable production contracts. If it won the contract for Phase I of the Blue Spider Project, this could lead to a $500 million production program spread out over 20 years. This project was an attempt to develop new, longer-life materials for the army's Spartan missile, which was exhibiting fatigue failures in the field (Kerzner 2009).

Gary, a PhD in mechanical engineering, was recently promoted to senior scientist, responsible for all R&D activities performed in the mechanical engineering department. Henry, the director of engineering, appointed Gary to head up the proposal team. This would provide him with a great opportunity to further develop his management skills and be a stepping stone to further advancement at Parks.

As Gary was working on the proposal, he identified a problem. There was a requirement that all components be able to operate normally over a temperature range of −65°F to 145°F. Current testing indicated that Parks' design would not function above 130°F. Gary felt that, based on his technical expertise, it would be impossible to meet the material specification requirements. Henry told Gary to claim that the material could exceed specifications. Gary's reaction was immediate: "That seems unethical to me. Why don't we just tell them the truth?" Henry replied: "The truth doesn't always win proposals. I picked you to head up this effort because I thought that you'd understand. I could have just as easily selected one of our many moral project managers. I'm considering you for program manager after we win the program. If you're going to pull this conscientious crap on me like the other project managers do, I'll find someone else. Look at it this way; later we can convince the customer to change the specifications. After all, we'll be so far downstream that he'll have no choice."

Ethical issues

Henry's approach violated *honesty*. Gary felt that it would be better to tell their customer the truth right from the start of the project. This situation was made worse by Henry's direction to tell the customer that the material could exceed specifications. Henry's approach also violated *integrity*. With Gary's educational knowledge and technical background, coupled with his knowledge of Parks' capability, he knew that no new design would meet the objective. Finally, this approach violated *fidelity*. How could he be faithful to his client in light of the inaccurate statements he was being asked to make?

On the other hand, Henry's approach would support *survive, sustain, and thrive*, both for Parks, which could win a huge contract, and for Gary, who could attain a higher position with more authority. Could Gary convince himself that his approach supported *responsibility*? After all, he did tell Henry that he thought that lying about the material's ability to meet temperature specifications was unethical. Was it Gary's responsibility to tell the client as well?

Alternatives

Gary could choose to tell the client that current testing indicated the material did not meet specifications. He could also tell the client that based on his technical expertise, it would be impossible to meet the material specification requirements. Since the first alternative was not based on actual material development and testing, Gary could choose to tell the client that current testing indicated that the material did not meet specifications. He could then tell the client that they had some ideas for material changes that may meet specification requirements. However, they would not know for certain until testing was conducted at the end of this contract phase.

Gary could also take Henry's approach and lie to the customer. This would be the safest route if Gary wanted to keep his job. Also, this was Gary's first R&D project, and Henry was experienced at it. Gary could perhaps assume that this is typical of the way that companies approach R&D contracts.

Resolution

In the end, Gary was willing to sacrifice his moral and ethical beliefs in order to be the project manager. Parks won the initial contract. Gary was eventually informed that the actual testing performed at the expense of Parks showed that the new material would not meet the specifications. They were now in discussions with the main client trying to understand the need for performance at temperatures above 130°. They were also asking for more time to develop a better design that would meet this standard.

CASE 16.6: Selecting a Supplier (Domestic, Developed Country, Emerging Market)

Bob was elated to be appointed project manager for a new commodity that was just beginning to be developed. One of his first tasks was to select a global supplier. His sponsor was Tony, the purchasing director, and the multidisciplinary supplier selection team would include representatives from purchasing, engineering, supplier technical assistance, and the office of general council. Engineering and manufacturing stakeholders had also been identified. This was a great opportunity for increased executive and cross-organizational exposure. This project was also aligned with a recently announced corporate strategy to develop a broader supply base with a primary objective of reducing commodity costs.

Bob felt that the team was very strong and worked well together, with everyone pulling their own weight. The team developed a list of supplier capability selection criteria, including on-site engineering support, technical design, product validation, manufacturing, on-site assembly (resident engineering) support, quality assurance, logistics, terms and conditions compliance, and cost.

The team then set out to identify potential suppliers in three categories: domestic, developed country, and emerging market. The list was narrowed to two potential suppliers in each category. The team conducted extensive on-site visits and capability evaluations on a scale of 0 (none) to 3 (exceptional) for each of the 6 and again narrowed the list to the best choice in each category. The team then developed Table 16.3, an evaluation matrix for the best supplier in each category. The domestic supplier and developed country supplier received a score of 68 and 54, respectively. The emerging market supplier was a distant third with a total score of only 28.

The team and stakeholders met with the sponsor and presented the evaluation matrix and abundant back-up information. At the end of the presentation, the sponsor stated that the evaluation criteria were slanted to favor a domestic supplier. The team was asked to go back to the drawing board and come up with a way to award the business to the emerging market supplier. Although the stakeholders supported the team's approach and recommendation, the sponsor pushed back, stating that the corporate objective was to source a significant amount of business to emerging markets. He stated that this was the company's vision for the future. He thought

TABLE 16.3 Evaluation matrix for global supplier decision.

Capability	Multiplier (1–3)	Supplier Capability Score in Each Region		
		Domestic	Developed Country	Emerging Market
On-site engineering	3	3	1	0
Technical design	3	3	3	2
Product validation	3	3	3	1
Manufacturing	3	3	3	2
On-site assembly	3	3	1	0
Quality assurance	3	3	3	1
Logistics	3	3	2	1
T and C's	2	2	2	2
Cost	1	1	2	3
Score		68	54	28

the commodity selected was low risk, and it was time for the organization to stop fighting the inevitable. Tony stated: "Emerging Markets is where the future is. We need to put the old ways of doing business behind us and embrace the 21ˢᵗ century!" He directed them to figure out a way to meet this objective.

Ethical issues

The team felt betrayed. Tony violated *honesty*. Why did he let the team do a structured analysis when he already knew what answer he wanted? Tony also violated *caring*. Bob felt Tony was using the team as a means to an end: to be able to state that the decision was based on the results of a cross-functional team. If the supplier later failed to deliver, his explanation would be that engineering and manufacturing were represented throughout the entire decision-making process.

Tony also violated *respect for others*. It seemed that he did not respect the team. He led them on and had them spend a lot of time and money developing what he called a slanted analysis. Tony also violated *pursuit of excellence*. It was clear to the team that the best emerging market supplier would not currently be able to meet the company's needs. The commodity would likely struggle through the entire design, development, manufacturing, and assembly process. He was convinced that the outcome would adversely impact the company's brand image once the commodity was in the field. Finally, Tony violated *responsible citizenship* by pushing the team to move business to a non-democratic country that was known to violate human rights, place workers in unsafe work environments, and have little regard for intellectual property rights.

Alternatives

It was now clear that management wanted the commodity to be sourced to an emerging market. However, there was no way to slant the matrix toward the emerging market unless almost all the weight was assigned to manufacturing cost. The team was unwilling to change their methodology. They did consider a number of alternatives for involving an emerging market supplier in ways that that could reduce the potential risk.

One alternative would be to select a domestic or developed country supplier who had manufacturing facilities in an emerging market. During the team's trip, they passed many familiar corporate logos on their path to the selected suppliers. They would need to select a supplier who could provide on-site engineering and assembly support. They would also need to choose one who had strong product validation, quality assurance, and logistics capability. The cost would likely rise because only the commodity manufacturing would be done in the emerging market. This alternative may not meet the company's emerging market directive. Also, a supplier needed to be identified quickly, so that the sourcing decision could be made soon.

Another alternative would be to require the emerging market supplier to partner with an existing supplier to fill in the gaps in its capabilities. Over time, the emerging market supplier could meet all the requirements. This approach had been used before. The results were successful, but the capable partner was not in direct competition with the emerging market supplier nor was it interested in entering that business. In this case, the team had already identified two domestic and two developed country suppliers who wanted this business. Cost would certainly rise, and the team would likely not have the time to find a viable partner to develop this business arrangement. This alternative might also not meet the company's emerging market directive.

The third alternative would be to source the business to the emerging market supplier. This would require locating engineering, supplier technical assistance, and manufacturing personnel with the supplier. There were two significant risks associated with this alternative. The first was whether or not the supplier, even with the company's help, would be able to develop and manufacture the commodity to the company's standards. The other was identifying people who would be willing to relocate to the emerging market for an extended period of time. Also, this approach would reduce available staff at home to work on other product development projects.

Resolution

The company selected the emerging market supplier. It was the only alternative that met the company's directive. To further complicate matters, engineering, supplier technical assistance, and manufacturing personnel were limited to periodic visits even though people were identified who would be willing to locate in the emerging market during the project lifecycle. The design and development process was completed significantly behind schedule. Once it was determined that the supplier was struggling through the manufacturing process, the company decided to make this commodity a stand-alone option and identify it as "late availability." The base product needed to be modified late in the development process to make this commodity a stand-alone option. The company eventually reevaluated its emerging market strategy and decided, at least for a time, to select a domestic or developed country supplier who had manufacturing facilities in an emerging market.

19. *Activity*: Identify an instance in which senior management had already made an unstated decision but asked staff to go through an analysis to come up with a decision that turned out not to be what management wanted. What was the final resolution of the decision?

CASE 16.7:   The Dilemma of Becoming a Lean Organization (Becoming More Efficient and Eliminating One's Own Job)

Ted, an industrial engineer, was ideally suited for his next assignment. His company was under a lot of pressure to reduce their manufacturing costs. Ted had suggested that a good approach to accomplish this objective would be to apply lean manufacturing methodologies. Ted was requested to map out a process, identify a team, and select a project that would have an immediate impact on the organization. John, Ted's immediate supervisor, was actively involved in the project to ensure that the team would have all the support they needed.

The first project was successful. It eliminated two production workers who were reassigned to another work area that was experiencing significant bottlenecks. John and Ted updated the initial process to document changes. The next lean manufacturing project selected was to eliminate the significant bottleneck. This project was also successful. Not only did it eliminate the two people reassigned from the first project, but it also eliminated three others as well. This time, John terminated the five production workers. John also told Ted that since the process was now well documented and had proved successful, his own job was no longer needed.

Ethical issues

Ted's proposal to use lean manufacturing methodologies to reduce cost supported *survive, sustain, and thrive*. It also supported *pursuit of excellence* because implementation of his projects would help the company be the best it could be. Ted believed that he had proved his *fidelity* to the company and his profession by volunteering a way to improve the company's operations based on his skills and training. Ted was shocked when he and others were terminated. He did not see it coming. Was it *fair* for the company to terminate people whose jobs were eliminated because of efficiency improvements? Could not the company find new jobs for these people and expand their business? Could not the company use Ted's skills to plan new work so they would be efficient from the start?

The company violated *caring*. The employees that were terminated were good workers. It was not their fault that the company had implemented inefficient processes. The company also violated *honesty*. Although management was not technically dishonest, John certainly did not disclose that the company was planning to terminate people as a result of this initiative.

Alternatives

The company could have redeployed the displaced workers temporarily while waiting for other openings they could fill. The company could also have retained Ted. As an industrial engineer, he could have continued to implement lean methodologies or worked on many other projects to help his company.

John's concern was that, once Ted realized that manufacturing workers were being terminated, Ted would lose his motivation to implement projects. John felt that the company needed to continue aggressive implementation to improve the bottom line.

Resolution

The company's only objective seemed to have been to reduce cost. Using Ted's process, John continued to implement projects to further reduce manufacturing staff.

Exercises

Complete Chapter Activities

16.1 Describe a situation (other than the ones identified in this chapter) in which dilemmas, paradoxes, inconsistencies, or differing expectations may result in a decision with ethical ramifications.

16.2 Describe a situation in which you made an association regarding things that you believed went together but later recognized did not.

16.3 Describe a situation in your experience where in-group favoritism was ethically justifiable.

16.4 Describe a situation in your experience where in-group favoritism was not ethically justifiable.

16.5 Describe a situation in which you have observed a conflict of interest in your surroundings at work or in local government.

16.6 Describe a situation where a euphemism was used to describe a situation or objective that may have led to an ethically colorless decision.

16.7 Describe an organizational situation in which you or a colleague received a small gift from someone. What do you think the giver was hoping to accomplish?

16.8 Describe a situation where you poorly estimated the consequences to others of a decision you had made. What was the impact on others? What was their reaction to the decision?

16.9 Describe a situation in which you know a lie was told to you or information was withheld but you wish you had been correctly informed.

16.10 Describe a situation in which you felt it was ethically correct to withhold information or lie. How do you think the other person would have felt if he had known the truth?

16.11 Describe a situation where you allowed some leeway or told a white lie that had ethical implications. How did you justify this decision to yourself?

16.12 Describe a situation where you became aware of a potentially suspect activity. What action, if any, did you take? What made the decision difficult?

16.13 Describe a situation where you experienced time pressure to complete an assignment, which led to cutting corners that undermined other ethical values.

16.14 Describe a situation where you were faced with cost pressure that had an impact on some other measure in your organization. How did you deal with the trade-offs associated with the pressure?

16.15 Describe a situation where you faced peer pressure. How did you deal with the pressure? Why did you choose to deal with the situation in this way?

16.16 Describe a situation where you had to make a decision between work and family commitments. How did you deal with the pressure?

16.17 Describe a situation where you faced competitive pressure. How did you deal with the pressure? Why did you choose to deal with the situation in this way?

16.18 Would you do anything other than squeeze suppliers if your company was making a profit but not hitting its profit target 2 years in a row?

16.19 Read the following case study based on an article from the *Wall Street Journal*. Identify and describe the ethical issues involved.

16.20 Discuss the ethical issues associated with Mark Hurd's firing by HP. Following is an overview of the story and outside comments.

Case

In 2005, Mark Hurd was appointed CEO of Hewlett-Packard (H-P). On September 22, 2006, Hurd was appointed to the additional role of chairman. During his tenure, H-P passed Dell as the world's leading personal computer maker and diversified the company away from its legacy printing business to include services, servers, and software. In April 2010, H-P acquired Palm. Under Hurd's leadership, the company grew profits and provided investors with a much higher return than the broader market. On August 6, 2010, however, Hurd was forced to resign from HP.

HP's code of conduct, which Hurd had publicly championed in 2006 following a boardroom scandal, has a simple test to decide whether an action is appropriate: "Before I make a decision, I consider how it would look in a news story" (Worthen and Lublin 2010). It appears that Hurd was forced to resign because he failed the test: "The H-P board asked for Mr. Hurd's resignation in large part because of the conflict between his actions and the code of conduct."

H-P conducted an internal investigation in late June when a woman named Jodie Fisher, who had been working as a contractor, sent Hurd a letter alleging sexual harassment. "H-P said that its board determined that Mr. Hurd didn't violate H-P's sexual harassment policy, but other irregularities were uncovered. The investigation found that Mr. Hurd submitted inaccurate expense reports that the company says concealed his relationship with Ms. Fisher, who assisted on H-P-sponsored events. The investigation also found that Mr. Hurd didn't disclose a close personal relationship with the contractor, and that the woman was paid at times when there was no legitimate purpose." Mr. Hurd agreed to pay back the approximately $20,000 in expenses, which were at issue. "Sadly, Mark's conduct undermined the standards we expect of our employees, not to mention the standards to which the CEO must be held, and the board decision was unanimous," said Mark Andreessen, an H-P director.

Alternatives

The *Wall Street Journal* article stated that "corporate governance experts were split on whether H-P's board acted properly in forcing Mr. Hurd's resignation. Corporate expense offenses can be regarded as relatively minor and can typically be settled if an executive pays back the amount, some experts said."

"Joseph Grundfest, a professor at Stanford University's Law School and a former member of the Securities and Exchange Commission, said H-P directors also could have considered a private or public reprimand and financial penalties. 'When all this comes out you'll have directors of other companies saying we would have dealt with it differently,' he said."

"Indeed, Charles Elson, head of the Weinberg Center for Corporate Governance at the University of Delaware's business school, praised H-P directors for forcing out Mr. Hurd rather than accepting his offer to repay the disputed expense money."

"In allowing recompense for expense-account abuses, 'Who's to say something like this wouldn't have happened again in a more serious way?' Mr. Elson asked. 'Once the trust is damaged, you can't go forward.'"

"Some governance experts went further, saying the board should have fired Mr. Hurd for cause. By falsifying expense accounts, he 'committed a serious, career-ending error and there should be some financial consequences,' said Nell Minow, editor of Corporate Library, a governance-research firm in Portland, Maine."

"At a different company, such offenses might have resulted in a reprimand or financial penalty. But H-P tried to renew its reputation as a leader in corporate governance standards after the probe of the 2006 board scandal, in which private investigators hired by a former H-P chairman were accused of using false pretenses to obtain phone records of directors and reporters."

"Over the ensuing years many H-P employees were dismissed for violating the company's policies, said people familiar with the matter. In some cases, people were terminated for offenses that would have been dealt with more leniently prior to Mr. Hurd's arrival, one of these people said."

"Mr. Hurd, in a statement Friday, said: 'As the investigation progressed, I realized there were instances in which I did not live up to the standards and principles of trust, respect and integrity that I have espoused at H-P and which have guided me throughout my career.'"

"Typically, these people say, employees terminated for offenses under the code of conduct weren't given any severance. Mr. Hurd, by contrast, negotiated an exit package that may be worth more than $35 million, including a cash payment of $12.2 million."

16.21 Read the following material regarding whistle blowing. Then, find a news item or case study involving whistle blowing. Describe the case, identify the ethical issues, describe the alternatives, and describe the resolution.

Whistle Blowing/Speaking Out

According to Merriam-Webster, a whistle blower is one who reveals something covert or who informs against another. An appropriate moral motive for whistle blowing is *preventing harm to others*. Revealing information becomes ethical when it will save someone from being hurt or mistreated or when it will keep someone's rights from being violated (Guy 1990). A whistle blower will likely perceive less risk if the employee is not already in good standing. However, if an employee is in good standing and has exhibited strong performance, the company may decide to overlook wrongdoing.

Based upon in-depth interviews with thirty recent graduates of the Harvard MBA program, less than a third believed that their organizations respected or encouraged whistle blowing (Badaracco and Webb 1995). A similar number were not sure how their organizations would treat whistle blowers. The remainder believed that whistle blowing was dangerous. Several of the interviewees reported that no action was taken after the unethical activity had been exposed.

Whistle blowers also experience moral conflict (Dozier and Miceli 1985). It arises when people recognize that their inclination to act may lead to a violation of the fundamental norms of their reference group. Blowing the whistle on your own company is seen as a violation of *loyalty*. In professional settings, there is a conflict between collegial loyalty and responsibility to the public. Based on the authors' studies, whistle blowing is more likely to take place where support or other cues from the environment would be perceived as valued by the organization. If the organizational climate is supportive, one would expect "highly moral" people to blow the whistle. However, whistle blowing is an uncommon phenomenon. According to limited empirical evidence, this is the result of many opportunities during the decision-making process for one to decide that the cost is not worth the benefits of sticking his or her neck out.

References

Badaracco, J. J. (1992). Business ethics: Four spheres of executive responsibility. *California Management Review*, 34(3), 65–78.

Badaracco, Jr., J. L. and Webb, A. P. (1995). Business ethics: A view from the trenches. *California Management Review*, 25(10), 15–21.

Banaji, M. R., Brazerman, M., and Chugh, D. (2003). How (un)ethical are you? *Harvard Business Review*, 81(12), 1–8.

Caputo, I. (2009, August 18). Activists focus on conflicts of interest among doctors with ties to industry. Retrieved September 8, 2010, from *The Washington Post*: http://www.washington-post.com/wp-dyn/content/article/2009/08/17/

Cialdini, R. B., Petrova, P. K., and Goldstein, N. J. (2004). The hidden costs of organizational dishonesty. *Sloan Management Review*, 45(3), 67–73.

Clark, F., Dean, G., and Oliver, K. (2003). *Corporate Collapse, Accounting, Regulatory and Ethical Failure*. Cambridge, U.K.: Cambridge University Press.

Dana, J. and Lowenstein, G. (2003). A social science perspective on gifts to physicians from industry. *JAMA*, 290(2), 252–255.

Dlouhy, J. A. (2010, August 31). Policy aims for divide between regulators, drillers. Retrieved September 8, 2010, from chron Business: http://www.chron.com/disp/story.mpl/business/7178864.html

Dougherty, C. and Lichtblau, E. (December 15, 2008). Settling bribery case to cost Siemens $1.6 billion. Retrieved September 19, 2010, from *The New York Times*: http://www.nytimes.com/2008/12/16/business/worldbusiness/16siemens.html

Dozier, J. B. and Miceli, M. P. (1985). Potential predictors of whistle-blowing: A prosocial behavior perspective. *Academy of Management Review*, 10, 823–836.

Driver Fatigue is an important cause of road crashes. (July 29, 2010). Retrieved October 13, 2010, from SmartMotorist.com: http://www.smartmotorist.com/traffic-and-safety-guideline/driver-fatigue-is-an-important-cause-of-road-crashes.html

Edited. (2010, January 27). Intellectual property theft. Retrieved October 22, 2010, from LockLizard: http://www.locklizard,com/intellectualproperty_theft.htm

Editor. (2010, February 21). IIPA report documents copyright theft in 39 countries. Retrieved October 22, 2010, from AGIPNEWS8054: http://www.ag-ip-news.com/getArticle. asp?Art_ID=8054&lang=english

Erez, A., Elms, H., and Fong, E. A. (2005). Lying, cheating, stealing: Groups and the Ring of Gyges. *Academy of Management Annual Meeting*, pp. 1–37. Honolulu, HI: Academy of Management.

Fleming, P. and Zyglidopoulos, S. C. (2008). The escalation of deception in organizations. *Journal of Business Ethics*, 81, 837–850.

Gordon, A. K. and Kaplar, M. E. (2004). The enigma of altruistic lying: Perspective differences in what motivates and justifies lie telling within romantic relationships. *Personal Relationships*, 11, 489–507.

Greene, N. (n.d.). Space shuttle challenger disaster—A NASA tragedy. Retrieved October 11, 2010, from About.com:Space/Astronomy: http://space.about.com/cs/challenger/a/challenger_2.htm

Greenhouse, S. and Rosenbloom, S. (2008, December 23). Wal-Mart settles 63 lawsuits over wages. Retrieved October 19, 2010, from *The New York Times*: http://www.nytimes.com/2008/12/24/business/24walmart.html

Gullo, K. (2009, March 10). Hitachi settles price-fixing case for $31 million (update 2). Retrieved September 19, 2010, from Bloomberg: http://www.bloomberg.com/apps/news?pid=2107000 1&sid=a1ZHoYNn6uAE

Guy, M. E. (1990). *Ethical Decision Making in Everyday Work Situations*. Westport, CT: Quorum Books.

Healzer, J. M., Howard, S. K., and Gaba, D. M. (1998). Attitudes towards production pressure and patient safety: A survey of anesthesia residents. *Journal of Clinical Monitoring and Computing*, 14, 145–146.

Kavilanz, P. (2010, August 16). Drug recalls surge. Retrieved October 13, 2010, from CNNMoney. com: http://money.cnn.com/2010/08/16/news/companies/drug_recall_surge/index.htm

Kerzner, Ph.D., H. (2009). *Project Management Case Studies (The Blue Spider Project)*. Hoboken, NJ: John Wiley & Sons, Inc.

Keyes, R. (2006). *The Quote Verifier: Who Said What, Where, and When*. New York: Macmillan.

Liker, J. K. (2004). *The Toyota Way*. New York: McGraw-Hill.

Mazur, T. C. (n.d.). Lying. Retrieved April 17, 2010, from Markkula Center for Applied Ethics, Santa Clara University: www.scu.edu/ethics/publications/iie/v6n1/lying.html

Moore, D. A., Cain, D. M., Lowenstein, G., and Bazerman, M. H. (2005). *Conflicts of Interest Challenges and Solutions in Business, Law, Medicine, and Public Policy*. New York: Cambridge University Press.

New Rules Staff. (2000, November 1). Wal-Mart charged with predatory pricing. Retrieved October 27, 2010, from A Program of the Institute for Local Self-Reliance: http://www.newrules.org/retail/news.walmart-charged-with-predatory-pricing

Owens, J. (2010, August). 2010 working mother 100 best companies. Retrieved September 29, 2010, from 2010 Working Mother 100 Best Companies: http://www.workingmother.com/BestCompanies/2010/08/

PEW Charitable Trusts. (2008, October 1). Conflicts of interest for doctors (fall 2008 trust magazine briefing). Retrieved September 8, 2010, from The PEW Charitable Trusts: http://www.pewtrusts.org/news_room_detail.aspx?id=43398

Phillpott, T. (2010, August 31). After a half billion bad eggs get released, the FDA reveals filthy conditions of Wright County Egg. Retrieved October 13, 2010, from Gross profits: http://www.grist.org/article/2010-08-31-after-a-half-billion-bad-eggs-get-fda-reveals-filthy-conditions-/

Shannon, V. (2008, January 14). Amazon.com is challenging French competition law. Retrieved October 27, 2010, from *The New York Times*: http://www.nytimes.com/2008/01/14/technology/14iht-amazon.4.9204272.html

SmartMotorist.com. (2010, July 29). *Driver Fatigue is an important cause of road crashes*. Retrieved 10 13, 2010, from SmartMotorist.com: http://www.smartmotorist.com/traffic-and-safety-guideline/driver-fatigue-is-an-important-cause-of-road-crashes.html

Tenbrunsel, A. E. and Messick, D. M. (2004). Ethical fading: The role of self-deception in unethical behavior. *Social Justice Research*, 17, 223–236.

Tenenbaum, D. (2001). How long is the average work week in the U.S. Retrieved October 21, 2010, from The Why Files: http://www.libraryspot.com/know/workweek.htm

Trevino, L. K. (1986). Ethical decision making in organizations: A person-situation interactionist model. *Academy of Management Review*, 11, 601–617.

Triandis, H. C., Carnevale, P., Gelfand, M., Robert, C., Wasti, S. A., Probst, T., Kashima, E. S., Dragonas, T., Chan, D., Chen, X. P., Kim, U., De Dreu, C., De Vliert, E. V., Iwao, S., Ohbuchi, K., and Scmitz, P. (2001). Culture and deception in business negotiations: A multilevel analysis. *International Journal of Cross Cultural Management*, 1, 73–90.

Wikipedia. (2010, October 15). Frank Serpico. Retrieved October 15, 2010, from Wikipedia: http://en.wikipedia.org/wiki/Frank_Serpico

Wikipedia. (2010, October 9). Work-life balance. Retrieved October 21, 2010, from Wikipedia: http://en.wikipedia.org/wiki/Work-life_balance

Withgott, J. and Brennan, S. (2009). *Essential Environment: The Science Behind the Stories*, 3rd edn. San Francisco, CA: Pearson.

Worthen, B. and Lublin, J. S. (2010, August 8). Mark Hurd neglected to follow H-P code. *Wall Street Journal*.

Worthen, B. J. (2010, August 8). *Mark Hurd Neglected to Follow H-P Code*. Retrieved June 06, 2011, from The Wall Street Journal: http://online.wsj.com/article/SB10001424052748704268004575417800832885086.html

Xiang, Q. Y. (2010, December 17). *Serum Fluoride Level and Children's*. Retrieved June 06, 2011, from National Institute of Health: http://www.eastcountymagazine.org/sites/default/files/fluoride%201Q%20study.pdf

Chapter 17

Strategic Direction, Planning, and Decision Making

17.1 Goal and Overview

The goal of this chapter is to develop a broad perspective as to how companies and individuals should plan, develop, and refine their strategies.

In earlier chapters, we introduced the tools of multi-criteria decision making, decision trees, and risk analysis. These have been used by organizations to evaluate, select, and refine the best strategies among a clearly defined set of alternatives. DuPont has used influence diagrams and decision trees to develop its global product and manufacturing strategy (Krumm and Rolle 1992). Phillips Petroleum has used decision trees to develop investment strategies in oil fields (Walls et al. 1995). Pharmaceutical companies routinely use them to evaluate drug development, investment, licensing, and partnership strategies for specific drugs. Power companies use multi-criteria decision tools in making strategic decisions (Kidd and Prabhu 1990; Keeney and McDonalds 1992; Keeney et al. 1995).

These methodologies are used when clear decision alternatives have been defined. In this chapter we describe a broader prerequisite framework for developing a strategic plan that helps identify alternatives for evaluation. The chapter begins with a discussion of the decision elements that make up a strategy. These include defining the scope of the business, its range of products and services, and its stakeholder relationships. We then discuss the elements of a SWOT analysis—indentifying an organization's strengths, weaknesses, opportunities, and threats. Next, we introduce a number of simple tools that consulting groups such as Strategic Decision Group use to organize thinking about specific strategic options. In large organizations, strategy development requires close interaction between the analysts developing the plan's specifics and the ultimate decision makers. We describe a dialogue decision process that a number of companies have adopted to ensure these two groups stay aligned. The final section discusses the process of scenario planning and its role in addressing broad uncertainties that cannot be readily contained within a probability distribution at the end of a decision tree.

This chapter is only an introduction to strategic planning; entire textbooks have been written on this topic and most business schools offer a complete course on it. Some of the classic books in the field of strategic planning and thinking are those by Porter (1998 and 2008), Hamel and Prahalad (1994) and Fahey and Randall (2001). Our primary interest is to provide a basic overview of strategic planning in order to clarify how decision-making tools that are the core of this text fit within the broader process.

17.2 Strategic Planning

Creating and implementing a strategy is a highly complex task that takes its cues from the competitive environment. The goal is to develop and implement multiple courses of action that are complementary and integrated. A business strategy is comparable not to a single cord but to a cable made up of the intertwining cords. It includes strategies for marketing, manufacturing, supply chain, technology, organizational structure, and so on.

The competitive landscape an organization confronts today is likely quite different from the one in which it will earn tomorrow's success. Change may come in the form of technologies, products, customer preferences, economic environment, regulations, and ways of communicating. It may include new competitors from anywhere on the globe or old competitors who have developed new strategies. Further, the changing environment is becoming more complex as it becomes more global and as change itself accelerates. Hewlett-Packard (HP), for example, struggles to adjust to a steady stream of new products from more than a dozen traditional competitors in each of its many product lines for home, office, small business, medium business, and large enterprise. These products not only include its well-established printing business, but also its laptops, desktops, workstations, and networking equipment. HP must constantly develop and implement a comprehensive dynamic strategy for growth while maintaining satisfactory sales and financial performance results to satisfy Wall Street each year.

The best-laid plans are likely to be thwarted not only by competitors, but also by evolving technology, shifting regulations, fluctuating macroeconomic variables, and customers whose tastes and needs may change. Thus, to implement strategy as circumstances change, managers must capture new information, make midcourse corrections, and get the timing right, because entering the market too early can often be just as costly as entering it too late.

Companies and industries can be destroyed as a result of changes outside their immediate control. Case in point: the impact of the Internet on traditional media. Newspapers and magazines struggle to survive in printed form today. Bookstores must compete with Internet retailers. On the other hand, new industries have developed as a result of the physical complexity of the telecommunications industry. Marketing and advertising strategy requires a whole new perspective when there are more than 500 million participants on Facebook and a billion people who use a search engine regularly.

Strategic projects represent the core of corporate growth, change, and wealth creation. They are major investments, often involving high uncertainty; they offer intangible benefits and promise attractive long-term financial outcomes (Buckley 1998). "They are the vehicles through which a sound vision gets implemented and realized" (Schoemaker et al. 1992). Strategic projects, in addition to motivating the creation, acquisition, and development of competencies (Foss 1997), also comprise a collection of diverse options (Amram and Kulatilaka 1999). Typical strategic decisions include entering or exiting markets, investing in new technology, building manufacturing capacity, and forming strategic partnerships.

The ability to make fast, widely supported, and high-quality strategic decisions on a regular basis is the cornerstone of effective strategy. To use the language of contemporary strategy thinking, strategic decision making is the fundamental dynamic capability within excellent firms. Executives from a variety of firms echo this perspective. John Browne, CEO of British Petroleum, stated, "No advantage and no success is ever permanent. The winners are those who keep moving" (Prokesh 1997). Likewise, Michael Dell, CEO of Dell, commented, "The only constant in our business is that everything is changing. We have to be ahead of the game" (Narayandas 1996). Anticipating and responding to a series of shifting advantages is challenging, however. It requires effective strategic decision making at several levels: at the unit level, to improvise business strategy; at the multi-business level, to create collective strategy and cross-business synergies; and at the corporate level, to articulate major shifts in strategic focus. For example, in 2010 the CEO of IBM identified business analytics as a core strength that will drive a significant share of IBM's growth. He sees the enormous expansion of information resulting from the global explosion in Internet usage. Yet companies are struggling to mine value out of this goldmine of customer preferences and buying behavior that the Internet is capturing. IBM thus aims to be the goldminer of choice for a wide range of organizations.

Brown and Eisenhardt (1997) examined top-management teams and their decisions in 12 entrepreneurial firms in Silicon Valley. They also examined 12 European, Asian, and North American multi-business firms (6 dominant and 6 modestly successful ones) in the broader context of strategy. They studied decision speed, conflict over goals and key decision areas, executive power, and politics. They found that the most effective strategic decision makers made choices that are fast, of high quality, and widely supported. How do effective decision makers accomplish this? Four approaches have emerged from this and other research (Eisenhardt 1999):

Build collective understanding. Effective strategic decision makers use extensive, real-time information about internal and external operations to enhance the ability of the executive team to see threats and opportunities sooner and more accurately. These executives build a collective intuition that allows them to move quickly and accurately as opportunities arise. They avoid accounting-based information, which tends to lag behind the realities of the business. They also do not rely on specific predictions of the future because these are likely to be wrong.

Stimulate internal conflict. It is critical that executives stimulate intense discussion about future scenarios and corporate strategy. Conflicting opinions stimulate innovative thinking, create a better understanding of options, and improve decision effectiveness. Without conflict, decision makers usually miss opportunities to question assumptions and overlook key elements of a decision. Stimulating conflict improves the quality of strategic thinking without sacrificing significant time.

Maintain the pace. Effective strategic decision makers maintain a disciplined pace that drives the decision process to a timely conclusion without pushing decision speed. They launch the decision-making process promptly, keep up the energy surrounding the process, and cut off debate at the appropriate moment. In this way, they promote strategic decision-making momentum.

Defuse political behavior. Politicking often involves managers using information to their own advantage. It distorts the information base, creates unproductive conflict, and wastes time that leads to poor strategic decision making. Delineating common goals, clarifying areas of responsibility, and employing humor can defuse politicking and minimize interpersonal conflict. An awareness of common goals that stresses collective success as well as common competitors gives managers a sense of shared fate. A balanced power structure shows a clear area of responsibility for each decision maker and dispels the assumption that the various managers need to engage in politicking. In addition, humor tends to strengthen the collaborative outlook and helps to create a positive mood. Alan Mulally, an outsider from Boeing, was appointed CEO of Ford in 2006. He is credited with dramatically reducing the destructive political culture that had prevented Ford from developing and maintaining a single consistent global strategy. In 2010, Ford made near-record profits and gained market share in a weak automotive market.

17.3 Elements of Strategic Decisions

A strategy has three distinct general characteristics: scope, posture, and goals (Fahey and Randall 2001). An organization specifies the *scope* of its strategy through the choice of products or solutions the company offers and the customers it seeks to serve. For example, should a telecommunication company acquire a cable TV company? Should a computer manufacturer deliver IT services? Should a U.S. biotechnology company enter emerging markets? The scope of the strategy specifies what a company offers to diverse customers in various geographic regions.

The *posture* of the strategy determines how a company competes in order to attract, win, and retain customers. For example, should a cable TV company upgrade its equipment to make it more customer-friendly, or should it instead lower prices? Should an automobile company add more functionality and features, such as more attractive built-in navigation systems, or focus on cost? The posture determines how a company will differentiate its offerings.

The *goal* of the strategy specifies the measures it intends to pursue. Is it seeking to grow in terms of volume or market share? Is it seeking to be the lowest cost producer? Is it seeking to be the leader in technology or fuel economy? The goal defines the results that a company desires and foresees.

Large organizations make strategic decisions at both the business and corporate levels. Business-level planning determines the boundaries of the business and decides how the business should compete in its area. At the corporate level, however, management decides which businesses offer the most opportunity for the corporation. Corporate management can also influence the competitiveness of its business units by developing strategies that leverage the strengths of individual units to help other business units. General Electric (GE) was excellent at leveraging its GE Capital business unit to fund purchases from GE Energy and GE Technology and Infrastructure to the benefit of each of its business units.

17.3.1   Business Scope

Organizations make the following business-scope decisions:

What product or product groups does the organization want to provide and what customers or customer needs does it want to serve?

What geographic regions does it want to reach?

What stakeholders does it want to involve in shaping and executing its product-market scope?

What assets, capabilities, and technologies does it possess or can it develop to serve its product-customer segments?

These questions compel an organization to systematically and carefully assess what business it is in, where opportunities exist within the marketplace in question, and what capacity the organization has, or can create, to use these opportunities.

17.3.2   Product-Market Scope

Every organization continually adds or deletes products as it seeks new customers and withdraws its offerings from specific customer groups. However, breadth and complexity of the relevant issues and questions about product-market scope are distinctly different at the corporate and business-unit levels.

At the corporate level, the principal challenge is to identify those broad businesses in which the corporation can generate value-adding opportunities with potential synergies between business units. GE struggled for years to determine and maximize the value of NBC to GE's broader strategy. They eventually sold it off. At the business-unit level, the strategic product-market scope decisions are as follows:

The range of products to offer

The categories of customers to serve

The customer's wants and needs to satisfy

The geographic markets to serve

The breadth of product varies significantly from company to company. For example, Intel's product-market scope focuses relatively narrowly on the use of chips in personal computers and business servers; it reported record quarterly profits 2011. In contrast, GE is greatly diversified, with more than 10 businesses that operate in a variety of product markets, ranging from finance to airplane engines. Many pharmaceutical companies are strictly drug developers and marketers, unlike Abbott Laboratories that describes itself as a global health care company. Its products include medicines, nutritional supplements, diagnostic instruments, and surgical devices. Ford decided to narrow its range of products by selling Jaguar, Land Rover, and Volvo as it pursues a Ford brand-centric strategy. HP felt a need to expand beyond products to include IT support services. To jumpstart this effort it bought and merged EDS into its product-service portfolio.

17.3.3 Geographic Scope

The business unit and corporate strategy must consider the regional, international, and global context of the business. Geographic-scope decisions present the following issues and questions:

What national or regional markets represent the best opportunities for the organization's current and future products and services?

How can the organization's products be customized or adapted for each customer group and geographic market?

There is great diversity among companies as to the geographic breadth of their business. While some companies solely serve their home countries or regions, others such as Toyota and Johnson & Johnson aggressively target all key markets around the globe. General Motors (GM) was the first U.S. car company to aggressively pursue markets in China and Russia, yet it is only in 2011 that it is seriously beginning to market its core brand, Chevrolet, on a global basis. At the same time, many of the car companies in emerging markets are exploring whether to become global competitors. Tata of India, for example, bought Jaguar.

Since deregulation, many power companies that were once strictly local or regional have expanded beyond their original territories. The same is true of many of the health systems that run hospitals. And while many of the small airlines, such as Spirit, have decided to limit their services to relatively few routes, Southwest made the decision to expand service by purchasing AirTran. In retail, Wal-Mart once consistently placed its stores outside urban areas and only recently started placing smaller stores in urban locations.

17.3.4 Vertical Business Scope and Stakeholders

Vertical scope describes the stages in the vertical chain in which a corporation participates. Modern businesses tend to focus on one aspect of the total supply chain for a product or service. This is generally their core competency. Every organization needs to decide

- What aspect of the business the organization will provide itself?
- What core aspects it will contract out but control?
- What aspects will involve ongoing relationships?

When Ford was created, it was a vertically integrated company beginning with raw steel production. Now it buys all its steel. In contrast, Proctor & Gamble still controls much of its business from product design, manufacture, and up through marketing. IBM takes a third path; it is vertically integrated in some ways but not others. It does basic research on the components of a computer, from central processing to storage devices, which it then manufactures itself or contracts out. Once the computers are installed, it provides companies with a wide range of services that network and support the IT infrastructure. It then proceeds to sell services that relate to how computers use information to help management make more analytic decisions. For their part, Apple and Dell have greater stakes in distribution. Apple decided to expand its role in the marketing of its products by setting up its own Apple stores. Dell focuses on managing its supply chain and delivering a computer assembled to meet a customer's specific needs. Among high-end clothing and sports shoe companies, Nike and Adidas focus strictly on design and marketing. And in the airliner industry, Boeing has struggled in recent years as it shifted from what was largely a go-it-alone strategy to one that integrates a large number of suppliers as partners in the production of the Dreamliner.

As companies define their core, they need to invest in it as well as arrange to obtain the other elements of their product or service. They need to identify

Which stakeholders can affect attainment of the organization's goals and how they can do so?

With which stakeholders the organization can align itself to enhance goal attainment and how it can do so?

Once the key stakeholders are identified, the organization must decide on the nature of the relationship. Microsoft and Yahoo established an Internet-search partnership in 2010 to compete more effectively against Google in search and online advertising. The deal gives Microsoft a 10 year license to integrate Yahoo's search technology into its existing search platforms, while Yahoo will become the "relationship sales force" for advertisers in both companies. In another instance of Internet partnership, eBay bought Skype for $2.6 billion in 2006 so that customers would be able to discuss their transactions in real time. The acquisition failed to provide eBay the profit increase that it sought, however, and Skype was eventually sold off at a loss (Musil and Skillings 2009).

Toyota and other Japanese companies have developed integrated networks of suppliers that include partial ownership. Often this analysis leads to an acquisition. Alternatively, GM and Ford each decided that various automotive components were not part of their core business and thus divested themselves of Delphi and Visteon as stand-alone suppliers. When Ford decided to pursue leadership in telematics, it formed a temporary partnership with Microsoft. This issue of core and shared responsibility is also relevant to public services. Many small cities, for instance, have their own police departments but share a regional dispatch system.

For some issues, companies even work closely with their competitors in trade associations or in sharing data. In 2011, for example, major drug makers, including Pfizer Inc, GlaxoSmithKline PLC, AstraZeneca, Johnson & Johnson, Novartis AG, Sanofi-Aventis, and Abbott Laboratories, established a shared database for their clinical trials for Alzheimer's and Parkinson's diseases. This was an effort to speed the development of new medicines to treat these brain disorders (Richwine 2011).

17.4 Situation Assessment: SWOT Analysis

The first step in strategy development is usually a situation assessment or SWOT (strengths, weaknesses, opportunities, and threats) analysis (Adams 2005; Bensoussan and Fleisher 2008). A SWOT analysis helps an organization identify both the strengths and weaknesses that it brings to the competitive fray and the opportunities and threats posed by its environment. An effective strategy calls for the organization to seize the opportunities, minimize the threats, leverage the strengths, and correct the weaknesses.

The SWOT analysis provides insight into an organization's resources and capabilities within its competitive environment, identifying the key factors related to achieving its objectives. A SWOT analysis groups information into two main categories: internal factors and external factors.

TABLE 17.1: Generic SWOT matrix.

Strengths	*Weaknesses*
What does the organization do better than anyone else? What advantages does the organization have? What unique resources does the organization have access to? What do others see as the organization's strengths?	What could the organization improve? What should the organization avoid? What are others likely to see as weaknesses? What factors lose sales?
Opportunities	*Threats*
Where are the good opportunities facing the organization? What trends could the organization take advantage of? How can the organization turn its strengths into opportunities?	What obstacles does the organization face? What is the competition doing that the organization should be worried about? Is changing technology threatening the organization's position? Could any of the organization's weaknesses seriously threaten its business?

The internal factors may be viewed as strengths or weaknesses depending upon their impact on the organization's objectives. What may represent strengths with respect to one objective may be weaknesses for another. The factors may include all the four P's—product, price, place (or distribution), and promotion—as well as personnel, finance, manufacturing capabilities, and so on.

The external factors are the opportunities and threats presented by the external environment. These may include macroeconomic matters, technological change, legislation and socio-cultural changes, as well as changes in the marketplace or a company's competitive position.

The SWOT analysis begins by conducting an inventory of internal strengths and weaknesses in the organization. It then notes the external opportunities and threats that may affect the organization, based on the overall environment. The results are often presented in the form of a matrix (see Table 17.1). The primary purpose of the SWOT analysis is to identify and assign each significant factor, whether positive or negative, to one of the four categories, allowing the organization to take an objective look at itself.

Strengths: Strengths describe the positive attributes, tangible and intangible, that are internal to the organization and add value or offer the organization a competitive advantage. These are all within the organization's control. What does it do well? What resources does it have? What advantages does it have over its competition? The organization may want to evaluate its strengths by area, such as marketing, finance, manufacturing, and organizational structure. Strengths include tangible assets such as available capital, equipment, credit, established customers, existing channels of distribution, copyrighted materials, patents, information and processing systems, and other valuable resources within the business. Strengths include the positive attributes of the people in the organization. These include their knowledge, social network, reputations, or other skills. Strengths may be intangible and include business processes such as innovative product development and efficient supply chains. Market share and brand recognition are potential market strengths. Apple is strong in innovative product development that involves integrating technologies. Wal-Mart is a leader in inventory management and tracking store purchases. Google has a leading-edge search engine and personnel to keep it state-of-the-art. Coca-Cola has a strong brand and market share.

1. **Activity:** Develop a career strategy using a SWOT analysis by first listing your strengths.

2. **Activity:** What are the strategic strengths of your organization?

_____

Weaknesses: Weaknesses are factors that are within the organization's control that detract from its ability to obtain or maintain a competitive edge. Which areas might it improve? Weaknesses might include an organization's lack of expertise, limited resources, lack of access to skills or technology, inferior service offerings, or poor location. These are factors that are under the organization's control but, for a variety of reasons, are in need of improvement to effectively accomplish its business objectives. Weaknesses capture the negative aspects internal to the business that detract from the value the organization offers or that place it at a competitive disadvantage. These are areas it needs to enhance in order to compete with its best competitor. The more accurately it identifies its weaknesses, the more valuable the SWOT will be for its assessment.

3. **Activity:** List your career weaknesses:

_____

4. **Activity:** What are the strategic weaknesses of your organization?

_____

Opportunities: An organization must survey the landscape to identify opportunities that would allow it to prosper by growing its business, increasing its market share, or improving its profit margins. Can it leverage the growth of the Internet? IBM is positioning itself to provide advanced data analysis to companies that are overwhelmed by the customer information the Internet can provide. How does globalization offer opportunities to reduce cost or sell more? Many U.S. universities see opportunities to offer their name-brand education around the globe. Harvard Business School is providing training in teaching the case method to faculty in leading Chinese business schools. How does the push for sustainable energy sources affect your organization? If you are a university, do you have courses on these subjects? Lastly, does the explosive growth in the senior citizen population offer any business opportunities for your organization?

5. **Activity:** Assess your career opportunities:

_____

6. **Activity:** What opportunities are available to your organization?

_____

Threats: What factors are potential threats to an organization's business? Threats include factors beyond the organization's control that could place its business strategy, or the business itself, at risk. Even if the organization cannot control them, it may be able to implement a contingency plan to mitigate their impact.

A threat is a challenge created by an unfavorable trend or development that may lead to deteriorating revenues or profits. Competition is always a threat. Other threats may include price increases by suppliers; governmental regulation; economic downturns; devastating media or press coverage; a shift in consumer behavior that reduces your sales; or the introduction of a "leap-frog" technology that may make your products, equipment, or services obsolete. Traditional universities are threatened by the growth of online degree programs. Newspapers and magazines are under attack by the ubiquitous (and free) Internet.

7. **Activity:** Assess your career threats.

_____

8. Activity: What threats face your organization?

The final step in this preparatory phase of strategy development is to compare the threats and weaknesses with external opportunities and threats. Which of your organization's strengths align with growth opportunities? What weaknesses will hold you back from pursuing them? Can your strengths be leveraged to address potential threats? Lastly, do your weaknesses exacerbate a potential deteriorating situation that would occur should the threats materialize?

17.4.1 Uncle Sam's Organic: SWOT Analysis

Uncle Sam's Organic (USO) is a medium-sized organic juice specialty company that was incorporated a few years ago. Its natural and organic juices and other drinks are created without any added sugar, preservatives, or artificial colors. USO handles all aspects of product development, manufacturing, and marketing of its health-conscious juices. Its offices and plant are in the heart of the agriculture area of California so that all its primary raw materials are shipped to its plant from within a 6 hour drive. Its brands include natural sports drinks, organic fruit juices, and non-alcoholic celebratory beverages. USO's products are distributed to independent organic food outlets and supermarkets that advertise high-quality organic products. These are all within 500 miles of its headquarters and operations.

USO has successfully penetrated a number of markets in the surrounding region and is now working on a strategy to expand. It is excited by growing interest in organic foods of all types. As USO develops its growth strategy, it will need to articulate the range of alternatives in each of the following areas:

Product-market scope: USO currently produces four main lines of products, focusing on oranges, lemons, grapes, and pomegranates as the main ingredients. All its end products have a relatively short lifecycle and thus require refrigeration. It has priced its products at the high end, so its primary customer base is upper-middle class suburbans. USO is considering the following possibilities:

- Expand beyond four core ingredient flavors to other ingredients readily available in California.
- Use the core four product groups to expand into drinks that do not require refrigeration.
- Expand into non-drinkable products such as spreads and jams.

Geographic scope: The current market scope is limited to 500 miles; the company has no interest in international markets at this time. Expansion could take on the following forms:

- Regional markets in suburbs up to 1000 miles away
- A national roll-out in suburban stores around the country
- A national roll-out through Internet sales
- High-end neighborhoods in major urban areas

Organization and stakeholder scope: Currently, USO buys all its raw materials on seasonal contracts, develops new products, and then produces and bottles the product themselves. It handles all its own marketing with some assistance from a local advertising firm that also helps with package design. Its products are sold primarily in small stores that emphasize organic products and one regional food market chain that advertises its large organic foods section. There is no formal long-term relationship between USO and the stores that sell its products.

Organic foods–stakeholder relationships

- Maintain status quo with seasonal contracts as needed
- Sign long-term contracts with a number of farms
- Develop a partnership with a farm cooperative that would share in the profits

Production

- Maintain ownership of production and expand production to other regions as demand grows
- Outsource all production
- Maintain local production but outsource production in other parts of the United States

Distribution channels

- Maintain basic strategy of selling primarily to independents
- Sell to regional supermarket chains that have a substantial organic food section
- Partner with a national organic food store chain to sell their products

USO conducted a SWOT analysis (Table 17.2) to begin developing its business strategy. The company's major strengths are the taste and quality of its products, which have attracted a loyal following including a number of high-profile individuals. Another is the breadth and depth of its executive and employee knowledge of the organic food market. Its major weaknesses are mostly related to its current size. The opportunities for growth are primarily reflections of the burgeoning

TABLE 17.2: USO SWOT matrix.

Strengths	*Weaknesses*
Quality and appeal: products win in taste tests	Lack of a broad market base
Numerous high-profile individuals strongly believe in their products	No presence in major store chains
High-quality farm sources	Limited capacity
Specialty stores like it—it sells and brings repeat customers	Young company that may have difficulty expanding quickly
Strong financial backing	Narrow product lineup that cannot justify a substantial increase in its marketing budget
Executives have a wide range of experience in the organic foods market	
USO employees are highly active in the social networking world of organic foods consumers	
Opportunities	*Threats*
Organic foods is a high growth area	Major store chains expanding their organic foods section
Target customers have high disposable incomes	Growth of several national brands with wide array of organic products
Internet opportunities for creative marketing	Existing competition always looking to expand
American organic food buyers prefer U.S. products because of closer supervision	Other start-up companies generated by healthy economic growth nationwide
Consumers willing to pay a premium leads to high profit margins	Volatile pricing of key organic ingredients
	Competition for growers
	Internet sales of competitors

interest in organic foods nationwide and the nature of its customer base. The threats are related to the existence of major competitors, potential problems with farm sources, and the attractiveness of the business opportunity that is attracting increased competition.

17.5 Basic Tools: Decision Hierarchy and Strategy Table

Scope definition and SWOT analysis are not ends in themselves (Hill and Westbrook 1997; Pickton and Wright 1998). They are only first steps in developing alternative strategies to be evaluated quantitatively. Each alternative strategy ultimately must include a set of related decisions that are consistent with a strategic theme. Strategic consulting groups use two tools, decision hierarchy and strategy tables, to organize the information associated with the development of alternatives. The results of this strategy framing are then fed into an analytic tool such as a decision tree or multi-criterion decision analysis.

Decision hierarchy: A decision hierarchy (Figure 17.1) establishes the boundaries of the strategic analysis. It consists of three layers. The top layer contains policy decisions already made and facts that cannot be changed. The middle layer details strategy decisions under consideration and is the primary focus of the strategic analysis. The bottom layer contains the tactical decisions for the future. The categorization allows a strategy team to navigate the evaluation of their decision problem by becoming aware of the decisions behind them, those immediately in front of them, and those on the horizon.

Policy decisions at the top of the hierarchy are givens that an organization or individual has already decided upon and is uninterested or unable to change. For example, a luxury vehicle strategy may have as a given the existing car dealership network. Wal-Mart, for most of its existence, operated under the assumption that it would not locate its stores in urban areas. Such givens may include financial factors, such as the company already having lined up financing to support expansion, or the company being unable to assume any significant increase in debt. Existing relationships may fall within this area. Toyota has a large Japanese supply chain network that it is unlikely to modify because of long-standing relationships and intertwined ownership. A drug company might list its patents and expiration dates as givens. Core corporate principles or legal restrictions should

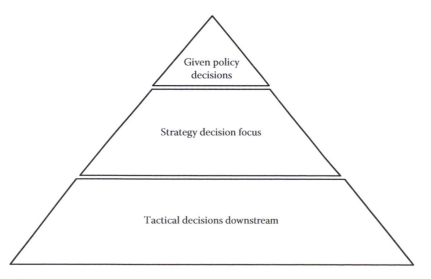

FIGURE 17.1: Decision hierarchy.

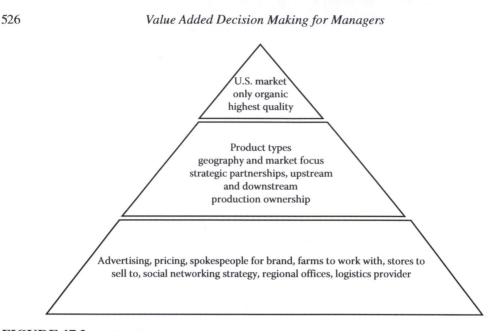

FIGURE 17.2:   USO decision hierarchy.

also be included. European companies, for example, are barred from using layoffs to adjust their workforce to reduce production in their home markets, and American companies may have union contracts that have a similar impact.

The strategy decision level includes all the major decision categories under consideration and is the focus of the analysis. It includes all scope decisions discussed earlier. The decision details associated with this level are laid out in a strategy table.

Tactical decisions are subordinate decisions that will be addressed once the strategic decisions have been made. These decisions will be part of the implementation strategy plan. Although the tactical decisions are usually not dealt with explicitly at the moment, it is important to acknowledge what they are so as to clarify what will need to be done to follow through on the chosen strategy.

USO developed a decision hierarchy (Figure 17.2) for its strategy development. They are an organic-based beverage company committed to high-quality products and are focused only on the United States. The decision areas are product scope, geography, partnerships, and production ownership. There are many decisions they will need to make once the strategic decisions have been made, as listed at the bottom of Figure 17.2.

Strategy table: A strategy table enables decision makers to define and organize a complex set of choices. The resulting decisions help create a coherent strategy. Each column in the table lists a different strategy decision category. For USO, the primary decision categories are investment level, product range, channel distribution, farm source, and geography. The leftmost column lists different strategy themes.

The strategy planning team develops roughly three to seven significantly different alternative strategies that define the dimensions of the solution to the problem. The strategies are developed by linking one or more of the choices listed under each category in each column of the strategy table. The strategy development team is challenged to

Find creative, fresh alternatives that go beyond variations of business as usual

Find significantly different alternatives that cover the complete range of possibilities

Not squander resources on evaluating undoable, unacceptable alternatives

TABLE 17.3: USO strategy matrix.

Theme	Investment Level	Product Range	Geography	Farm Sources	Distribution Channels	Production
Grow incrementally and on its own	Low	None	Expand one region at a time	Annual contracts	Partner with specialty organic stores	In-house
Grow rapidly and on its own	High	More core ingredients but same product types	Expand nationally	Long-term contracts	Supermarkets with organic marketing focus	Contract some production
Grow rapidly with strategic partners		Wide range of products	Expand into urban markets	Partner with growers	Supermarkets with organic food sections but no marketing focus	

Challenge the common perception of what is acceptable and what is not, and what is possible and what is not

Look at the problem from different perspectives, such as those of a corporate and stakeholder perspective

USO started with three strategies in the first column, "Theme." "Grow rapidly with strategic partners" might combine the following decisions: investment level (low), product range (wide), geography (national), farm sources (partners), distribution channels (partner with specialty stores and sell to supermarkets with large organic food sections), and production (contract some production). A strategy is portrayed by highlighting or linking the companion decisions across the table (Table 17.3).

17.6   Strategy Development Steps for Large Organizations

In its strategy practice, the Strategic Decision Group developed a six-step structured process, later adapted by GM in the 1980s (Barabba 1995), called dialogue decision process. It builds upon the principles of decision analysis and deals with major, one-of-a-kind decisions that cut across organizations within a company. This tool may be modified to handle many other types of decisions as well.

In developing strategies, a company usually creates a cross-functional team that includes individuals with varied expertise who will address opportunities or problems. Too often, however, there is not enough formal communication between decision makers and members of the team. In many cases, the team analyzes a given situation, creates a solution, and then presents the solution to the decision makers. Little if any formal communication takes place between decision makers and the project team members during the process. All too frequently, the team presents only a single strategy in response to the known issues and the information that has been gathered and

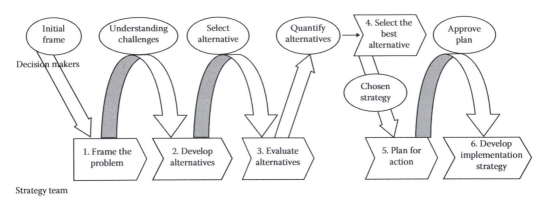

FIGURE 17.3: Decision dialogue process.

analyzed. At this point, the decision makers can approve, disapprove, or send the problem back for more work.

This traditional process encourages advocacy rather than collaboration: the team is under pressure to get it right the first time or its work will be dismissed. This motivates the team to develop conventional solutions and hide any potential problems. The decision makers, for their part, have only limited opportunity to use their own collective wisdom and experience to collaborate in creating a great solution.

Dialogue decision process, on the other hand, is designed to bring value-creating ideas to the surface, combine them into attractive, doable strategies, and, at the same time, build organizational commitment to successfully implement the chosen alternative. A key to the success of this six-step process (Figure 17.3) is regular dialogue between the decision makers and project or strategy team (Kusnic and Owen 1992; Bodily and Allen 1999).

The decision board consists of everyone who must agree to the decision. This board initiates the strategy development effort. It reviews the business assessments and alternatives and then evaluates the alternatives along with the implementation plan. Finally, the board selects the strategy and allocates the resources for implementation.

The strategy or project team is composed of people who have profound knowledge as well as a stake in implementing the decisions. The team conducts the business assessment, develops alternatives, evaluates the alternatives, and develops a plan of action. The dialogue between the decision board and strategy team is critical for knitting together the various aspects of strategic management leadership.

Step 1: Frame the Problem

In this step, the team identifies the purpose, scope, and perspective of the problem. It specifies the boundaries of problem, the dimensions of the solution, the sources of information that will help determine the solution, and the criteria to be used in the choice. The team also establishes a filter to classify and discard irrelevant information early in the process. The purpose of this identification is to

Develop a shared understanding of the opportunity or challenge in question

Reflect the different perspectives of the group members

Bring to the surface any unstated assumptions that could affect the project

Explicitly formulate and communicate the problem

The principal processes and deliverables of assessing business situations are issue raising, strategy vision statements, decision hierarchies, and influence diagrams.

Step 2: Develop alternatives

In Step 2, the strategy team presents alternatives to the decision board, asking the following questions:

Should we consider any additional alternatives?

Have we presented any alternative that you would not consider?

Step 3: Evaluate alternatives

The strategy team evaluates the alternatives by quantifying the value, risk, timing, and trade-offs associated with each. The MAUT or AHP can be applied when the problem includes multiple objectives, while a decision tree or simulation can be used when the problem includes uncertainty. The team determines the trade-offs among the financial measures, non-financial measures, time, and risks. The team then communicates to the decision makers and stakeholders the results of their evaluation as well as the benefits and drawbacks of the various alternatives. Only then do they offer their insights regarding why one alternative is preferable.

Step 4: Select the best alternative

The decision board chooses among the alternatives after reviewing the benefits and risks and the insights provided by the strategy team. The decision board weighs strategic and organizational considerations along with the financial comparisons of alternatives. In the decision phase, the decision board must step up to these challenges:

Choosing a particular alternative that involves making difficult trade-offs

Communicating the frame in order to make sure that all stakeholders understand the full picture

Clarifying the insights that were shared so that all stakeholders see the reason for the choice

Explaining the robustness of the choice so that the stakeholders can understand why other, less flexible, alternatives were not chosen

Communicating enthusiasm for and commitment to the chosen alternative

Obtaining organizational ownership of and commitment to the decision

Step 5: Plan for action and implementation

In this step, the strategy team develops a vision of the company's growth as a result of implementing the chosen alternative. It designs the implementation plan and designates leaders for the implementation.

Step 6: Implement action plan

17.6.1 Therapharma Case

To illustrate the steps of the dialogue decision process, we present a case study of a composite pharmaceutical company called Therapharma (Bodily and Allen 1999), adapted from actual situations. The senior management of Therapharma has initiated an effort to develop a new strategy, as the managers were concerned that the company was not taking full advantage of its opportunities within the pharmaceutical industry.

Step 1: Frame the problem

Therapharma identified the challenges they faced and made a clear plan for filling gaps in information, developing alternatives, and assessing value trade-offs. The company identified the following issues:

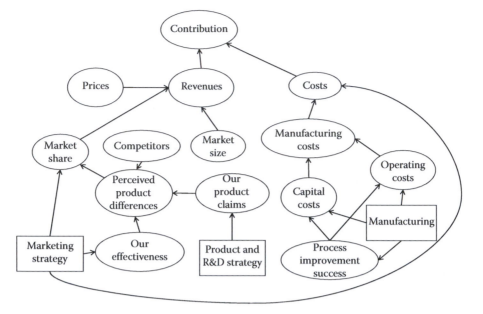

FIGURE 17.4: Therapharma influence diagram.

Elderly people constitute the fastest growing population segment and also have many chronic diseases.

Therapharma's market share is expected to decline because its product line is poorly matched to the fastest-growing therapeutic classes.

Therapharma faces a 3–7 year gap in product launches.

Therapharma's German sales force is 20% of the size needed to cover its products effectively.

The U.S. prescription business will continue to contribute most of the corporate cash flow.

Therapharma then identified the challenges it would have to address in its strategy. It used an influence diagram (Figure 17.4) to frame the challenges and develop a shared understanding. The value of a new product to Therapharma is influenced by the number of competitors, its market share, and whether the product will be approved and reach market introduction. Patent protection and strength of managed care buyers are important for Therapharma, as they impact the retail price and their product's long-term value. Therapharma specified the following challenges:

Find new products in fast-growing therapeutic classes that would fill the gap in product introductions

Maximize profits by increasing its sales force and product support capabilities in key Far Eastern and European markets

Better assess when, and with what probability, new drugs would be discovered

Work with managed care providers to increase the clinical value and price positions of products

Reduce drug development time

Expand Therapharma's capability to market over-the-counter (OTC) medicines

Step 2: Develop alternatives

Therapharma's strategy team held brainstorming sessions that focused on the challenges identified in Step 1 in order to identify alternatives. At the end of each session, the team compiled a strategy table (Table 17.4). The team created five strategies and listed the names of each in the first

TABLE 17.4: Therapharma's strategy matrix.

Strategy	International Focus	Domestic Marketing	Licensing and Joint Ventures	Generics	R&D Strategy
Short-term profit improvement	Worldwide presence	Maintain current sales force levels	Out-licensing limited to low potential drugs and focused in-licensing with joint research agreements	Stay out and promote trademark to prevent share erosion	Current
Current	Build critical sales force mass in top 12 countries	Expand sales force to maintain current office-based coverage and increase hospital and key physician coverage			Concentrate on key product classes
Focused reallocation	Build, acquire companies, or joint venture to build critical mass in Germany, Japan, etc.		Aggressive out-licensing to achieve foreign potential and in-licensing to add domestic products	Selectively enter to use manufacturing capacity	Aggressively license to exploit strengths
Joint venture and license focus					
Broad-based expansion	Acquire company in Germany and out-license in Japan, etc.	Maintain current sales force levels and increase advertising	Aggressive licensing and key joint venture	Cautious entry strategy based on marketing strength	Expand entire program

column, with the choices that could be made in each decision area in the other five columns. For example, the focused reallocation and its associated decisions are in solid rectangles. The strategy includes the following choices in each of the five areas:

International focus: build or acquire companies, or create a joint venture to bolster critical sales-force mass in certain countries

Domestic marketing: maintain domestic sales-force levels and increase advertising

Licensing: limit the licensing to outside firms of low-potential drugs

Generics: stay out of the business of producing generic drugs and promote trademarks to prevent share erosion

R&D: concentrate R&D on key therapeutic areas

Step 3: Evaluate alternatives

The team created influence diagrams for major products to identify the sources of uncertainty and developed spreadsheet models of the individual products. The model incorporated 10 years of cash flow plus the continuing value to Therapharma at the end of 10 years. These models were then consolidated into the overall financial model. The team assessed the probability distributions for the uncertainties, using both data and expert judgment to establish the range of possibilities and their relative likelihood.

The team calculated the net present value (NPV) using the spreadsheet model for each alternative. They demonstrated total uncertainty of each alternative using probability distribution of their NPV (Figure 17.5). The focused-reallocation alternative had the least uncertainty and risk as it had the narrowest NPV range. On the other hand, this alternative had lower NPV values than the broad-based-expansion alternative. The decision board can use the comparison to account for risk/value trade-offs in evaluating and choosing alternatives.

Step 4: Select the best alternative

The decision board considered strategic and organizational factors along with the financial comparisons of alternatives. The board rejected the momentum and the short-term-profit-improvement strategies since these strategies were dominated by others. The focused-reallocation strategy was better than others due to low risk:

It concentrated on therapeutic areas in which Therapharma already had major franchises.

It avoided the risk of failure related to negotiating and implementing a joint venture.

It emphasized new products that played to Therapharma's strengths.

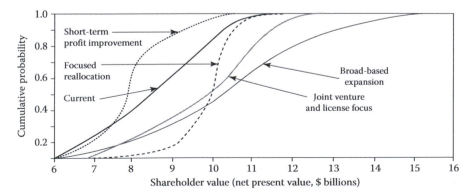

FIGURE 17.5:   Therapharma risk profiles.

The decision board also discussed implementation issues associated with the best alternative. The critical implementation issues included a major expansion of OTC medicine programs, abandoning three therapeutic areas, and the impact of doing so upon the R&D organization.

Steps 5 and 6: Plan for action and implementation

The board decided that Therapharma would focus on four therapeutic areas and OTC medicines. They realized that the company would need to advertise directly to consumers in order to promote the OTC medicines along with some of its prescription medicines. The board decided that two of the potential OTC products could best be marketed by coordinating a direct sales force to educate physicians and an advertising campaign directed toward consumers. Another major product that they wished to market would require a level of advertising strength that Therapharma would not be able to develop quickly. Therefore, they elected to launch and market this product by forming a joint venture with an OTC company that was known as a leader in advertising.

Additionally, the team decided to phase out three therapeutic area programs by reassigning personnel and licensing the products to companies that would value them. The implementation plan also included business process changes that would improve Therapharma's ability to begin marketing the drug while it was still in the process of development.

The plan of action (Step 6) was easy for Therapharma because they had used the dialogue decision process. All members of senior management had bought into the focused-reallocation strategy during Steps 1–4. All stakeholders participated in the process and had been able to contribute ideas.

17.7 Scenario Planning

When a situation involves several random events, a decision tree is an appropriate tool for incorporating uncertainty of each over a specific range. With multiple random events represented in the tree, the corresponding probabilities simply multiply. Sensitivity analysis expands the range even further by exploring whether the decision would change as a result of modest variation in key parameters. The ultimate objective is to maximize the expected value, which is a probabilistically weighted sum of different outcomes. In scenario planning, we explore a level of collective uncertainty not easily captured in decision trees. The different scenarios represent significantly different streams of connected uncertain events. No weighted sum can reflect this diversity.

The concept of divergent scenarios is easy to understand in a political context. An election, for example, can set in motion a whole range of outcomes, not all of which may be simple to plan for. The election of George W. Bush rather than of Al Gore in 2000 triggered a whole series of random events. Similarly, the loss of the Democratic Party's majority in the House of Representatives in 2010 will influence unpredictably the legislative agenda for the next 2 years. For someone planning for the election of 2012, no amount of sensitivity analysis can capture the divergence of tax policies that would result with a Republican president and Republican control of both houses of Congress, versus a Democratic president with a majority in one chamber. Business decisions affected by tax policies would need to explore both scenarios. This is the essence of scenario planning.

Scenario planning, also called scenario thinking or scenario analysis, is a strategic planning method in which an organization anticipates a range of possible developments in order to better prepare for the future (Schoemaker 1995; Garvin and Levesque 2006; Koehler and Harvey 2007; Lindgren and Bandhold 2009). It attempts to describe a group of distinct futures, all of which are plausible, without assigning probabilities to the likelihood of any specific scenario. As a group process, scenario planning encourages knowledge exchange and cultivating a deeper understanding of issues central to the business in question. The goal is to map out a small number of possible alternative futures, craft narratives to describe these scenarios, and develop options for the organization that would enable it to adapt within these depictions.

Scenario planning differs from such tools as traditional strategic planning, contingency planning, and sensitivity analysis. Traditional strategic planning assumes that there will be one best answer to a strategic question. Scenario planning, on the other hand, entertains multiple possibilities. Contingency planning examines only a single uncertainty, such as, "What will happen if we fail to develop a specific product?" It presents a base case and examines one particular exception or contingency, whereas scenario planning investigates the joint impact of various uncertainties. Sensitivity analysis investigates the impact of change in one or two variables, but keeps all the other variables constant. This is not realistic, however, because any large change in one variable means that other variables are unlikely to remain constant (Matheson and Matheson 1999). If the change in interest rate is large, for example, other variables, such as inflation, money supply, and exchange rate, will change as well. Scenario analysis, on the hand, changes several variables at a time. Moreover, unlike simulation modeling which is heavily numbers-driven, scenario planning involves subjective interpretation as well as objective analysis. Finally, scenario planning organizes the future possibilities into narratives that are easier to grasp and use than great volumes of data.

Scenario planning was originally conceived for the U.S. military during World War II and was first applied to business planning in the late 1960s. Planners at Royal Dutch/Shell have used scenarios since the early 1970s to generate and evaluate its strategic options (Schoemaker et al. 1992; Garvin and Levesque 2006). As a result, Shell has consistently been better prepared than other petrochemical companies for futures radically different from the present. Scenario planning helped Shell anticipate overcapacity in the tanker business and Europe's petrochemicals before its competitors and better prepare it for the ensuing drop in demand.

17.7.1 Developing Scenarios

The genesis of scenario planning in an organization may result from a number of factors (Schoemaker 1995, 2007). It may be a reaction to a wildly inaccurate forecast with regard to financial matters or scientific or technological breakthroughs either of which could be over- or underestimated. Perhaps executives did not appreciate the motivational bias of medical researchers who seek funding and predict imminent breakthroughs in the treatment of cancer, AIDS, or Alzheimers. Conversely, who would have predicted the speed at which Facebook would reach 500 million users or that Twitter would follow a similar growth trajectory? Over-prediction leads to premature investments that may be off the mark. Under-prediction results in lost opportunities to be a market leader.

Other triggers involve costly surprises. Most companies now plan for random rapid spikes in the cost of oil, but that was not the case 10 years ago. Similarly, how many companies have paid attention to the monopoly China has acquired in the production of rare earth metals? They may have been caught by surprise at China's willingness to use threats of cutoffs in its contentious interaction with Japan in 2010. Were European countries ready when Russia reduced supplies of gas to Europe as a result of a dispute with Ukraine over the gas pipeline that passed through it?

Scenario planning is one element in the broader concept of strategic planning. The first step is to define an appropriate timeframe for the range of changes to explore with scenarios. Next, a diverse team should identify relevant trends that will influence decisions. These should be coupled with an identification of key random events or random variables that could influence the trend. The trends and variables form the basis for several distinct scenarios used to create plans that can be implemented or modified depending upon which scenario ultimately unfolds (Figure 17.6).

One of the easiest trends to predict is the age distribution of the U.S. population in 10 or 20 years. The timeframe is a function of the scope of the decisions at hand. If you are in the business of building and providing assisted living facilities or cancer units, a time horizon of 20 years or longer may be appropriate, with developments phased in as the target population grows increasingly older and in need of such facilities. It will be one or two decades before the real impact of the baby boom generation is felt in this market. In contrast, if you are in the business of servicing the travel interests of baby boomers, your time horizon could be only 5–10 years, because travel is of immediate interest

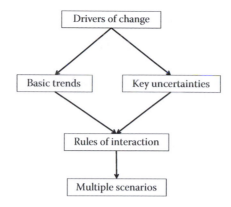

FIGURE 17.6: Scenario planning process.

as boomers approach retirement. If your business relates to social networking, the time horizon of the scenarios you develop should be less than 5 years.

The next task is to identify relevant trends. These trends should be verified by a diverse group of knowledgeable people, the better to understand the difference between actual trends and what are no more than spikes in activity. One can envision a trend in the growth of hybrid cars, for instance, but it is too early to call the production of electric vehicles a trend, and fleets of hydrogen cars are just a pipe dream for now. The development of wind farms may be considered a trend, but a comeback for nuclear energy in the United States can only be seen on planning documents. In 2011, the medical value of the human genome project or stem cell research is still closer to science fiction than reality.

As these trends are incorporated into multiple scenarios, it is important to ask, "Is this trend sustainable? Could it be stopped? Might it be reversed?" In the 1970s the United States was expanding its manned flight space program. Each trip to the moon was an exciting adventure. Futurists imagined manned flights to even more distant celestial bodies. But were the budgets and the public attention sustainable? The United States has not been back to the moon since 1972. In 2011 all NASA-controlled manned flights might cease with the retirement of the space shuttle fleet.

In the late 1980s and early 1990s, Russia was trending toward increasing democratization, but since then, many elements of democracy have been curtailed or rolled back. China has launched a massive increase in college education opportunities, but is this growth sustainable? There are numerous reports that the Chinese economy is unable to absorb these college graduates and meet their expectations for professional jobs. Could growing mass disillusionment put a halt to near-term growth? Could the college population decline? To explore these questions and develop scenarios, experts in Chinese society would need to explain the multiple sociological drivers of this growth.

One interesting question for decision makers in industry or government is whether their organization can directly influence a trend. Courtney et al. (1997) note that the ideal strategy in the presence of significant uncertainty is for companies to shape their future. Apple's phenomenal growth has been fueled by its key product lines. Intel has maintained its lead in chip design and manufacturing for more than two decades. This shaping of trends is not limited to high-tech industries; Wal-Mart has shaped mass retailing for as long as Intel has led the computer chip industry.

Sometimes trend viewers connect phenomenon too closely. They assume that growth in one area automatically leads to a parallel impact in a related area as night must follow day. As business and personal computers became widely available, for example, prognosticators envisioned a paperless society. Instead, the development and marketing of low-cost, high-quality printers has fueled demand for paper.

Lastly, it should be noted that both stability and little or no progress are also trends. This is true of the conflict between Israelis and Palestinians and between North and South Korea. For decades it was true of the economy of Africa as a whole. Similarly, the trend line for the relationship between the United States and Canada is flat.

A parallel task to trend identification involves identifying critical random events and random variables. With regard to the aging population discussed earlier, a critical random variable is the age of retirement. There is significant uncertainty about this variable because of the recession of 2007 and 2008. Dramatic reductions in housing values reduced the net worth of these individuals. In addition, the stock market has barely returned to where it was in 1999. Thus, anticipated returns on portfolio investment never materialized, leaving many without enough funds to retire as planned.

With regard to the electric car, substantive growth in this vehicle segment will depend on several unknowns. The first is a breakthrough in battery technology that allows for much longer travel between recharging. The second relates to how much costs per unit decrease as the industry gains manufacturing experience. These are the internal random variables, and then there are external factors such as the price of oil that would influence demand for electric vehicles. The outcome of the 2012 presidential election is also a factor, as President Obama has demonstrated a willingness to push for substantial funding in support of this research.

17.7.2   Trends and Random Events

In this section we suggest a number of trends and critical random events in diverse areas. In each case, the reader is expected to question whether the trend actually exists and suggest other relevant trends and random events.

17.7.3   State and Local Government Services

The recession of 2007 and 2008 has increased costs of state governments as they deal with the unemployed and increasing numbers of families below the poverty line. Lost jobs and declining consumer purchases have caused revenues to decline. Local governments have also been hit with large declines in property values and a growing list of foreclosed homes that produce no tax revenue. Several trends that result are as follows:

T_1: There is increasing pressure to bring revenue and costs into alignment at the state level so as to balance the budget annually without using short-term accounting gimmicks.

T_2: Local governments will continue to cut services including police and other emergency services.

U_1: The key uncertainty is whether elected officials will raise taxes substantially to balance their budgets.

9. Activity: Critique whether or not the aforementioned trends are actual trends. Identify other trends and random events that will critically impact local and state government services.

_____

_____

17.7.4   Energy

A number of factors have contributed to turmoil in the energy industry. There have been wide fluctuations in the price of oil. Worldwide, there is strong scientific evidence in support of global warming that has elevated environmental concerns in many developing countries.

T_3: There is growing global investment in implementing sustainable power generation plants fueled by alternative energy sources. This is coupled with a significant increase in global research into alternative energy technologies.

T_4: There is a steady flow of discovery of new sources of oil and natural gas around the world that can be recovered and processed with substantial profit margins.

T_5: China will continue to consume an ever-growing share of natural energy resources.

U_2: A primary uncertainty is how quickly a smart grid will develop with more efficient electric power transmission capability.

U_3: Will the U.S. government invest significant dollars in research into alternative energy technology, and will it subsidize the cost of delivering power from alternative energy sources?

10. **Activity**: Critique whether or not the aforementioned are actual trends. Identify other trends and random events that will critically affect the energy industry in the United States.

_____

_____

17.7.5   U.S. Economy

By 2011, the U.S. economy had not recovered from the interrelated shocks to its financial and housing segments. Millions of individuals have remained unemployed for more than a year. Foreclosures have not yet peaked and continue to depress housing values. There are no short-term forecasts of dramatic improvement for many who continue to suffer from the worst economic crisis in the nation since the Great Depression.

T_6: Foreclosures continue to rise.

T_7: Unemployment and underemployment decline slowly.

T_8: There is a continued surplus of single family housing.

T_9: There are fewer high-paying jobs for the less educated.

T_{10}: The United States continues to fall in the international rankings of student performance in mathematics.

U_4: When will the U.S. economy begin sustainable growth that would provide substantial job growth?

U_5: When will the housing foreclosure rate return to normal levels?

17.7.6   Health Care Industry

The health care industry includes hospitals, medical practitioners and their organizations, insurance agencies, pharmaceuticals and bio-medical researchers, medical equipment manufacturers, state governments, and the federal government. Each has its own interest in the continuing growth of this sector of the U.S. economy. However, the trends and uncertainties listed in the following are not equally relevant to the strategic plans of each of these groups.

T_{11}: Health care spending will continue to become a larger share of the GDP.

T_{12}: Expensive new technologies will continue to increase the cost of health care.

T_{13}: The percentage of deaths caused by cancer will continue to grow.

T_{14}: The number of elderly people will continue to grow, the average lifespan will continue to increase, and the number of people with chronic illnesses will grow.

T_{15}: Cost of health insurance will rise significantly faster than inflation.

T_{16}: Medical records will be increasingly computerized.

T_{17}: Co-pays continue to increase slowly.

U_5: Will the 2010 health care law requiring almost universal health insurance coverage be repealed or eviscerated?

U_6: Will there be changes that create increased competition for health care insurance across state lines?

U_7: Will President Obama be reelected with a Democratic majority in the Senate?

U_8: Will there be a major restructuring of the income tax code that will affect the deductibility of health care costs?

U_9: Will there be a standard universal computerized hospital medical record implemented within the next 5 years?

Two divergent scenarios for the next 10 years are presented. One describes unrestricted growth in health care expenditures. The other describes a significant decrease in the rate of growth to a rate close to inflation.

Scenarios

1. Growth in health care expenditures continues at its current rate with individuals, companies, and government costs increasing proportionately. Key elements of the health care law are repealed. There is no substantial change in the fee-for-service medical system. There are no changes in medical litigation. Expensive medical procedures continue to grow with little review of their relative cost effectiveness.

2. The growth in expenditures for health care is contained to within 1% or 2% of inflation. The health care law remains in place and is implemented as planned. The adoption of universal medical record systems leads to an increased use of standardized best practices. Almost all organizations providing health benefits significantly increase the share of costs borne by the individual. There is increased competition for all health insurance dollars within every state. There is a cap placed on medical litigation.

11. **Activity**: Critique whether or not the aforementioned are actual trends. Identify other trends and random events that will critically affect the health care industry in the United States.

---

12. **Activity**: Create another health care industry scenario that is internally consistent and plausible.

Exercises

Complete Chapter Activities

17.1 Develop a SWOT analysis of your career.

a. List your strengths.

b. List your career weaknesses.

      c. Assess your career opportunities.

      d. Assess your career threats.

17.2 Develop a SWOT analysis of your organization.

      a. List your organization's strengths.

      b. List your organization's weaknesses.

      c. Assess your organization's opportunities.

      d. Assess the threats your organization faces.

Elements of Scenario Planning Trends and Uncertainties

17.3 Describe trends and key uncertainties with regard to the Internet.

17.4 Describe trends and key uncertainties with regard to social networking.

17.5 Describe trends and key uncertainties with regard to college education.

17.6 Describe trends and key uncertainties with regard to K-12 education.

17.7 Describe trends and key uncertainties with regard to the global economy.

17.8 Describe trends and key uncertainties with regard to global politics.

17.9 Describe trends and key uncertainties with regard to television.

References

Adams, J. (2005). Analyze your company using SWOTs. *Supply House Times*, 48, 26–28.

Amram, M. and Kulatilaka, N. (1998). Uncertainty: The new rules for strategy. *Journal of Business Strategy*, 20(3), 25–29.

Barabba, V. P. (1995). *Meeting of the Minds: Creating the Market-Based Enterprise*. Boston, MA: Harvard Business School Press.

Bensoussan, B. E. and Fleisher, C. S. (2008). *Analysis without Paralysis: 10 Tools to Make Better Strategic Decisions*. Upper Saddle River, NJ: Pearson Education, Inc.

Bodily, S. E. and Allen, M. S. (1999). A dialogue process for choosing value-creating strategies. *Interfaces*, 29(6), 16–28.

Brown, S. L. and Eisenhardt, K. M. (1997). The art of continuous change: Linking complexity and time-paced evolution in relentlessly shifting organizations. *Administrative Science Quarterly*, 42, March, 1–34.

Buckley, A. (1998). *International Investment – Value Creation and Appraisal: A Real Options Approach*. Denmark: Copenhagen Business School Press.

Courtney, H., Kirkland, J., and Viguerie, P. (1997). Strategy under uncertainty. *Harvard Business Review*, 75(6), 66–79.

Eisenhardt, K. M. (1999). Strategy as strategic decision making. *Sloan Management Review*, 40(3), 65–72.

Fahey, L. and Randall, R. M. eds. (2001). *The Portable MBA in Strategy*, 2nd edn. New York: Wiley.

Foss, N. J. (1997). Resources and strategy: Problems, open issues and ways ahead. In: *Resources, Firms and Strategies*, Foss, N. J., ed. Oxford: Oxford University Press.

Garvin, D. A. and Levesque, L. C. (2006). A note on scenario planning. *Harvard Business School*, July 31, Case No. 9-306-003.

Hamel, G. and Prahalad, C. K. (1994). *Competing for the Future*. Boston, MA: Harvard Business School Press.

Hill, T. and Westbrook, R. (1997). SWOT analysis: It's time for a product recall. *Long Range Planning*, 30(1), 46–52.

Keeney, R. L. and McDaniels, T. L. (1992). Value-focused thinking about strategic decisions at BC hydro. *Interfaces*, 22(6), 94–109.

Keeney, R. L., McDaniels, T. L., and Swoveland, C. (1995). Evaluating improvements in electric utility reliability at British Columbia Hydro. *Operations Research*, 43, 933–947.

Kidd, J. B. and Prabhu, S. P. (1990). A practical example of a multi-attribute decision aiding technique. *Omega, The International Journal of Management Science*, 18, 139–149.

Koehler, D. J. and Harvey, N. eds. (2007). *Blackwell Handbook of Judgment and Decision Making*. Malden, MA: Blackwell Publishing.

Krumm, F. V. and Rolle, C. F. (1992). Management and application of decision and risk analysis in Du Pont. *Interfaces*, 22(6), 84–93.

Kusnic, M. W. and Owen, D. (1992). The unifying vision process: Value beyond traditional decision analysis in multiple-decision-maker environments. *Interfaces*, 22(6), 150–166.

Lindgren, M. and Bandhold, H. (2009). *Scenario Planning: The Link between Future and Strategy*. Hampshire, England: Palgrave Macmillan.

Matheson, D. and Matheson, J. E. (1999). Outside-in strategic modeling. *Interfaces*, 29(6), 29–41.

Musil, S., and Skillings, J. E. (2009). Sold! Ebay Jettisons Skype in $2 billion deal. http://News.Cnet.Com/8301-1035_3-10322833-94.html

Narayandas, D. (1996). Dell Computer Corporation. *Harvard Business School*, Case 9-596-058.

Pickton, D. W. and Wright, S. (1998). What's SWOT in strategic analysis? *Strategic Change*, 7(2), 101–109.

Porter, M. E. (1998). *Competitive Strategy: Techniques for Analyzing Industries and Competitors*. New York: Free Press.

Porter, M. E. (2008). *On Competition: Updated and Expanded Edition*. Cambridge, MA: Harvard Business School Press.

Prokesh, S. (1997). Unleashing the power of learning: An interview with British Petroleum's John Browne. *Harvard Business Review*, 75, September–October, 166.

Richwine, L. (2011). Drugmakers to share data to speed brain research. http://News.Yahoo.Com/S/Nm/20100611/H1_Nm/Us_Drugs_Alzheimers

Schoemaker, P. J. H. (1995). Scenario planning: A tool for strategic thinking. *Sloan Management Review*, 36(2), Winter, 25–36.

Schoemaker, P. J. H. (2007). Forecasting and scenario planning: The challenge of uncertainty and complexity. In: *Blackwell Handbook of Judgment and Decision Making*, Koehler and Harvey, N., eds. Malden, MA: Blackwell Publishing.

Schoemaker, P. J. H., Cornelius, A. J. M., and van der Heijden, K. (1992). Integrating scenarios into strategic planning at Royal Dutch/Shell. *Planning Review*, 20, 41–46.

Walls, M. R., Morahan, G. T., and Dyer, J. S. (1995). Decision analysis of exploration opportunities in the onshore US at Phillips Petroleum Company. *Interfaces*, 25, 39–56.

Appendix A: Instructions for Downloading the DecisionTools Suite

To download the DecisionTools Suite, copy and paste the link below into your browser. Click on the book's title and you will be prompted to answer a question about the book's index. Please note: *You must have the book in hand in order to download the software.*

Simply type in the requested answer, fill out the following form, and download your software.

The software will be delivered in a ZIP file format. You must extract the contents of the ZIP folder first before attempting to install the software. If you attempt to install the software from within the ZIP folder, it will not work.

http://www.palisade.com/bookdownloads/chelst

Appendix B: Instructions for Downloading Logical Decisions

To download Logical Decisions, copy and paste the link below into your browser.

When you have installed the software, enter one of the following names and keys, depending on whether you have downloaded version 6.2 or version 7.x:

Value Added Decision Making 6.2

00UPKJ-TW85AY-6DZTB8-V0AYKB

Value Added Decision Making 7.0

0JK3UN-DKU6C6-C4ZRBF-GG8FG6

http://www.logicaldecisionsshop.com/catalog/index.php?main_page=product_info&cPath=2&products_id=14

Index